"十二五"普通高等教育本科国家级规划教材

# Theory of Machines and Mechanisms

## (Bilingual Edition)

## Second Edition



# 机械原理

## （英汉双语）

## 第2版

主　编　张　颖　张春林

副主编　何岑成　赵自强

参　编　李志香　马　超　冯欣欣

王泽洋　赵嘉珩

机 械 工 业 出 版 社

本书是按照高等工科教育逐步与国际教育接轨的要求编写的机械原理课程英汉双语教材。在满足国内教学基本要求的基础上，体现了我国机械原理课程教学的现状和特色，吸取了国外同类教材的特点。全书内容从机构分析、机构设计、机构系统设计到机械动力学，遵循以设计为主线，加强对基本概念、基本理论、基本方法的理解，以理论与工程实践相结合为指导思想，对传统机械原理的内容进行了整合，结合现代科学技术的发展，增加了新内容，删减了过于陈旧的内容。

全书共有 13 章。第 1 章介绍机械的基本概念；第 2 ~ 4 章主要介绍机构的结构分析、运动分析和力分析；第 5 ~ 9 章主要介绍常用机构的设计；第 10 章介绍空间机构和机器人机构；第 11 章介绍机构系统的设计；第 12 和 13 章主要介绍机械系统的运转及速度波动的调节，机械的平衡。创新设计的思想融于各章内容中。

本书的英文内容，在反映中文内容的前提下，采用典型的科技英语表现方式和通俗易懂的词汇，简便易读。同时在附录中增加了英语常用符号与公式读法，供广大读者参考。

本书可作为高等工科学校机械类专业的机械原理教材，特别适合作为英汉双语教材，也可以作为机械工程人员的参考用书。

**图书在版编目（CIP）数据**

机械原理：英汉对照/张颖，张春林主编. —2 版.
—北京：机械工业出版社，2016. 11（2024. 6 重印）
“十二五”普通高等教育本科国家级规划教材
ISBN 978-7-111-54883-6

Ⅰ. ①机… Ⅱ. ①张… ②张… Ⅲ. ①机械原理-高等学校-教材-英、汉 Ⅳ. ①TH111

中国版本图书馆 CIP 数据核字（2016）第 222736 号

机械工业出版社（北京市百万庄大街 22 号 邮政编码 100037）
策划编辑：余 皞 责任编辑：余 皞 李 超 王晓艳 杨 洋
责任校对：任秀丽 封面设计：张 静
责任印制：郜 敏
中煤（北京）印务有限公司印刷
2024 年 6 月第 2 版第 7 次印刷
184mm × 260mm · 24. 5 印张 · 618 千字
标准书号：ISBN 978-7-111-54883-6
定价：69. 80 元

电话服务 网络服务
客服电话：010-88361066 机 工 官 网：www. cmpbook. com
010-88379833 机 工 官 博：weibo. com/cmp1952
010-68326294 金 书 网：www. golden-book. com
**封底无防伪标均为盗版** 机工教育服务网：www. cmpedu. com

# Preface to the Second Edition

This book has been written for mechanical engineering students who are beginning to study the theory of machines and mechanisms or professional English.

Since the book was published in 2012, some universities take it as Chinese textbook of theory of machines and mechanisms for undergraduates, some universities take it as English textbook of theory of machines and mechanisms for undergraduates, some universities take it as bilingual education textbook of theory of machines and mechanisms, some universities take it as foreign students textbook of theory of machines and mechanisms, and some universities take it as international class students textbook of theory of machines and mechanisms. Such students will have acquired a set of mechanical engineering knowledge and the ability to use English language to express themselves in spoken and written forms.

This edition has been influenced by changes in engineering education, suggestions by colleagues and users, and the author's checkup.

The major revisionary works for publishing this new edition are as follows.

Effort has been made to remove all sorts of errors and misprints as far as possible.

We adopt a simple mathematical manner for easy comprehension by students in Chapter 5, but there are not any examples to describe the new methods. So many examples have been added in Chapter 5.

Geneva wheel with 6 slots will be more helpful to understand the action ratio, so we use a Geneva wheel with 6 slots instead of Geneva wheel with 4 slots.

Some of examples have been added in Chapter 13 too.

Two-dimensional codes are added in this edition to describe the mechanisms animated figures.

I acknowledge the efforts of professor He Cencheng of Beijing University of Technology for revisionary and compiled works for this edition.

Zhao Jiaheng of Beijing Institute of Technology compiled some examples.

Finally, I make an affectionate acknowledge to my PhD doctors team for their support.

I am grateful to all readers who pointed out errors and mistakes of the previous edition and gave valuable suggestions.

Zhang Chunlin

# 第2版前言

本教材是为机械工程专业的学生学习机械原理课程或专业英语编写的。

本教材于2012年出版后，一些学校选其作为本科生机械原理课程的中文教材，教学时讲授中文，英语内容作为参考；一些学校选其作为机械原理课程的英文教材，教学时讲授英文，中文内容作为参考；一些开设双语教学的学校，选其作为本科生的英汉双语教学机械原理课程教材，教学时有时用英文讲授，有时用中文讲授，中、英文内容可用来对比和参考。一些学校选其作为外国留学生的机械原理课程教材，也有一些学校选其作为国际班的机械原理课程教材。本教材不但使学生获得机械原理的基本知识，还可以提高学生英语的口语和写作能力。

经过几年的使用，根据作者自查和用户意见反馈，对其进行了修改、补充和完善，如在平面连杆机构的综合内容中，引入了新方法，但缺少例题佐证。所以，本次修改增加了相应例题；在说明槽轮机构的运动特性时，采用了六槽槽轮代替四槽槽轮，这样就容易理解主动销轮和从动槽轮的转角关系，更容易理解槽轮机构的运动系数。在机械平衡设计一章中也增加了例题。

本次修订还更正了一些文字错误和语法错误，并引入了二维码技术，读者只要用手机扫描插图旁的二维码，就可以观看相应机构的动画，增强了本教材的可读性。

北京工业大学的何岑成老师参与了本次修订工作和英语教材的编审工作，北京理工大学的赵嘉珩为本教材增编了例题。

感谢我的博士生团队为本书的修订所做的工作。

由于编者水平有限，书中难免存在错误和疏漏之处，欢迎广大读者批评指正。

张春林

# Preface to the First Edition

Along with the rapid development of science and technology, the higher education will speed up internationalization. So it is necessary to compile a compatible textbook which not only has Chinese feature, but also makes for globalization. This is a new object that the teaching reform of China's higher education is faced with. On the one hand, the direct adoption of the foreign classical textbooks does not suit the actual state of our teaching in Chinese universities; on the other hand, the Chinese language textbooks cannot be brought in line with the international practice. Therefore, we will compile a bilingual textbook on "theory of machines and mechanisms" with the support of China Machine Press.

This book is based on the basic teaching requirement of the "theory of machines and mechanisms" issued by the Ministry of Education. The basic concepts, basic principles and basic skills are reinforced, and we take the content of synthesis or design as a main clue to cultivate consciousness and capability of innovation for mechanical engineering. We hope that this book can include the features inheriting from the traditional textbooks, and can be in line with some foreign textbooks.

1. Descriptions about "theory of machines and mechanisms" in the Chinese book

1) This book is compiled with analysis of mechanisms, design of mechanisms, design of mechanism systems and dynamics of mechanisms.

2) It highlights the analytical methods, such as kinematic analysis of mechanisms, force analysis of mechanisms, synthesis of mechanisms, cam mechanisms and balancing design, etc., and de-emphasizes graphical methods.

3) Some of the traditional contents of "theory of machines and mechanisms" are modified, integrated and expanded in this book.

The concepts of mechanism and machine are defined in Chapter 1, and they are up-to-date ideas. Composition principle of mechanisms is expanded to creative design of mechanisms. A new method which is used to determine the self-locking of a mechanism has been bring up, in which we will use the total reaction force acting on the friction cone or friction circle to determine self-locking phenomenon of a mechanism instead of use of efficiency. The concept is clear and the method is simple and practical. It highlights design as the main clue. By using the less complicated geometry algebra method instead of traditional coordinate transformation method in the analytical design of four-bar mechanism and cam mechanism, we can simplify mathematical design methods easily. This textbook also provides readers with creative design of mechanisms and design of mechanism systems to implement the overall content from simple to complicated; the contents of modified gears, helical gears, bevel gears, worm gears and intermittent mechanisms are simplified; the kinetostatic analysis of mechanism and force analysis including friction and self-locking are merged into a force analysis of mechanisms.

4) This textbook attempts to be different from the traditional versions by adopting color rendered figures and three-dimensional figures illustrations, which are clear and easy to read.

5) The exercises are divided into four parts: reading, comprehension, practice, and self-test. Furthermore, there are some analog test papers. These problems can help students to understand the

basic concepts and to prepare them to cope with real engineering problems in practice.

6) As a bilingual textbook, a lot of described contents are simplified to reduce its length. It shortens traditional Chinese text to keep concise and compact to highlight consistent context and accurate terminology.

2. Descriptions about "theory of machines and mechanisms" in the English book

1) The main purpose of the bilingual textbook is to help students study the contents of "theory of machines and mechanisms", rather than only English.

2) The English version is not the strict translation of the Chinese version, but their contents are coincident basically.

3) All the figures, tables and formulas between the Chinese version and English version correspond with each other; they have a good comparability.

4) The professional and conventional vocabularies are adopted from foreign textbooks; so the book has a good credibility.

5) Basic English grammar of science and technology makes reading easy.

6) The reading and comprehension which include the basic concepts are adopted from foreign textbooks.

7) Some concepts between the Chinese version and English version are a little different, such as redundant degree of freedom and local degree of freedom, redundant constraint and virtual constraint. This will benefit students to some extent.

The authors of the Chinese version are as follows.

Professor Zhang Chunlin (Chapters 1, 2 and 3), Zhao Ziqiang (Chapters 4, 5 and 6), Li Zhixiang (Chapters 7 and 8), Ma Chao (Chapters 9, 10, 11, 12 and 13).

Zhang Ying of Beijing University of Technology compiled and translated Chapters 1, 2, 3, 4, 5, 6, 7, 8, 9. Wang Zeyang compiled and translated Chapters 10 and 11. Feng Xinxin compiled and translated Chapters 12 and 13. Zhang Ying thoroughly compiled and edited the English Version and Professor Zhang Chunlin compiled and edited the entire bilingual text.

Acknowledgments

The authors would like to express their thanks for the cooperation of all those individuals who generously made this item available.

Sincere thanks also go to Professor Ge Wenjie of Northwestern Polytechnical University, who provided a part of 3D graphics. Some of the graphics are adopted from the other authors. We would like to express our thanks to them.

Finally, we would like to make an affectionate acknowledgment to China Machine Press for the support and we acknowledge the efforts of editorial staff for bringing out the new bilingual version in an excellent format.

It is hard to avoid errors and mistakes in the first edition, because the knowledge of English of authors is limited; the authors invite criticisms and suggestions from all readers for further improvement of the book, so that the errors can be corrected in future versions.

Zhang Chunlin

Ever Bright Garden, Beijing

# 第1版前言

随着科学技术的飞速发展和教育逐渐走向世界，编写具有中国特色、又逐步与发达国家接轨的教材，是中国高等学校教学改革所面临的新课题。直接引入国外教材的方法不适合国内的教学现状，单纯的中文教材不能满足逐步与国际接轨的需要。因此，在机械工业出版社的支持下，作者开始尝试编写中英文双语机械原理教材。

本教材的指导思想是在教育部颁布的教学基本要求基础上，加强基本理论、基本方法和基本技能的培养；以设计为主线，突出创新意识和创新设计能力的培养；既反映中国同类教材的内容与特色，又逐步与国外同类教材接轨。

1. 关于本教材中文部分的编写说明

1）本教材按照机构分析、机构设计、机构系统设计和机构动力学设计的体系编写。

2）教材内容中淡化图解法、强化解析法，如机构运动分析、力分析、连杆机构综合、凸轮机构设计、平衡设计等，淡化了图解法，突出了解析法。

3）本教材对传统机械原理课程内容进行了修改、整合与拓展。在绪论中重新定义了机构与机器的基本概念，使其更具有时代特点；把机构组成原理拓展到机构创新设计；放弃传统的利用效率判断机构自锁的方法，改为利用总反力作用在摩擦锥或摩擦圆之内来判断机构的自锁现象，概念清晰，方法实用，把摩擦与自锁机构设计相结合，强化了设计主线；在连杆机构和凸轮机构的解析法设计中，放弃了传统的坐标变换法，采用了简单的几何代数法，简化了数学运算过程。

本教材还增加了创新设计、机构系统设计的内容，完善了从简单机构设计到复杂机构系统设计的整体内容；同时简化了齿轮机构中的变位齿轮、斜齿轮、锥齿轮、蜗杆机构以及其他常用机构的内容。把机构动态静力分析、摩擦与自锁、考虑摩擦的力分析等内容整合为机构的力分析。

4）为了突出教材的新颖性，学习了国外同类教材的插图风格，重新绘制了平面彩色渲染图和一些三维图形，代替了传统的黑白线条图形，层次分明，有益于对教材内容的理解。

5）改革了配套习题的体系与内容。本教材的配套习题集内容分为阅读、理解、设计和自测四个组成部分，习题集最后辅以部分模拟试卷，形成一套完整的课后训练系统。在加强基本理论、基本概念的基础上，采用与工程实践相结合的题目，突出了理论与实践相结合的训练目的，使学生逐步学习解决工程设计中的问题。

6）由于本书为双语教材，为了不增加篇幅，简化了传统中文教材的叙述内容，使论述更加简练。

2. 关于本教材英文部分的说明

1）编写英汉双语教材，其目的是通过英文内容的学习，掌握机械原理的基本知识，为本专业领域内进行国际交流奠定基础。

2）英文内容没有采取严格的汉语翻译内容，而是强调基本内容的一致性和英文词义的准确性。

3）教材体系和内容与对应的中文教材相一致，如图、表、公式都有严格的对应，具有可比性。

4）编写中尽量采用国外同类教材中的常用标准专业词汇，具有可信性。

5）尽量采用常用的词汇和简单的初级科技语法，容易理解，具有可读性。

6）本书配套习题集中的阅读与理解部分，以基本概念与基本理论为主，参考大量原版教材及文献。

7）该书中，有些中英文的概念存在差别，有益于中外教材内容的接轨。如该书第2章中，中文教材中的局部自由度、虚约束，在英文教材中为多余自由度、多余约束。保留这些差别是有益的。

参加本书中文编写的教师有：张春林（第1章，第2章，第3章），赵自强（第4章，第5章，第6章），李志香（第7章，第8章），马超（第9章，第10章，第11章，第12章，第13章）。

参加英文编写的教师有：张颖（第1章，第2章，第3章，第4章，第5章，第6章，第7章，第8章，第9章），王泽洋（第10章，第11章），冯欣欣（第12章，第13章）。

本书英文部分由北京工业大学的张颖负责统稿。全书由张春林负责统稿。另外，在编写过程中，西北工业大学的葛文杰教授提供了部分3D图形，还有一些老师提供了部分图形，编者表示感谢。

由于编者水平有限，书中难免存在错误和疏漏之处，特别是英文部分难免存在中国式的英文表述，敬请广大读者批评指正。

**张春林**

于北京光大花园

CONTENTS

# 目 录

**Preface to the Second Edition (第 2 版前言)**

**Preface to the first Edition (第 1 版前言)**

**Chapter 1 Introduction 绪论** ······ 1

1. 1 General Information 机械总论 ······ 2

1. 2 Teaching Content and Object of the Course 机械原理课程的研究对象与内容 ······ 6

1. 3 Purpose of This Course 学习机械原理课程的目的 ······ 8

**Chapter 2 Structural Analysis of Planar Mechanisms 平面机构的结构分析** ······ 11

2. 1 Kinematic Chain and Mechanisms 运动链与机构 ······ 12

2. 2 Schematic Diagram of Mechanisms 机构运动简图 ······ 16

2. 3 Degree of Freedom of Planar Mechanisms 机构自由度的计算 ······ 22

2. 4 Mechanism Analysis and Innovation 机构分析与创新 ······ 30

**Chapter 3 Kinematic Analysis of Planar Mechanisms 平面机构的运动分析** ······ 41

3. 1 Introduction 平面机构运动分析概述 ······ 42

3. 2 Velocity Analysis with Instantaneous Center 用速度瞬心法对机构进行速度分析 ······ 44

3. 3 Kinematic Analysis by Graphical Method 用相对运动图解法对机构进行运动分析 ······ 48

3. 4 Kinematic Analysis by Algebraic Method 用解析法对机构进行运动分析 ······ 60

**Chapter 4 Force Analysis of Planar Mechanisms 平面机构的力分析** ······ 67

4. 1 Introduction 平面机构力分析概述 ······ 68

4. 2 Force Analysis Including Inertia Forces in Mechanisms 计入惯性力的机构力分析 ······ 68

4. 3 Force Analysis Including Friction in Mechanisms 计入摩擦的机构力分析 ······ 76

4. 4 Friction and Design of Self-Locking Mechanisms 摩擦与自锁机构的设计 ······ 88

**Chapter 5 Synthesis of Planar Linkages 平面连杆机构及其设计** ······ 95

5. 1 Characteristics and Types of Planar Linkages 平面连杆机构的特点与基本型式 ······ 96

5. 2 Fundamental Features of Four-bar Linkages 平面连杆机构的基本性质 ······ 104

5. 3 Synthesis of Four-Bar Linkages 平面连杆机构的设计 ······ 112

**Chapter 6 Design of Cam Mechanisms 凸轮机构及其设计** ······ 137

6. 1 Introduction 凸轮机构概述 ······ 138

6. 2 Basic Types of Follower Motion and Design 从动件的运动规律及其设计 ······ 144

6.3 Cam Profile Synthesis 凸轮轮廓曲线的设计 …… 156
6.4 Sizes of Cam Mechanisms 凸轮机构基本尺寸的设计 …… 164
6.5 Computer-Aided Design of Cam Mechanisms 计算机辅助凸轮设计 …… 170

**Chapter 7 Design of Gear Mechanisms 齿轮机构及其设计 …… 177**
7.1 Classification of Gear Mechanisms 齿轮机构的分类 …… 178
7.2 Fundamental Law of Gearing 齿廓啮合基本定律 …… 180
7.3 Involute Properties and Involute Tooth Profiles 渐开线齿廓及其啮合特点 …… 182
7.4 Nomenclatures of Standard Spur Gear and Gear Sizes
渐开线标准直齿圆柱齿轮的基本参数和几何尺寸 …… 186
7.5 Meshing Drive of Standard Spur Gears 渐开线直齿圆柱齿轮机构的啮合传动 …… 192
7.6 Forming and Undercutting of Gear Teeth 渐开线圆柱齿轮的加工及其根切现象 …… 202
7.7 Nonstandard Spur Gears 变位齿轮概述 …… 208
7.8 Parallel Helical Gears 平行轴斜齿圆柱齿轮机构 …… 212
7.9 Worm and Worm Gears 蜗杆传动机构 …… 220
7.10 Bevel Gears 锥齿轮机构 …… 226

**Chapter 8 Design of Gear Trains 轮系及其设计 …… 233**
8.1 Classification of Gear Trains 轮系及其分类 …… 234
8.2 Ratio of Ordinary Gear Trains 定轴轮系传动比的计算 …… 236
8.3 Ratio of Epicyclic Gear Trains 周转轮系传动比的计算 …… 242
8.4 Ratio of Combined Gear Trains 混合轮系传动比的计算 …… 246
8.5 Some Considerations for Design of Planetary Gear Train 周转轮系设计中的若干问题 …… 250
8.6 Introduction of Miscellaneous Planetary Trains 其他类型的周转轮系简介 …… 258

**Chapter 9 Introduction of Screws, Hook's Couplings and Intermittent Mechanisms
螺旋机构、万向联轴器和间歇运动机构简介 …… 265**
9.1 Screw Mechanisms 螺旋机构 …… 266
9.2 Universal Joints 万向联轴器 …… 268
9.3 Ratchet Mechanisms 棘轮机构 …… 270
9.4 Geneva Mechanisms 槽轮机构 …… 274
9.5 Indexing Cam Mechanisms 凸轮式间歇运动机构 …… 278
9.6 Intermittent Gear Mechanisms 不完全齿轮机构 …… 280

**Chapter 10 Spatial Mechanisms and Robotic Mechanisms
空间连杆机构及机器人机构概述 …… 285**
10.1 Introduction of Spatial Mechanisms 空间连杆机构概述 …… 286
10.2 Introduction of Robotic Mechanisms 机器人机构概述 …… 292

**Chapter 11 Design of Mechanism Systems 机构系统设计 …… 299**
11.1 Introduction of Mechanism Systems 机构系统设计概述 …… 300
11.2 Harmonization Design of Mechanism Motions 机构系统的运动协调设计 …… 302
11.3 Combined Methods of Mechanism Systems 机构系统的组合方法 …… 304

**Chapter 12 Fluctuation and Regulation in Speed of Machines 机械系统的运转及速度波动的调节** …… 313

12.1 Operating Analysis of Machinery 机械运转过程分析 …… 314

12.2 Equivalent Kinetic Model of Mechanism Systems 机械系统的等效动力学模型 …… 318

12.3 Kinetic Equations of Mechanism Systems 机械系统的运动方程及其求解 …… 326

12.4 Periodic Speed Fluctuation and Regulation in a Machine 周期性速度波动及飞轮设计 …… 330

12.5 Aperiodic Speed Fluctuation and Regulation in a Machine 非周期性速度波动及其调节 …… 338

**Chapter 13 Balance of Machinery 机械的平衡设计** …… 343

13.1 Introduction 机械平衡概述 …… 344

13.2 Balance Design of Rigid Rotors 刚性转子的平衡设计 …… 346

13.3 Balance Test of Rigid Rotors 刚性转子的平衡试验 …… 356

13.4 Balance of Planar Mechanisms 平面机构的平衡简介 …… 360

**Appendix 附录一** …… 362

**Appendix 附录二** …… 372

**References 参考文献** …… 379

# Chapter 1

# Introduction

# 第1章

# 绪论

## 1.1 General Information

A variety of machineries, such as textile machinery, printing machinery, food machinery, transportation machinery, mining machinery, construction machinery, engineering machinery, agricultural machinery, forestry machinery, packaging machinery, metallurgical machinery, fluid machinery, service machinery, weapons, robots and so on, have played an important role in nowadays economic development and national defense. They have promoted the development of human society. How to improve old pattern of machinery and design new machinery are essential tasks in the field of mechanical engineering.

**1. The Concept of Machinery**

Along with the progress of human society, the concept of machinery is developed and improved gradually. Human beings began to use stone tools from the primitive society at first, such as stone axes and stone knives. With the development of technology, our ancestor invented some simple wooden tools, such as lever, windlass, human power waterwheel, animal power waterwheel and so on. On the basis of these simple tools, some complicated devices were invented, such as water-powered roller and windmills. In the 18th century, after the British industrial revolution, human invented steam engine, internal combustion engine and electromotor, therefore the foundation of modern mechanical industry was established. After the computer was invented, the automatic control technology, information technology, and the transducer technology were used in machinery, which made the machines realize automation and intelligent. New technology and equipments, such as robots, numerical control machines, high-speed vehicles and aircrafts, heavy machinery, micro machinery and so on, have promoted the human society's prosperity and progress.

In different period, the definition of machinery is different, as well. In general, machinery is a device which can realize the mechanical movement. For example, screwdrivers, hammers, pliers, scissors are machinery and tanks, planes, cars, ships, robots, and other senior complicated equipment are also machinery. However, in modern society, people used to define the simplest machinery with no power as tools, such as lever, pliers, scissors, wheelbarrow, and so on.

In mechanical engineering, when talking about specific machinery, it is called a machine.

(1) Mechanism  A mechanism is a device, which produces specific mechanical motions. Thus, the function of a mechanism is to transmit and modify a motion. We often use a drawing with some simple lines and symbols to describe the mechanism, and it is called kinematical diagram.

(2) Machine  A machine is a device which produces specific mechanical motions, and it can transmit or modify mechanical energy, materials and information. A machine may be a mechanism or a combination of mechanisms capable of transmitting or modifying motion and mechanical energy.

For instance, vehicles, tanks, missiles, planes, ships and so on are all machines. But the television is not a machine, because it has nothing to do with mechanical movement.

Fig. 1-1a shows an internal combustion engine which transforms thermal energy into mechanical energy. The sliding motion of piston 1 will be transformed to the series rotation of crank shaft 3 by coupler 2. This kind of mechanism is called a linkage.

## 1.1 机械总论

各种各样的机械，如纺织机械、印刷机械、食品机械、交通运输机械、矿山机械、建筑机械、工程机械、农业机械、林业机械、包装机械、冶金机械、流体机械、服务机械、兵器与机器人等，在经济建设和国防建设领域中发挥了巨大作用，促进了人类社会的发展。改造原有机械、设计新机械是机械工程领域中的重要任务。了解机械、设计机械是机械工程专业教学中的重要内容 。

**1. 机械的概念**

机械的概念是伴随人类社会的不断进步逐渐发展与完善的。从早期原始社会人类使用的诸如石斧、石刀等最简单的石制工具，到杠杆、辘轳、人力脚踏水车、兽力汲水车等简单木制工具，发展到较复杂的水力驱动的水碾和风力驱动的风车等都是简单机械。18 世纪英国工业革命后，人类发明的蒸汽机、内燃机、电动机等复杂的机械，奠定了现代机械的基础。计算机发明后，自动控制技术、信息技术、传感技术融入机械中，机械实现了自动化和智能化。机器人、数控机床、高速运载工具，重型机械、微型机械等大量先进机械加速了人类社会的繁荣和进步，机械进入机电结合的新时代。

不同历史时期，人们对机械的定义也有所不同。从广义角度讲，凡是能实现机械运动的装置都是机械。例如：螺钉旋具、锤子、钳子、剪子等简单工具是机械，汽车、坦克、飞机、舰船、各类加工机床、机械手、机器人、复印机、打印机等高级复杂装备也是机械。无论其结构和材料如何，只要是实现机械运动的装置，就称之为机械。在现代社会中，人们常把最简单的、没有动力源的简单机械称为工具或器械，如杠杆、钳子、剪子、手推车等最简单的机械常称为工具。

工程中，谈到具体的机械时，常使用机器这个名词，泛指时则用机械来统称。

（1）机构　机器的重要特征是执行机械运动，工程上把机器中执行机械运动的装置称为机构。为研究方便，常用简单的符号和线条表示机构的组成情况和运动情况，并称之为机构运动简图。

（2）机器　机器是执行机械运动的装置，用来变换或传递能量、物料或信息。机器的重要特征是执行机械运动，同时完成能量的转换，或物料、信息的传递。汽车、坦克、导弹、飞机、轮船、车床、起重机、织布机、印刷机、包装机等大量具有不同外形、不同性能和用途的装置都是具体的机器。电视机不是机器，因其功能与机械运动无关。

图 1-1a 所示的四冲程内燃机，是一个把热能转换为机械能的机器。该内燃机中，活塞 1 做往复移动并通过连杆 2 推动曲轴 3 连续旋转。这种把活塞移动转化为曲轴连续转动的装置称为连杆机构。

凸轮 7 转动，驱动推杆 8 往复移动，该机构称为凸轮机构。再通过杠杆 9，驱动气门 10 的开启，控制进、排气阀的运动，保证缸体 11 按顺序吸进燃气和排出废气。四冲程内燃机中，活塞往复移动四次，曲轴转动两周，进气阀和排气阀各启闭一次，所以凸轮转速为曲轴转速的一半。也就是说，在曲轴和凸轮轴之间要设置减速齿轮 4、5、6，该机构称为齿轮机构。齿轮机构实现了高速转动到低速转动的运动变换。

综上所述，机构是组成机器的主体。为表明机器的组成和运动情况，常用机构运动简图来表示。

Cam 7 and follower 8 are called cam mechanism. The opening and closing of valve 10 can be controlled by lever 9. This can ensure that cylinder 11 takes in the gas and exhales the waste gas. In a four-stroke engine, it takes four strokes of the piston to complete one cycle. That is to say, the circle is completed in two revolutions of the crank shaft, and the valves work at a time. Therefore, the engine requires a gear mechanism to reduce the speed of the crankshaft, such as gears 4, 5, 6.

There are four strokes in the engine. They are intake stroke, compression stroke, power stroke and exhaust stroke, but only one power stroke works, so the speed of the crank shaft fluctuates cyclically. A flywheel must be mounted on the crank shaft.

Fig. 1-1b shows the skeleton diagram of the four-stroke engine. The design can be simplified by use of skeleton diagram.

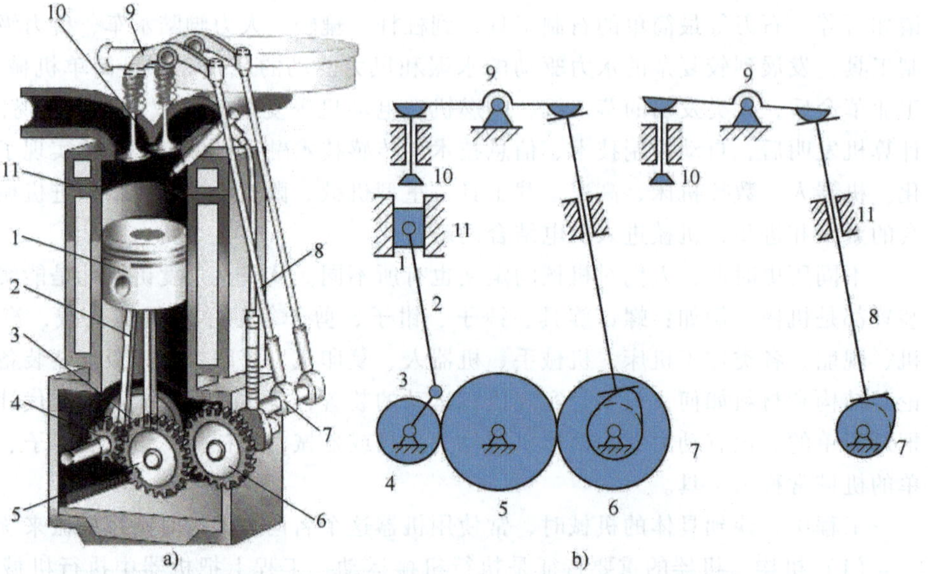


Fig. 1-1 Internal combustion engine and scheme(内燃机及其机构简图)
1—piston(活塞) 2—coupler(连杆) 3—crank shaft(曲轴) 4、5、6—speed reducing gear(减速齿轮) 7—cam(凸轮) 8—follower(推杆) 9—lever(杠杆) 10—valve(气门) 11—frame(缸体)

Although there are various types of machines, the mechanisms consisting of machines are finite. Therefore, the theory of mechanism is very important to design machines.

(3) Machinery In the viewpoint of kinematics, mechanism and machine have no difference, so mechanisms and machines are generally called machinery in mechanical engineering. There is no energy transforming and modifying in a mechanism, for example, mechanical watch is a mechanism but not a machine, because it can not transform energy.

**2. Composition of Machines**

A machine consists of prime power, transmission system and working system. A modern machine also contains a control system. Fig. 1-2 shows the block diagram of composition of a machine.

Fig. 1-3b shows a diagram of an automatic gate. Fig. 1-3a shows the composition of the driver. The prime power is electromotor 3 which has a high speed. But gate 2 needs a low speed to work, so we have to install gear mechanisms and chain mechanisms to modify the speed of gate 2. The reducer 4 and chain driver 5 are transmission system. Gate 2 consisting of the parallel linkages is called working system. The control system and sense system are not illustrated in the gate.

图 1-1b 所示为该内燃机的机构简图。为表达清楚，另一套凸轮机构单独画出。使用机构简图对内燃机进行分析和设计时，简化了设计工作。由于缸体中进气、压缩、燃烧、排气交替进行，导致曲轴运动速度不均匀，所以在曲轴的另一端还要安装调速飞轮。上述各机构协调动作，才能满足内燃机的工作要求。

机器的种类很多，其功能、形状、结构、尺寸等也各不相同，但组成不同机器的机构种类却是有限的，仅有十余种。因此，要研究用有限的机构组成的无限的机器，就必须掌握机构的种类、工作原理及设计方法。

(3) 机械　机构与机器都是实现机械运动的装置，所以从运动学的观点看，两者是一样的；不同的是，机构没有能量的转换和信息的传递。例如：机械表是机构而不是机器，因为弹簧能量没有转换为机械能输出，其作用仅为克服各运动件的摩擦阻力。

从机械运动的观点看问题，机构与机器没有本质区别，工程中将机构与机器统称为机械。

**2. 机器的组成**

根据机器的定义，机器中要有动力源，称为原动机；机器中还要有机械运动的传递装置或机械运动形态的变换装置，常将它们称为机械传动系统和工作执行系统，统称为机械运动系统；现代机器还有控制系统。图 1-2 所示为常见机器组成示意图。

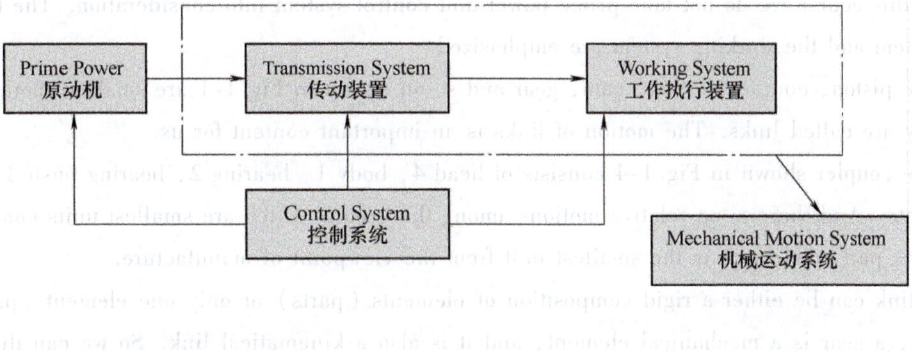


Fig. 1-2　Composition of common machine（常见机器组成示意图）

图 1-3 所示为电动大门示意图，其驱动器 1 的内部组成如图 1-3a 所示。原动机为电动机 3，其转速很高，而大门 2 的开启速度较低，所以要经过齿轮传动机构组成的减速器 4 和链传动机构 5 把电动机 3 的转速降下来，图示的减速器和链传动机构就是速度变换机构。由许多平行四边形机构和滚轮组成的大门 2 称为工作执行机构。

工程中应用的机械运动系统大都具有减速机构。但也有一些现代机器没有减速机构，直接用可控电动机驱动工作执行机构。

In mechanical engineering, most of machines have a transmission system, however, a variety of modern machines use control motor instead of transmission system.

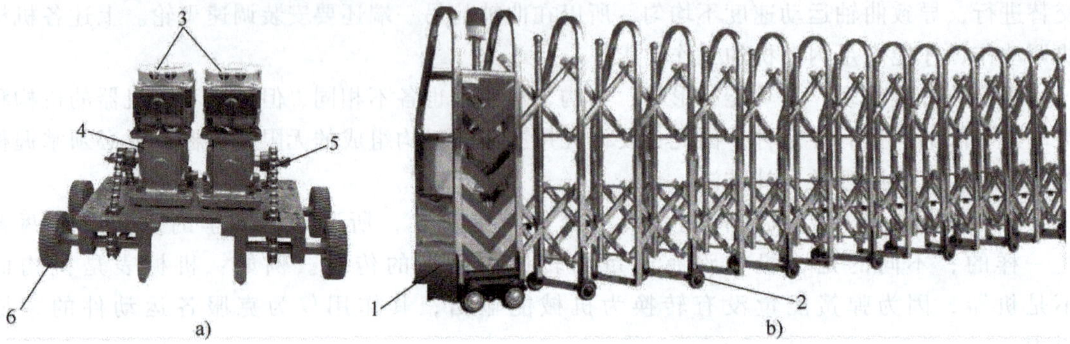


Fig. 1-3 Automatic gate (电动大门示意图)
1—driver(驱动器) 2—gate (大门) 3—electromotor(电动机) 4—reducer(减速器)
5—chain driver(链传动机构) 6—rollers(机构滚轮)

## 1.2 Teaching Content and Object of the Course

**1. Research Object of the Course**

The research object of the course is the theory of machinery, and machinery contains mechanisms and machines. Therefore, the theory of mechanisms and machines is the important content of this course.

In this course we do not take prime power and control system into consideration. The transmission system and the working system are emphasized.

The piston, coupler, crank, cam, gear and so on shown in Fig. 1-1 are smallest motion units, and they are called links. The motion of links is an important content for us.

The coupler shown in Fig. 1-4 consists of head 4, body 1, bearing 2, bearing bush 3, bolt 5, nut 6, etc. And there is no relative motions among these parts which are smallest units composing a link. The part or element is the smallest unit from the viewpoint of manufacture.

A link can be either a rigid composition of elements (parts) or only one element (part), for instance, a gear is a mechanical element, and it is also a kinematical link. So we can distinguish between link and element easily.

**2. Content of This Course**

As mentioned earlier, a mechanism transmits and modifies a motion and a machine is a mechanism or combination of mechanisms, thus this content of this course is about mechanisms. The study of mechanisms involves its structural analysis, motion analysis, force analysis, dimension synthesis and design of mechanism system.

Generally, the content of the course is divided into three parts. They are structural analysis of mechanisms, kinematics of mechanisms and dynamics of machinery respectively.

(1) Structural analysis of mechanisms It involves the composition of mechanism, kinematical diagram, calculation of degree of freedom and analysis of mechanism.

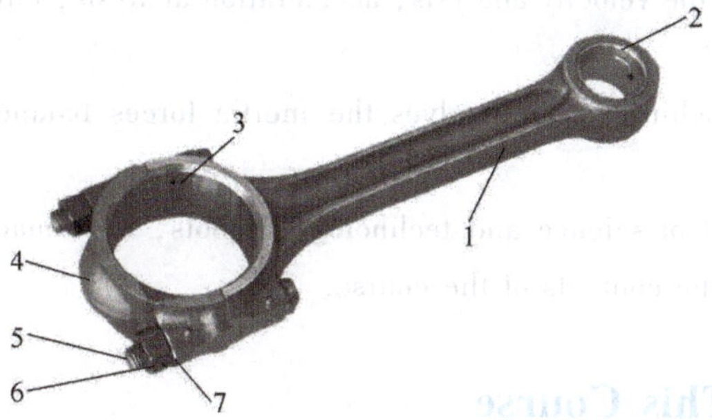


Fig. 1-4　Constitute of the coupler（连杆的组成）
1—body of the coupler（连杆体）　2—bearing（小端轴承）　3—bearing bush（轴瓦）
4—head of the coupler（连杆头）　5—bolt（螺栓）　6—nut（螺母）　7—washer（垫圈）

## 1.2　机械原理课程的研究对象与内容

**1. 机械原理的研究对象**

机械原理是研究有关机械基本理论的课程，其研究对象为机械。而机械又是机器和机构的总称，所以，机械原理是研究机器和机构基本理论的课程。

从机器的组成情况看，原动机是把其他形式的能量转化为机械能的机器，为机器的运转提供动力。机械原理的研究对象不涉及原动机的选择，也不涉及机器的控制系统。机器的传动机构和工作执行机构才是机械原理的研究重点。

图 1-1 所示内燃机中，活塞做往复直线运动，连杆做平面运动，齿轮、凸轮做定轴转动，它们都是最小运动单元，这里把组成机器的最小运动单元称为构件。构件的运动轨迹和运动规律也是机械原理研究的对象。

这里把组成构件的最小单元称为机械零件，简称零件。零件是制造后没有经过组装的物体，因而是组成机器的最小制造单元。构件可以是若干零件的刚性组合体，如图 1-4 所示的连杆是由连杆体、连杆头、大端的轴瓦、小端轴承、螺栓、螺母、垫圈等多个零件刚性连接而成的组合体，各零件之间不能相对运动；构件也可以是单个零件，如一个齿轮可能是一个零件，也是一个构件。机械零件的设计问题将在后续课程中进行讲述。

**2. 机械原理的研究内容**

机构是机器中执行机械运动的主体，或者说机构是组成机器的要素。因此，机械原理的主要研究内容是机构，即研究机构的种类、组成、分析方法、运动、受力、设计以及系统设计等内容。

一般说来，常把机械原理的内容分成三大部分：

（1）机构结构学　研究机构组成、机构简图的画法、机构自由度的计算以及机构的结构分析等。

（2）机构运动学　研究机构运动时的位置、速度、加速度，构件之间的作用力，按照工作要求设计机构的运动学尺寸等。

(2) Kinematics of mechanisms  It deals with the relative motions of links and forces acting on the links, thus it involves the velocity analysis, acceleration analysis, force analysis and synthesis of mechanisms.

(3) Dynamics of machinery  It involves the inertia forces balancing and the fluctuation in speed and regulation.

With the development of science and technology, robots, micromachines, bio-simulation machines and so on expand the contents of the course.

## 1.3 Purpose of This Course

**1. Purpose**

The machinery is closely related to human life, economic construction, and national defense construction, so mechanical industry can reflect a industry level of a country and a level of science and technology. Although there are a lot of machines, the mechanisms consisting machines are finite. So on the basis of finite mechanisms to design some new machine is a creative work. It is also the purpose of this course.

This course is the basis of some other subsequent courses, such as Machinery Design, Advanced Kinematics and Dynamics of Mechanisms, Mechanical Manufacture, etc. So the purpose of this course is also to lay the foundation for the further machine design.

**2. Conclusion**

1) Machinery is very important to develop the economy.

2) Theory of Mechanisms and Machines is very useful to design machines.

3) Mechanism is a device which can transmit and modify motion.

4) The kinematical diagram is a simple form to describe the mechanisms.

5) A machine may be a mechanism or combination of mechanisms capable of transmitting or modifying motion and mechanical energy.

6) Machinery is the generic term of mechanisms and machines.

7) A link is the smallest kinematical unit of a machine from the point of view of the movement.

8) An element is the smallest manufacturing unit of a machine.

9) Theory of Machines and Mechanisms is a basic course, and it is a part of machinery design.

（3）机构动力学 研究机构运转过程中惯性力的平衡以及速度波动的调节等。

随着科学技术的飞速发展和各学科之间的融合与渗透，机械的内容不断丰富，机器人、小型机械、微型机械、仿生机械、生物机械的出现，使机械原理的研究内容不断拓展。

## 1.3 学习机械原理课程的目的

**1. 目的**

机械与人类生活、经济建设、国防建设密切相关。机械工程的发展程度代表一个国家工业基础的强弱和科学技术的发展水平。机械种类众多，但组成各类机械的机构种类有限，因此学会各类机构的设计以及各类机构的组合设计，是设计各类新机械的理论基础。

机械原理是后续课程的基础，也是机械设计过程中的一个重要环节。只有掌握机械原理的基本知识，才能设计新机械。学习本课程的目的是为设计各类机械奠定理论和技术基础。

**2. 结论**

1）机械很重要。

2）机械原理很重要。

3）机构是执行机械运动的装置。

4）机构运动简图是机构的简化表达形式。

5）机器是执行机械运动的装置，且能转换能量，或传递物料与信息。

6）机械是机构与机器的总称。

7）构件是机器中的最小运动单元。

8）零件是机器中的最小制造单元。

9）机械原理不但是机械设计的组成内容，而且是后续课程的学习基础。

# Chapter 2

# Structural Analysis of planar Mechanisms

# 第2章

# 平面机构的结构分析

## 2.1 Kinematic Chain and Mechanisms

### 1. Link

A part is a minimum unit in a machine from the standpoint of manufacture, such as a rigid link, a nut, a bolt, a gear and so on. A link is a minimum unit in a machine from the standpoint of movement, such as a coupler of an engine. A link or a member may be a single part, such as a gear, a cam, or an assembly of rigidly connected parts. There is no relative motion among the parts in a link.

Fig. 2-1a shows a single-cylinder four-stroke-cycle engine, and Fig. 2-1b is its the schematic diagram. Fig. 2-1d is a coupler. The coupler consists of several parts which are connected rigidly (Fig. 2-1c).

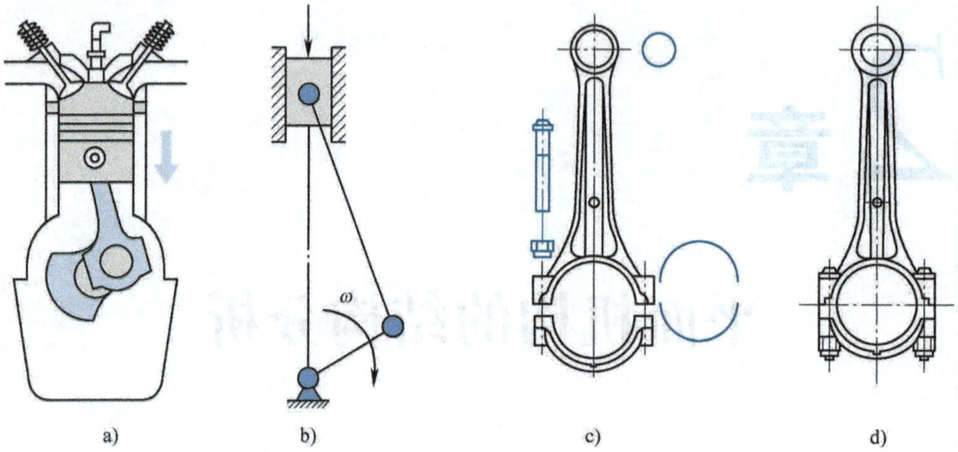


Fig. 2-1 Coupler of the internal combustion engine(内燃机中的连杆)

### 2. Kinematic Pair

A kinematic pair is a joint that permits relative motion. Two rigid links placed together so as to permit relative motion will build a kinematic pair. Kinematic pairs can be classified according to nature of contact and nature of relative motion.

(1) Kinematic pairs according to nature of relative motion According to the feature of relative motion between two links, kinematic pairs may be classified as turning pairs and sliding pairs.

1) Turning pair. When one link has a turning or revolving motion relative to the other link, these two links constitute a turning pair or a revolving pair. Fig. 2-2 shows some turning pairs and their symbols.

If one link is fixed, it is called a frame, such as link 2 in Fig. 2-2b. A shaft turning inside a bearing is a turning pair.

2) Sliding pair. If two links have a sliding motion relative to each other, they form a sliding pair or prismatic pair. Fig. 2-3 shows some sliding pairs and their symbols.

A rectangular rod in a rectangular hole in a prism is a sliding pair.

(2) Kinematic pairs according to nature of contact According to the feature of contact between

## 2.1 运动链与机构

**1. 构件**

构件是指组成机器的最小运动单元。图 2-1a 所示为一内燃机，其机构简图如图 2-1b 所示。其中箱体、活塞、连杆和曲轴都是构件。构件可能是一个零件，也可能是几个零件的刚性组合。内燃机中的连杆结构如图 2-1c 所示，该构件就是由几个零件刚性组合在一起的(图 2-1d)。

当不考虑构件的自身弹性变形时，则视之为刚性构件。本教材在不作特殊说明时，均指刚性构件。

在机构简图中，常用简单直线或曲线表示构件。图 2-1b 中的曲柄、连杆用简单直线表示，活塞用方形滑块表示。

**2. 运动副**

两构件之间具有相对运动的连接称为运动副。

两构件既然被运动副连接起来，就必须要保持接触，而且在接触过程中还要能够产生相对运动，因此，可以按两构件的接触方式和相对运动方式对运动副进行分类。

(1) 按两构件之间的相对运动方式分类　两构件之间的相对运动只有转动和移动，其他运动形式可以看作转动和移动的合成运动。

1) 转动副。两构件之间的相对运动为转动的运动副称为转动副。

图 2-2a 所示为构件 2 固定、构件 1 转动的转动副，对应的简图如图 2-2b 所示；图 2-2c 所示为连接两运动构件的转动副，对应的简图如图 2-2d 所示。

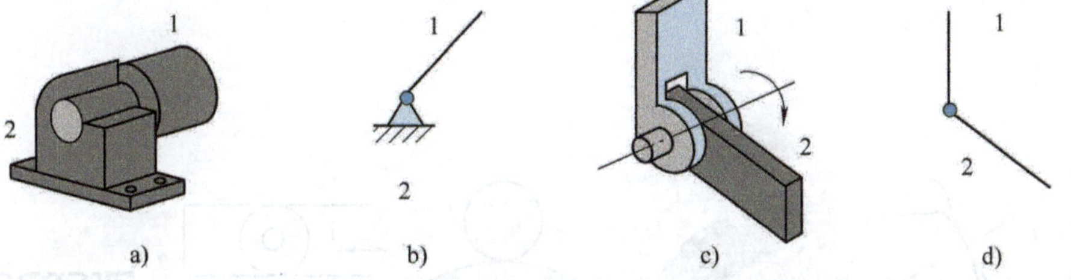


Fig. 2-2　Turning pairs（转动副）

2) 移动副。两构件之间的相对运动为移动的运动副称为移动副。

图 2-3a、b 所示是构件 1 相对构件 2 移动的移动副。若两构件均是运动构件，其运动副简图如图 2-3c 所示；若其中某一构件固定，则其运动简图如图 2-3d 所示。

(2) 按两构件的接触方式分类　两构件之间的接触方式共有三种，即面接触、点接触和线接触。

1) 低副。两构件之间是面接触的运动副称为低副。

two links, kinematic pairs may be classified as lower pairs and higher pairs.

1) Lower pair. When a pair has surface or area contact between two links, it is known as a lower pair. The kinematic pairs showed in Fig. 2-2 and Fig. 2-3 are lower pairs.

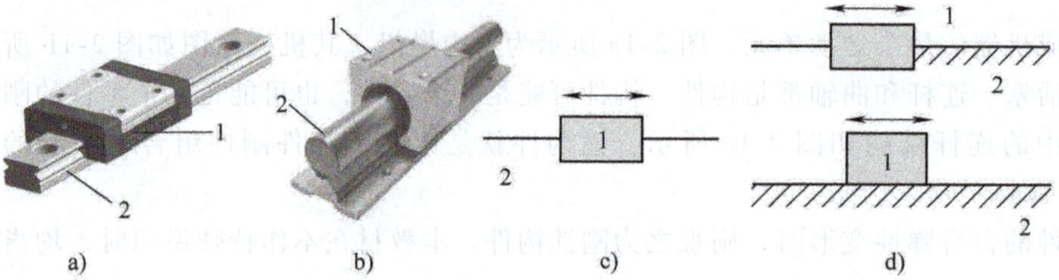


Fig. 2-3 Sliding pairs(移动副)

2) Higher pair. When a pair has a point or line contact between two links, it is known as a higher pair. Fig. 2-4 shows some higher pairs and their symbols.

Meshing gears, cam and follower, ball and roller on a surface form higher pairs.

(3) Kinematic pair element The geometric forms of contact in a pair, such as point, line or surface, are known as pair elements.

A pair is made up of two elements, one on each link being joined, such as the outside cylindrical surface of shaft 1 in Fig. 2-2 and inner cylindrical surface of bearing 2 in Fig. 2-2.

A kinematic pair can be closed by form or by force.

**3. Kinematic Chain**

A kinematic chain is an assembly of links in which the relative motions of the links are possible. Kinematic chains can be classified as closed chains and unclosed chains. If every link in a kinematic chain has at least two pair elements, and links form a closed loop, this kinematic chain is called a closed chain. A closed chain at least has one loop. Fig. 2-5 shows some closed kinematic chains.

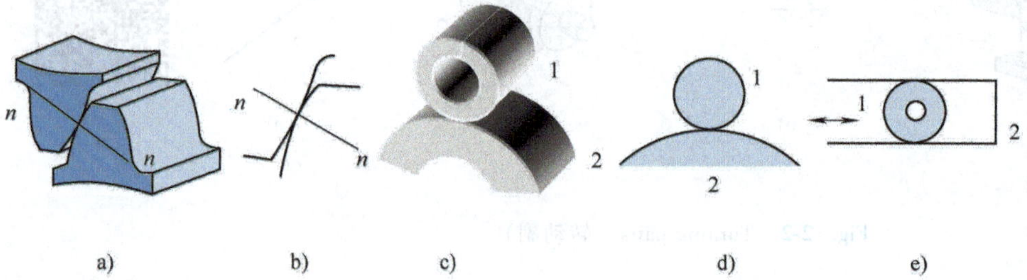


Fig. 2-4 Higher pairs(高副)

Fig. 2-6 shows some unclosed kinematic chains. The first and the last link have only one pair element in the unclosed chain.

If the relative motion of the links in the assembly is impossible, the assembly of links is called a structure or superstructure. Fig. 2-7 shows some structures. A structure may be considered as a link.

由于在承受同等作用力时，面接触具有较小的压强，所以称其为低副。图 2-2 所示的转动副中，转轴 1 与轴承座 2 的接触面是圆柱面；图 2-3 所示的移动副中，滑块 1 与导轨 2 之间也是面接触，它们都是低副。

2）高副。两构件之间是点或线接触的运动副称为高副。

在承受同等作用力时，点或线接触的运动副中具有较大的压强，所以称之为高副。图 2-4a中轮齿与轮齿接触时，从端面看是点接触，从空间看是线接触，称之为齿轮高副，其对应的运动副简图如图 2-4b 所示。图 2-4c 中滚子 1 与凸轮 2 接触时，从端面看是点接触，从空间看是线接触，称之为凸轮高副，其对应的运动副简图如图 2-4d 所示。图 2-4e 中一滚子在槽面内移动，按相对运动是移动副，按其接触性质则为高副。也就是说，移动副有时是低副接触，有时是高副接触。

（3）运动副元素　在研究运动副时，经常涉及两构件在运动副处的表面形状。把两构件在运动副处的点、线、面接触部分称为运动副的元素。

图 2-2 所示的转动副中，轴 1 的外圆柱面是轴 1 上的运动副元素，轴承座 2 的内圆柱面是轴承座 2 的运动副元素。图 2-3a 所示的移动副中，运动副元素为接触平面；图 2-3b 所示的移动副中，运动副元素为圆柱面。2-4a 所示轮齿形成的运动副中，各自的轮廓曲线是轮齿的运动副元素。图 2-4c 所示的凸轮高副中，各自的轮廓线则是相应的运动副元素。因此，高副的运动简图一般用其对应的曲线表示。单一构件的运动副连接处经常使用运动副元素表示。

**3. 运动链**

若干个构件通过运动副连接起来并可做相对运动的构件系统称为运动链。

若运动链中的各构件构成了首尾封闭的系统，则称之为闭链。闭链中每个构件上至少有两个运动副元素。图 2-5a、b 所示的运动链为闭链，图 2-5c 所示为含有两个运动副元素的构件。

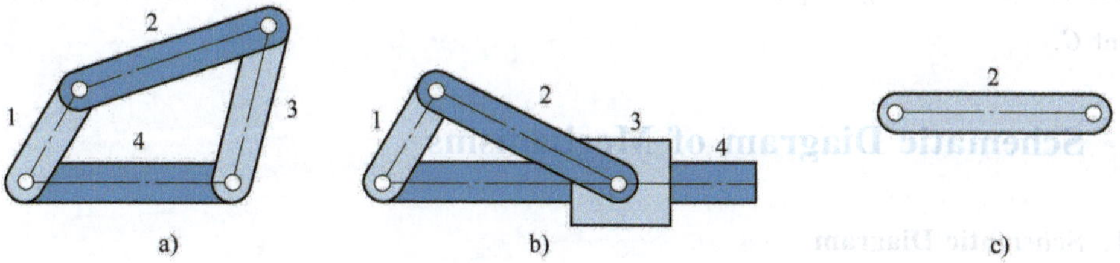


Fig. 2-5　Closed kinematic chains（闭链）

若各构件之间没有形成首尾封闭的系统，则称之为开链。开链中首尾构件仅含有一个运动元素。图 2-6 所示的运动链为开链，构件 3、4 只含有一个运动元素。开链在机器人领域中有广泛应用。

图 2-7 所示的构件系统中，各构件间均不能做相对运动。因此，它们不是运动链，而是桁架，该系统在运动中只相当于一个运动单元，即是一个构件。

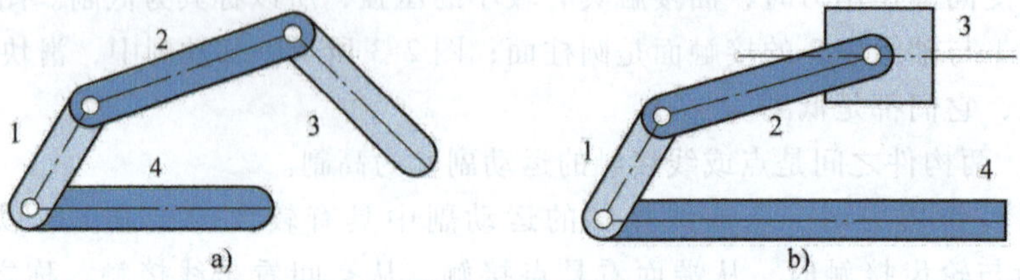


Fig. 2-6 Unclosed kinematic chains(开链)

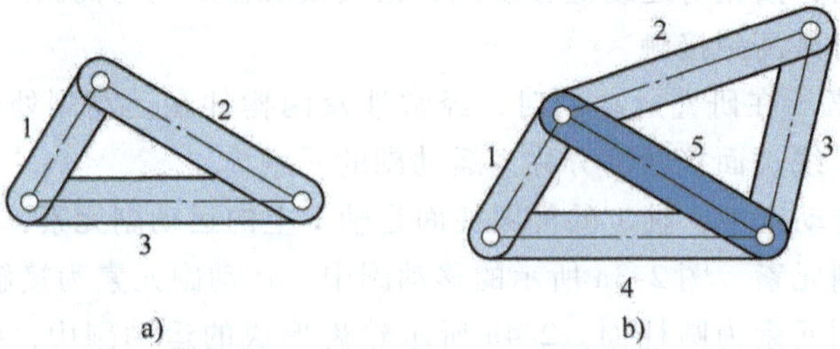


Fig. 2-7 Structures(桁架)

**4. Mechanism**

If one link of a kinematic chain is fixed to the ground, the kinematic chain becomes a mechanism. The mechanism can be classified as planar and spatial mechanism. In planar mechanism, all links have parallel motion to each other, and in spatial mechanism, some of links have spatial motion. If all the pairs in a mechanism are lower pairs, the mechanism is called a lower pair mechanism. If a mechanism has one or more higher pairs, the mechanism is called a higher pair mechanism.

Fig. 2-8 shows some lower pair mechanisms.

Fig. 2-9 shows a higher pair mechanism in which links 1 and 2 are connected by a higher pair at point $C$.

## 2.2 Schematic Diagram of Mechanisms

**1. Schematic Diagram**

To take the trouble to show the links of a mechanism in their true shapes would be a laborious and unnecessary chore. Only a few dimensions of the links of a mechanism are relevant to mechanism analysis and synthesis.

A simple diagram in which the links and pairs are represented by some simple lines and pair symbols to describe the composition of a mechanism is called a schematic diagram of mechanisms.

The schematic diagram takes one or two forms: a schematic diagram and scaled schematic diagram. A schematic diagram is proportional but not exactly to scale, while a scaled schematic diagram requires a "stripped-down" stick diagram, "stripped-down" stick diagram is usually used for further motion analysis and force analysis.

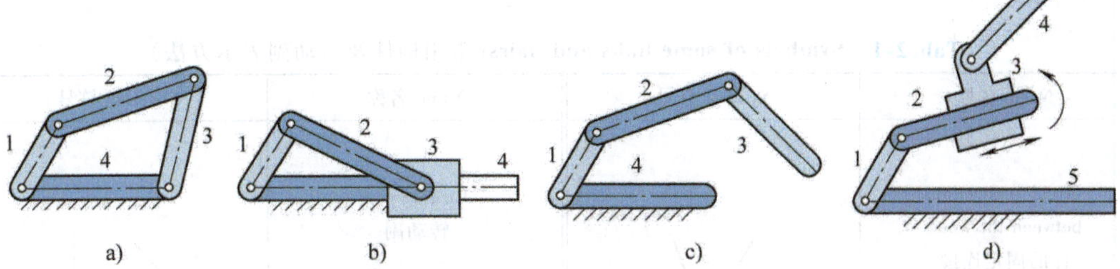

Fig. 2-8 Mechanisms in which all the pairs are lower pairs（低副机构）

**4. 机构**

在运动链中，若选定某个构件为机架，则该运动链成为机构。

机架是固定不动的构件，如安装在车辆、船舶、飞机等运动物体上的机构。机架相对于该运动物体是固定不动的。

机构中各构件的运动平面若互相平行，则称为平面机构；若机构中至少有一个构件不在相互平行的平面上运动，或至少有一个构件能在三维空间中运动，则称为空间机构。

固定图 2-5a、b 所示运动链中的构件 4 后，可得到图 2-8a、b 所示的闭链机构；固定图 2-6a所示开链中的构件 4 后，可得到图 2-8c 所示的开链机构。

完全由低副连接而成的机构，称为低副机构。连杆机构是常用的低副机构。

机构中只要含有一个高副，就称之为高副机构。图 2-9 所示机构在 $C$ 处用高副连接，故称为高副机构。齿轮机构、凸轮机构是常用的高副机构。

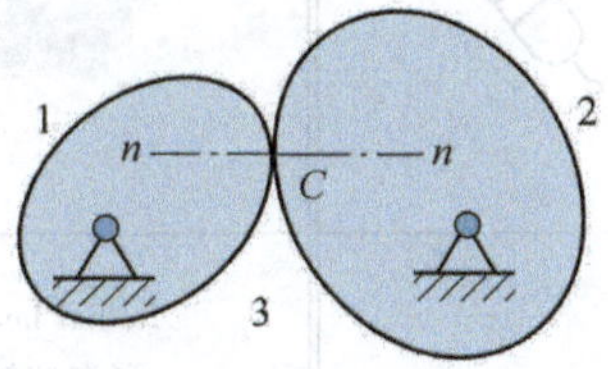

Fig. 2-9 Mechanism including higher pair（高副机构）

## 2.2 机构运动简图

**1. 机构运动简图**

机械设计与分析过程中，用简单的线条表示构件，用图形符号表示运动副，这样描述机构的组成和运动情况，概念清晰、简单实用。这种用简单的线条和运动副的图形符号表示机构的组成情况的简单图形称为机构简图。如按比例尺画出，则称之为机构运动简图，否则为机构示意图。机构运动简图所反映的主要信息是：机构中构件的数目、运动副的类型和数目、各运动副的相对位置即运动学尺寸。而对于构件的外形、断面尺寸、组成构件的零件数目及连接方式，在画机构运动简图时均不予考虑。

## 2. Symbols of Common Used Links and Pairs

The special symbols used in a schematic diagram of mechanisms are listed in Tab. 2-1.

**Tab. 2-1 Symbols of some links and pairs(常用构件及运动副表示方法)**

| Name 名称 | Symbol 符号 | Name 名称 | Symbol 符号 |
|---|---|---|---|
| Rigid connection between the links 杆的固定连接  |  | Turning pair 转动副 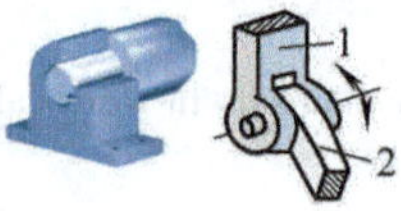 | 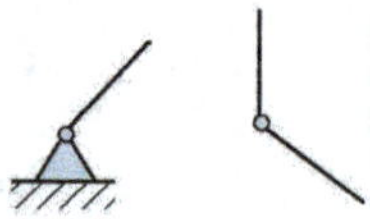 |
| Binary link 二副构件  |  | Prismatic pair 移动副  | 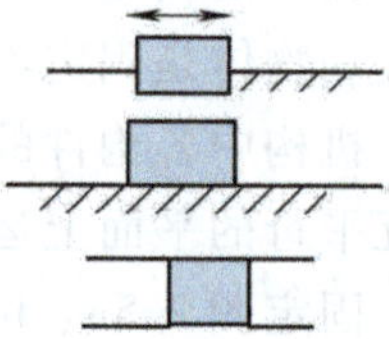 |
| Ternary link 三副构件 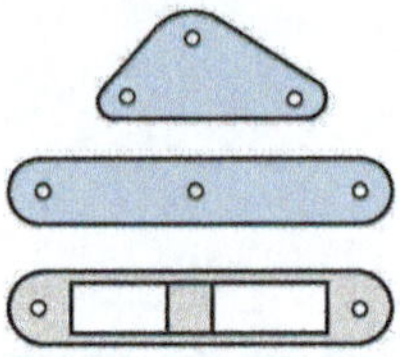 | 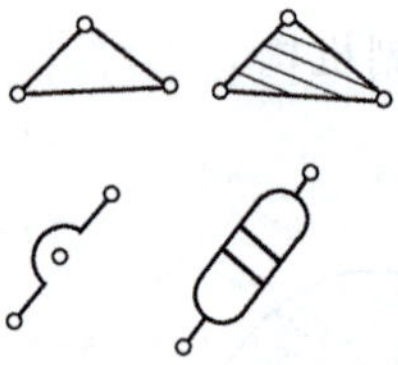 | Electric motor 电动机  | 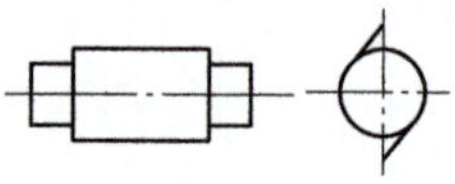 |
| Helical pair 螺旋副  | 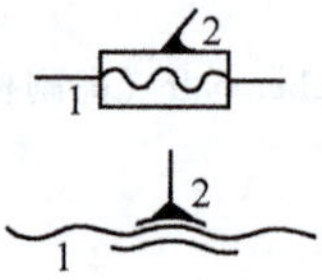 | Radial bearing 深沟球轴承  | 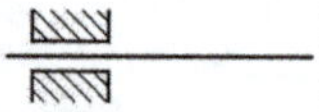 |
| Thrust bearing 推力轴承 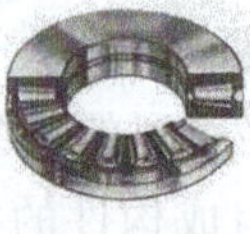 | 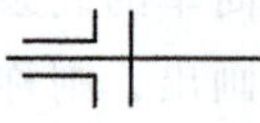 | Pinion and rack 齿轮齿条机构  | 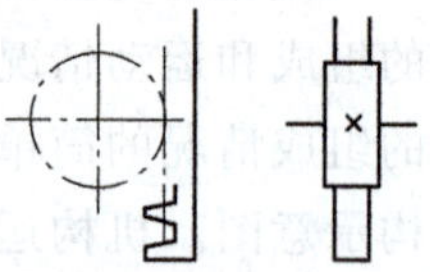 |

（续）

| Name 名称 | Symbol 符号 | Name 名称 | Symbol 符号 |
|---|---|---|---|
| Cam mechanism 凸轮机构 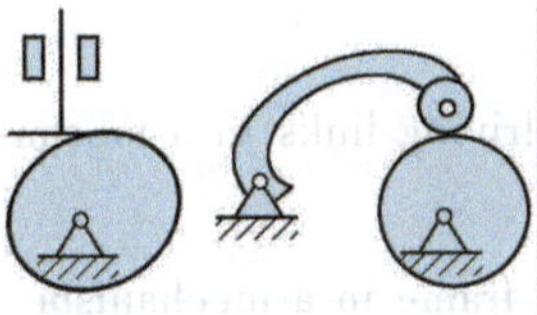 | 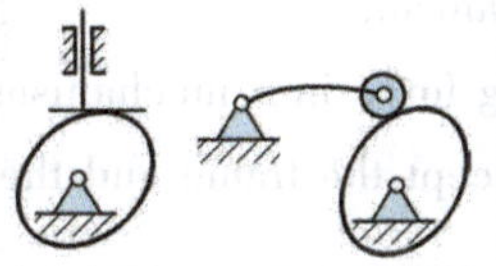 | Bevel gear 锥齿轮  | 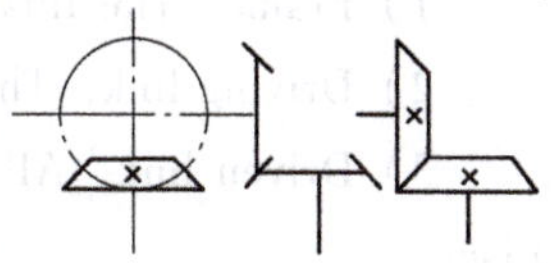 |
| Belt drive 带传动  | 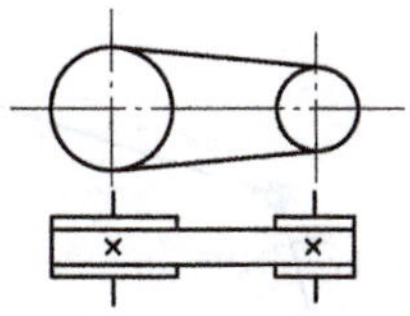 | Worm gear 蜗杆传动  | 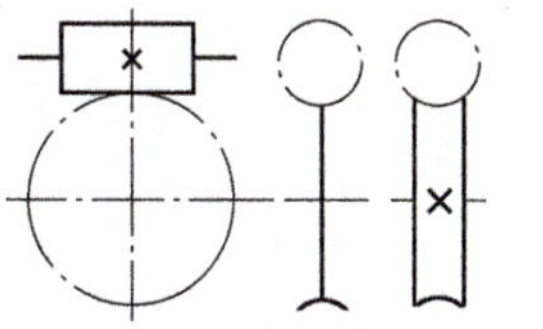 |
| Chain drive 链传动 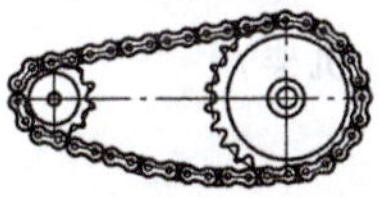 | 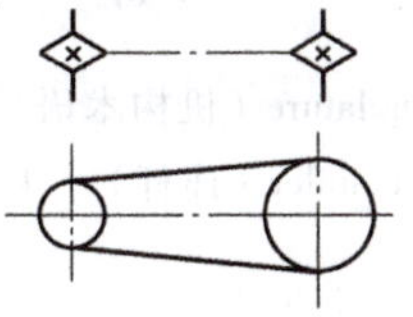 | Ratchet wheel 棘轮机构  | 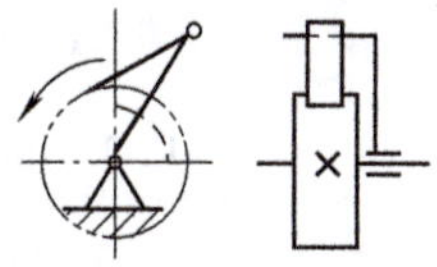 |
| External gear 外啮合齿轮  | 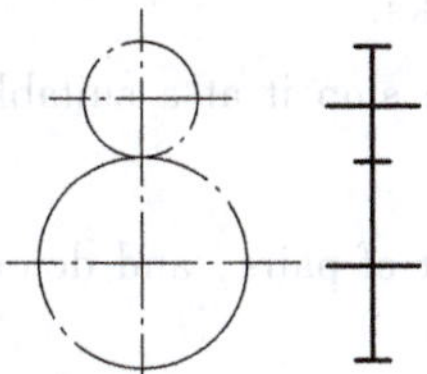 | Coupling 联轴器  | 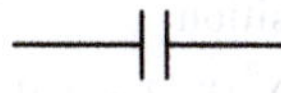 |
| Internal gear 内啮合齿轮 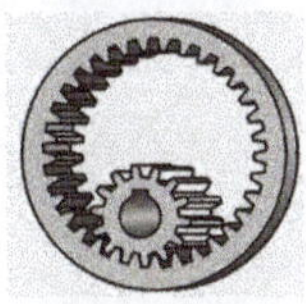 | 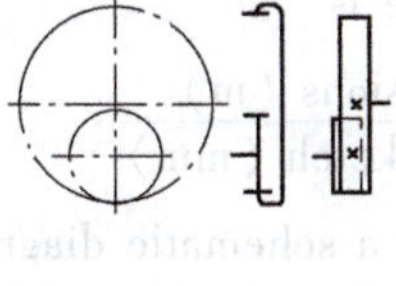 | Brake 制动器 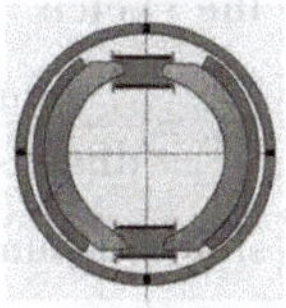 | 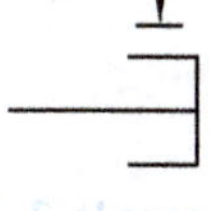 |

机构运动简图应与原机构具有相同的运动特性，因此须按一定的长度比例尺来画。长度比例尺 $\mu_l$ 采用如下定义：

$$\mu_l = \frac{\text{运动尺寸的实际长度}}{\text{图上所画的长度}}\left(\frac{\text{m}}{\text{mm}}\text{或}\frac{\text{mm}}{\text{mm}}\right)$$

严格按照比例尺正确画出的机构运动简图，可作为图解法运动分析与机构设计的依据。

**2. 常用构件与运动副画法**

在画机构运动简图时，必须对构件和运动副的画法进行一些规定。具体规定可参阅表 2-1给出的常用构件和运动副的符号。

### 3. The Procedure of Drawing a Schematic Diagram of a Mechanism

(1) Mechanism nomenclature When mentioning a mechanism, the following terms are often used.

1) Frame. The link which is fixed in a mechanism.

2) Driving link. The link acted by the driving force in a mechanism.

3) Driven link. All the other moving links except the frame and the driving links in a mechanism.

4) Coupler or connected rod. Links which are not connected with the frame in a mechanism. They are shown in Fig. 2-10.

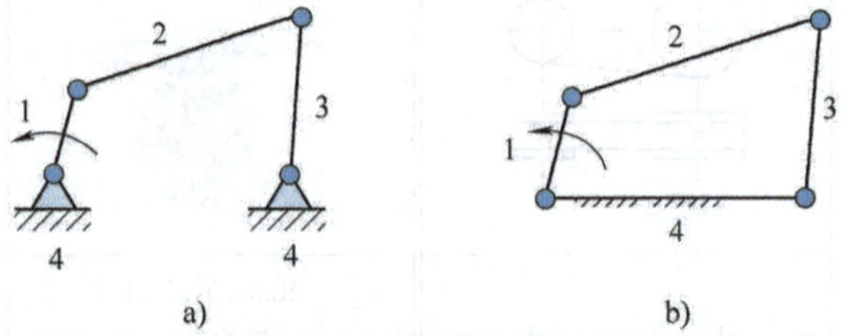


Fig. 2-10 Mechanism nomenclature (机构术语)

1、3—link connected frame (连架杆) 2—coupler (连杆) 4—frame (机架)

(2) The procedure of drawing a schematic diagram

1) Find out the driving links and the driven links.

2) Run the mechanism slowly for a while, then stop it at a suitable position, and observe its composition.

3) Find out the number of links and the number of pairs, and determine the type of pairs from input link to the output link.

4) The plane on which most links move can be selected as a drawing plane.

5) The dimensions between two pairs and the other kinematic dimensions must be measured, then select proper scale to draw the sketch. The scale is

$$\mu_l = \frac{\text{actual dimensions (m)}}{\text{dimensions in sketch (mm)}}$$

**Example 2-1** Fig. 2-11a shows a pump. Draw a schematic diagram of the pump.

**Solution**

The eccentric disk is a driving link, and it rotates about the fixed pin $A$. The eccentric disk is connected with toroid of the coupler by a turning pair at $B$. Line $AB$ represents eccentric disk 1 in Fig. 2-16. The coupler is connected with the slider by a turning pair at $C$. The line $BC$ represents coupler 2 in Fig. 2-11b. Slider 3 and frame 4 form a sliding pair, and its guide line is through the rotating center of the eccentric disk. The kinematic diagram is shown in Fig. 2-11b.

**Example 2-2** Fig. 2-12a shows a shaper. Draw a schematic diagram of the shaper.

**Solution**

The shaper consists of gear 1, gear 2, block 3, rocker 4, link 5, slide bar 6 and frame 7. Gear

**3. 机构运动简图画法**

（1）机构的基本术语　在图 2-10a、b 中，固定不动的构件 4 称为机架，与机架相连接的构件 1、3 称为连架杆，不与机架相连接的杆件 2 称为连杆。连杆是把一个连架杆的运动传递到另一个连架杆的传动构件，此类机构也称为连杆机构。

施加驱动力的构件，称为原动件或主动件。原动件一般是某个连架杆，如图 2-10 中的构件 1 为原动件，其运动方向可用图示箭头表示。原动件是设计人员根据机构运动要求自行确定的，原动件确定后，其余构件均为从动件。

（2）机构运动简图的具体画法

1）找出原动件和从动件。

2）使机构缓缓运动，观察其组成情况和运动情况。

3）沿主动件到从动件的传递路线找出构件数目和运动副的数目与种类。

4）选择大多数构件所在平面为投影面。

5）测量各运动副之间的尺寸，用运动副表示各构件的连接，选择适当的比例尺画出机构运动简图。

绘制机构运动简图是工程技术人员的一种基本技能。

**例 2-1**　画出图 2-11a 所示泵的机构运动简图。

**解**　图 2-11a 所示的泵，由偏心轮 1、连杆 2、滑块 3 和机架 4 组成。偏心轮 1 为主动件，各构件的连接关系如下：

偏心轮 1 与机架 4 在 $A$ 点处以转动副连接，偏心轮 1 与连杆 2 在 $B$ 点处以转动副连接，$AB$ 为构件 1。构件 2 与构件 3 在 $C$ 点处以转动副连接，$BC$ 为构件 2。构件 3 与机架 4 用移动副连接。选择合适投影面和比例尺，测量出 $AB$ 与 $BC$ 的尺寸以及滑块运动方向线偏离 $AB$ 的转动中心 $A$ 的距离。画出的机构运动简图如图 2-11b 所示。

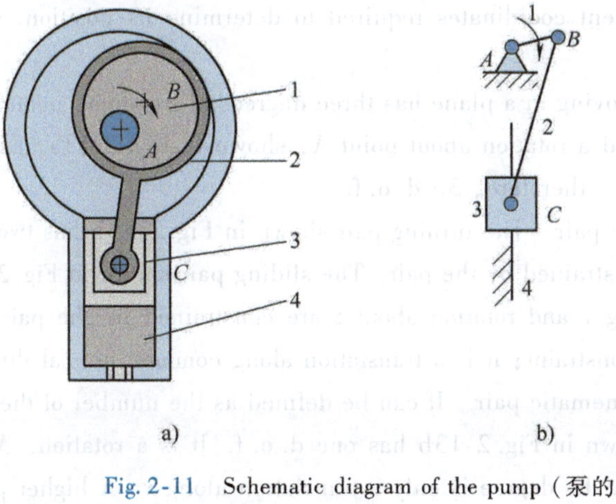


Fig. 2-11　Schematic diagram of the pump（泵的机构运动简图）

1—eccentric disk（偏心轮）　2—coupler（连杆）　3—slider（滑块）　4—frame（机架）

**例 2-2**　画出图 2-12a 所示牛头刨床的机构运动简图。

**解**　图 2-12a 所示的牛头刨床由小齿轮 1、大齿轮 2、滑块 3、摆杆 4、连杆 5、滑枕 6 和机架 7 组成，各构件间的连接关系如下：

小齿轮 1、大齿轮 2 与机架 7 在 $O_1$、$O_2$ 处以转动副连接，两齿轮以高副连接。大齿轮 2

1 and 2 rotate about the fixed pins $O_1$ and $O_2$. Block 3 rotates about pin $A$ which is fixed to gear 2, and it translates along the rocker guide. Link 5 is connected with rocker 4 and slide bar 6 by turning pairs at $C$ and $D$, and the ram translates along the frame. The schematic diagram of the shaper is illustrated in Fig. 2-11b.

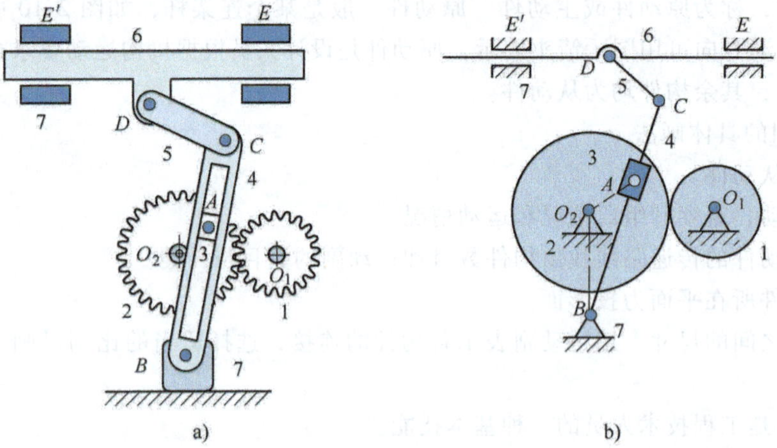


Fig. 2-12 Shaper and its schematic diagram（牛头刨床及其机构运动简图）
1、2—gear（齿轮） 3—block（滑块） 4—rocker（摆杆）
5—link（连杆） 6—slide bar（滑枕） 7—frame（机架）

## 2.3 Degree of Freedom of Planar Mechanisms

### 1. Gruebler's Equation

(1) Degree of freedom of a link  Degree of freedom is also called the mobility, and it can be defined as the number of independent coordinates required to determine its position. Sometimes it can be written as d. o. f.

An unconstrained rigid link moving in a plane has three degrees of freedom, namely both translations corresponding to $x$ and $y$ and a rotation about point A, shown in Fig. 2-13a, and such separate links would have a cluster of $n$, therefore, $3n$ d. o. f.

(2) Constraints of a kinematic pair  The turning pair shown in Fig. 2-13b has two constraints. The translation along $x$ and $y$ is constrained by the pair. The sliding pair shown in Fig. 2-13c has two constraints too; the translation along $y$ and rotation about $z$ are constrained by the pair. The higher pair shown in Fig. 2-13d has one constraint; it is a translation along common normal direction.

(3) Degree of freedom of a kinematic pair  It can be defined as the number of the independent relative motion. A turning pair shown in Fig. 2-13b has one d. o. f. It is a rotation. A sliding pair shown in Fig. 2-13c has one d. o. f. Its degree is only a translation along $x$. A higher pair shown in Fig. 2-13d has two d. o. f. They are a translation along the common tangent $t$—$t$ and a rotation about the contact point $A$.

(4) Degree of freedom of a planar mechanism  In a planar mechanism, the fixed link has zero degree of freedom; each moving link has 3 d. o. f., each lower pair has 2 constraints and each higher pair has 1 constraint. We suppose that there are moving links $n$, lower pairs $p_l$, higher pairs $p_h$, then the degree of freedom in a planar mechanism is as follows:

和滑块3在$A$点处以转动副连接，滑块3与摆杆4以移动副连接，摆杆4分别与机架7和连杆5以转动副在$B$点、$C$点处连接，连杆5与滑枕6以转动副在$D$点处连接，滑枕6与机架7在$E$点、$E'$点处以移动副连接。分别测量齿轮节圆半径，距离$O_1O_2$、$O_2A$、$O_2B$、$BC$、$CD$以及滑枕导路方向到$B$点的距离，选择投影面和比例尺，画出机构运动简图，如图2-12b所示。

用机构运动简图表示机构的组成情况，简单实用。后续内容将使用机构运动简图进行机构的分析与设计。

## 2.3 机构自由度的计算

**1. 平面机构自由度的计算公式**

（1）构件自由度　构件自由度是指自由运动的构件所具有的独立运动的数目。

图2-13a所示自由运动的构件在平面内有3个自由度，即沿$x$、$y$方向的移动和绕$A$点的转动。$n$个构件在平面内则有$3n$个自由度。

（2）运动副的约束　构件之间用运动副连接后，其相对运动就会受到约束。把这种运动副对构件运动产生的约束称为运动副约束。图2-13b所示的转动副中，沿$x$、$y$方向的移动受到约束；图2-13c所示的移动副中，约束了沿$y$方向的移动和绕$z$轴的转动（$z$轴垂直纸面，图中未画）；图2-13d所示的高副中，约束了沿两曲线公法线方向的移动。平面运动副提供的最大约束数目为2。

（3）运动副自由度　连接构件的运动副所具有的独立运动数目称为运动副自由度。

设平面运动副提供的约束数目为$C$，则该运动副的自由度为$3-C$。

图2-13b所示转动副的自由度为1，即绕$z$轴的转动；图2-13c所示移动副的自由度为1，即沿$x$轴的移动；图2-13d所示高副的自由度为2，即沿公切线$t$—$t$的移动和绕切点$A$的转动。

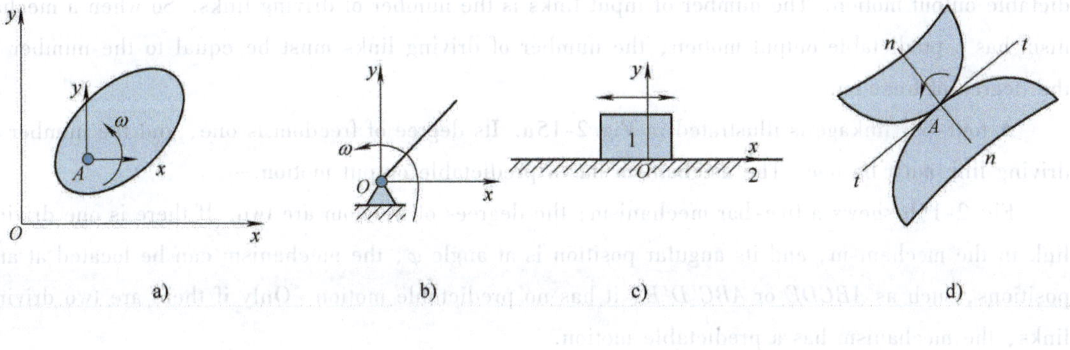


Fig. 2-13　Constrains of pairs（运动副的约束）

（4）平面机构的自由度及其计算

1）机构的自由度。机构只有实现确定的运动，才能满足特定的功能要求。机构具有确定运动时，所具有的独立运动参数的数目，称为机构的自由度。

$$F = 3n - 2p_l - p_h \tag{2-1}$$

Where $F$ is the number of degrees of freedom in a planar mechanism; $n$ is the number of all the moving links in a mechanism; $p_l$ is the number of the lower pairs in a mechanism; $p_h$ is the number of the higher pairs in a mechanism.

**Example 2-3** Determine the degrees of freedom of the mechanisms shown in Fig. 2-14.

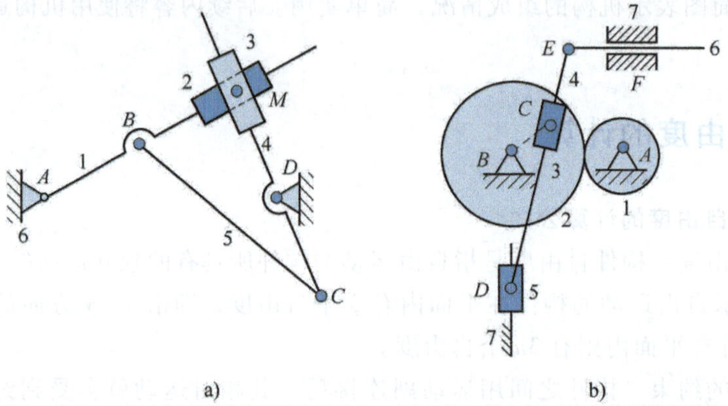


Fig. 2-14 The caculation of degree of freedom (自由度计算)

**Solution**

Fig. 2-14 a: $n = 5$, $p_l = 5$, $p_h = 0$

$$F = 3n - 2p_l - p_h = 3 \times 5 - 2 \times 7 - 0 = 1$$

Fig. 2-14 b: $n = 6$, $p_l = 8$, $p_h = 1$

$$F = 3n - 2p_l - p_h = 3 \times 6 - 2 \times 8 - 1 = 1$$

### 2. Conditions Having Predictable Motion in a Mechanism

The degree of freedom of a mechanism is the number of independent coordinates to define its position, and is also the number of input links which need to be provided in order to create a predictable output motion. The number of input links is the number of driving links. So when a mechanism has a predictable output motion, the number of driving links must be equal to the number of the degree of freedom.

A four-bar linkage is illustrated in Fig. 2-15a. Its degree of freedom is one, and the number of driving link must be one. The mechanism has a predictable output motion.

Fig. 2-15b shows a five-bar mechanism; the degrees of freedom are two. If there is one driving link in the mechanism, and its angular position is at angle $\varphi$, the mechanism can be located at any positions, such as $ABCDE$ or $ABC'D'E$, it has no predictable motion. Only if there are two driving links, the mechanism has a predictable motion.

### 3. Points for Attention When Calculating Degree of Freedom

When we determine the degree of freedom of a mechanism, sometimes it may be not the same with the actual number of degree of freedom. The following attention must be considered when calculating the degree of freedom.

(1) Redundant degree of freedom  Sometimes, one or more links of a mechanism may be moved without causing any motion to the other links of the mechanism. Such a link is said to have a

2）机构自由度的计算。一个平面低副提供 2 个约束，设机构中有 $p_l$ 个低副，则提供 $2p_l$ 个约束。一个平面高副提供 1 个约束，设机构中有 $p_h$ 个高副，则提供 $p_h$ 个约束。机构中各运动副提供的约束总数为 $2p_l+p_h$。因此，机构的自由度 $F$ 为

$$F=3n-2p_l-p_h \tag{2-1}$$

式中，$n$ 为机构中活动构件的数目；$p_l$ 为机构中低副的数目；$p_h$ 为机构中高副的数目。式 (2-1) 即为计算平面机构自由度的一般公式。

**例 2-3**　计算图 2-14a 所示双曲线画规机构和图 2-14b 所示牛头刨床机构的自由度。

**解**　图 2-14a 中：活动构件数 $n=5$，低副数 $p_l=5$，高副数 $p_h=0$，故

$$F=3n-2p_l-p_h=3\times5-2\times7-0=1$$

该机构的自由度为 1，说明该机构具有 1 个独立运动参数。

图 2-14b 中：活动构件数 $n=6$，低副数 $p_l=8$，高副数 $p_h=1$，故

$$F=3n-2p_l-p_h=3\times6-2\times8-1=1$$

该机构的自由度也为 1，说明该机构具有 1 个独立运动参数。

机构自由度是机构的固有属性，只要机构中的构件数、运动副数和种类确定后，其自由度就确定了，所需独立运动的参数也就确定了。

**2. 机构具有确定运动的条件**

机构具有确定运动是指：当给定机构原动件的运动时，该机构中的其余运动构件也都随之做相应的确定运动。

如果机构中的自由度等于原动件的数目，则该机构具有确定运动。因此，机构是否具有确定的运动，与机构的自由度及给定的原动件数目有关。

图 2-15a 所示的四杆机构中，机构自由度为 1，若给定 1 个原动件，则该机构有确定运动。如给定构件 1 的角位移 $\varphi$，则其余构件的位置都是完全确定的。如果原动件 $AB$ 运动到 $AB'$，则该机构由 $ABCD$ 位置运动到唯一的位置 $AB'C'D$。

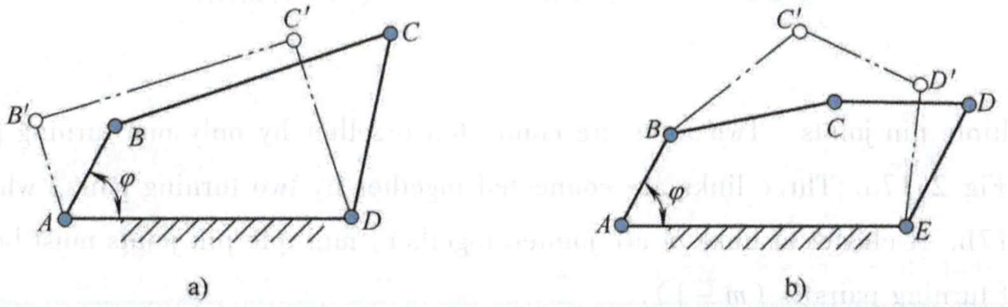


Fig. 2-15　Conditions of causing definite and predictable motions（机构具有确定运动的条件）

图 2-15b 所示的五杆机构中，机构自由度为 2。当 $AB$ 为原动件的给定位置时，其余构件的位置并不能确定。很明显，当原动件占据位置 $AB$ 时，其余构件既可分别占有位置 $BC$、$CD$、$DE$，也可占有位置 $BC'$、$C'D'$、$D'E$，还可以占有其他位置。但若再给定一个原动件位置，如给定 $DE$ 为另一原动件的位置，即同时给定 2 个原动件位置，则不难看出该五杆机构中各构件的运动便完全确定了。

从以上两例中可看出，只有当给定的原动件数目与机构的自由度相等时，才可使机构具有确定的运动。

如果机构自由度等于或大于 1，能否具有确定运动，取决于原动件数目是否等于自由度。当自由度小于 1 时，该机构蜕变为桁架，此时已经不是机构了。

redundant degree of freedom. For example, in the cam mechanism shown in Fig. 2-16a, its degree of freedom is as follows:

$$F = 3n - 2p_l - p_h = 3 \times 3 - 2 \times 3 - 1 = 2$$

But in this mechanism, roller 2 connected to the follower 3 with a turning pair can rotate about its axis without causing any movement to the other links. It is a redundant rotation from the standpoint of movement. When meeting a problem like this, you may weld roller 2 to follower 3, and the motion of follower 3 would remain unchanged (Fig. 2-16b). If we "weld" roller 2 to follower 3, the degree of freedom of the cam mechanism is:

$$F = 3n - 2p_l - p_h = 3 \times 2 - 2 \times 2 - 1 = 1$$

To reduce the wear between two contact surfaces, the sliding friction is often replaced by rolling or rolling-sliding friction.

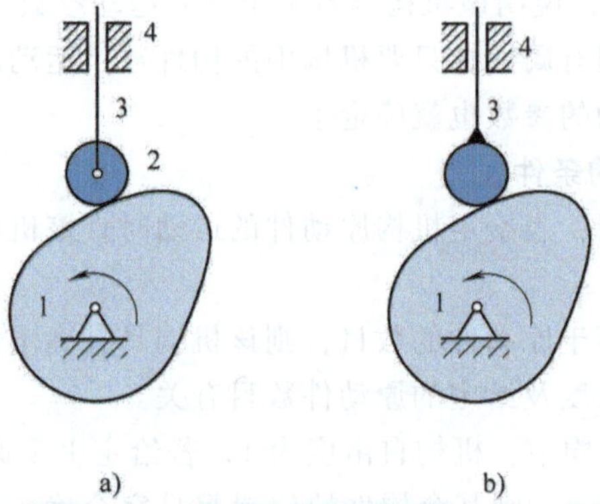


Fig. 2-16 Partial degree of freedom(局部自由度)

(2) Multiple pin joints Two links are connected together by only one turning pair, which is illustrated in Fig. 2-17a. Three links are connected together by two turning pairs, which is illustrated in Fig. 2-17b. A cluster of links $m$ are jointed together, multiple pin joints must be needed, and the number of turning pairs is $(m-1)$.

Some of the multiple pin joints are illustrated in Fig. 2-18.

(3) Redundant constraints Sometimes, a mechanism may have one or more redundant constraints which do not effect the movement of links, or a mechanism may have one or more links which do not introduce any extra constraint. For example, the parallel-crank mechanism shown in Fig. 2-19 has 5 links, but the function of link 5 is not affected. It is a redundant link with two turning pairs.

To find out the redundant constraints is not difficult, but to find out the redundant links sometimes is very difficult. We can put forward some proposals to find out the redundant constraints and links.

1) Two links are connected by several turning pairs and their axes are coincident. There is only

**3. 计算机构自由度的注意事项**

在计算平面机构的自由度时，有时会出现计算出的自由度与机构的实际运动不一致的现象，对其进行分析并寻求解决方法是必要的。

(1) 局部自由度　在某些机构中，某个构件所产生的相对运动并不影响其他构件的运动，把这种不影响其他构件运动的自由度称为局部自由度。

图 2-16a 所示凸轮机构的自由度为

$$F=3n-2p_l-p_h=3\times3-2\times3-1=2$$

显然，该机构只需要一个主动件就有确定的运动，滚子的转动与推杆的运动规律无关，其作用仅仅是把滑动摩擦转换为滚动摩擦。滚子 2 绕自身轴线的转动不影响机构运动，称为局部自由度。处理方法是把滚子 2 固化在支承滚子的构件 3 上，去掉局部自由度，如图 2-16b 所示。其自由度为

$$F=3n-2p_l-p_h=3\times2-2\times2-1=1$$

局部自由度经常出现在用滚动摩擦代替滑动摩擦的场合，这样可减小机构的磨损。

(2) 复合铰链　两个以上的构件在同一处以转动副连接，则形成复合铰链。

图 2-17a 所示的两个构件 1 和 2 在一处用转动副连接时，仅需一个转动副。当图 2-17b 所示的三个构件用转动副连接时，则有两个转动副。$m$ 个构件在一起用转动副连接时，则有 $m-1$ 个转动副。

在计算机构自由度时，必须注意正确判别复合铰链，否则会发生计算错误。图 2-18 给出了一些典型的三个构件连接的复合铰链示例。

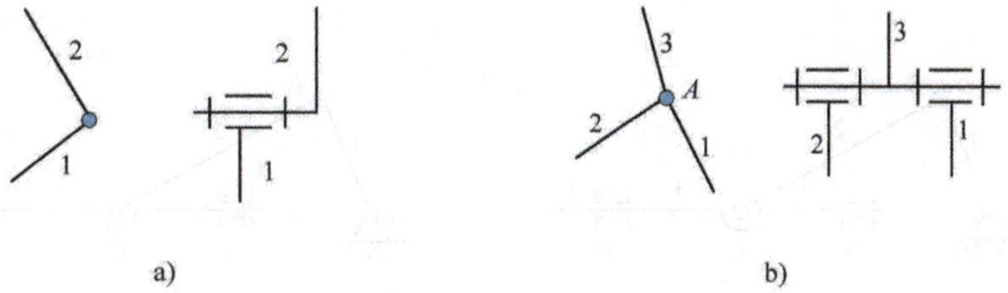


Fig. 2-17　Multiple pin joints (复合铰链)

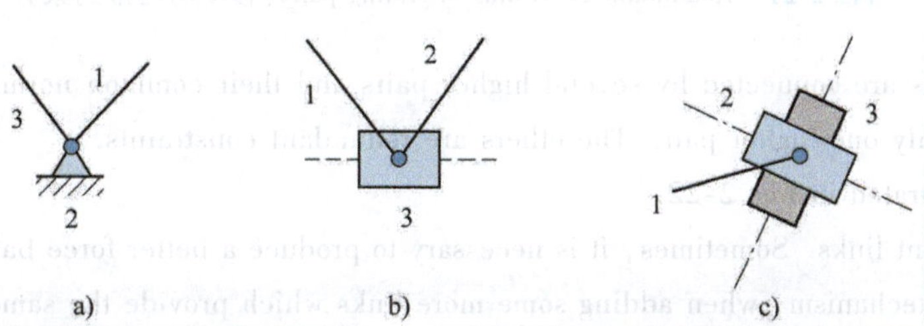


Fig. 2-18　Examples of multiple pin joints (复合铰链的示例)

(3) 虚约束　对机构运动不起限制作用的约束称为虚约束。图 2-19 中实线所示的平行四边形机构，其自由度 $F=1$。

若在构件 2 和机架 4 之间与 $AB$ 或 $CD$ 平行地铰接一构件 $EF$，且 $EF$ 的尺寸等于构件 $AB$ 和 $CD$ 的尺寸，此时，机构自由度为

$$F=3n-2p_l-p_h=3\times4-2\times6-0=0$$

one turning pair. The others are redundant constraints.

This is illustrated in Fig. 2-20.

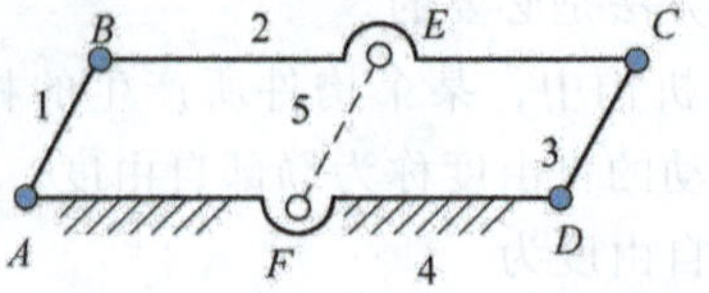


Fig. 2-19 Redundant constrain in the parallel-crank mechanism(平行四边形机构的虚约束)

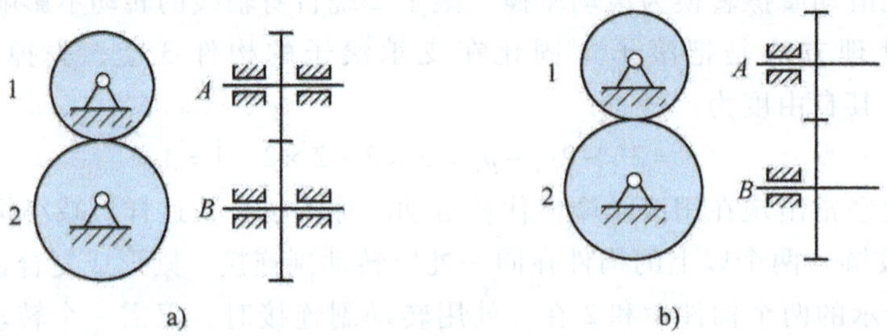


Fig. 2-20 Redundant constrains of turning pairs(转动副的虚约束)

2) Two links are connected by several sliding pairs and their guide lines are parallel. There is only one sliding pair. The others are redundant constraints.

This is illustrated in Fig. 2-21.

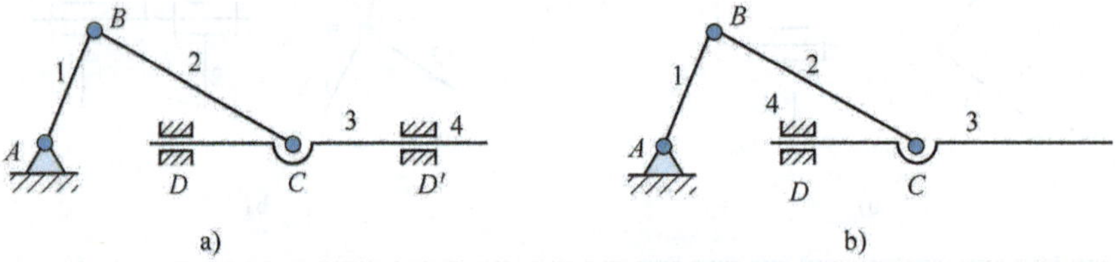


Fig. 2-21 Redundant constrains of sliding pairs(移动副的虚约束)

3) Two links are connected by several higher pairs and their common normal lines are coincident. There is only one higher pair. The others are redundant constraints.

This is illustrated in Fig. 2-22.

4) Redundant links. Sometimes, it is necessary to produce a better force balance or to improve structure of the mechanism, when adding some more links which provide the same constrained function.

Fig. 2-23a shows a planetary gear train, and there is a redundant gear $z_{2'}$.

Fig. 2-23b shows a ellipse trammel, and there is a redundant link, such as link $OA$. In this mechanism, one of the two blocks can also be a redundant link.

Fig. 2-24 shows another type of redundant links. The redundant link is produced by connecting two equidistance points.

**Example 2-4** Calculate the degree of freedom of the mechanism shown in Fig. 2-25.

很明显，该计算结果与实际情况是不相符的。因为在连接构件 $EF$ 前，构件 2 上 $E$ 点的轨迹是以 $F$ 为圆心、以 $EF=AB$ 为半径的圆弧，$EF$ 没有起到对构件 2 的约束作用，是虚约束。

这说明虚约束会影响计算自由度的正确性。处理手段是将机构中构成虚约束的构件连同其所附带的运动副一概去掉不计。

机构中的虚约束不会影响机构的运动情况，但却能改善机构的受力情况并增加机构的刚度。从机构运动的角度看，虚约束是多余的，但从机械结构的角度看，虚约束又是必要的。

虚约束的类型较多，比较复杂，在计算自由度时要特别注意。为便于判断，将常见的几种虚约束形式简述如下：

1）两构件在多处用转动副连接，且各转动副的轴线重合，这时只有一处转动副起作用，其余转动副均为虚约束。图 2-20a 所示的齿轮机构中，每根轴处都有两个转动副。计算机构自由度时，每根轴上仅计一个转动副，其余为虚约束，如图 2-20b 所示。

2）两构件在多处用移动副连接，且各移动副的导路平行，这时只计一个移动副，其余为虚约束。图 2-21a 所示的机构中，构件 3 与机架 4 用两个移动副 $D$、$D'$ 连接，且导路平行，计算机构自由度时，仅考虑一个移动副，其余为虚约束。

3）两构件在多处用高副连接，且各高副的公法线重合，这时只计一个高副约束，其余为虚约束。图 2-22 所示的机构中，圆形构件与框架在 $A$、$B$ 两处形成两个高副，且各高副处的公法线重合，计算机构自由度时，仅考虑一个高副，其余为虚约束。

4）不起约束作用的构件将导致虚约束，在计算机构自由度时要去掉该构件。图 2-23a 所示的轮系机构中，齿轮 $z_1$、$z_2$、$z_3$ 和 H 组成一个具有确定运动的轮系机构，为平衡行星齿轮 2 的惯性力，在其对称方向又安装了一个行星轮，该行星轮连同支承该齿轮的转动副为虚约束，计算自由度时应该去掉。

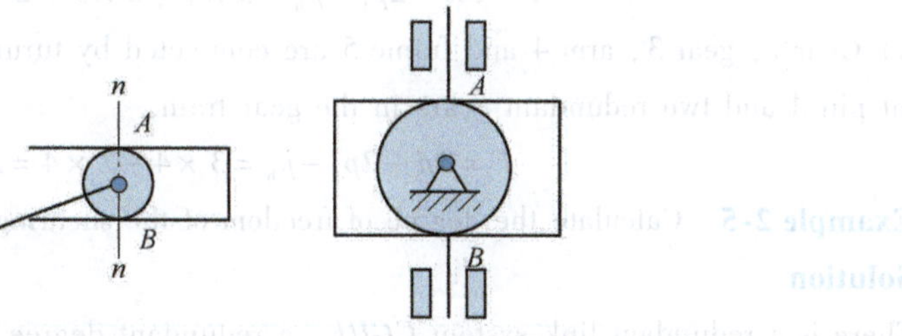


Fig. 2-22 Redundant constrains of higher pairs（高副机构的虚约束）

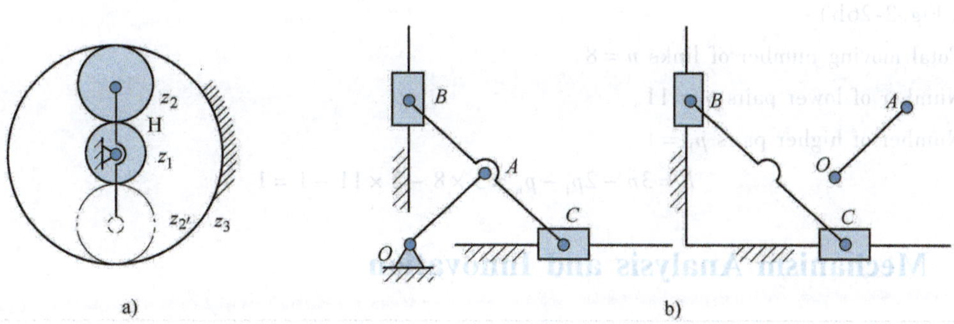


Fig. 2-23 Redundant constraints（虚约束）

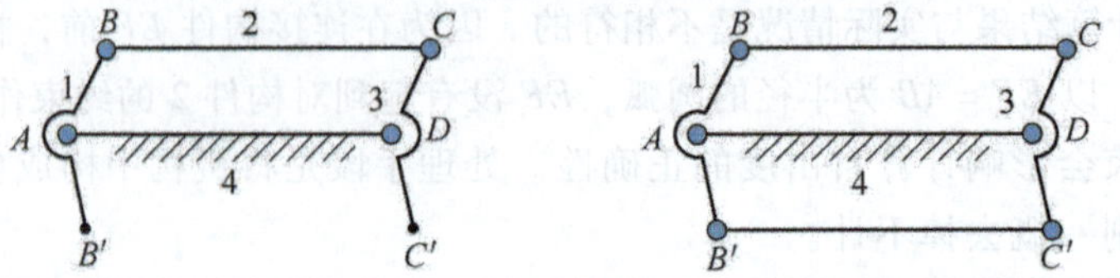


Fig. 2-24 Redundant constrain produced by connecting two equidistance points
(连接等距点产生的虚约束)

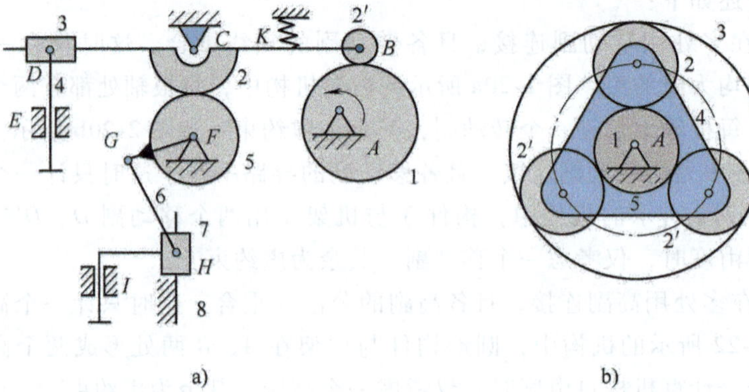


Fig. 2-25 Degree of freedom of the complex mechanism (复杂机构的自由度)

**Solution**

a) There is a redundant degree of freedom in a cam pair and a redundant constraint at position $I$ or $H$, $n=7$, $p_l=9$, $p_h=2$

$$F=3n-2p_l-p_h=3\times 7-2\times 9-2=1$$

b) Gear 1, gear 3, arm 4 and frame 5 are connected by turning pairs. There are three turning pairs at pin $A$ and two redundant gears in the gear train.

$$F=3n-2p_l-p_h=3\times 4-2\times 4=2$$

**Example 2-5** Calculate the degree of freedom of the shearing mechanism shown in Fig. 2-26a.

**Solution**

There is a redundant link system $CGHF$, a redundant degree of freedom in the cam pair and a multiple pin hinges at point $C$ in the mechanism. After we consider these points of attention, we have (Fig. 2-26b):

Total moving number of links $n=8$

Number of lower pairs $p_l=11$

Number of higher pairs $p_h=1$

$$F=3n-2p_l-p_h=3\times 8-2\times 11-1=1$$

## 2.4 Mechanism Analysis and Innovation

### 1. Link Group Analysis

(1) Driving link The driving link may rotate about its axis or translate along a guideline, and

图 2-23b 所示机构中，$\overline{AB}=\overline{AC}=\overline{OA}$，没有 $OA$ 之前，$A$ 点的运动轨迹是以 $O$ 为圆心、$OA$ 为半径的圆。加装 $OA$ 后，$A$ 点的轨迹没改变，因此 $OA$ 为虚约束。在计算自由度时应该去掉带有两个转动副元素的构件 $OA$。这类约束的判断比较复杂，一般要经过几何证明。

5）若两构件上两点间距离在运动过程中始终保持不变，当用运动副和构件连接该两点时，则构成虚约束。如图 2-24 所示机构中，$B'$、$C'$两点之间的距离不随机构的运动而改变，杆件 $B'C'$连同转动副元素 $B'$、$C'$为虚约束，计算机构自由度时必须将其去掉。

正确处理虚约束是计算机构自由度的难点。

**例 2-4**　计算图 2-25 所示机构的自由度。

**解**　图 2-25a 中的弹簧 $K$ 对计算机构自由度没有影响，滚子 2′有一个局部自由度，构件 7 与机架 8 在平行的导路上组成两个移动副，其中之一为虚约束。通过分析可知，运动构件 $n=7$，低副 $p_l=9$，高副 $p_h=2$，机构自由度为

$$F=3n-2p_l-p_h=3\times7-2\times9-2=1$$

图 2-25b 所示的轮系机构中，齿轮 2′为虚约束，太阳轮 1、齿轮 3、系杆 4 及机架 5 共 4 个构件在 $A$ 处组成转动副，构成复合铰链。$A$ 处的转动副实际为 3 个。通过分析可知，该轮系 $n=4$，$p_l=4$，$p_h=2$，机构自由度为

$$F=3n-2p_l-p_h=3\times4-2\times4-2=2$$

**例 2-5**　计算图 2-26 所示剪床机构的自由度。

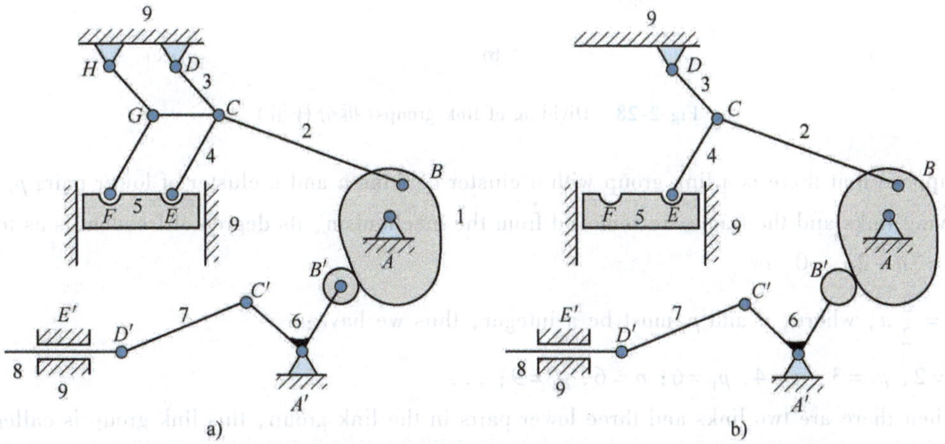


Fig. 2-26　Degree of freedom of the shearing mechanism（剪床机构的自由度）

**解**　图 2-26a 中，由于 $C$、$G$ 两点等距，构件 $GC$ 为虚约束，杆组 $FGH$ 为不起作用的重复约束。运动副 $C$ 处为复合铰链，$B'$处为局部自由度。将图 2-26a 所示机构等效为图 2-26b 所示机构后，可知 $n=8$，$p_l=11$，$p_h=1$，该机构的自由度为

$$F=3n-2p_l-p_h=3\times8-2\times11-1=1$$

## 2.4　机构分析与创新

### 1. 杆组分析

（1）原动件　原动件由运动副连接机架和一个杆件组成，自由度为 1，常做定轴转动或往复移动，如图 2-27 所示。

it has one degree of freedom. Fig. 2-27 shows two kinds of driving links.

(2) Link group  Any mechanism consists of driving links, driven links and a frame. The number of driving links is equal to the number of degrees of freedom. Therefore, if the driving links and the frame which has zero degree of freedom are removed from the mechanism, the degree of freedom of the link group which is left must be zero.

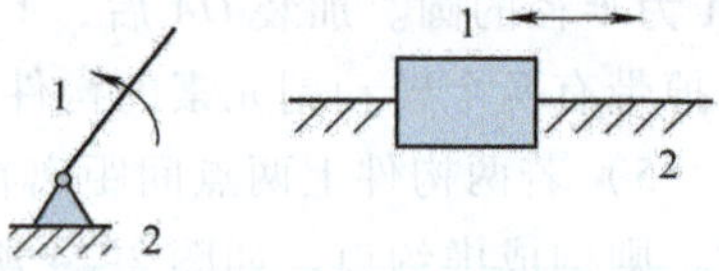

Fig. 2-27  Driving links(原动件)

For example, Fig. 2-28 shows a one-degree-of-freedom mechanism, and the driving link is crank *AB*. After the driving link *AB* and the frame have been removed from the mechanism, the degree of freedom of the link group *BCDEF* is zero.

The link group *BCDEF* may be divided into two link groups, each link group consists of three pairs and two links, and this link group can not be divided any more. They are called basic link groups, which are illustrated in Fig. 2-28c.

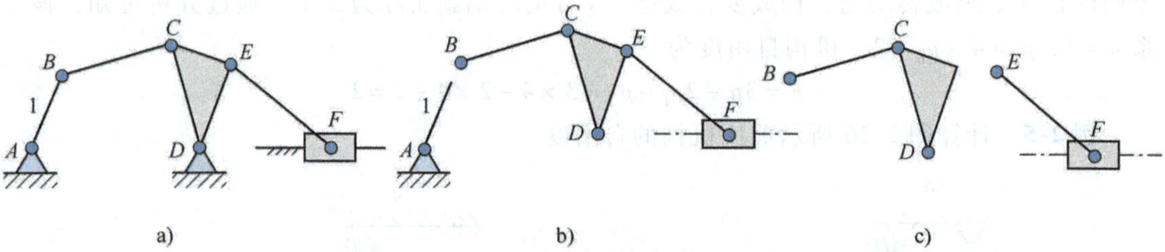

Fig. 2-28  Dividing of link groups(拆分杆组)

Suppose that there is a link group with a cluster of links $n$ and a cluster of lower pairs $p_1$, when the driving links and the frame are removed from the mechanism, its degree of freedom is as follows:

$F = 3n - 2p_1 = 0$, or

$p_1 = \frac{3}{2}n$, where, $n$ and $p_1$ must be a integer, thus we have:

$n = 2$, $p_1 = 3$; $n = 4$, $p_1 = 6$; $n = 6$, $p_1 = 9$; ...

When there are two links and three lower pairs in the link group, this link group is called class Ⅱ link group. There is one pair which connects two links in the link group and two pairs which will connect the other links. The class Ⅱ link groups are illustrated in Fig. 2-29.

When there are four links and six lower pairs in the link group, the link group is called class Ⅲ link group. There are three pairs which connect links in the link group and three pairs which will connect the other links. The common class Ⅲ link groups are illustrated in Fig. 2-30.

Another link group in which there are four links and six lower pairs is illustrated in Fig. 2-31. There are four pairs which connect links in the link group and two pairs which will connect the other links. We called this link group as class Ⅳ link group. This kind of link group is used rarely.

**2. Principle of Mechanism Composition**

Any mechanism can be designed by connecting basic link group in which the degree of freedom is zero with the driving link and the frame. This is illustrated in Fig. 2-32.

This is a good method to create a new mechanism.

（2）杆组　前述已知，机构具有确定运动时，该机构的自由度等于原动件的数目。如果去掉原动件，则剩余部分杆件系统的自由度为零，并称之为杆组。本书仅考虑低副连接而成的杆组。

把自由度为零且不能再分割的杆组称为基本杆组。图 2-28a 所示机构中，其自由度为 1。去掉原动件 *AB* 后，相当于减少 1 个自由度，则图 2-28b 所示的剩余杆件系统 *BCDEF* 的自由度一定为零。自由度为零的杆件系统 *BCDEF* 还可以进一步拆分为图 2-28c 所示的自由度为零的杆组 *BCD* 和 *EF*。这两个杆组都是由两个构件和三个低副组成的杆组，已不能再进行拆分。

由于杆组自由度为零，则有

$$3n-2p_1=0$$

其中构件数 $n$ 和运动副数 $p_1$ 都必须是整数。$n$ 和 $p_1$ 满足下列关系：

$$p_1=\frac{3}{2}n$$

把 $n=2$、$p_1=3$ 的杆组称为Ⅱ级杆组。Ⅱ级杆组有一个内接副（指连接杆组内部构件的运动副）、两个外接副（与杆组外部构件连接的运动副）。内接副和外接副可以是转动副，也可以是移动副。常见的Ⅱ级杆组如图 2-29 所示。图 2-29 中运动副 *B* 为杆组的内接副，运动副 *A*、*C* 为外接副。

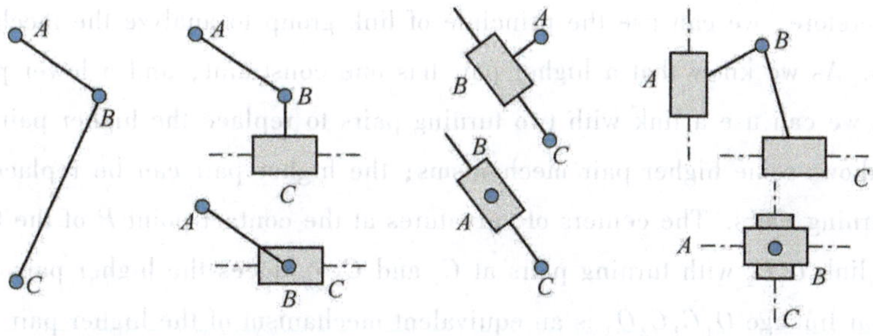

Fig. 2-29　The common class Ⅱ link groups（常见的Ⅱ级杆组）

当 $n=4$、$p_1=6$ 时，如果杆组中含有三个内接副，则称之为Ⅲ级杆组。如有四个内接副，则称之为Ⅳ级杆组。图 2-30 所示为常见的Ⅲ级杆组。图中的运动副 *A*、*B*、*C* 为内接副，运动副 *D*、*E*、*F* 为外接副。

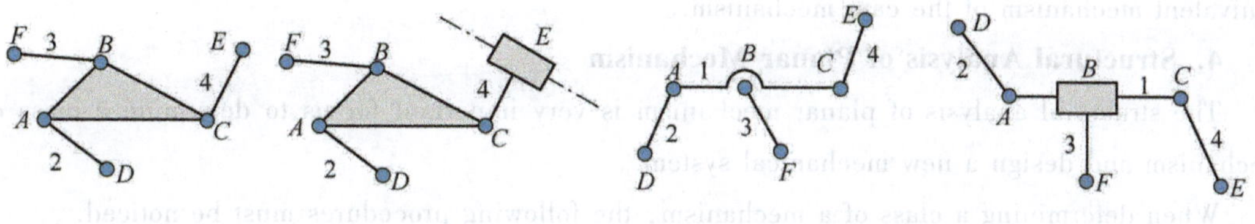

Fig. 2-30　The common class Ⅲ link groups（常见的Ⅲ级杆组）

图 2-31 所示为常见的Ⅳ级杆组。Ⅳ级杆组中有四个内接副和两个外接副。四级杆组应用较少，本书不进行讨论。

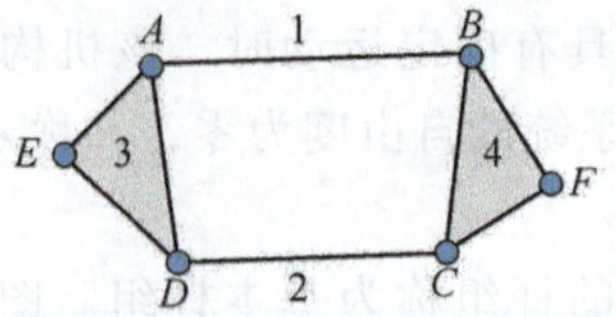


Fig. 2-31 The common class Ⅳ link group (常见的Ⅳ级杆组)

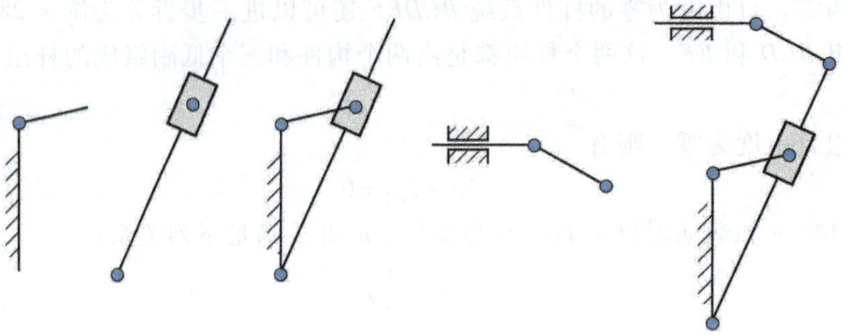


Fig. 2-32 Shaper mechanism design (牛头刨床的组合过程)

**3. Replacement of Higher Pair by Lower Pairs**

When a mechanism including a higher pair must be analyzed, we can replace the higher pair by lower pairs. Therefore, we can use the principle of link group to analyze the mechanism connected with lower pairs. As we know that a higher pair has one constraint, and a lower pair has two constraints, so that we can use a link with two turning pairs to replace the higher pair.

Fig. 2-33 shows some higher pair mechanisms; the higher pair can be replaced by one binary link with two turning pairs. The centers of curvatures at the contact point $P$ of the two profiles lie at $C_1$ and $C_2$; the link $C_1C_2$ with turning pairs at $C_1$ and $C_2$ replaces the higher pair.

The four-bar linkage $O_1C_1C_2O_2$ is an equivalent mechanism of the higher pair mechanism. It is illustrated in Fig. 2-33a.

If one of the profiles at the contact point becomes a point, the turning pair is located at this point. This is shown in Fig. 2-33b. The linkage $O_2C_2C_1$ is the equivalent mechanism.

If one of the profiles at the contact point becomes rectilinear, the curvature radius of such a profile at the contact point will be infinite. The turning becomes a sliding pair, and the other is located at the curvature center of the profile. This is shown in Fig. 2-33c. The linkage $O_1C_1C_2O_2$ is the equivalent mechanism of the cam mechanism.

**4. Structural Analysis of Planar Mechanism**

The structural analysis of planar mechanism is very important for us to determine a class of a mechanism and design a new mechanical system.

When determining a class of a mechanism, the following procedures must be noticed.

1) Remove the redundant degree of freedom and redundant constraints.

2) The higher pairs are replaced by lower pairs.

3) Calculate the degree of freedom, and determine the driving links.

4) Find out the class II link groups first and remove them from the mechanism. If there is not

**2. 机构的组成原理**

任何复杂的平面机构都可看成是把基本杆组连接到原动件和机架上组成的。

图 2-32e 所示牛头刨床的主运动机构就是在图 2-32a 所示的原动件上连接不同的Ⅱ级杆组（图 2-32b、c、d）所构成的。

**3. 高副低代简介**

前面讨论了仅含低副的杆组，对含有高副的机构进行分析时，则可采用低副代替高副的方法进行变通处理，简称高副低代。

高副低代是一种运动上的代换，其代换原则为：

1）代换前后保持机构的自由度不变。

2）代换前后保持机构的运动关系不变。

一个平面高副引入 1 个约束，而一个低副引入 2 个约束，欲使自由度不变就须使它们引入的约束数不变。最简单的代换方式是用带有两个低副的一个构件来代换一个平面高副，满足替代原则 1。又因为平面高副所引入的约束是限制两高副元素沿接触点的公法线方向做相对移动，所以高副低代的要点是找出两高副元素接触点处的公法线和曲率中心。只要将代换后的两个低副分别置于两个曲率中心，便可满足替代原则 2。

图 2-33 给出了几种典型高副接触的代换图例。图 2-33a 中，两高副构件各自绕 $O_1$ 和 $O_2$ 转动，其接触点为 $P$。过接触点 $P$ 作两曲线的公法线，并确定其曲率中心 $C_1$ 和 $C_2$，则含有两个转动副的构件 $C_1C_2$ 代替了高副 $P$。显然，$C_1C_2$ 构件有 3 个自由度，两端的转动副共有 4 个约束，总的约束为 1，与 $P$ 处高副约束数目相等。铰链四杆机构 $O_1C_1C_2O_2$ 为该高副机构的等效代替机构。$O_1C_1$、$O_2C_2$ 分别代表原高副机构的构件 1 和 2。用低副机构代替高副机构后，会比高副机构增加一个含有两个转动副的构件。

当其中一个高副曲线的曲率半径为零，即出现尖点时，其曲率中心即在该尖点处。图 2-33b 所示的直动尖顶从动件盘形凸轮机构即是此类实例。曲柄滑杆机构 $O_2C_2C_1$ 即是该高副机构的替代机构。

如果其中一个高副曲线的曲率半径无穷大，即高副曲线为直线，其曲率中心在无穷远，绕无穷远点的转动即演化为直线移动，该转动副演化为移动副。图 2-33c 所示的摆动平顶从动件盘形凸轮机构即是此类实例，导杆机构 $O_2C_2C_1O_1$ 即是该高副机构的替代机构。

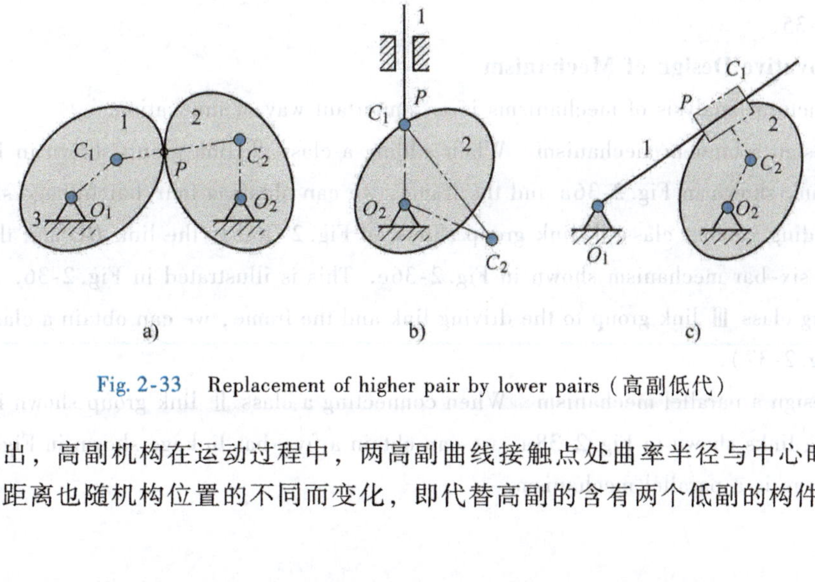


Fig. 2-33　Replacement of higher pair by lower pairs（高副低代）

需要指出，高副机构在运动过程中，两高副曲线接触点处曲率半径与中心时刻在变化，该两点间的距离也随机构位置的不同而变化，即代替高副的含有两个低副的构件长度和位置

any class II link group, the class III link group must be considered.

5) The last links which have been left must be the driving links, and they are equal to the number of the degrees of freedom.

**Example 2-6** Determine the class of the shearing mechanism as shown in Fig. 2-26.

**Solution**

The redundant links, redundant constraints, redundant degrees of freedom and multiple pin joints are considered first (Fig. 2-34), which was accomplished in Example 2-5.

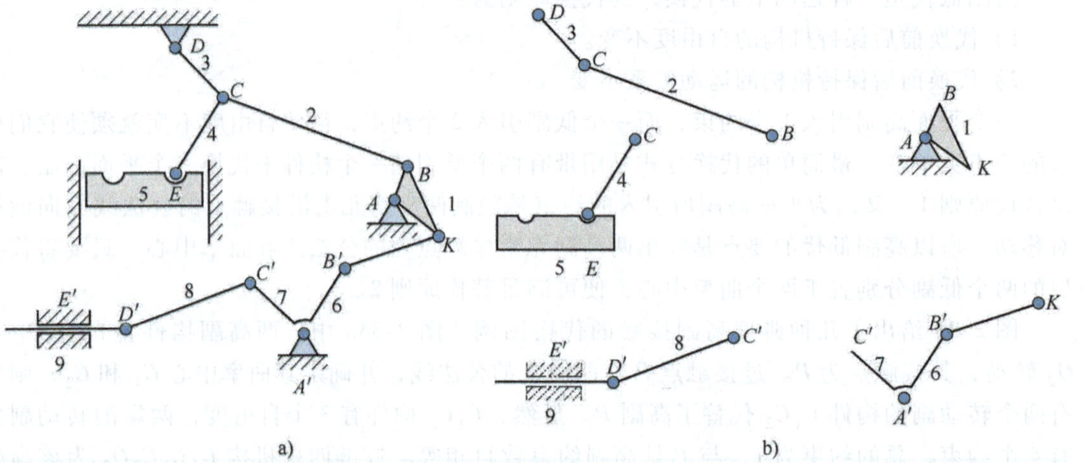


Fig. 2-34 Mechanism analysis (机构的分析)

There is one degree of freedom in the shearing mechanism, and link 1 is the driving link. This mechanism consists of four class Ⅱ link groups, so it is a class Ⅱ mechanism.

**Example 2-7** Determine the class of the shaper mechanism shown in Fig. 2-35.

**Solution**

The degree of freedom of the shaper mechanism is one, and link $AB$ is the driving link. There is no class Ⅱ link group, only a class Ⅲ link group, so it is a class Ⅲ mechanism. This is illustrated in Fig. 2-35.

**5. Innovative Design of Mechanism**

The structural analysis of mechanisms is an important way of innovation.

(1) Design a tandem mechanism  When adding a class Ⅱ link group shown in Fig. 2-36b to the driving link shown in Fig. 2-36a and the frame, we can obtain a four-bar linkage shown in Fig. 2-36c. If adding another class Ⅱ link group shown in Fig. 2-36d to the link $DC$ and the frame, we can obtain a six-bar mechanism shown in Fig. 2-36e. This is illustrated in Fig. 2-36.

If adding class Ⅲ link group to the driving link and the frame, we can obtain a class Ⅲ mechanism (see Fig. 2-37).

(2) Design a parallel mechanism  When connecting a class Ⅱ link group shown in Fig. 2-38b to two driving links shown in Fig. 2-38a, we can obtain a five-bar linkage shown in Fig. 2-38c. The five-bar linkage is a parallel mechanism.

也时刻在变化。说明高副机构的位置不同，其代替机构的尺寸也不同，所以高副低代是瞬时替代关系。

**4. 平面机构的结构分析**

平面机构结构分析的主要任务是判定机构的级别。机构的级别是按照机构中所含基本杆组的最高级别来决定的。最高级别为Ⅱ级杆组组成的机构称为Ⅱ级机构，最高级别为Ⅲ级杆组组成的机构称为Ⅲ级机构。

把机构分解为原动件和基本杆组，并确定机构级别的过程称为机构的结构分析。

机构结构分析的一般步骤为：

1）计算机构的自由度并确定原动件。同一机构中，原动件不同，机构的级别可能不同。

2）高副低代，去掉局部自由度和虚约束。

3）从远离原动件的部位开始拆杆组，首先考虑Ⅱ级杆组，拆下的杆组是自由度为零的基本杆组，最后剩下的原动件数目与自由度相等。

**例 2-6**　图 2-26 所示的剪床机构中，凸轮 1 为原动件，对该机构进行结构分析。

**解**　该机构的自由度为 1。

高副低代，去掉局部自由度和虚约束，如图 2-34a 所示。

从远离原动件的位置处开始拆分杆组。共拆下 4 个Ⅱ级杆组，没有Ⅲ级杆组。最后剩下 1 个原动件。

杆组的最高级别为Ⅱ级杆组，该机构为Ⅱ级机构。

**例 2-7**　对图 2-35 所示的以曲柄 *AB* 为原动件的牛头刨床机构进行杆组分析。

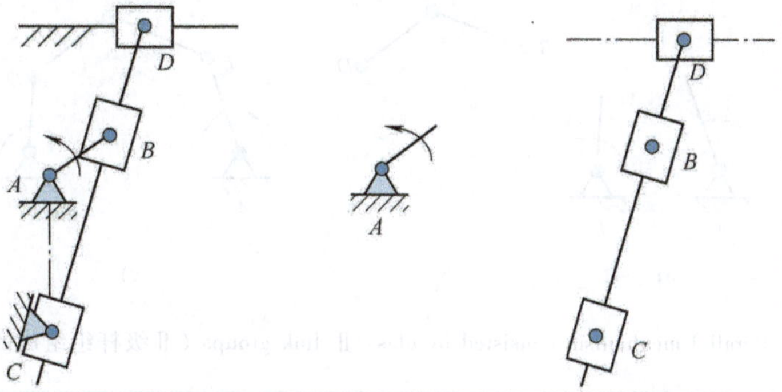


Fig. 2-35　Shaper mechanism analysis（牛头刨床机构的分析）

**解**　该机构的自由度为 1，没有局部自由度和虚约束。

该机构不含Ⅱ级杆组，仅有 1 个原动件和 1 个Ⅲ级杆组。该机构为Ⅲ级机构。

**5. 机构创新**

机构组成与分析是机构创新的重要途径。

（1）设计串联机构　把杆组的外接副连接到原动件和机架上，可以组成串联机构，在此基础上，再把其他杆组的外接副连接到前述机构的运动构件和机架上，可组成更加复杂的机构。

把图 2-36b 所示的Ⅱ级杆组的外接副 *B* 和 *D* 连接到图 2-36a 所示的原动件和机架上，组成图 2-36c 所示的四杆机构。再把图 2-36d 所示的Ⅱ级杆组的外接副 *E* 和 *F* 分别连接到四杆

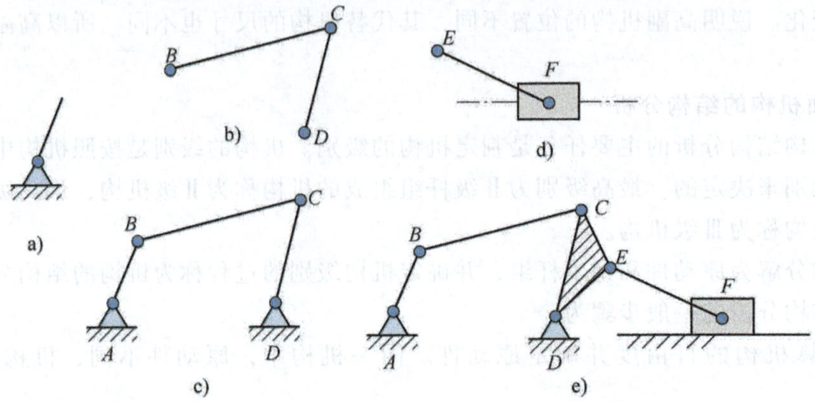


Fig. 2-36 Series mechanism consisted by class Ⅱ link groups (Ⅱ级杆组组成的串联机构)

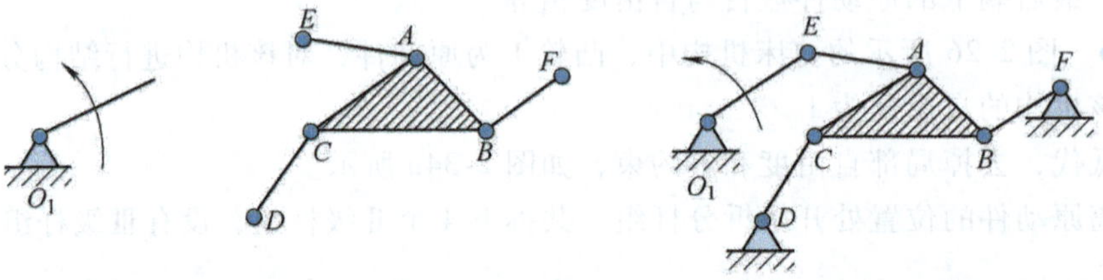


Fig. 2-37 Series mechanism consisted by class Ⅲ link groups (Ⅲ级杆组组成的串联机构)

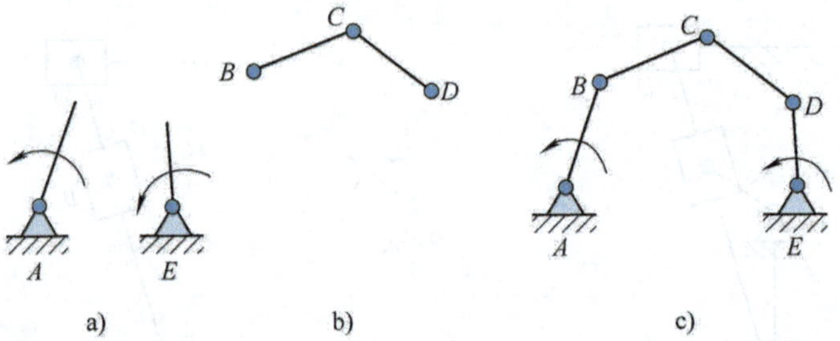


Fig. 2-38 Parallel mechanism consisted by class Ⅱ link groups (Ⅱ级杆组组成的并联机构)

If connecting a class Ⅲ link group to three driving links, we can obtain another parallel mechanism (see Fig. 2-39). This kind of mechanism can be used to a parallel robot.

Various link groups may be used to create many mechanisms.

机构的构件 $CD$ 和机架上，又组成图 2-36e 所示的六杆机构。该机构可近似实现滑块的等速运动，在机械工程中有重要应用。

图 2-37 所示是Ⅲ级杆组组成串联机构的过程。Ⅲ级杆组的运动副 $A$、$B$、$C$ 为内接副，$E$、$F$、$D$ 为外接副。将外接副 $E$ 连接到原动件上，其余两个外接副连接到机架上，就组成了图 2-37c 所示的Ⅲ级串联机构。

（2）设计并联机构　把杆组的外接副连接到原动件上可以组成并联机构。并联机构在机器人领域有广泛应用。图 2-38b 所示的Ⅱ级杆组的两个外接副 $B$ 和 $D$ 连接到图 2-38a 所示的两个原动件上，组成了图 2-38c 所示的 2 自由度的五杆并联机构。该机构可实现 $C$ 点的复杂运动轨迹。

图 2-39 所示的Ⅲ级杆组的 3 个外接副 $D$、$E$、$F$ 连接到 3 个原动件上，组成了 3 自由度并联机构。该机构可作为平面并联机器人的主体机构，也可以应用到微机械中。

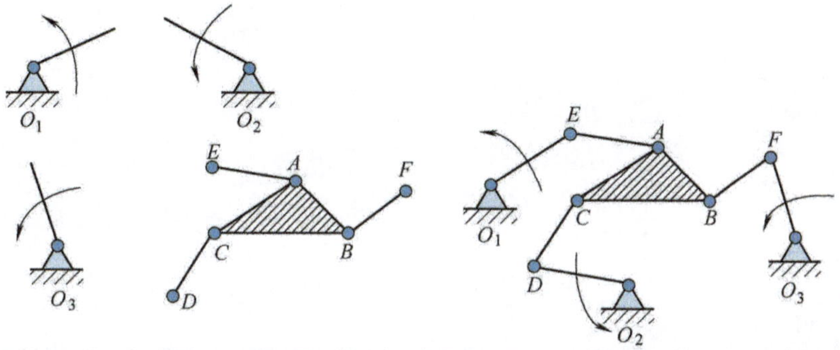


Fig. 2-39　Parallel mechanism consisted by class Ⅲ link groups（Ⅲ级杆组组成的并联机构）

杆组的结构众多，利用上述方法可以设计出许多新机构。杆组组成原理是机械创新设计的有效途径。

# Chapter 3

# Kinematic Analysis of Planar Mechanisms

# 第3章

# 平面机构的运动分析

## 3.1 Introduction

When the dimensions of a mechanism and the speed of the driving link in the mechanism are known, it is required to find out the positions, velocities and accelerations of the other links without regard to the forces which affect these motions. This process is called the kinematic analysis of the mechanism.

**1. Purpose of Kinematic Analysis**

(1) The workspace of a mechanism is necessary by means of analysis of positions or tracing path  Fig. 3-1a shows a internal combustion engine, in which the stroke of the piston can be used to design the length of the cylinder, and path of the coupler can be used to design the internal dimensions of the engine block.

(2) Determine the velocities and accelerations of links to investigate the working characteristics of a machine  Fig. 3-1b shows a shaper mechanism. The ram in the working stroke demands constant velocity approximately and the variation of accelerations is as little as possible. So the velocity analysis is very important to design a shaper mechanism.

(3) Motion analysis is needed for the dynamic force calculation  Once a position analysis is done, the next step is to determine the velocities of driven links or tracing points of interest in the mechanism, and also as a step on the way to determine the accelerations which are needed for the dynamic force analysis.

**2. Methods of Motion Analysis**

Motion analysis in mechanism can be determined either analytically or graphically. With the invention of computers, it becomes convenient to make use of analytical methods. However, a graphical method is more direct and it can not be neglected.

(1) Graphical method  The graphical method is divided into two types. They are instantaneous center method and relative motion method.

1) Instantaneous center method. The instantaneous center is a center of rotation of a moving body relative to another body. It is also a coincident point which has the same absolute velocity. Therefore, once the instantaneous centers of a mechanism have been found, they can be used to do a very rapid graphical velocity analysis of the mechanism.

2) Relative motion method. According to the principle of relative motions, we can establish the velocity equation and acceleration equation, and these equations can be solved by using vectors which represent velocities and accelerations. This is a manual graphical method, which is often used as a check on the more complete and accurate method.

(2) Analytical method  We first establish a vector loop (or loops) around the mechanism, in which the links are represented as position vectors, then we can take the derivatives with respect to time to find the velocity and acceleration.

(3) Experimental method  We may install the displacement sensor, speed sensor or acceleration sensor in the machine to measure the displacements, speeds and accelerations, which are required. The experimental method is a conventional method to analyze the performance of machines.

## 3.1 平面机构运动分析概述

平面机构的运动分析是根据给定的原动件运动规律，求解其他从动件上某些点的位置、速度和加速度，以及这些构件的角位置、角速度和角加速度的过程。

**1. 平面机构运动分析的目的**

(1) 求解机构中某些点的运动轨迹或位移，确定机构的运动空间 图 3-1a 所示内燃机中的曲柄滑块机构，滑块的运动行程$\overline{C_1C_2}$是设计活塞缸长度尺寸的依据，连杆 $BC$ 的运动轨迹是设计内燃机箱体尺寸的依据。

(2) 求解某些构件的速度、加速度，了解机构的工作性能 图 3-1b 所示的牛头刨床机构中，要求滑枕在工作行程中接近等速运动，其加速度的变化要小，才能提高加工质量。因此对刨床进行运动分析就成为刨床设计的重要内容。

(3) 为力分析做前期工作 现代机械的运转正在向高速化发展，惯性力的影响不能忽略。而构件的惯性力与其加速度成正比，惯性力矩与其角加速度成正比，所以运动分析又为机构的力分析奠定了基础。

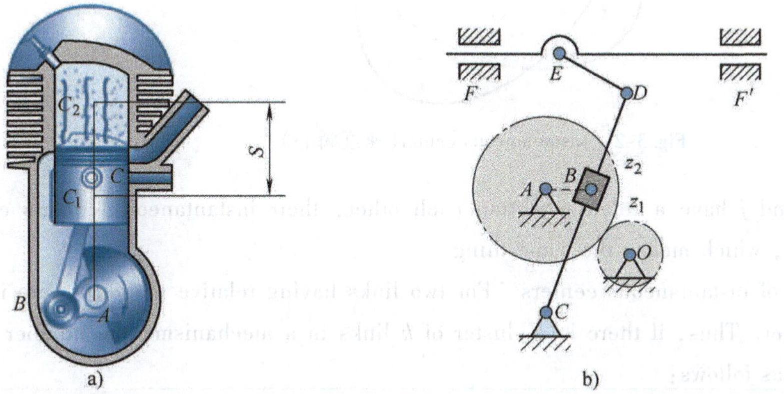


Fig. 3-1 Kinematic analysis of mechanisms (机构的运动分析)

**2. 运动分析的方法**

机构运动分析的方法主要有图解法、解析法及实验法。图解法又可分为速度瞬心法与相对运动图解法。在解析法中，矩阵法因其简便实用而常被人们采用。

(1) 图解法

1) 速度瞬心法。利用瞬心是两个构件的瞬时转动中心，又是两构件绝对速度相等的重合点原理，求解从动构件的角速度或某些点的速度。利用速度瞬心法只能进行速度分析。

2) 相对运动图解法。利用理论力学中的相对运动原理，把速度方程和加速度方程转换为几何矢量方程，用作图的方法求解构件的角速度、角加速度或某些点的速度及加速度。

(2) 解析法 在建立机构运动学模型的基础上，采用数学方法求解构件的角速度、角加速度或某些点的速度及加速度。本书采用矩阵法进行机构的运动分析。

(3) 实验法 通过位移、速度、加速度等各类传感器对实际机械的位移、速度、加速度等运动参数进行测量，实验法是研究已有机械运动性能的常用方法。

## 3.2 Velocity Analysis with Instantaneous Centers

### 1. Concept of Instantaneous Center of Velocity

(1) Instantaneous center of velocity An instantaneous center of velocity is a center of rotation of a moving link relative to another link. If a link is in motion relative to a fixed link, the center is called as an absolute center; otherwise it is called as a relative center.

There is plane body 1 having plane motion relative to another reference plane body 2. At any instant, the relative velocities of two points $A$ and $B$ on the body 1 and the body 2 are $v_{A_2A_1}$ and $v_{B_2B_1}$ respectively in the direction as shown in Fig. 3-2. The instantaneous center $P_{12}$ lies on the intersection of two velocities $v_{A_2A_1}$ and $v_{B_2B_1}$.

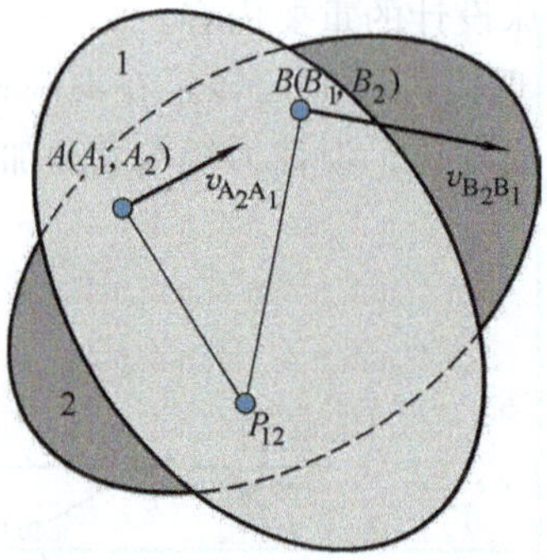


Fig. 3-2 Instantaneous center(速度瞬心)

If two bodies $i$ and $j$ have a relative motion each other, their instantaneous centers can be named either $P_{ij}$ or $P_{ji}$, which means the same thing.

(2) The number of instantaneous centers For two links having relative motion, there is only one instantaneous center. Thus, if there is a cluster of $k$ links in a mechanism, the number of instantaneous centers is as follows:

$$N = C_k^2 = \frac{k(k-1)}{2}$$

(3) Locating instantaneous centers The following rules are used when locating instantaneous centers.

1) Two links are connected by a kinematic pair. If two links are connected by a pivot joint, the center of the pivot is the instantaneous center(see Fig. 3-3a, and Fig. 3-3b). If two links have sliding contact, the instantaneous center lies at infinity in a direction perpendicular to the path of the motion of the slider(see Fig. 3-3c). If two links have pure rolling contact, the instantaneous center is the point of contact, this is because the two points of contact on the two bodies have the same linear velocity and there is no relative motion at the contact point(see Fig. 3-3d). If two links have rolling and sliding contact, the instantaneous center lies somewhere on the common normal of the contact point(see Fig. 3-3e).

These centers are called primary centers, and they can be located by inspection.

2) Two links having relative motion are not connected by kinematic pair. This instantaneous

## 3.2　用速度瞬心法对机构进行速度分析

**1. 瞬心的基本概念**

(1) 瞬心　在任一瞬时，两个做平面相对运动的构件都可以看作为绕一个瞬时重合点做相对转动。这个瞬时重合点又称为瞬时转动中心，简称为瞬心。这两个构件在该重合点处的绝对速度相等，所以瞬心又称为等速重合点或同速点。当这两个构件之中有一个构件固定不动时，瞬心处的绝对速度为零，这时的瞬心称为绝对瞬心。当两个构件都在运动时，其瞬心称为相对瞬心。

如图 3-2 所示，构件 1 和 2 做相对平面运动，两构件在重合点 $A$ 处的相对速度为 $v_{A_2A_1}$，在重合点 $B$ 处的相对速度为 $v_{B_2B_1}$，两相对速度垂直线的交点为两构件的瞬心 $P_{12}$。若构件 1 和 2 都在运动，则 $P_{12}$ 为相对速度瞬心；若有一构件固定不动，则 $P_{12}$ 为绝对速度瞬心。因两构件做相对运动，故 $P_{ij}$ 与 $P_{ji}$ 代表同一个瞬心。

(2) 平面机构瞬心的数目　假设机构中含有 $k$ 个构件，其中既包含运动构件也包含机架，每两个构件之间有一个瞬心，则全部瞬心的数目 $N$ 为

$$N = C_k^2 = \frac{k(k-1)}{2}$$

当构件数 $k$ 较多时，找出全部瞬心是一项比较繁琐的工作。所以，瞬心法通常适用于构件数较少的简单机构的运动分析。

(3) 瞬心位置的确定　可以把瞬心分为两类：其一是两个构件之间直接用运动副连接时的瞬心；其二是两个构件之间没有用运动副连接时的瞬心。下面分别讨论如何确定这两类瞬心的位置。

1) 两个构件用运动副连接时的瞬心位置。

① 两个构件用转动副连接时的瞬心位置。图 3-3a、b 所示的构件 1 和 2 由转动副连接，显然，铰链中心点就是两个构件的瞬心 $P_{12}$。

② 两个构件用移动副连接时的瞬心位置。图 3-3c 所示的构件 1 和 2 的相对移动速度方向与导路方向平行，瞬心 $P_{12}$ 位于垂直导路方向的无穷远点。

③ 两构件用平面高副连接时的瞬心位置。平面高副分为纯滚动高副和滚动兼滑动的高副。图 3-3d 所示为纯滚动高副，两构件在接触点处的相对速度为零。该接触点即为瞬心 $P_{12}$。

图 3-3e 中，构件 1 和 2 之间组成高副，高副廓线在接触点处的相对速度为 $v_{12}$，其方向沿高副廓线在接触点处的切线方向。而瞬心则位于过接触点且与 $v_{12}$ 方向相垂直的法线 $n—n$ 上。至于瞬心 $P_{12}$ 位于法线上的哪一点，则还需要由其他条件来确定。

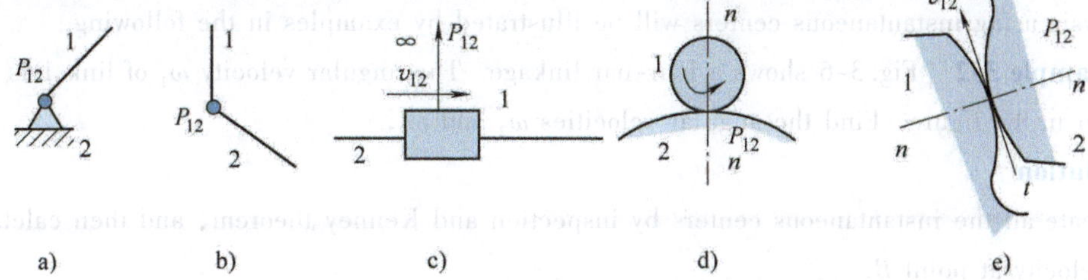


Fig. 3-3　Primary instantaneous centers of two links linked by kinematic pair
(两构件用运动副连接时的瞬心位置)

center can be determined by Kennedy theorem.

Any three bodies in plane motion will have exactly three instantaneous centers, and they will be on the same straight line. This is known as Kennedy theorem.

Consider a three-bar mechanism involving a higher pair shown in Fig. 3-4, the primary centers $P_{12}$ and $P_{13}$ can be located easily. It is found that point $C$ is not the instantaneous center, because the two velocities $v_{C_2}$ and $v_{C_3}$ of the coincident point $C$ are in different directions. The velocities $v_{C_2}$ and $v_{C_3}$ of the instantaneous center will be the same. Only this center lies on the point $P_{23}$ which is the intersection of the line joining $P_{12}$, $P_{23}$, and the common normal line $n$—$n$. That is to say, if three plane bodies have relative motions, their instantaneous centers must be lie on a straight line. We can use this rule to find the remaining instantaneous centers which are not obvious from inspection. The following example describes the procedure in detail.

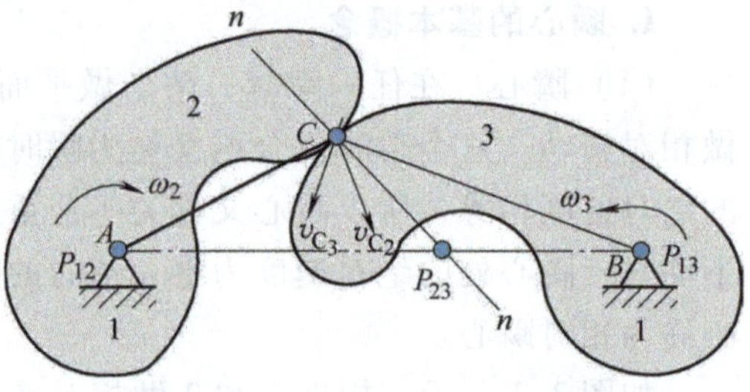


Fig. 3-4 Kennedy theorem (三心定理)

**Example 3-1** Fig. 3-5 shows a four-bar linkage and a slider-crank mechanism. Find all the instantaneous centers by graphical method.

**Solution**

Fig. 3-5a: The number of instantaneous centers is given by:

$$N = \frac{k(k-1)}{2} = \frac{4(4-1)}{2} = 6$$

The primary centers $P_{14}$, $P_{12}$, $P_{23}$, and $P_{34}$ can be located by inspection directly, then they have been labeled in Fig. 3-5a. The unknown centers are $P_{13}$ and $P_{24}$, and they can be located by Kennedy theorem.

The instantaneous center $P_{13}$ can be located by using links 1, 2, 3 and links 1, 4, 3, because the three centers $P_{12}$, $P_{23}$, $P_{13}$ are on the same straight line, $P_{14}$, $P_{34}$, $P_{13}$ are on the same straight line, then $P_{13}$ lies at the intersection of lines joining $P_{12}$, $P_{23}$, and $P_{14}$, $P_{34}$ by Kennedy theorem.

The instantaneous center $P_{24}$ can be located also by Kennedy theorem.

Fig. 3-5b: The number of instantaneous centers is 6 too.

The procedure is illustrated in Fig. 3-5b.

**2. Velocity Analysis with Instantaneous Centers**

When the scaled kinematic diagram of a linkage has been drown; and the angular velocity of a link is known, it is required to find the angular velocities of the other links. The techniques of velocity analysis using instantaneous centers will be illustrated by examples in the following.

**Example 3-2** Fig. 3-6 shows a four-bar linkage. The angular velocity $\omega_1$ of link 1 is known, as shown in the figure. Find the angular velocities $\omega_2$ and $\omega_3$.

**Solution**

Locate all the instantaneous centers by inspection and Kenney theorem, and then calculate the linear velocity at point $B$.

$$v_{P_{12}} = v_B = \omega_1 L_{P_{14}P_{12}}$$

2）没有用运动副连接的两构件之间的瞬心位置。两构件之间没有用运动副连接时，其瞬心位置可用三心定理来确定。

三心定理　做平面运动的三个构件有三个瞬心，且位于同一直线上，称这一结论为三心定理。图 3-4 所示的高副机构中，设构件 1、2、3 之间的相对运动为平面运动。根据瞬心计算公式可知，该机构的瞬心数目为 3。

其中，构件 1 和 2 之间由转动副 $A$ 连接，瞬心 $P_{12}$ 位于转动副 $A$ 的回转中心。构件 1 和 3 之间由转动副 $B$ 连接，其中心点 $B$ 即为瞬心 $P_{13}$ 的位置。按照三心定理，另外一个瞬心 $P_{23}$ 必须在过高副接触点 $C$ 处的公法线上，还必须位于瞬心 $P_{12}$ 与 $P_{13}$ 的连线上。

由于速度瞬心是两构件的绝对速度相等的重合点，也就是说两构件在该重合点处的速度大小与方向都必须相同，只有在图 3-4 所示的重合点 $P_{23}$ 处才能满足该条件。

如果 $P_{23}$ 点不在 $P_{12}$ 与 $P_{13}$ 的连线上，假定在图示的 $C$ 点，则构件 2 上 $C_2$ 点的速度为 $v_{C_2}$，方向垂直于 $P_{12}$ 点与 $C$ 点的连线。构件 3 上 $C_3$ 点的速度为 $v_{C_3}$，方向垂直于 $P_{13}$ 点与 $C$ 点的连线。只要 $C$ 点不在 $P_{12}$ 与 $P_{13}$ 的连线上，两构件在该点处的速度方向永远不会相同，只有当 $C$ 点落在 $P_{12}$ 与 $P_{13}$ 的连线上时，$v_{C_2}$ 与 $v_{C_3}$ 的方向才能相同，$C$ 点才能成为瞬心点。这就是说，瞬心 $P_{23}$ 必定位于 $P_{12}$ 与 $P_{13}$ 的连线上。

应用三心定理求解没有用运动副连接的两构件瞬心时非常方便。

**例 3-1**　确定图 3-5 所示四杆机构的全部瞬心。

**解**　图 3-5a 所示铰链四杆机构中，瞬心总数为

$$N=\frac{k(k-1)}{2}=\frac{4(4-1)}{2}=6$$

各构件的铰链中心分别为瞬心 $P_{14}$、$P_{12}$、$P_{23}$ 和 $P_{34}$。未知瞬心 $P_{13}$ 和 $P_{24}$ 可利用三心定理找出。

$P_{13}$ 在 $P_{12}$ 和 $P_{23}$ 的连线上，又在 $P_{14}$ 和 $P_{34}$ 的连线上，其交点为 $P_{13}$。

$P_{24}$ 在 $P_{12}$ 和 $P_{14}$ 的连线上，又在 $P_{34}$ 和 $P_{23}$ 的连线上，其交点为 $P_{24}$。

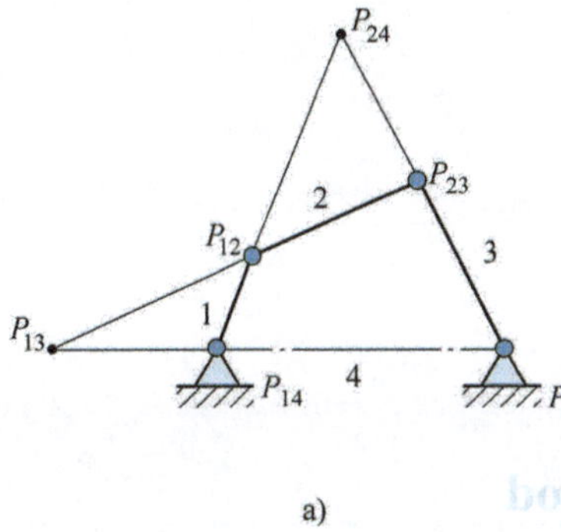


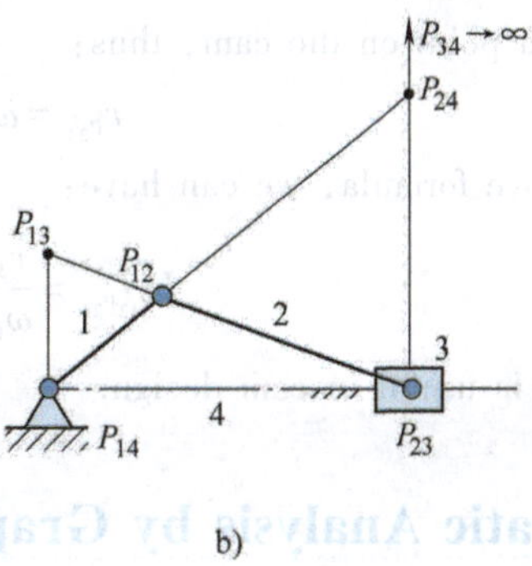


Fig. 3-5　Instantaneous centers for four-bar mechanisms（四杆机构的瞬心）

图 3-5b 所示铰链四杆机构中，瞬心总数为 6。

可直接找出的瞬心为 $P_{14}$、$P_{12}$、$P_{23}$ 和 $P_{34}$。未知瞬心 $P_{13}$ 和 $P_{24}$ 可利用三心定理找出。

$P_{13}$ 在 $P_{12}$ 和 $P_{23}$ 的连线上，又在 $P_{14}$ 和 $P_{34}$ 的连线上，其交点为 $P_{13}$。

$P_{24}$ 在 $P_{12}$ 和 $P_{14}$ 的连线上，又在 $P_{34}$ 和 $P_{23}$ 的连线上，其交点为 $P_{24}$。

Consider the $P_{12}$ to be on link 1 and obtain the velocity of the point $P_{12}$, then consider the $P_{12}$ to be on link 2 and obtain the velocity of point $P_{12}$, we have:

$$v_{P_{12}} = v_B = \omega_1 L_{P_{14}P_{12}} = \omega_2 L_{P_{12}P_{24}}$$

$$\omega_2 = \omega_1 \frac{L_{P_{14}P_{12}}}{L_{P_{24}P_{12}}}$$

The rotating direction can be determined by inspection. It is clockwise about the instantaneous center $P_{24}$.

The velocities of the instantaneous center $P_{13}$ will be the same. Imagine point $P_{13}$ to be on link 1 and link 3, we have:

$$v_{P_{13}} = \omega_1 L_{P_{13}P_{14}} = \omega_3 L_{P_{34}P_{13}}$$

$$\omega_3 = \frac{L_{P_{13}P_{14}}}{L_{P_{13}P_{34}}}\omega_1$$

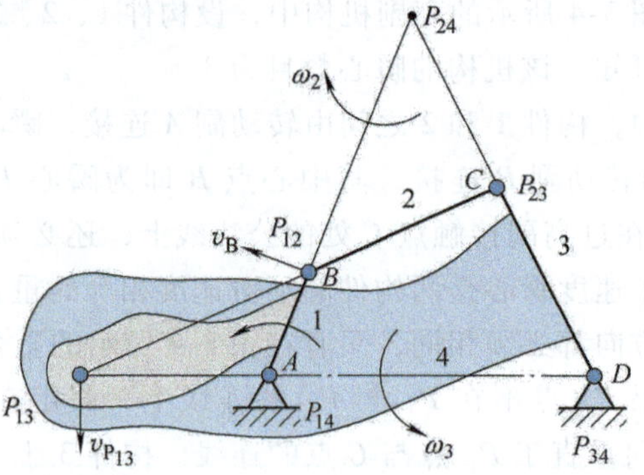

Fig. 3-6 Application of instantaneous centers for four-bar mechanism (瞬心法在铰链四杆机构速度分析中的应用)

The rotating direction can be determined by inspection. It is counterclockwise about the instantaneous center $P_{34}$.

**Example 3-3** Fig. 3-7 shows a cam mechanism with a flat follower. The angular velocity of cam 1 is known and it is required to find the velocity of the follower 3.

**Solution**

Locate instantaneous centers $P_{12}$, $P_{13}$ and $P_{23}$.

Since the follower 3 is a slider having rectilinear translation, each point on it has the same velocity. So the instantaneous center $P_{23}$ can be considered as a point on the slider. Therefore, the velocity of instantaneous center $P_{23}$ is the same as that of slider. The instantaneous center $P_{23}$ can also be considered as a point on the cam, thus:

$$v_{P_{23}} = \omega_2 L_{P_{12}P_{23}} = v_3$$

From the above formula, we can have:

$$L_{P_{12}P_{23}} = \frac{v_3}{\omega_2} = \frac{ds/dt}{d\varphi/dt} = \frac{ds}{d\varphi}$$

This formula is useful in cam design.

## 3.3 Kinematic Analysis by Graphical Method

When we study engineering mechanics, there are two different cases.

Case 1: Knowing the velocity and acceleration of one point, determine the velocity and acceleration of another point in the same link.

Case 2: Knowing the velocity and acceleration of one point of a link, determine the velocity and acceleration of the same point in different link.

**2. 用速度瞬心法进行机构的速度分析**

利用速度瞬心法对机构进行速度分析时，要选择一个适当的比例尺 $\mu_l$ 画出机构运动简图，找出机构的全部瞬心并标注在机构简图上。利用瞬心是两构件重合点处的同速点和瞬时转动中心的概念，找出待求构件的速度。

**例 3-2**　已知图 3-6 所示铰链机构 $ABCD$ 的尺寸和位置及构件 1 的角速度 $\omega_1$，用速度瞬心法求解构件 2、3 的角速度 $\omega_2$、$\omega_3$。

**解**　找出该机构的全部速度瞬心并标注在机构运动简图上。

已知构件 1 的角速度，待求角速度的构件要同构件 1 联系起来。利用瞬心 $P_{12}$ 求 $\omega_2$，瞬心 $P_{12}$ 是构件 1 和 2 的同速点。

$P_{12}$ 在构件 1 上，有

$$v_{P_{12}} = v_B = \omega_1 L_{P_{14}P_{12}}$$

$P_{12}$ 在构件 2 上，有

$$v_{P_{12}} = v_B = \omega_1 L_{P_{14}P_{12}} = \omega_2 L_{P_{24}P_{12}}$$

所以

$$\omega_2 = \omega_1 \frac{L_{P_{14}P_{12}}}{L_{P_{24}P_{12}}}$$

扩大构件 1 和 3，其同速点为 $P_{13}$，利用瞬心 $P_{13}$ 求 $\omega_3$。

$$v_{P_{13}} = \omega_1 L_{P_{13}P_{14}} = \omega_3 L_{P_{34}P_{13}}$$

$$\omega_3 = \frac{L_{P_{13}P_{14}}}{L_{P_{13}P_{34}}}\omega_1$$

其方向可由其速度方向判别。

**例 3-3**　已知图 3-7 所示凸轮机构中的凸轮角速度 $\omega_2$，用速度瞬心法求从动件速度 $v_3$。

**解**　画出机构简图并标注全部速度瞬心。$P_{23}$ 为凸轮 2 与推杆 3 的速度瞬心，即两个构件的同速点，则有：$v_{P_{23}} = \omega_2 L_{P_{12}P_{23}} = v_3$，推杆的运动方向可由瞬心处的速度方向决定。

从上式可找出以下关系：$L_{P_{12}P_{23}} = \dfrac{v_3}{\omega_2} = \dfrac{ds/dt}{d\varphi/dt} = \dfrac{ds}{d\varphi}$，该关系式在凸轮机构的设计中有重要应用。

当机构的构件数较少时，应用瞬心法进行速度分析是比较简便快捷的。

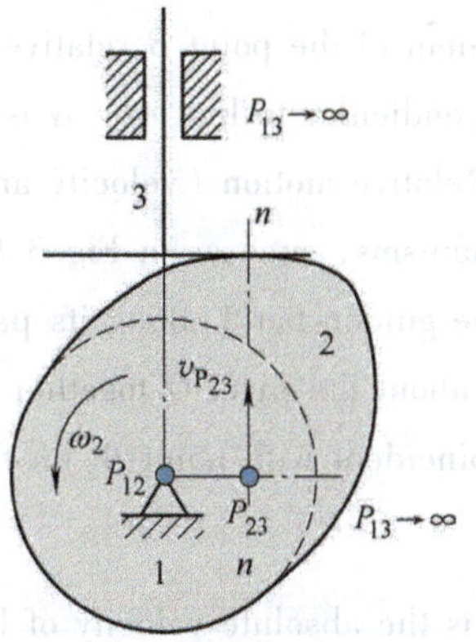


Fig. 3-7　Application of instantaneous centers for cam mechanism（瞬心法在凸轮机构中的应用）

## 3.3　用相对运动图解法对机构进行运动分析

机构运动分析时，常会遇到两类问题：其一是已知某个构件上一点的速度和加速度，求该构件上另外一点的速度和加速度；其二是两个做平面相对运动的构件之间，存在一个瞬时重合点，其中一个构件在这个重合点处的速度和加速度是已知的，求解另外一个构件在该点处的速度和加速度。

要解决这两类问题，首先要建立两点之间的速度和加速度矢量方程，然后通过求解矢量方程来得到未知点的速度和加速度。下面就来讨论如何针对上述两类问题建立速度和加速度

**1. Principles of Relative Motions**

(1) Relative motion (velocity and acceleration) of two points on the same link Let us consider a body which has plane motion (see Fig. 3-8). The motion of the body is the combination of the translation, in which all the points move in the same way as point $A$ and the rotation about the point $A$ with an angular velocity $\omega$. The absolute velocity of point $B$ related to the point $A$ is:

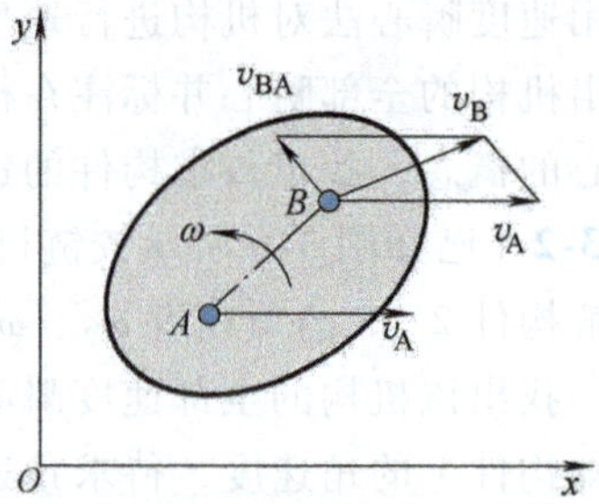


Fig. 3-8 Relative velocity of two points on a link (同一构件上两点之间的速度关系)

$$\boldsymbol{v}_B = \boldsymbol{v}_A + \boldsymbol{v}_{BA} \tag{3-1}$$

Where $\boldsymbol{v}_A$ is the absolute velocity of the base point $A$. $\boldsymbol{v}_B$ is the absolute velocity of point $B$. $\boldsymbol{v}_{BA}$ is the velocity of point $B$ relative to Point $A$. Its magnitude is $\boldsymbol{v}_{BA} = \omega l_{AB}$; its direction is perpendicular to line $AB$ and as the same as its angular velocity $\omega$.

We can easily differentiate the velocity vector equation to obtain an expression for the acceleration of point $B$.

$$\boldsymbol{a}_B = \boldsymbol{a}_A + \boldsymbol{a}_{BA}^n + \boldsymbol{a}_{BA}^t \tag{3-2}$$

Where $\boldsymbol{a}_A$, $\boldsymbol{a}_B$ are the absolute acceleration at the base point $A$ and point $B$ in the same link. $\boldsymbol{a}_{BA}^n$ is the relative normal acceleration component of point $B$ relative to point $A$. Its magnitude is: $\boldsymbol{a}_{BA}^n = \boldsymbol{v}_{BA}^2 / l_{AB} = \omega^2 l_{AB}$, and the direction is always from $B$ toward $A$. $\boldsymbol{a}_{BA}^t$ is the relative tangential acceleration component of the point $B$ relative to point $A$. Its magnitude is: $\boldsymbol{a}_{BA}^t = \alpha l_{AB}$, and the direction is always perpendicular to line $AB$. $\alpha$ is the angular acceleration of the body.

(2) Relative motion (velocity and acceleration) of two coincident points on different links In many mechanisms, such as in Fig. 3-9, constraint of relative motion is achieved by guiding the slider 2 on the guider-bar 1 along its path. The slider 2 is reciprocated along the guider-bar 1, and they rotate about the pivot $O$ together with an angular velocity $\omega_1$. Consider point $B_1$ on the link 1 which is coincident with point $B_2$ on the slider 2, we have:

$$\boldsymbol{v}_{B_2} = \boldsymbol{v}_{B_1} + \boldsymbol{v}_{B_2B_1}$$

Where $\boldsymbol{v}_{B_1}$ is the absolute velocity of link 1 at point $B_1$. $\boldsymbol{v}_{B_2}$ is the absolute velocity of link 2 at the point $B_2$. $\boldsymbol{v}_{B_2B_1}$ is the velocity of point $B_2$ on link 2 relative to point $B_1$ on link 1 at the coincident point $B$. Its direction is along the guider-bar.

The velocity vector equation is differentiated with respect to time; we can obtain an expression for the acceleration of coincident point $B$.

$$\boldsymbol{a}_{B_2} = \boldsymbol{a}_{B_1} + \boldsymbol{a}_{B_2B_1}^k + \boldsymbol{a}_{B_2B_1}^r$$

Where $\boldsymbol{a}_{B_1}$ is the absolute acceleration of coincident point $B$ on the link 1, and $\boldsymbol{a}_{B_2}$ is the absolute acceleration of coincident point $B$ on the link 2. $\boldsymbol{a}_{B_2B_1}^k$ is the Coriolis acceleration of coincident point $B$. Its magnitude is given by $\boldsymbol{a}_{B_2B_1}^k = 2\boldsymbol{v}_{B_2B_1}\omega_1$, and its direction can be obtained by rotating the relative velocity $\boldsymbol{v}_{B_2B_1}$ through 90° in the direction of the angular velocity $\omega_1$. $\boldsymbol{a}_{B_2B_1}^r$ is the relative acceleration of coincident point $B$. Its direction is along the guider-bar.

矢量方程。

**1. 相对运动图解法的基本原理**

（1）同一构件上两点之间的速度和加速度的关系　做平面运动的物体，任意一点的运动都可以看成是随同基点的平动以及绕基点的转动的合成。图 3-8 所示为做平面运动的刚体，已知基点 $A$ 的速度为 $\boldsymbol{v}_A$，则该刚体上任意一点 $B$ 的速度为

$$\boldsymbol{v}_B=\boldsymbol{v}_A+\boldsymbol{v}_{BA} \tag{3-1}$$

式中，$\boldsymbol{v}_A$ 是 $A$ 点的绝对速度，方向已知；$\boldsymbol{v}_B$ 为 $B$ 点的绝对速度，方向未知；$\boldsymbol{v}_{BA}=\omega l_{AB}$，是 $B$ 点相对于 $A$ 点的相对速度，其方向垂直于 $AB$。

速度为矢量，具有方向和大小两个物理量。每个速度方程可求解两个未知数。为求解方便，列速度方程时尽量使未知数位于等号两侧。

同一构件上两点之间的运动关系可以概括为牵连运动是移动、相对运动是转动的运动关系。

$B$ 点与 $A$ 点之间的加速度关系可以表达为

$$\boldsymbol{a}_B=\boldsymbol{a}_A+\boldsymbol{a}_{BA}^n+\boldsymbol{a}_{BA}^t \tag{3-2}$$

式中，$\boldsymbol{a}_{BA}^n=\boldsymbol{v}_{BA}^2/l_{AB}=\omega^2 l_{AB}$，是 $B$ 点相对于 $A$ 点的法向加速度，其方向由 $B$ 指向 $A$；$\boldsymbol{a}_{BA}^t=\alpha l_{AB}$，是 $B$ 点相对于 $A$ 点的切向加速度，其方向垂直于 $A$ 和 $B$ 两点的连线；$\omega$ 和 $\alpha$ 分别为该构件的角速度和角加速度。

加速度为矢量，具有方向和大小两个物理量。每个加速度方程可求解两个未知数。为求解方便，列加速度方程时尽量使未知数位于等号两侧。

（2）两构件重合点处的速度和加速度矢量关系　如图 3-9 所示，构件 1 和 2 用移动副连接且构件 1 绕 $A$ 点转动，两构件在重合点 $B$ 处的运动关系可用理论力学中的牵连运动是转动、相对运动是移动来描述。

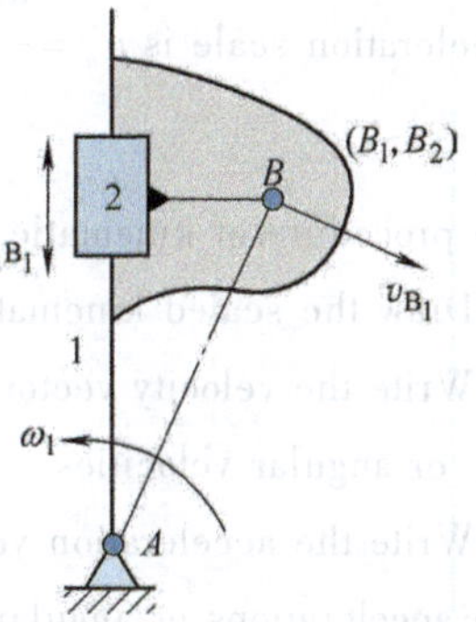


Fig. 3-9　Relative velocity of coincident point on separate links（两构件重合点处的运动关系）

该重合点处的速度矢量关系为

$$\boldsymbol{v}_{B_2}=\boldsymbol{v}_{B_1}+\boldsymbol{v}_{B_2B_1}$$

式中，$\boldsymbol{v}_{B_2}$为构件 2 上 $B$ 点的绝对速度，一般不知道其方向；$\boldsymbol{v}_{B_1}$为构件 1 上 $B$ 点的绝对速度，其方向垂直于 $AB$；$\boldsymbol{v}_{B_2B_1}$为构件 2 上 $B$ 点相对构件 1 上 $B$ 点的相对速度，其方向平行于导路。$\boldsymbol{v}_{B_2B_1}$与$\boldsymbol{v}_{B_1B_2}$的大小相等，方向相反。

两构件在重合点 $B$ 处的加速度关系为

$$\boldsymbol{a}_{B_2}=\boldsymbol{a}_{B_1}+\boldsymbol{a}_{B_2B_1}^k+\boldsymbol{a}_{B_2B_1}^r$$

式中，$\boldsymbol{a}_{B_2}$为构件 2 上 $B$ 点的绝对加速度。$\boldsymbol{a}_{B_1}$为构件 1 上 $B$ 点的绝对加速度，$\boldsymbol{a}_{B_1}=\boldsymbol{a}_{B_1}^n+\boldsymbol{a}_{B_1}^t$；如构件 1 等速转动，$\boldsymbol{a}_{B_1}=\boldsymbol{a}_{B_1}^n=\omega_1^2 l_{AB}$，方向由 $B$ 指向 $A$。$\boldsymbol{a}_{B_2B_1}^r$为构件 2 上 $B$ 点相对构件 1 上 $B$ 点的相对加速度，其方向与导路方向平行。$\boldsymbol{a}_{B_2B_1}^k$为构件 2 上 $B$ 点相对构件 1 上 $B$ 点的科氏加速度，其方向为把$\boldsymbol{v}_{B_2B_1}$沿$\omega_1$方向转过 90°，$\boldsymbol{a}_{B_2B_1}^k=2\boldsymbol{v}_{B_2B_1}\omega_1$。

The absolute acceleration of point $B_1$ can be divided into two components. They are:

$$a_{B_1} = a_{B_1}^n + a_{B_1}^t$$

When the straight guider rotates with a constant angular velocity, then, $a_{B_1} = a_{B_1}^n = \omega_1^2 l_{AB}$, its direction is toward to its rotating center $A$.

(3) Velocity image and acceleration image When we know the velocities or accelerations at two different points on a link, the velocity or acceleration of the third point can be determined by drawing their images.

While drawing the images, the following points should be kept in mind:

1) The velocity image or acceleration image of a link is a scaled reproduction of the link shape in the velocity diagram or acceleration diagram.

2) The order of the letters in the velocity image or acceleration image is the same as in the link configuration.

**2. Graphical Method of Relative Motions**

Any vector equation, such as velocity equation or acceleration equation, can be solved by vector addition or subtraction. Of course, the velocities or accelerations must be represented by directed line segments.

Velocity scale is $\mu_v = \dfrac{\text{velocity represented}}{\text{drawing distance}}\left(\dfrac{\text{m/s}}{\text{mm}}\right)$

Acceleration scale is $\mu_a = \dfrac{\text{acceleration represented}}{\text{drawing distance}}\left(\dfrac{\text{m/s}^2}{\text{mm}}\right)$

The procedure of kinematic analysis of planar mechanism is as follows:

1) Draw the scaled kinematic diagram.

2) Write the velocity vector equation and draw the velocity diagram, then find out the unknown velocities or angular velocities.

3) Write the acceleration vector equation and draw the acceleration diagram, then find out the unknown accelerations or angular accelerations.

**Example 3-4** Fig. 3-10a shows a four-bar linkage; all the dimensions of the links and angular position of the driving link $AB$ are known. When the crank $AB$ rotates counterclockwise with an angular velocity $\omega_1$, determine the angular velocities $\omega_2$, $\omega_3$, the velocity of the point $E$ on the link 2 and angular accelerations $\alpha_2$, $\alpha_3$.

**Solution**

1) Draw the scaled skeleton with $\mu_l$ (see Fig. 3-10a).

2) Velocity analysis. The velocity of point $B$ is as follows:

$$v_B = v_{B_1} = v_{B_2} = \omega_1 l_{AB}$$

The velocity vector equation of point $C$ relative to point $B$ is written as follows:

当两构件以相同的角速度转动且有相对移动时，其重合点处必有科氏加速度。为求解方便，列上述方程时尽量使未知数分布在等号两侧。

(3) 速度和加速度矢量图　在绘制速度和加速度矢量图时，应注意以下两点：

1) 选择合适的比例尺。

2) 矢量方向一定要与实际方向一致。

**2. 相对运动图解法**

通过引入速度比例尺 $\mu_v$，把速度矢量转化为长度矢量，即可用图解法求解未知速度。

$$\mu_v = \frac{\text{实际速度(m/s)}}{\text{图中的长度(mm)}}$$

通过引入加速度比例尺 $\mu_a$，把加速度矢量转化为长度矢量，即可用图解法求解未知加速度。

$$\mu_a = \frac{\text{实际加速度}(\text{m/s}^2)}{\text{图中的长度(mm)}}$$

具体步骤如下：

1) 选长度比例尺 $\mu_l$ 画出机构运动简图。

2) 列出速度矢量方程，标注出速度的大小与方向的已知与未知情况。

3) 列出加速度矢量方程，标注出加速度的大小与方向的已知与未知情况。

如果知道同一构件上两点的速度或加速度，求解第三点的速度或加速度，可利用在速度多边形或加速度多边形上作出对应构件的相似三角形的方法求解第三点的速度或加速度。该方法称之为影像法。

下面举例说明相对运动图解法的具体应用。

**例 3-4**　在图 3-10a 所示机构中，已知曲柄 $AB$ 以逆时针方向等速转动，其角速度为 $\omega_1$，求构件 2、3 的角速度 $\omega_2$、$\omega_3$ 和角加速度 $\alpha_2$、$\alpha_3$，及构件 2 上 $E$ 点的速度和加速度。

**解**

1) 选长度比例尺 $\mu_l$ 画出图 3-10a 所示的机构运动简图。

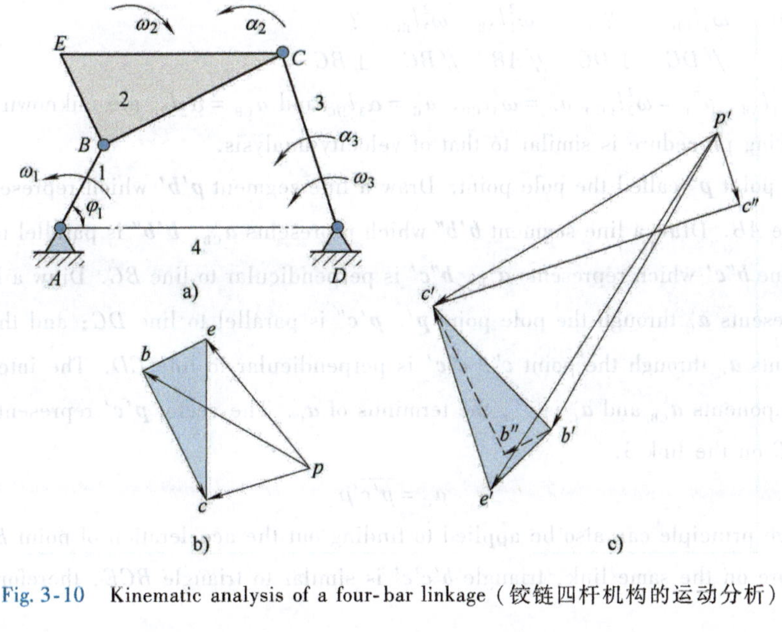


Fig. 3-10　Kinematic analysis of a four-bar linkage（铰链四杆机构的运动分析）

| | $\boldsymbol{v}_C =$ | $\boldsymbol{v}_B$ | $+ \boldsymbol{v}_{CB}$ |
|---|---|---|---|
| magnitude | ? | $\omega_1 l_{AB}$ | ? |
| direction | $\perp DC$ | $\perp AB$ | $\perp BC$ |

From Fig. 3-10 we have $\boldsymbol{pc} = \boldsymbol{pb} + \boldsymbol{bc}$

Choose a convenient velocity scale $\mu_v = \frac{v_B}{\overline{pb}}$, then $\overline{pb} = \frac{v_B}{\mu_v}$

Locate a point $p$ which is called the pole point and the origins of the absolute velocities are at it.

Draw a line segment $pb \perp AB$ and line $pc \perp DC$ through the pole point $p$, and then draw a line $bc \perp BC$. The intersection of line $pc$ and line $bc$ is the point $c$(see Fig. 3-10b). The segment $pc$ indicates the velocity of the point $C$ on the link 3, and $bc$ indicates the velocity of the point $C$ relative to point $B$ on the link 2, therefore:

$$v_C = \mu_v \overline{pc},\ v_{CB} = \mu_v \overline{bc}$$

$$\omega_2 = \frac{v_{CB}}{L_{BC}} = \frac{\mu_v \overline{bc}}{L_{BC}},\ \omega_3 = \frac{v_C}{L_{DC}} = \frac{\mu_v \overline{pc}}{L_{DC}}$$

From Fig. 3-10b, $v_{CB}$ is directed from $b$ toward $c$, thus in Fig. 3-10a, the link 2 is rotating clockwise about point $B$. Similarly, the direction of $\omega_3$ is counterclockwise about point $D$.

To find the velocity of point $E$, the image principle can be used. The velocity of point $E$ could have been obtained simply by making a triangle $bce$ similar to the triangle $BCE$ after points $b$ and $c$ had been located. Then connect $p$ to $c$, we have:

$$v_E = \mu_v \overline{pe}$$

3) Acceleration analysis. The acceleration of point $B$ is known.

$$a_B = a_B^n = \omega_1^2 l_{AB}$$

Its direction is from $B$ toward $A$, and then choose an acceleration scale $\mu_a$.

The acceleration vector equation of point $C$ relative to $B$ is written as follows:

| | $\boldsymbol{a}_C^n$ | $+ \boldsymbol{a}_C^t$ | $= \boldsymbol{a}_B^n$ | $+ \boldsymbol{a}_{CB}^n$ | $+ \boldsymbol{a}_{CB}^t$ |
|---|---|---|---|---|---|
| magnitude | $\omega_3^2 l_{CD}$ | ? | $\omega_1^2 l_{AB}$ | $\omega_2^2 l_{BC}$ | ? |
| direction | $//DC$ | $\perp DC$ | $//AB$ | $//BC$ | $\perp BC$ |

Where $a_B^n = \omega_1^2 l_{AB}$, $a_{CB}^n = \omega_2^2 l_{BC}$, $a_C^n = \omega_3^2 l_{CD}$. $a_C^t = \alpha_3 l_{DC}$ and $a_{CB}^t = \alpha_2 l_{BC}$ are unknown.

The drawing procedure is similar to that of velocity analysis.

Locate a point $\boldsymbol{p'}$ called the pole point. Draw a line segment $\boldsymbol{p'b'}$ which represents $\boldsymbol{a}_B^n$. $\boldsymbol{p'b'}$ is parallel to line $AB$. Draw a line segment $\boldsymbol{b'b''}$ which represents $\boldsymbol{a}_{CB}^n$. $\boldsymbol{b'b''}$ is parallel to line $BC$, and then draw a line $\boldsymbol{b''c'}$ which represents $\boldsymbol{a}_{CB}^t$. $\boldsymbol{b''c'}$ is perpendicular to line $BC$. Draw a line segment $\boldsymbol{p'c''}$ which represents $\boldsymbol{a}_C^n$ through the pole point $\boldsymbol{p'}$. $\boldsymbol{p'c''}$ is parallel to line $DC$; and then draw a line which represents $\boldsymbol{a}_C^t$ through the point $\boldsymbol{c''}$. $\boldsymbol{c''c'}$ is perpendicular to line $CD$. The intersection of two tangential components $\boldsymbol{a}_{CB}^t$ and $\boldsymbol{a}_C^t$ is $\boldsymbol{c'}$, the terminus of $\boldsymbol{a}_C$. The vector $\boldsymbol{p'c'}$ represents the acceleration of point $C$ on the link 3.

$$a_C = \overline{p'c'}\mu_a$$

The image principle can also be applied to finding out the acceleration of point $E$. Since points $B$, $C$ and $E$ are on the same link, triangle $b'c'e'$ is similar to triangle $BCE$, therefore, $\boldsymbol{a}_E$ is repre-

2）速度分析。因为机构中 $B$ 点的速度为已知，可从 $B$ 点开始进行速度分析，构件 2 上 $C$ 点与 $B$ 点为同一构件上的两点，故有

$$v_{\mathrm{B}} = v_{\mathrm{B}_1} = v_{\mathrm{B}_2} = \omega_1 l_{\mathrm{AB}}$$

| | $\boldsymbol{v}_{\mathrm{C}}$ = | $\boldsymbol{v}_{\mathrm{B}}$ + | $\boldsymbol{v}_{\mathrm{CB}}$ |
|---|---|---|---|
| 大小 | ? | $\omega_1 l_{\mathrm{AB}}$ | ? |
| 方向 | $\perp DC$ | $\perp AB$ | $\perp BC$ |

该速度矢量方程有两个已知方向的未知数，可用矢量加法求得 $\boldsymbol{v}_{\mathrm{C}}$ 和 $\boldsymbol{v}_{\mathrm{CB}}$。

上述速度矢量方程可通过引入速度比例尺转化为下列矢量方程。该方程可用高等数学中的矢量运算求解，即

$$\boldsymbol{pc} = \boldsymbol{pb} + \boldsymbol{bc}$$

选择速度比例尺，$\mu_v = \dfrac{v_{\mathrm{B}}}{\overline{pb}}$，则 $\overline{pb} = \dfrac{v_{\mathrm{B}}}{\mu_v}$。

任选一点 $p$，简称极点，作线段 $pb \perp AB$，代表速度 $_{\mathrm{B}}$。过 $p$ 点作 $CD$ 的垂直线，代表 $\boldsymbol{v}_{\mathrm{C}}$ 的方向线。过 $b$ 点作 $BC$ 的垂直线，代表 $\boldsymbol{v}_{\mathrm{CB}}$ 的方向线，交点即为 $c$ 点。线段 $pc$ 代表速度 $\boldsymbol{v}_{\mathrm{C}}$，线段 $bc$ 代表速度 $\boldsymbol{v}_{\mathrm{CB}}$。

$$v_{\mathrm{C}} = \mu_v \overline{pc} \qquad v_{\mathrm{CB}} = \mu_v \overline{bc}$$

已知构件 $BC$ 上两点的速度后，可以用影像法求解该构件上另一点 $E$ 的速度。在图 3-10b所示的速度多边形中，以 $bc$ 为边，作 $\triangle bce$ 与构件 $BCE$ 相似，即 $\triangle bce \cong \triangle BCE$，可在速度多边形中直接求得点 $e$，线段 $pe$ 代表速度 $\boldsymbol{v}_{\mathrm{E}}$。作相似三角形时要注意保持速度多边形和机构中构件的字母顺序的一致性。

构件 2 和 3 的角速度很容易求出来。

$\omega_2 = \dfrac{v_{\mathrm{CB}}}{L_{\mathrm{BC}}} = \dfrac{\mu_v \overline{bc}}{L_{\mathrm{BC}}}$，根据 $\boldsymbol{bc}$ 的方向可判别构件 2 的角速度方向为顺时针方向。

$\omega_3 = \dfrac{v_{\mathrm{C}}}{L_{\mathrm{DC}}} = \dfrac{\mu_v \overline{pc}}{L_{\mathrm{DC}}}$，根据 $\boldsymbol{pc}$ 的方向可判别构件 3 的角速度方向为逆时针方向。

3）加速度分析。因为机构中 $B$ 点的加速度为已知，可从 $B$ 点开始进行加速度分析，构件 2 上 $C$ 点与 $B$ 点为同一构件上的两点，故有

| | $\boldsymbol{a}_{\mathrm{C}}^{\mathrm{n}}$ + | $\boldsymbol{a}_{\mathrm{C}}^{\mathrm{t}}$ = | $\boldsymbol{a}_{\mathrm{B}}^{\mathrm{n}}$ + | $\boldsymbol{a}_{\mathrm{CB}}^{\mathrm{n}}$ + | $\boldsymbol{a}_{\mathrm{CB}}^{\mathrm{t}}$ |
|---|---|---|---|---|---|
| 大小 | $\omega_3^2 l_{\mathrm{CD}}$ | ? | $\omega_1^2 l_{\mathrm{AB}}$ | $\omega_2^2 l_{\mathrm{BC}}$ | ? |
| 方向 | $/\!/DC$ | $\perp DC$ | $/\!/AB$ | $/\!/BC$ | $\perp BC$ |

式中，$a_{\mathrm{B}}^{\mathrm{n}} = \omega_1^2 l_{\mathrm{AB}}$，$a_{\mathrm{CB}}^{\mathrm{n}} = \omega_2^2 l_{\mathrm{BC}}$，$a_{\mathrm{C}}^{\mathrm{n}} = \omega_3^2 l_{\mathrm{CD}}$，均为已知数。$a_{\mathrm{C}}^{\mathrm{t}} = \alpha_3 l_{\mathrm{DC}}$，$a_{\mathrm{CB}}^{\mathrm{t}} = \alpha_2 l_{\mathrm{BC}}$，为待求的值。

上述加速度矢量方程可通过引入加速度比例尺转化为下列矢量方程。该方程可用高等数学中的矢量运算求解，矢量方程表达为

$$\boldsymbol{p'c''} + \boldsymbol{c''c'} = \boldsymbol{p'b'} + \boldsymbol{b'b''} + \boldsymbol{b''c'}$$

矢量加法的具体过程为：任选极点 $\boldsymbol{p'}$，作 $\boldsymbol{p'b'} /\!/ AB$，$\boldsymbol{p'b'}$代表 $\boldsymbol{a}_{\mathrm{B}}^{\mathrm{n}}$，过 $b'$点作 $\boldsymbol{b'b''} /\!/ BC$，$\boldsymbol{b'b''}$代表 $\boldsymbol{a}_{\mathrm{CB}}^{\mathrm{n}}$；过 $b''$点作 $BC$ 的垂直线，代表 $\boldsymbol{a}_{\mathrm{CB}}^{\mathrm{t}}$的方向线；过 $p'$作 $\boldsymbol{p'c''} /\!/ DC$，$\boldsymbol{p'c''}$代表 $\boldsymbol{a}_{\mathrm{C}}^{\mathrm{n}}$；过 $\boldsymbol{c''}$点作 $DC$ 的垂直线，代表 $\boldsymbol{a}_{\mathrm{C}}^{\mathrm{t}}$ 的方向线，交点 $c'$即为所求。$\boldsymbol{p'c'}$代表 $C$ 点的加速度 $\boldsymbol{a}_{\mathrm{C}}$。

sented in the acceleration diagram by the vector from $p'$ to $e'$.

$$a_E = \overline{p'e'}\mu_a$$

$a^t_{CB} = \overline{b''c'}\mu_a$, $\alpha_2 = \frac{a^t_{CB}}{l_{BC}}$, put the $\boldsymbol{a}^t_{CB}$ on the point $C$, we know that the angular acceleration of link 2 is counterclockwise.

**Example 3-5** Fig. 3-11a shows a guider-bar mechanism; all the dimensions of the links and angular position of the driving link $AB$ are known. When the crank $AB$ rotates counterclockwise with an angular velocity $\omega_1$, determine the angular velocities $\omega_2$, $\omega_3$ and the angular acceleration $\alpha_2$ and $\alpha_3$.

**Solution**

1) Draw the scaled skeleton with $\mu_l$ (see Fig. 3-11a).

2) Velocity analysis. Let us imagine that the link 3 can be expanded on the point $B$. The point $B$ is a coincident points of point $B_1$ which is on the crank 1, point $B_2$ which is on the slider 2 and $B_3$ which is on the rocker 3. Point $B$ is the instantaneous center of link 1 and link 2, so we have:

$$v_{B_1} = v_{B_2} = \omega_1 l_{AB}$$

The velocity equation relating point $B_2$ and $B_3$ is:

| | $\boldsymbol{v}_{B_3}$ | = $\boldsymbol{v}_{B_2}$ | + $\boldsymbol{v}_{B_3B_2}$ |
|---|---|---|---|
| direction | $\perp BD$ | $\perp AB$ | // guider line |
| magnitude | ? | $\omega_1 l_{AB}$ | ? |

Where, the direction of $\boldsymbol{v}_{B_2}$ is perpendicular to line $AB$ and to the right, $\boldsymbol{v}_{B_3B_2}$ is parallel to the guider line and its magnitude is unknown, and $\boldsymbol{v}_{B_3}$ is perpendicular to line $BD$ and its magnitude is unknown too.

Choose a velocity scale $\mu_v$ and a pole point $p$. Beginning at the pole $p$, we draw $\boldsymbol{pb}_2$ which represents $\boldsymbol{v}_{B_2}$, then from its terminus draw a line of indefinite length which represents the direction of $\boldsymbol{v}_{B_3B_2}$. Draw a line which is perpendicular to line $BD$ from the pole $p$, the intersection of the two lines is point $b_3$ (see Fig. 3-11c). Then we have:

$$v_{B_3} = \mu_v \overline{pb_3},\ v_{B_3B_2} = \mu_v \overline{b_2b_3},\ \omega_3 = \frac{v_{B_3}}{l_{BD}},\ \omega_2 = \omega_3$$

The direction of $\omega_3$ is clockwise.

3) Acceleration analysis. The acceleration equation of point $B_3$ relative to $B_2$ is:

$$\boldsymbol{a}_{B_3} = \boldsymbol{a}_{B_2} + \boldsymbol{a}^k_{B_3B_2} + \boldsymbol{a}^r_{B_3B_2}$$

The acceleration of point $B$ on the link 3 can be divided into normal component and tangential component.

$\boldsymbol{a}_{B_3} = \boldsymbol{a}^n_{B_3} + \boldsymbol{a}^t_{B_3}$, so we have:

| | $\boldsymbol{a}^n_{B_3}$ | + $\boldsymbol{a}^t_{B_3}$ | = $\boldsymbol{a}^n_{B_2}$ | + $\boldsymbol{a}^k_{B_3B_2}$ | + $\boldsymbol{a}^r_{B_3B_2}$ |
|---|---|---|---|---|---|
| direction | $B \to D$ | $\perp BD$ | $B \to A$ | $\perp$ guider | // guider |
| magnitude | $\omega_3^2 l_{BD}$ | ? | $\omega_1^2 l_1$ | $2v_{B_3B_2}\omega_2$ | ? |

Choose an acceleration scale $\mu_a$ to transform the accelerations $\boldsymbol{a}^n_{B_2}$, $\boldsymbol{a}^n_{B_3}$, $\boldsymbol{a}^k_{B_3B_2}$ into length vectors. Locate a pole point $p'$. Draw a line segment $\boldsymbol{p'b'_2}$ which represents $\boldsymbol{a}^n_{B_2}$. $\boldsymbol{p'b'_2}$ is parallel to $AB$

$a_C^t = \overline{c''c'}\mu_a$，$\alpha_3 = \dfrac{a_C^t}{l_{DC}}$，其方向由 $c''c'$ 的方向判别，为逆时针方向。

$a_{CB}^t = \overline{b''c'}\mu_a$，$\alpha_2 = \dfrac{a_{CB}^t}{l_{BC}}$，其方向由 $b''c'$ 的方向判别，为逆时针方向。

连杆上 $E$ 点的加速度也可用加速度影像法直接求出来。

$a_E = \mu_a \overline{p'e'}$，方向如图 3-10c 所示。

**例 3-5**　图 3-11 所示机构中，已知曲柄 $AB$ 以逆时针方向等速转动，其角速度为 $\omega_1$，求构件 2、3 的角速度 $\omega_2$、$\omega_3$ 和角加速度 $\alpha_2$、$\alpha_3$。

**解**　1）选长度比例尺 $\mu_l$ 画出图 3-11a 所示的机构运动简图。

2）速度分析。构件 1 上 $B_1$ 点的速度为

$$v_{B_1} = v_{B_2} = \omega_1 l_{AB}$$

列速度方程时必须与 $B_1$ 点联系起来，才能使矢量方程的未知数少于 2。因此扩大构件 3，如图 3-11b 所示。此时，$B$ 点为构件 1、2、3 的重合点，可用 $B_1$、$B_2$、$B_3$ 表示重合点 $B$ 的位置。

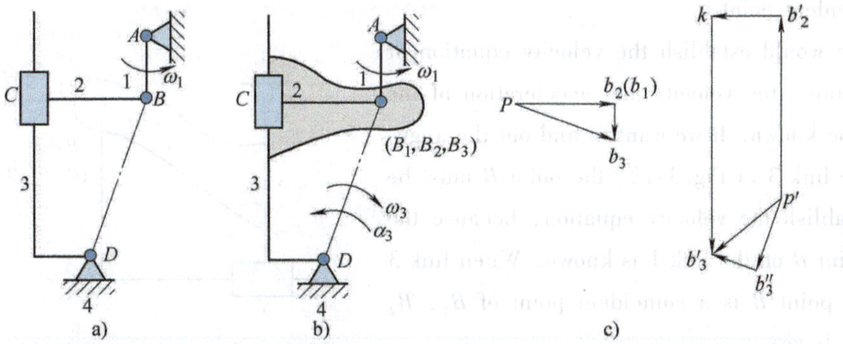


Fig. 3-11　Kinematic analysis of a guide-bar mechanism（导杆机构的运动分析）

构件 3、2 在重合点 $B$ 的速度矢量方程为

$$\boldsymbol{v}_{B_3} = \boldsymbol{v}_{B_2} + \boldsymbol{v}_{B_3B_2}$$

| | $\boldsymbol{v}_{B_3}$ | $\boldsymbol{v}_{B_2}$ | $\boldsymbol{v}_{B_3B_2}$ |
|---|---|---|---|
| 方向 | $\perp BD$ | $\perp AB$ | // 导路 |
| 大小 | ? | $\omega_1 l_{AB}$ | ? |

选速度比例尺 $\mu_v$，把 $v_{B_2}$ 转化为长度 $pb_2$。

任选一极点 $p$ 作矢量加法，则

$$\boldsymbol{pb_3} = \boldsymbol{pb_2} + \boldsymbol{b_2b_3}$$

$v_{B_3} = \mu_v \overline{pb_3}$，$v_{B_3B_2} = \mu_v \overline{b_2b_3}$，$\omega_3 = \dfrac{v_{B_3}}{l_{BD}}$，$\omega_2 = \omega_3$，其方向为顺时针方向。

3）加速度分析。两构件在重合点 $B$ 处的加速度关系为

$$\boldsymbol{a}_{B_3} = \boldsymbol{a}_{B_2} + \boldsymbol{a}_{B_3B_2}^k + \boldsymbol{a}_{B_3B_2}^r$$

$$\boldsymbol{a}_{B_3} = \boldsymbol{a}_{B_3}^n + \boldsymbol{a}_{B_3}^t$$

| | $\boldsymbol{a}_{B_3}^n$ | $+\boldsymbol{a}_{B_3}^t$ | $=\boldsymbol{a}_{B_2}^n$ | $+\boldsymbol{a}_{B_3B_2}^k$ | $+\boldsymbol{a}_{B_3B_2}^r$ |
|---|---|---|---|---|---|
| 方向 | $B \to D$ | $\perp BD$ | $B \to A$ | $\perp$ 导路（指左） | // 导路 |
| 大小 | $\omega_3^2 l_{BD}$ | ? | $\omega_1^2 l_1$ | $2v_{B_3B_2}\omega_2$ | ? |

and toward $A$. Draw a line segment $b_2'k'$ which represents $a^k_{B_3B_2}$. $b_2'k'$ is perpendicular to the guider line toward the left, and then from the point $k'$ draw a line which is parallel to the guider line. It represents $a^r_{B_3B_2}$. Draw a line segment $p'b_3''$ which represents $a^n_{B_3}$ through the pole point $p'$. $p'b_3''$ is parallel to $BD$ and toward $D$, then draw a line which is perpendicular to line $BD$ through the point $b_3''$. It represents $a^t_{B_3}$. The intersection of two lines is $b_3'$. The vector $b_3''b_3'$ represents the tangential acceleration of point $B$ on the link 3.

The acceleration polygon is shown in Fig. 3-11c.

$$\alpha_3 = \frac{a^t_{B_3}}{l_{BD}} = \frac{\overline{b_3''b_3'}\mu_a}{l_{BD}}$$

The direction of the angular acceleration $\alpha_3$ can be determined by $a^t_{B_3}$ (see Fig. 3-11a).

**3. Some Key Points of Motion Analysis**

1) The Coriolis acceleration of coincident points on two different links must be discriminated correctly. This Coriolis component of acceleration will always be presented when there is a velocity of slip associated with any member which also has an angular velocity. The Coriolis acceleration will be produced at coincident point.

2) When we would establish the velocity equation or acceleration equation, the velocity and acceleration of the base point must be known. If we want to find out the angular velocity of the link 3 in Fig. 3-12, the point $B$ must be considered to establish the velocity equation, because the velocity of the point $B$ on the link 1 is known. When link 3 is expanded, the point $B$ is a coincident point of $B_1$, $B_2$ and $B_3$. Then we have:

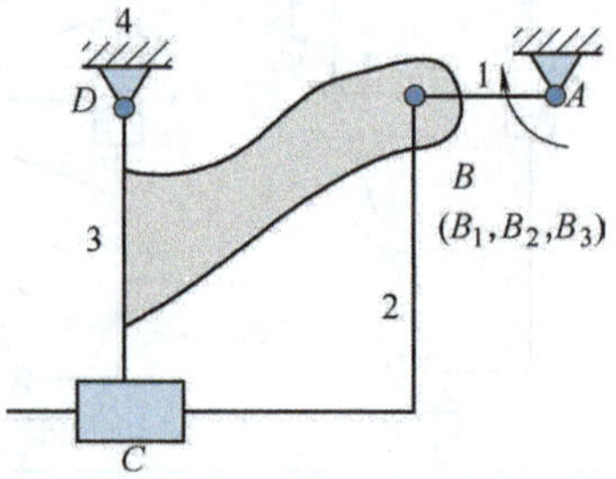

Fig. 3-12 Expanded link(构件的扩大)

$$v_{B_3} = v_{B_2} + v_{B_3B_2}$$

Therefore, we can determine $\omega_3$ from the velocity equation.

3) When the mechanism is at its limited positions, the velocity polygon or acceleration polygon becomes simple, but sometimes it is difficult to analysis.

Fig. 3-13a shows a four-bar linkage in which the crank and coupler are in collinear. In the guider-bar mechanism shown in Fig. 3-13b, the crank is perpendicular to the rocker. Their velocity polygon and acceleration polygon are very simple.

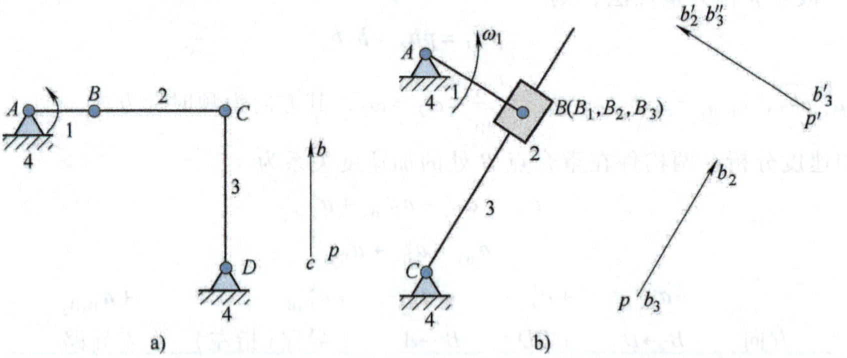

Fig. 3-13 Kinematic analysis in limited positions (特殊位置的运动分析)

选加速度比例尺 $\mu_a$，把 $a^n_{B_2}$、$a^n_{B_3}$、$a^n_{B_3B_2}$ 转化为相应长度。

任选一点 $p'$ 作矢量加法，则

$$p'b_3' = p'b_2' + b_2'k + kb_3'$$

$$p'b_3'' + b_3''b_3' = p'b_2' + b_2'k + kb_3'$$

$p'b_3''$ 代表 $B_3$ 点的法向加速度 $a^n_{B_3}$，$b_3''b_3'$ 代表 $B_3$ 点的切向加速度 $a^t_{B_3}$，$pb_2'$ 代表 $B_2(B_1)$ 点的法向加速度 $a^n_{B_2}$，$b_2'k$ 代表重合点 $B$ 点的科氏加速度 $a^k_{B_3B_2}$，$kb_3'$ 代表重合点 $B$ 点的相对加速度 $a^r_{B_3B_2}$。

加速度多边形如图 3-11c 所示。

$\alpha_3 = \dfrac{a^t_{B_3}}{l_{BD}} = \dfrac{b_3''b_3'\mu_a}{l_{BD}}$，其方向由 $a^t_{B_3}$ 判断，如图 3-11b 所示。

**3. 机构运动分析中应注意的若干问题**

1）正确判别科氏加速度。科氏加速度的存在条件是两构件以相同的角速度共同转动的同时，还必须做相对运动，其重合点存在科氏加速度。

2）建立速度或加速度矢量方程时，一定要从已知速度或加速度的点开始列方程，另一个构件与该点不接触时，可采用构件扩大的方法重合到该点，这样就可以建立两重合点的速度方程或加速度方程。如图 3-12所示机构中，若想求出构件 3 的速度或角速度，只要把构件 3 按图示扩大，即可列出简单的速度方程和加速度方程，从而实现求解的目的。

$$\boldsymbol{v}_{B_3} = \boldsymbol{v}_{B_2} + \boldsymbol{v}_{B_3B_2}$$

在速度方程中，要注意构件 3 在 $B$ 点的速度方向垂直于 $BD$，构件 2 在 $B$ 点的速度等于构件 1 在 $B$ 点的速度，且可直接求解。构件 3、2 在 $B$ 点的相对速度方向平行于导路方向。

重合点选取得当，可使求解过程大大简化。

3）机构在极限位置、共线位置等特殊位置时，其速度和加速度多边形变得简单。图 3-13a所示铰链四杆机构的曲柄与连杆共线，图 3-13b 所示的导杆机构中，导杆 $BC$ 处于极限位置。

4）液压机构的运动分析可转化为相应的导杆机构进行。图 3-14a 所示的摆动液压缸机构可转化为图 3-14b 所示的导杆机构，然后再用相对运动图解法进行运动分析。

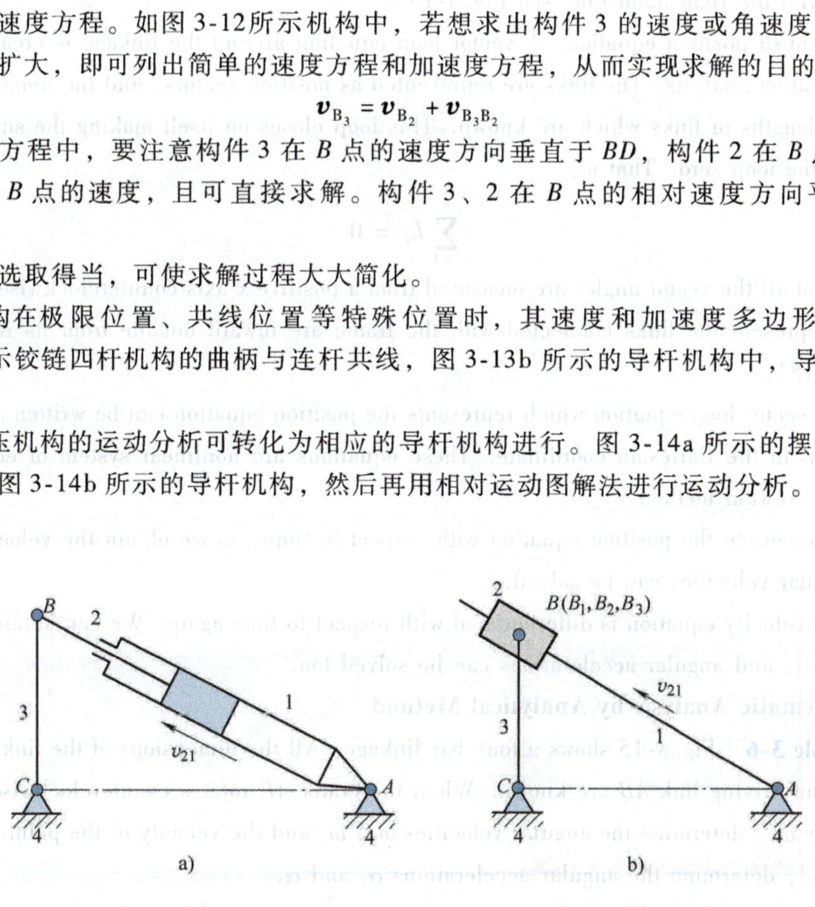


Fig. 3-14　Kinematic analysis of hydraulic mechanism（摆动液压缸机构运动分析）

4) Hydraulic mechanism can be transformed into a relating guider-bar mechanism. Fig. 3-14a shows a hydraulic mechanism. It can be transformed into a guider-bar mechanism shown in Fig. 3-14b; they are equivalent mechanisms.

## 3.4 Kinematic Analysis by Algebraic Method

### 1. Fundamental Law of Analytical Method

The procedure of analytical method is that the position equation must be established first, then differentiate the position equation with respect to time, we obtain the velocity equation. In the end, the velocity equation is differentiated with respect to time. We can obtain the acceleration equation.

An approach to motion analysis creates a vector loop around the mechanism. This loop closes on itself making the vector sum around the loop zero. The links are represented as position vectors, and the lengths of the vectors are the lengths of the links which are known.

The procedure of motion analysis is as follows:

1) Establish a coordinate system in which the origin of the coordinate system is coincident with the rotating center of the driving link and the $x$ axis is along the frame of the mechanism. The $y$ axis is satisfied with the right hand role (see Fig. 3-15).

2) Establish position equation. A vector loop equation around the linkage is created in an approach to position analysis. The links are represented as position vectors, and the lengths of the vectors are the lengths of links which are known. This loop closes on itself making the sum of the vectors around the loop zero. That is:

$$\sum_{i=1}^{n} L_i = 0$$

Note that all the vector angles are measured from a positive $x$ axis counterclockwise and the vectors which represent the links connected with the frame are toward outside from its rotating center (see Fig. 3-15).

3) The vector loop equation which represents the position equation can be written as two projective equations in the Cartesian coordinate. These equations are nonlinear system of equations. We can solve by Newton method.

4) Differentiate the position equation with respect to time, so we obtain the velocity equation, and the angular velocities can be solved.

5) The velocity equation is differentiated with respect to time again. We can obtain the acceleration equation, and angular accelerations can be solved too.

### 2. Kinematic Analysis by Analytical Method

**Example 3-6** Fig. 3-15 shows a four-bar linkage. All the dimensions of the links and angular position of the driving link $AB$ are known. When the crank $AB$ rotates counterclockwise with an angular velocity $\omega_1$, determine the angular velocities $\omega_2$, $\omega_3$ and the velocity of the point $E$ on the link 2. In the end, determine the angular accelerations $\alpha_2$ and $\alpha_3$.

**Solution**

The Cartesian coordinate has been established first; then we create a closed vector loop around

## 3.4 用解析法对机构进行运动分析

**1. 解析法的基本知识**

解析法的实质是建立机构的位置方程 $s=s(\varphi)$、速度方程 $v=v(\varphi)$、加速度方程 $a=a(\varphi)$ 并求解的过程。

解析法的一般步骤为：

1）建立直角坐标系。一般情况下，坐标系的原点与原动件的转动中心重合，$x$ 轴通过机架，$y$ 轴的确定按直角坐标系法则处理。

2）建立机构运动分析的数学模型。把机构看作一个封闭环，构件尺寸看作矢量，连架杆的矢量方向指向与连杆连接的铰链中心。其余杆件的矢量方向可任意选定。最后列出的机构封闭矢量和应为零。即

$$\sum_{i=1}^{n} L_i = 0$$

3）求解位置方程。各矢量与 $x$ 轴的夹角以逆时针方向为正，把矢量方程中各矢量向 $x$、$y$ 轴投影，其投影方程即为机构的位置方程。该方程为非线性方程，可用牛顿法求解。

4）求解速度方程。位置方程中的各项对时间求导数，可得到机构的速度方程，从中解出待求的角速度或某些点的速度。

5）求解加速度方程。速度方程中的各项对时间求导数，可得到机构的加速度方程，从中解出待求的角加速度或某些点的加速度。

**2. 解析法在机构运动分析中的应用**

以下通过几个示例说明解析法的具体应用。

**例 3-6** 已知图 3-15 所示的铰链四杆机构中各构件的尺寸和原动件 1 的位置 $\varphi_1$ 以及角速度 $\omega_1$，求解构件 2、3 的角速度 $\omega_2$、$\omega_3$ 和角加速度 $\alpha_2$、$\alpha_3$。

**解**

1）建立直角坐标系 $Axy$，坐标原点通过 $A$ 点，$x$ 轴沿机架 $AD$。

2）封闭矢量环如图 3-15 所示，连架杆矢量外指（分别指向与连杆连接处的铰链中心），余者任意确定。封闭环矢量方程为

$$\boldsymbol{L}_1+\boldsymbol{L}_2-\boldsymbol{L}_3-\boldsymbol{L}_4=0$$

3）建立各矢量的投影方程。注意各矢量与 $x$ 轴的夹角以逆时针方向为正。

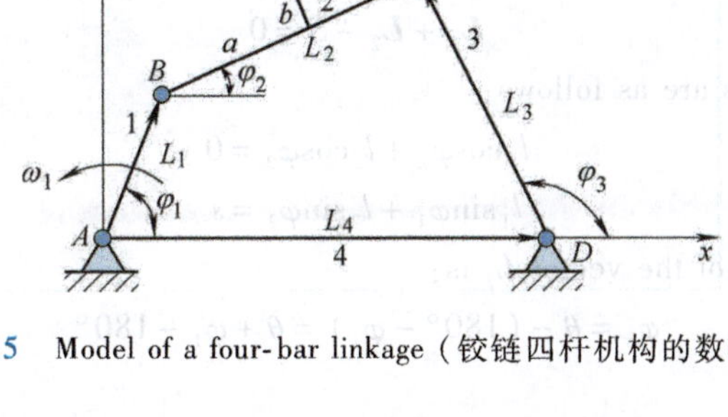


Fig. 3-15 Model of a four-bar linkage（铰链四杆机构的数学模型）

the linkage. The links are now drawn as position vectors which form a closed vector loop, and their directions are shown in Fig. 3-15.

The sum of the vectors is zero, thus:

$$\boldsymbol{L}_1 + \boldsymbol{L}_2 - \boldsymbol{L}_3 - \boldsymbol{L}_4 = 0$$

Write down the vector loop equations as projective equations in the Cartesian coordinate.

$$l_2\cos\varphi_2 - l_3\cos\varphi_3 = l_4 - l_1\cos\varphi_1$$
$$l_2\sin\varphi_2 - l_3\sin\varphi_3 = -l_1\sin\varphi_1$$

These position equations can be solved by Newton method; the $\varphi_2$ and $\varphi_3$ are known.

Differentiate the position equations with respect to time, and rewrite them in matrix, so we can obtain the velocity equations, and the angular velocities $\omega_2$, $\omega_3$ can be solved.

$$\begin{pmatrix} -l_2\sin\varphi_2 & l_3\sin\varphi_3 \\ l_2\cos\varphi_2 & -l_3\cos\varphi_3 \end{pmatrix}\begin{pmatrix} \omega_2 \\ \omega_3 \end{pmatrix} = \begin{pmatrix} l_1\omega_1\sin\varphi_1 \\ -l_1\omega_1\cos\varphi_1 \end{pmatrix}$$

We differentiate the velocity equations with respect to time again, so the acceleration equations are obtained, and the angular accelerations $\alpha_2$, $\alpha_3$ can be solved.

$$\begin{pmatrix} -l_2\sin\varphi_2 & l_3\sin\varphi_3 \\ l_2\cos\varphi_2 & -l_3\cos\varphi_3 \end{pmatrix}\begin{pmatrix} \alpha_2 \\ \alpha_3 \end{pmatrix} = -\begin{pmatrix} -l_2\omega_2\cos\varphi_2 & l_3\omega_3\cos\varphi_3 \\ -l_2\omega_2\sin\varphi_2 & l_3\omega_3\sin\varphi_3 \end{pmatrix}\begin{pmatrix} \omega_2 \\ \omega_3 \end{pmatrix} + \begin{pmatrix} l_1\omega_1^2\cos\varphi_1 \\ l_1\omega_1^2\sin\varphi_1 \end{pmatrix}$$

If the velocity or acceleration of a point on a link, such as point $E$, must be determined, we can first write position equation of this point, and then differentiate with respect to time two times.

The position equations of the point $E$ are as follows:

$$x_E = l_1\cos\varphi_1 + a\cos\varphi_2 + b\cos(\varphi_2 + 90°)$$
$$y_E = l_1\sin\varphi_1 + a\sin\varphi_2 + b\sin(\varphi_2 + 90°)$$

The velocity of point $E$ is:

$$v_E = \sqrt{x'^2_E + y'^2_E}$$

The acceleration of point $E$ is:

$$a_E = \sqrt{x''^2_E + y''^2_E}$$

A computer program can be employed to relieve the designers of many routine processes.

**Example 3-7** In the following mechanism of Fig. 3-16, the link 1 rotates at a constant angular velocity $\omega_1$; its direction is counterclockwise. When the link 1 is in the phase $\varphi_1$, determine the displacement, velocity and acceleration of link 3.

**Solution**

Establish the Cartesian coordinate system and a closed vector loop around the linkage, as shown in Fig. 3-16.

The sum of the vectors in the loop is zero, thus:

$$\boldsymbol{L}_1 + \boldsymbol{L}_2 - \boldsymbol{S} = 0$$

The position equations are as follows:

$$l_1\cos\varphi_1 + l_2\cos\varphi_2 = 0$$
$$l_1\sin\varphi_1 + l_2\sin\varphi_2 = s$$

The position angle $\varphi_2$ of the vector $L_2$ is:

$$\varphi_2 = \theta - (180° - \varphi_1) = \theta + \varphi_1 - 180°$$

$$l_1\cos\varphi_1+l_2\cos\varphi_2-l_3\cos\varphi_3-l_4=0$$
$$l_1\sin\varphi_1+l_2\sin\varphi_2-l_3\sin\varphi_3=0$$

该位置方程为非线性方程组，可用牛顿法解出构件 2、3 的角位移 $\varphi_2$、$\varphi_3$。

4）位置方程对时间求导数，可得到速度方程。两边求导并整理后得

$$-l_2\omega_2\sin\varphi_2+l_3\omega_3\sin\varphi_3=l_1\omega_1\sin\varphi_1$$
$$l_2\omega_2\cos\varphi_2-l_3\omega_3\cos\varphi_3=-l_1\omega_1\cos\varphi_1$$

写成矩阵方程为

$$\begin{pmatrix}-l_2\sin\varphi_2 & l_3\sin\varphi_3\\ l_2\cos\varphi_2 & -l_3\cos\varphi_3\end{pmatrix}\begin{pmatrix}\omega_2\\ \omega_3\end{pmatrix}=\begin{pmatrix}l_1\omega_1\sin\varphi_1\\ -l_1\omega_1\cos\varphi_1\end{pmatrix}$$

此方程为线性方程组，可用消元法求解出构件 2、3 的角速度 $\omega_2$、$\omega_3$。

5）速度方程再对时间求一次导数，可得加速度方程。即

$$\begin{pmatrix}-l_2\sin\varphi_2 & l_3\sin\varphi_3\\ l_2\cos\varphi_2 & -l_3\cos\varphi_3\end{pmatrix}\begin{pmatrix}\alpha_2\\ \alpha_3\end{pmatrix}=-\begin{pmatrix}-l_2\omega_2\cos\varphi_2 & l_3\omega_3\cos\varphi_3\\ -l_2\omega_2\sin\varphi_2 & l_3\omega_3\sin\varphi_3\end{pmatrix}\begin{pmatrix}\omega_2\\ \omega_3\end{pmatrix}+\begin{pmatrix}l_1\omega_1^2\cos\varphi_1\\ l_1\omega_1^2\sin\varphi_1\end{pmatrix}$$

此为线性方程组，可求解出构件 2、3 的角加速度 $\alpha_2$、$\alpha_3$。

要求构件 2 上 $E$ 点的速度或加速度，可写出 $E$ 点的位置坐标，然后求导数。即

$$x_E=l_1\cos\varphi_1+a\cos\varphi_2+b\cos(\varphi_2+90°)$$
$$y_E=l_1\sin\varphi_1+a\sin\varphi_2+b\sin(\varphi_2+90°)$$
$$v_E=\sqrt{x'^2_E+y'^2_E},\quad a_E=\sqrt{x''^2_E+y''^2_E}$$

**例 3-7** 对图 3-16 所示机构进行运动分析。已知机构的尺寸和原动件 1 的位置 $\varphi_1$、角速度 $\omega_1$，求构件 3 的位移、速度、加速度。

**解** 画出机构简图并建立图示的坐标系，建立矢量环。

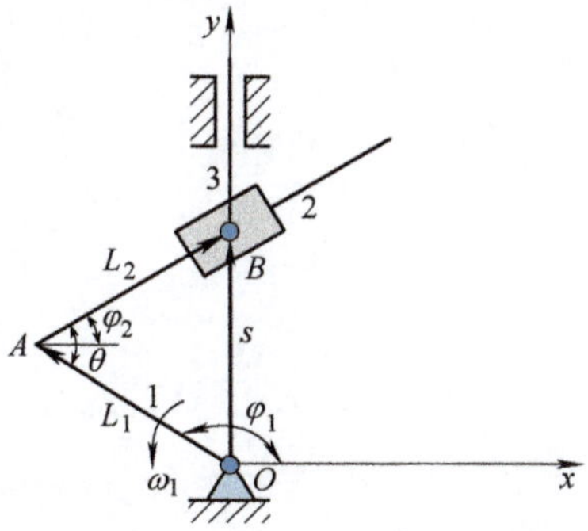


Fig. 3-16 Model of a four-bar linkage with a sliding pair
（含有移动副四杆机构的运动分析模型）

封闭矢量环方程为

$$\boldsymbol{L}_1+\boldsymbol{L}_2-\boldsymbol{s}=0$$

投影方程为

$$l_1\cos\varphi_1+l_2\cos\varphi_2=0$$
$$l_1\sin\varphi_1+l_2\sin\varphi_2=s$$

$\varphi_2=\theta-(180°-\varphi_1)=\theta+\varphi_1-180°$，将 $\varphi_2$ 代入上式可有

$$l_1\cos\varphi_1+l_2\cos(\varphi_1+\theta-180°)=0$$
$$l_1\sin\varphi_1+l_2\sin(\varphi_1+\theta-180°)=s$$

Then, we have:

$$l_1\cos\varphi_1 + l_2\cos(\varphi_1 + \theta - 180°) = 0$$
$$l_1\sin\varphi_1 + l_2\sin(\varphi_1 + \theta - 180°) = s$$

Rearrange the above equations, we have

$$l_1\cos\varphi_1 - l_2\cos(\varphi_1 + \theta) = 0$$
$$l_1\sin\varphi_1 - l_2\sin(\varphi_1 + \theta) = s$$

If we differentiate the position equations with respect to time two times, the velocity and acceleration of the link 3 can be solved easily.

**3. Summary**

The key point of kinematical analysis with algebraic method is how to establish the vector loop. Fig. 3-17 shows some closed vector loops in different mechanisms.

Fig. 3-17a shows an offset slider-crank mechanism; its closed vectors loop is *ABCDA*, not *ABC*. Fig. 3-17b shows a guider-bar mechanism; its closed vectors loop is *ABC*.

When a mechanism is in a limited position, we can draw its common position, and establish the vector loop equation, then substitute the special position, such as angle $\varphi_1 = 90°$. This makes it easy to solve (see the Fig. 3-17c).

$$l_1\cos\varphi_1 - l_2\cos(\varphi_1 + \theta) = 0$$
$$l_1\sin\varphi_1 - l_2\sin(\varphi_1 + \theta) = s$$

本例题求解过程比较简单。解上述位置方程可求出位移 $s$，对 $s$ 求导数可求出速度与加速度。由于 $l_2$ 是变量，$l_2$ 的一次导数是构件 2、3 的相对速度，二次导数为相对加速度。

**3. 解析法总结**

封闭矢量环的建立是解析法的关键问题。图 3-17 所示为一些机构的封闭矢量环。

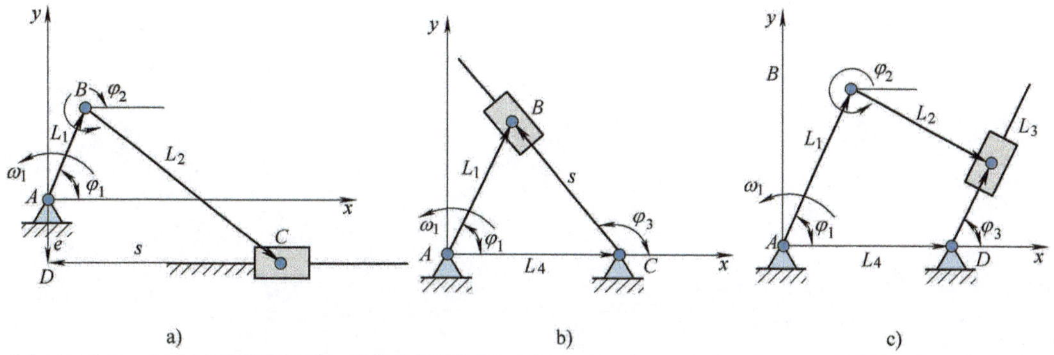


Fig. 3-17　Some closed vector loops in mechanisms（一些机构的封闭矢量环）

图 3-17a 所示的曲柄滑块机构，不能用 $ABC$ 建立封闭矢量环，要建成封闭矢量环 $ABCDA$，$\overline{AD} = e$，$\overline{DC} = s$，$s$ 为待求量。

图 3-17b 所示为摆动导杆机构的封闭矢量环及其坐标系的选择。

当机构处于特殊位置时，如 $\varphi_1 = 90°$，可按图 3-17c 所示的一般位置建立矢量环方程，代入特定角度后，可求解对应位置的速度与加速度。

# Chapter 4

# Force Analysis of Planar Mechanisms

# 第4章

# 平面机构的力分析

## 4.1 Introduction

A machine is a device that performs work and transmits energy by means of mechanical force from a power source to a driven load. So the forces acting on the mechanism is the foundation of calculating the strength and stiffness of the members. These forces are also theoretical foundation of calculating the mechanical efficiency.

The forces acting on the mechanism can be divided into two classes; they are external applied loads and the constraint forces. The external applied loads include the driving force and resistance etc. The constraint forces which occur in pairs are internal forces in the machine, but they can also be external forces on an individual link.

**1. Contents of Force Analysis**

1) Knowing the external applied loads, calculate the constraint forces which occur in pairs. Constraint forces can be used to design kinematic pairs.

2) Knowing the resistance acting on driven link, calculate the forces acting on driving link, or knowing the driving force, calculate the resistance acting on the driven link.

3) The force analysis is the theoretical foundation to calculate the efficiency.

4) Make use of the force analysis, we can design some self-locking mechanism.

Various assumptions must be made during the force analysis. A major assumption concerns dynamic or inertia forces. All machines have masses, and if an element of a machine is accelerating, there will be inertia force associated with this motion. However, if the magnitude of the inertia force is small compared to the external applied loads, it can be neglected while analyzing the mechanism. Such an analysis is known as static-force analysis.

When the inertia force due to the mass of the components is also considered, and the friction effects are negligible, it is called dynamic-force analysis.

In this chapter, we will discuss the static-force analysis considering the friction and dynamic-force analysis considering the inertia force.

**2. Methods of Force Analysis**

There are two general methods of performing a complete force analysis.

(1) Graphical force analysis  Graphical force analysis employs scaled free-body diagrams and vector graphics in the determination of unknown forces. An important advantage of the graphical approach is that it provides useful insight as to the nature of the forces in the physical system.

(2) Analytical force analysis  Analytical method for investigating static and dynamic force in machines employs mathematical models, and the mathematical base of the analytical approach lends itself well to computer implementation. Solutions can be obtained quickly and accurately for many positions of a mechanism.

## 4.2 Force Analysis Including Inertia Forces in Mechanisms

All machines have some acceleration links; dynamic forces are always present when the ma-

## 4.1 平面机构力分析概述

机构运动过程中，会受到各种力的作用。作用在机构上的力是计算各构件的强度、刚度及进行结构设计的重要依据，也是计算机械效率的必要条件。

作用在机构上的力，可分为外部施加于机构的力以及机构中各运动副的反作用力。外部施加的力主要包括作用在机构上的驱动力或驱动力矩、生产阻力等。机构中运动副的反力对整个机构系统来说是内力，但对一个分离出来的构件来说则是外力。

**1. 机构力分析的内容**

机构的力分析主要包含以下内容：

1）根据作用在机构中的已知外力，求解各运动副中的反力。运动副中的反力是运动副结构设计的依据。

2）已知作用在机构上的生产阻力，可求解出原动件上施加的驱动力；已知原动机的驱动力，可以求解出作用在从动件上的生产阻力。

3）机构的受力分析是计算机械效率的基础。

4）机构的受力分析还是设计自锁机构的基础。合理设计反作用力的方向，可使机构反行程自锁。自锁机构有重要用途。

**2. 机构力分析的方法**

机构力分析的方法有两种，即图解法和解析法。

在分析机构一个运动周期中的受力状态时，由于计算量大，使用解析法并利用计算机编制程序可缩短计算时间并提高计算精度。当对机构的某一具体位置进行力分析时，图解法也有其简单明了的优点。

## 4.2 计入惯性力的机构力分析

高速运转的机械中，惯性力的影响不能忽略。将惯性力看作为一般外力，施加在产生惯性力的构件上，该构件处于静平衡状态，这种力分析方法称为动态静力分析。

**1. 构件惯性力的确定**

图 4-1a 所示的曲柄滑块机构中，通过运动分析可以求得连杆 2 在质心 $s_2$ 处的加速度 $a_{s_2}$ 和角加速度 $\alpha_2$ 以及滑块 3 的加速度 $a_{C_3}$。$G_2$ 为构件 2 所受的重力，$J_{s_2}$ 为构件 2 绕质心的转动惯量，$F_{i_2}$ 和 $M_{s_2}$ 分别表示连杆 2 质心处的惯性力和惯性力矩。其值为

$$F_{i_2} = -m_2 a_{s_2}, \quad M_{s_2} = -J_{s_2}\alpha_2$$

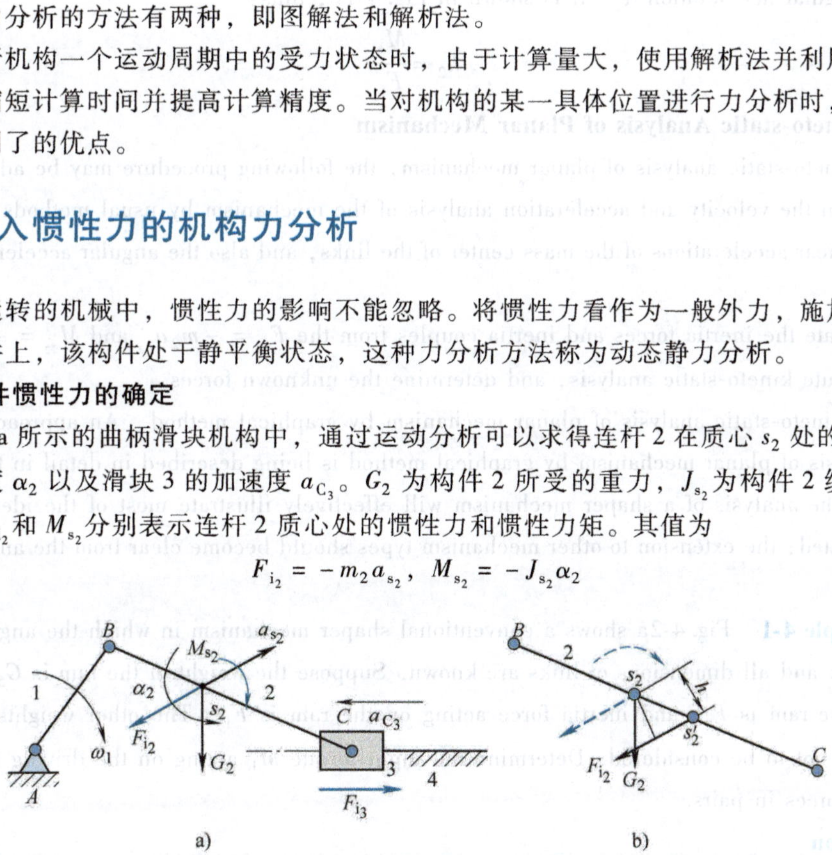


Fig. 4-1 Inertia force and inertia torque into resultant force（连杆的惯性力与惯性力矩的合成）

chines operate in a high speed. The dynamic force analysis can be expressed in a form very similar to static force analysis by using Dalember principle.

Dalember principle states that the inertia forces, torques, the external forces and torques on a link together give static equilibrium. Thus, a dynamic analysis problem is reduced to one requiring static analysis.

**1. Inertia Forces**

Fig. 4-1a shows a slider crank mechanism. Because of the constrained motion of the mechanism, accelerations, such as $a_{s_2}$, $\alpha_2$ of the individual link 2 and $a_{C_3}$ of the slider 3, may be determined first, then the inertia forces $F_{i_2}$ and torque $M_{s_2}$ acting on the link 2 can be calculated respectively.

$$F_{i_2} = -m_2 a_{s_2}$$

$$M_{s_2} = -J_{s_2}\alpha_2$$

Where $m_2$ is the mass of link 2, $a_{s_2}$ is the acceleration of mass center of the link 2, $\alpha_2$ is the angular acceleration of the link 2, and $J_{s_2}$ is the moment of inertia about a axis passing through the mass center of the link 2.

We can also obtain an offset inertia force acting on point $s_2'$. The equivalent inertia force is offset from the mass center $s_2$ so as to produce a same moment about the mass center that is opposite in sense to angular acceleration $\alpha_2$. It is shown in Fig. 4-1b, thus:

$$h_2 = \frac{M_{s_2}}{F_{i_2}}$$

**2. Kineto-static Analysis of Planar Mechanism**

For kineto-static analysis of planar mechanism, the following procedure may be adopted.

Perform the velocity and acceleration analysis of the mechanism by usual methods, and determine the linear accelerations of the mass center of the links, and also the angular accelerations of the links.

Calculate the inertia forces and inertia couples from the $F_{i_2} = -m_2 a_{s_2}$ and $M_{s_2} = -J_{s_2}\alpha_2$.

Prosecute kineto-static analysis, and determine the unknown forces.

(1) Kineto-static analysis of planar mechanism by graphical method An approach of kineto-static analysis of planar mechanism by graphical method is being described in detail in the following example. The analysis of a shaper mechanism will effectively illustrate most of the ideas that have been presented; the extension to other mechanism types should become clear from the analysis of this mechanism.

**Example 4-1** Fig. 4-2a shows a conventional shaper mechanism in which the angular velocity of the crank and all dimensions of links are known. Suppose the weight of the ram is $G_5$, resistance acting on the ram is $F_r$, and inertia force acting on the ram is $F_{i_5}$. The other weights and inertia forces need not to be considered. Determine the input torque $M_b$ acting on the driving link and the constraint forces in pairs.

**Solution**

The graphical analysis is shown in Fig. 4-2. First, consider connecting rod 4 in the absence of

构件质心 $s_2$ 上的惯性力 $F_{i_2}$ 和惯性力矩 $M_{s_2}$ 可以合成为图 4-1b 所示的一个惯性力，其大小不变，但力的作用点偏离了距离 $h$，其值为

$$h_2 = \frac{M_{s_2}}{F_{i_2}}$$

**2. 机构的动态静力分析**

机构动态静力分析的步骤是，首先求出各构件的惯性力，并把它们视为外力加在产生惯性力的构件上，然后将机构分解为若干个构件组，分别列出它们的力平衡方程，再逐一求解未知力及运动副反力。

力分析过程中，通常由二力杆开始，然后再考虑已知力作用的构件组。构件组中所含未知力的数目应该等于所能列出的力平衡方程的数目，以保证这些未知力能够顺利求解。

（1）用图解法进行机构的动态静力分析

**例 4-1** 图 4-2a 所示牛头刨床机构中，各构件的尺寸及原动件的角速度 $\omega_1$ 均为已知。刨头所受重力为 $G_5$，在图示位置刨头的惯性力为 $F_{i_5}$，刀具所受的生产阻力为 $F_r$。其余构件的重力及惯性力、惯性力矩均忽略不计。求机构各运动副中的反力及需要加在原动件上的平衡力矩 $M_b$。

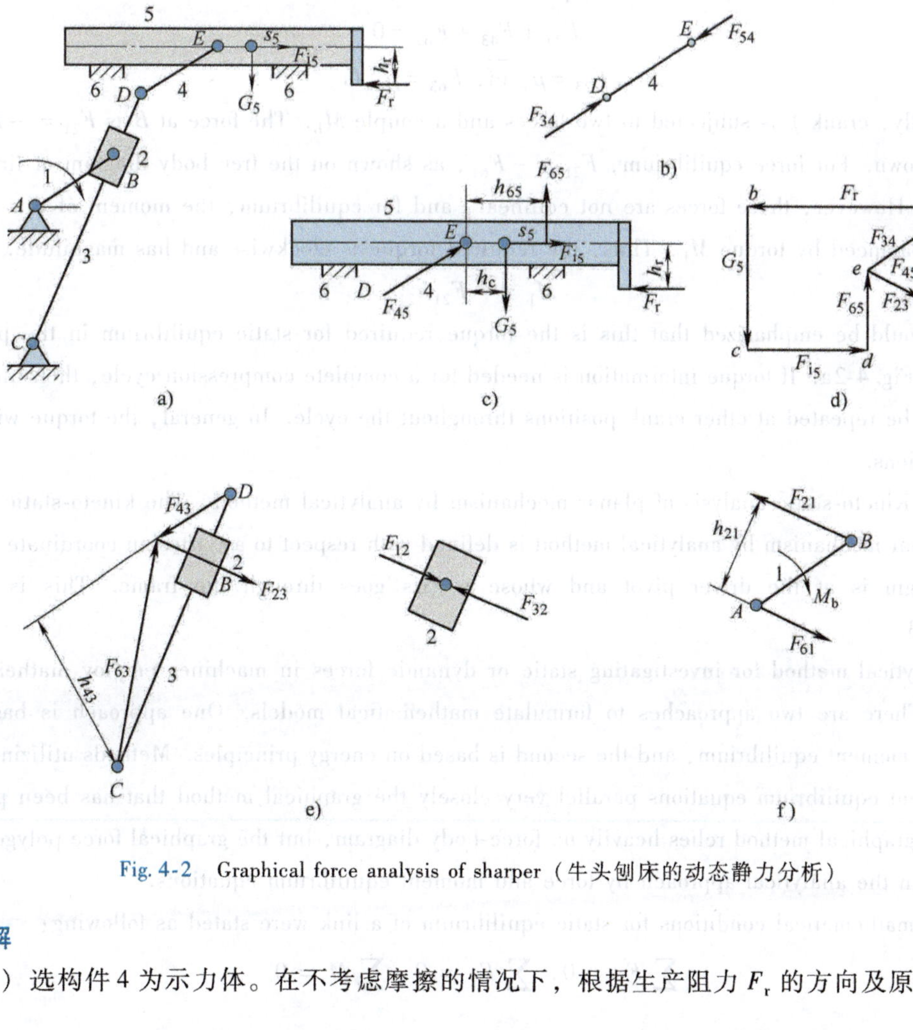


Fig. 4-2 Graphical force analysis of sharper（牛头刨床的动态静力分析）

**解**

1）选构件 4 为示力体。在不考虑摩擦的情况下，根据生产阻力 $F_r$ 的方向及原动件角速

the gravity and inertia forces. This link is acted on by two forces only, at point $D$ and $E$. Thus, link 4 is a two forces link loaded at each end as shown in Fig. 4-2b.

Take the ram 5 as a free body shown in Fig. 4-2c; it is a five forces link, with $F_r$, $G_5$, $F_{i_5}$, $F_{65}$ and $F_{45}$, and it is in equilibrium, thus

$$F_r + G_5 + F_{i_5} + F_{65} + F_{45} = 0$$

The $F_{65}$ and $F_{45}$ are unknown; they can be solved by a force polygon shown in Fig. 4-2d. We have:

$$F_{65} = \mu_F \overline{de},\ F_{45} = \mu_F \overline{ea}$$

The exact position of $F_{65}$ is not known, but we can take the moments about the axis passing through the point $E$. We have:

$$h_{65} = \frac{G_5 h_c + F_r h_r}{F_{65}}$$

Take the link 2 as a free body shown in Fig. 4-2e. It is a two forces link. $F_{12}$ and $F_{32}$ are normal to the guiding faces, and they have the same magnitude and opposite direction. Link 3 is a three forces link. When it is in equilibrium, the resultant of the three forces is zero, and all the lines of action of the forces intersect at the same point. We have:

$$F_{23} + F_{43} + F_{63} = 0$$

$$F_{23} = \mu_F \overline{ef},\ F_{63} = \mu_F \overline{fa}$$

Finally, crank 1 is subjected to two forces and a couple $M_b$. The force at $B$ is $F_{21} = -F_{12}$ and is now known. For force equilibrium, $F_{21} = -F_{61}$, as shown on the free body diagram of link 1 in Fig. 4-2f. However, these forces are not collinear, and for equilibrium, the moment of this couple must be balanced by torque $M_b$. Thus, the required torque is clockwise and has magnitude.

$$M_b = -F_{21} h_{21}$$

It should be emphasized that this is the torque required for static equilibrium in the position shown in Fig. 4-2a. If torque information is needed for a complete compression cycle, then the analysis must be repeated at other crank positions throughout the cycle. In general, the torque will vary with positions.

(2) Kineto-static analysis of planar mechanism by analytical method The kineto-static analysis of planar mechanism by analytical method is defined with respect to a Cartesian coordinate system whose origin is at the driver pivot and whose $x$ axis goes through the frame. This is shown in Fig. 4-3.

Analytical method for investigating static or dynamic forces in machines employ mathematical models. There are two approaches to formulate mathematical models. One approach is based on force and moment equilibrium, and the second is based on energy principles. Methods utilizing force and moment equilibrium equations parallel very closely the graphical method that has been presented. The graphical method relies heavily on force-body diagram, but the graphical force polygons are replaced in the analytical approach by force and moment equilibrium equations.

The mathematical conditions for static equilibrium of a link were stated as following:

$$\sum F_x = 0,\ \sum F_y = 0,\ \sum M = 0$$

度 $\omega_1$ 的方向判断，构件 4 为受压二力杆，$F_{34}$ 与 $F_{54}$ 的方向如图 4-2b 所示。

2）选图 4-2c 所示构件 5 为示力体，根据力平衡条件可得

$$F_r + G_5 + F_{i_5} + F_{65} + F_{45} = 0$$

式中，$F_r$、$G_5$、$F_{i_5}$ 的大小、方向均为已知，而 $F_{65}$ 的方向垂直于刨头导轨，仅大小未知。$F_{45}$ 与 $F_{54}$ 方向相反，大小未知。该力平衡方程中含有两个未知量，可以用图解法求解。选择适当的力比例尺 $\mu_F$，作力的封闭矢量多边形如图 4-2d 所示。由此可得

$$F_{65} = \mu_F \overline{\mathrm{de}},\ F_{45} = \mu_F \overline{\mathrm{ea}}$$

根据构件 5 上的力矩平衡条件，对 $E$ 点列力矩平衡方程，可求出 $F_{65}$ 的作用线的位置。

$$h_{65} = \frac{G_5 h_c + F_r h_r}{F_{65}}$$

3）选图 4-2e 所示的构件 2、3 为示力体。滑块 2 为二力杆，在不考虑移动副和转动副中的摩擦的情况下，$F_{32}$ 与移动副导路方向垂直。另一反力 $F_{12}$ 与 $F_{32}$ 等值、反向、共线且通过铰链中心 $B$ 点。构件 3 上受三个力，分别是 $F_{43}$、$F_{23}$、$F_{63}$。其中，$F_{43}$ 与 $F_{34}$ 大小相等，方向相反；$F_{23}$ 垂直于导路方向，大小为未知量。根据构件 3 的力矩平衡条件，可得

$$F_{23} + F_{43} + F_{63} = 0$$

封闭矢量多边形如图 4-2d 所示。

$$F_{23} = \mu_F \overline{\mathrm{ef}},\ F_{63} = \mu_F \overline{\mathrm{fa}}$$

4）取构件 1 为示力体。构件 1 受平衡力矩（即驱动力矩）$M_b$ 及运动副反力 $F_{61}$、$F_{21}$ 的作用。其中，$F_{21} = -F_{12} = F_{32}$，大小及方向均为已知。根据构件 1 的力矩平衡条件，对 $A$ 点列力矩平衡方程得

$$M_b = -F_{21} h_{21}$$

又根据力平衡条件

$$F_{61} = -F_{21}$$

（2）用解析法进行机构的动态静力分析　在建立直角坐标系的基础上，分别以各构件为示力体，将约束反力和各作用力分解为沿 $x$ 轴和 $y$ 轴的分力，列出各构件的平衡方程，然后联立求解，即可取得预期结果。下面以例 4-2 进行说明。

**例 4-2**　图 4-3a 所示的曲柄滑块机构中，已知曲柄和连杆的尺寸分别为 $L_1$、$L_2$，经过运动分析后已经知道各构件的运动参数。已知作用在滑块的生产阻力为 $F$，求各运动副的反力和作用在曲柄上的平衡力矩。

**解**　分别以构件 1、2、3 为示力体，标注各力的分量如图 4-3b、c、d 所示，按力系平衡条件列出力的平衡方程

$$\sum F_x = 0,\ \sum F_y = 0,\ \sum M = 0$$

对于图 4-3b 所示的构件 1：

$$F_{41x} + F_{21x} - m_1 a_{s_1x} = 0$$

$$F_{41y} + F_{21y} - m_1 a_{s_1y} = 0$$

$$M_1 - F_{21x} L_1 \sin\varphi_1 + F_{21y} L_1 \cos\varphi_1 - (-m_1 a_{s_1x}) r_1 \sin\varphi_1 + (-m_1 a_{s_1y}) r_1 \cos\varphi_1 - (-J_{s_1}\alpha_1) = 0$$

**Example 4-2** Fig. 4-3 shows a slider-crank mechanism used in a lot of machines, such as internal combustion engine and reciprocating compressors. All the dimensions of links, angular position of the crank and angular velocity of the crank are known. The kinematic analysis is first performed also. The center of mass and mass moment of inertia are designated by letters $s_i$ and $J_{s_i}$ respectively, where $i=1, 2, 3$ (see Fig. 4-3a). The resistance acting on the slider is $F_3$. Determine the driving torque $M_1$ and constraint forces in pairs.

**Solution**

As the mechanism is in static equilibrium, each link must also be in equilibrium individually. The location of the $xy$ coordinate system is in Fig. 4-3, and all the forces are expressed in term of $x$ and $y$ components. Since the mechanism is planar, a maximum of three independent equilibrium equations can be written for each free link.

Taking the crank as a free body, we have:

$$F_{41x}+F_{21x}+(-m_1 a_{s_1x})=0$$

$$F_{41y}+F_{21y}+(-m_1 a_{s_1y})=0$$

$$M_1-F_{21x}L_1\sin\varphi_1+F_{21y}L_1\cos\varphi_1-(-m_1 a_{s_1x})r_1\sin\varphi_1+(-m_1 a_{s_1y})r_1\cos\varphi_1-(-J_{s_1}\alpha_1)=0$$

Taking the connect rod as a free body, we have:

$$F_{12x}+F_{32x}+(-m_2 a_{s_2x})=0$$

$$F_{12y}+F_{32y}+(-m_2 a_{s_2y})=0$$

$$F_{32x}L_2\sin\varphi_2+F_{32y}L_2\cos\varphi_2-(-m_2 a_{s_2x})r_2\sin\varphi_2+(-m_2 a_{s_2y})r_2\cos\varphi_2-(-J_{s_2}\alpha_2)=0$$

Taking the slider as a free body, we have:

$$F_{23x}-F_{3x}+(-m_3 a_{C_3x})=0$$

$$F_{23y}+F_{43y}-F_{3y}=0$$

Substitute $F_{12x}=-F_{21x}$, $F_{12y}=-F_{21y}$, $F_{23x}=-F_{32x}$, $F_{23y}=-F_{32y}$ into the above equations, and there are eight unknowns. Note that the force $F_{43x}=0$, because the friction is negligible.

These equations are then conveniently arranged in matrix form. The matrix equation is easily solved by Gauss elimination using a computer.

The matrix equation is as follows:

$$\begin{bmatrix} -1 & 0 & 0 & 0 & 1 & 0 & 0 & 0 & 0 \\ 0 & -1 & 0 & 0 & 0 & 1 & 0 & 0 & 0 \\ L_1\sin\varphi_1 & -L_1\cos\varphi_1 & 0 & 0 & 0 & 0 & 0 & 0 & 1 \\ 1 & 0 & -1 & 0 & 0 & 0 & 0 & 0 & 0 \\ 0 & 1 & 0 & -1 & 0 & 0 & 0 & 0 & 0 \\ 0 & 0 & -L_2\sin\varphi_2 & -L_2\cos\varphi_2 & 0 & 0 & 0 & 0 & 0 \\ 0 & 0 & 1 & 0 & 0 & 0 & 0 & 0 & 0 \\ 0 & 0 & 0 & 1 & 0 & 0 & 0 & 1 & 0 \\ 0 & 0 & 0 & 0 & 0 & 0 & 0 & 0 & 0 \end{bmatrix} \begin{bmatrix} F_{12x} \\ F_{12y} \\ F_{23x} \\ F_{23y} \\ F_{41x} \\ F_{41y} \\ F_{43x} \\ F_{43y} \\ M_1 \end{bmatrix} =$$

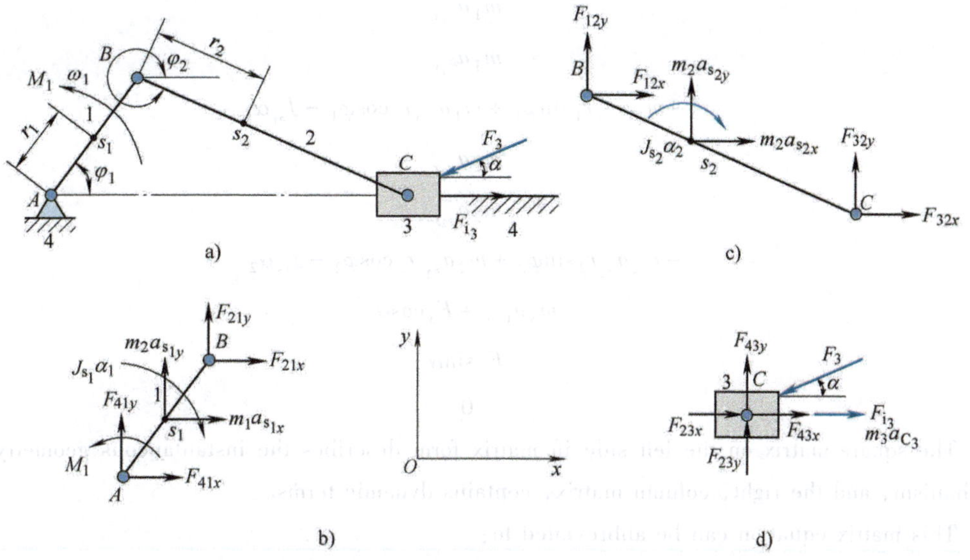


Fig. 4-3　Analytical force analysis（动态静力分析的解析法）

对于图 4-3c 所示的构件 2：

$$F_{12x}+F_{32x}+(-m_2a_{s_{2x}})=0$$

$$F_{12y}+F_{32y}+(-m_2a_{s_{2y}})=0$$

$$F_{32x}L_2\sin\varphi_2+F_{32y}L_2\cos\varphi_2-(-m_2a_{s_{2x}})r_2\sin\varphi_2+(-m_2a_{s_{2y}})r_2\cos\varphi_2-(-J_{s_2}\alpha_2)=0$$

对于图 4-3d 所示的构件 3：

$$F_{23x}-F_{3x}+(-m_3a_{C_{3x}})=0$$

$$F_{23y}+F_{43y}-F_{3y}=0$$

考虑到 $F_{12x}=-F_{21x}$，$F_{12y}=-F_{21y}$，$F_{23x}=-F_{32x}$，$F_{23y}=-F_{32y}$，则未知数的个数为 8，而方程的个数也为 8，故该方程组可解。

将其写成矩阵形式得

$$\begin{bmatrix} -1 & 0 & 0 & 0 & 1 & 0 & 0 & 0 & 0 \\ 0 & -1 & 0 & 0 & 0 & 1 & 0 & 0 & 0 \\ L_1\sin\varphi_1 & -L_1\cos\varphi_1 & 0 & 0 & 0 & 0 & 0 & 0 & 1 \\ 1 & 0 & -1 & 0 & 0 & 0 & 0 & 0 & 0 \\ 0 & 1 & 0 & -1 & 0 & 0 & 0 & 0 & 0 \\ 0 & 0 & -L_2\sin\varphi_2 & -L_2\cos\varphi_2 & 0 & 0 & 0 & 0 & 0 \\ 0 & 0 & 1 & 0 & 0 & 0 & 0 & 0 & 0 \\ 0 & 0 & 0 & 1 & 0 & 0 & 0 & 1 & 0 \\ 0 & 0 & 0 & 0 & 0 & 0 & 0 & 0 & 0 \end{bmatrix} \begin{bmatrix} F_{12x} \\ F_{12y} \\ F_{23x} \\ F_{23y} \\ F_{41x} \\ F_{41y} \\ F_{43x} \\ F_{43y} \\ M_1 \end{bmatrix} =$$

$$m_1 a_{s_1x}$$

$$m_1 a_{s_1y}$$

$$-m_1 a_{s_1x} r_1 \sin\varphi_1 + m_1 a_{s_1y} r_1 \cos\varphi_1 - J_{s_1}\alpha_1$$

$$m_2 a_{s_2x}$$

$$m_2 a_{s_2y}$$

$$-m_2 a_{s_2} r_2 \sin\varphi_2 + m_2 a_{s_2y} r_2 \cos\varphi_2 - J_{s_2}\alpha_2$$

$$m_3 a_{C_3x} + F_3 \cos\alpha$$

$$F_3 \sin\alpha$$

$$0$$

The square matrix on the left side in matrix form describes the instantaneous geometry of the mechanism, and the right, column matrix, contains dynamic terms.

This matrix equation can be abbreviated to:

$$\boldsymbol{A}\boldsymbol{F}_{ij} = \boldsymbol{B}$$

Matrixes $\boldsymbol{A}$ and $\boldsymbol{B}$ are known parameters matrix. $\boldsymbol{F}_{ij}$ is unknown forces matrix.

This method is suitable for dynamic analysis of any planar mechanisms.

## 4.3 Force Analysis Including Friction in Mechanisms

Whenever two connected links have relative motion, friction occurs at the contact surfaces, such as a pin joint. The friction produces heat and wear, which may eventually lead to bearing failure. It can also adversely affect the motion response of the mechanism. In addition, the presence of friction can substantially increase the energy requirements of a machine. If the wear becomes serious, the vibration and noise will occur, the performance of a machine can be effected also.

**1. Friction in Kinematic Pairs**

Prismatic pairs, turning pairs and helical pairs are common pairs in engineering.

(1) Friction in sliding pairs  The friction in sliding pairs can be divided into plane surface friction and V surface friction.

1) Friction on plane surface. A slider 1 is rest on a smooth plane surface 2 which is horizontal. Suppose that the slider 1 is sliding relative to the surface 2 to the right at a constant velocity $v_{12}$ acted by a driving force $F$ which is inclined at an angle $\alpha$ to the normal, as shown in Fig. 4-4. The direction of friction force $F_{21}$ acting on slider is always opposite to the relative motion, and the plane surface exerts a reaction force $N_{21}$ on the slider 1 which is normal to the plane surface. The friction $F_{21}$ and the normal reaction force $N_{21}$ can be combined into a single reaction force $R_{21}$ which is inclined at an angle $\varphi$ to the normal.

Therefore, the angle $\varphi$ of the resultant force $R_{21}$ from the normal direction is:

$$\tan\varphi = \frac{F_{21}}{N_{21}}$$

Since, $F_{21} = fN_{21}$, thus:

$$\begin{matrix} m_1 a_{s_1x} \\ m_1 a_{s_1y} \\ -m_1 a_{s_1x} r_1 \sin\varphi_1 + m_1 a_{s_1y} r_1 \cos\varphi_1 - J_{s_1}\alpha_1 \\ m_2 a_{s_2x} \\ m_2 a_{s_2y} \\ -m_2 a_{s_2} r_2 \sin\varphi_2 + m_2 a_{s_2y} r_2 \cos\varphi_2 - J_{s_2}\alpha_2 \\ m_3 a_{C_3x} + F_3 \cos\alpha \\ F_3 \sin\alpha \\ 0 \end{matrix}$$

该矩阵可简写为

$$\boldsymbol{A}\boldsymbol{F}_{ij} = \boldsymbol{B}$$

$\boldsymbol{A}$、$\boldsymbol{B}$ 矩阵均为已知参数矩阵，未知力矩阵 $\boldsymbol{F}_{ij}$ 可以非常容易求解。

## 4.3　计入摩擦的机构力分析

运动副中的摩擦是一种有害阻力。它不仅降低机械效率，还使运动副表面受到磨损，导致机械运转精度降低，引起机械振动和噪声的增大，缩短了机器的使用寿命。摩擦还会引起运动副发热并膨胀，可能导致运动副卡死，使机械运转不灵活等。

**1. 运动副中的摩擦**

工程中常用的运动副主要有移动副、转动副和螺旋副，以下分别讨论。

(1) 移动副中的摩擦　根据移动副的具体结构，常把移动副分为平面移动副、斜面移动副和槽面移动副。

1) 平面移动副中的摩擦。图 4-4 所示滑块 1 在总驱动力 $F$ 的作用下，相对平面 2 以速度 $v_{12}$ 等速移动。平面 2 给滑块 1 的作用力有法向反力 $N_{21}$ 和摩擦力 $F_{21}$，两者的合力 $R_{21}$ 为平面 2 给滑块 1 的总反力，$R_{21}$ 与法线方向的夹角为 $\varphi$。

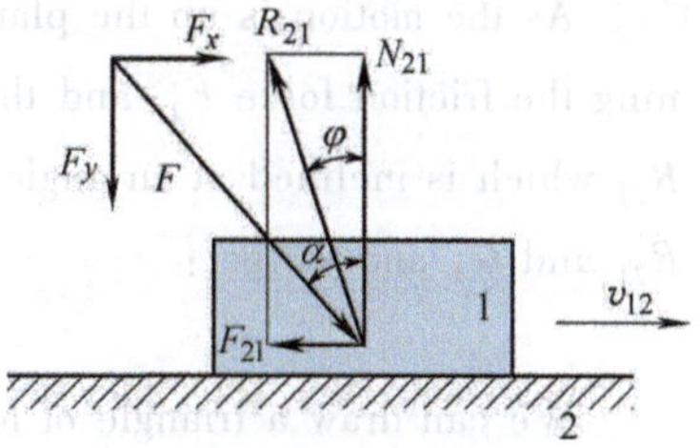


Fig. 4-4　Friction on the plane surface（平面中的摩擦）

摩擦力 $F_{21}$ 与法向反力 $N_{21}$ 之间的关系为

$$\tan\varphi = \frac{F_{21}}{N_{21}},\ F_{21} = fN_{21}$$

$$\tan\varphi = \frac{fN_{21}}{N_{21}} = f,\ \varphi = \arctan f$$

当滑块与平面的材料一定时，摩擦因数为定值，总反力与正压力方向夹角 $\varphi$ 为一恒定角度，称为摩擦角。构件 2 对构件 1 的总反力 $R_{21}$ 和构件 1 相对构件 2 的运动方向成 $(90° + \varphi)$ 角。

如果外加驱动力 $F$ 与法线之间的夹角为 $\alpha$，沿运动方向和法线方向的分量分别为 $F_x$、$F_y$，则两者关系为

$$\tan\alpha = \frac{F_x}{F_y}$$

根据平衡条件，$N_{21} = F_y$。联立求解上述方程后，可有

$$\tan\varphi = \frac{fN_{21}}{N_{21}} = f,\ \varphi = \arctan f$$

Where $f$ is defined as the coefficient of sliding friction, and its value depends on many factors, such as surface smooth, materials in contact, nature of lubricant, temperature, etc. , and for a given conditions, it is a constant. The angle $\varphi$ is known as the angle of friction.

The determination of direction of the resultant $R_{21}$ is as follows:

The direction of the resultant reaction $R_{21}$ will be inclined at an angle $(90° + \varphi)$ to the sense of the relative velocity $v_{12}$.

If the applied force $F$ is resolved into $x$ and $y$ components, thus:

$$\tan\alpha = \frac{F_x}{F_y}$$

The block 1 is in equilibrium under the action of these forces, so we have:

$$N_{21} = F_y$$

Simultaneously resolve the above equations:

$$F_x = \frac{\tan\alpha}{\tan\varphi} F_{21}$$

When $\alpha < \varphi$, $F_x < F_{21}$, let a infinite force $F$ be applied to the block to move on the surface, the block is unable to move forever. This phenomenon is called as self-locking.

When $\alpha = \varphi$, the block is at rest or has a uniform motion.

When $\alpha > \varphi$, the block is accelerated to move.

Therefore, the condition of self-locking in a plane pair can be described that the applied force acted in the block locates within the friction angle, namely $\alpha \leqslant \varphi$, the self-locking will occur.

2) Friction on inclined plane surface. A block of weight $G$ resting on a plane inclined at an angle $\alpha$ to the horizontal is acted on by a horizontal force $F_d$ which tends to move the body up the plane with a uniform velocity $v_{12}$. This is shown in Fig. 4-5a.

As the motion is up the plane, the friction $F_{21}$ would act downwards along the plane. Combining the friction force $F_{21}$ and the normal reaction force $N_{21}$ as before, we have the resultant force $R_{21}$ which is inclined at an angle $\varphi$ to the normal. The body is in equilibrium under the forces $F_{21}$, $R_{21}$ and $G$, and we get:

$$F_d + G + R_{21} = 0$$

We can draw a triangle of forces as shown in Fig. 4-5b, and obtain:

$$F_d = G\tan(\alpha + \varphi)$$

If the block moving up the inclined plane is not self-locking, then the incline angle of the plane surface must satisfy the following formula.

$$\alpha < 90° - \varphi$$

If the block moves down the plane with uniform velocity $v_{12}$, the resultant force $R_{21}$ is shown in Fig. 4-5c. The effective driving force is $G\sin\alpha$, and the resistance is as follows:

$$F_{21} = G\cos\alpha f = G\cos\alpha\tan\varphi$$

If the block can move down the inclined plane under action of the gravity $G$, we have:

$$G\sin\alpha \geqslant G\cos\alpha\tan\varphi \Rightarrow$$

$$F_x = \frac{\tan\alpha}{\tan\varphi} F_{21}$$

当 $\alpha < \varphi$ 时，如果滑块处于静止状态，无论力 $F$ 多大，驱动力 $F_x$ 都小于最大静摩擦力 $F_{21}$，滑块不能运动，这种现象称为自锁。如滑块处于运动状态，则滑块将减速运动到静止不动。

当 $\alpha = \varphi$ 时，如果滑块处于静止状态，则滑块仍然发生自锁；如滑块处于运动状态，则将做等速运动。

当 $\alpha > \varphi$ 时，滑块做加速运动。

因此，平面运动副的自锁条件可描述为当外加驱动力作用在摩擦角之内时，该运动副处于自锁状态，即 $\alpha \leqslant \varphi$ 为其自锁条件。

2）斜面移动副的摩擦。如果把图 4-4 所示的平面移动副导路倾斜 $\alpha$ 角度，则演化成为图 4-5 所示的斜面摩擦移动副。

图 4-5a 中，滑块受铅垂载荷 $G$，在水平力 $F_d$ 的作用下等速上升，斜面 2 给滑块 1 的正压力 $N_{21}$ 和摩擦力 $F_{21}$ 合成总反力 $R_{21}$ 后，滑块的力系平衡条件为

$$F_d + G + R_{21} = 0$$

作出图 4-5b 所示受力分析图后，可求出水平力 $F_d$ 和铅垂载荷 $G$ 之间的关系为

$$F_d = G\tan(\alpha + \varphi)$$

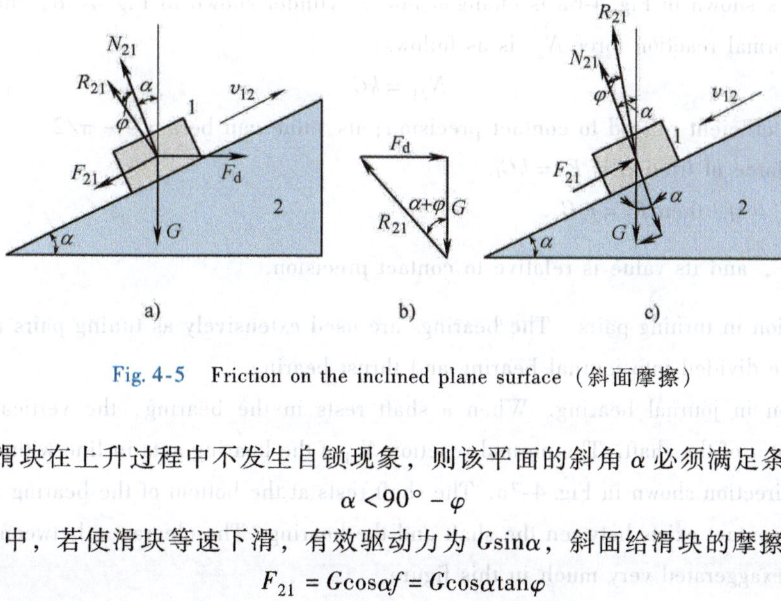


Fig. 4-5　Friction on the inclined plane surface（斜面摩擦）

若要求滑块在上升过程中不发生自锁现象，则该平面的斜角 $\alpha$ 必须满足条件：

$$\alpha < 90° - \varphi$$

图 4-5c 中，若使滑块等速下滑，有效驱动力为 $G\sin\alpha$，斜面给滑块的摩擦阻力为

$$F_{21} = G\cos\alpha f = G\cos\alpha\tan\varphi$$

滑块沿斜面下滑的条件为

$$G\sin\alpha \geqslant G\cos\alpha\tan\varphi \Rightarrow \tan\alpha \geqslant \tan\varphi \Rightarrow \alpha \geqslant \varphi$$

若要求滑块在铅垂载荷 $G$ 作用下发生自锁，则必须满足：$\alpha < \varphi$。

3）槽面摩擦。如果将图 4-6a 所示滑块作成图 4-6b 所示夹角为 $2\theta$ 的楔形滑块，并置于相应的槽面中，楔形滑块 1 在外力 $F$ 的作用下沿槽面等速运动。设两侧法向反力分别为 $N_{21}$，铅垂载荷为 $G$，总摩擦力为 $F_f$。

对于图 4-6a 所示的平面摩擦，有　　$F_f = fG$

$$\tan\alpha \geqslant \tan\varphi \Rightarrow$$

$$\alpha \geqslant \varphi$$

If the block can not move down the inclined plane, we have:

$$\alpha < \varphi$$

3) Friction on V-plane surface. If the block shown in Fig. 4-6a is changed into a V-block with a included angle $2\theta$ shown in Fig. 4-6b, and its weight is $G$, there are two normal reaction forces $N_{21}$, and from Fig. 4-6c, its magnitude is as follows:

$$N_{21} = \frac{G/2}{\sin\theta} = \frac{G}{2\sin\theta}$$

The total frictional force is:

$$F_f = 2F_{21} = 2N_{21}f$$

$$F_f = 2f\frac{G}{2\sin\theta} = \frac{f}{\sin\theta}G = f_v G$$

Where, $f_v = \frac{f}{\sin\theta}$, $f_v$ is a equivalent coefficient of friction, and in Fig. 4-6a, the frictional force is as:

$$F_f = fG$$

It is quite clear that $f_v > f$, that is to say, the frictional force on V-plane surface is greater than that of plane surface.

If the block shown in Fig. 4-6a is changed into a cylinder shown in Fig. 4-6d, and its weight is $G$. The total normal reaction force $N_{21}$ is as follows:

$$N_{21} = kG$$

Where, $k$ is coefficient related to contact precision; its value can be $k = 0 \sim \pi/2$.

The total force of friction is $F_f = kGf$.

Suppose $f_v = kf$, then $F_f = f_v G$.

$f_v = 1 \sim \frac{\pi}{2}$, and its value is relative to contact precision.

(2) Friction in turning pairs The bearings are used extensively as tuning pairs in machinery, and they can be divided into journal bearing and thrust bearing.

1) Friction in journal bearing. When a shaft rests in the bearing, the vertical load $G$ acts through the center of the shaft. The normal reaction $N_{21}$ of the bearing acts in line with $G$ in the vertically upward direction shown in Fig. 4-7a. The shaft rests at the bottom of the bearing at point $A$ and metal to metal contact exists between the shaft and the bearing. The clearance between the shaft and the bearing is exaggerated very much in this figure.

When a torque $M_d$ is applied to the shaft, it rotates clockwise with an angular velocity $\omega_{12}$. Friction at the line of contact with the bearing will cause it to roll up the right until a point $B$ is reached at which relative sliding takes place. This is shown in Fig. 4-7b. At this point, the friction force is $F_{21}$ which is tangential to the shaft reaction and opposite to the relative velocity $v_{12}$, the normal reaction $N_{21}$ passes through the center of the shaft. The normal reaction force $N_{21}$ and the friction force $F_{21}$ can be combined into a resultant force $R_{21}$ which is equal to the vertical load $G$ and opposite in direction. The vertical load $G$ and resultant force $R_{21}$ will be parallel and constitute a cou-

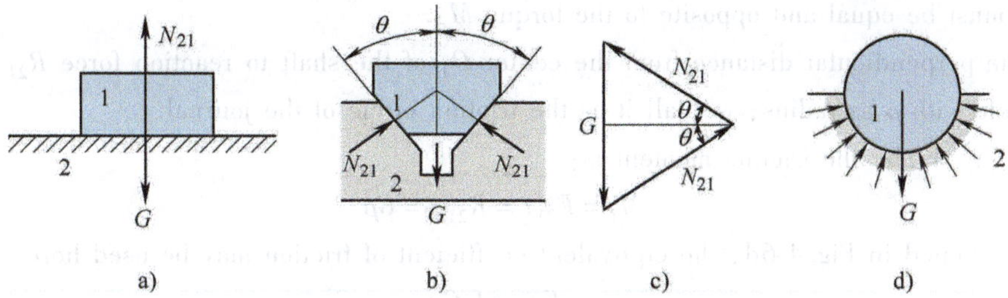


Fig. 4-6　Friction on V-plane surface（槽面摩擦）

对于图 4-6b 所示的槽面结构式的平面摩擦，有

$$F_f = 2F_{21} = 2N_{21}f$$

由图 4-6c 所示的力多边形可知

$$N_{21} = \frac{G/2}{\sin\theta} = \frac{G}{2\sin\theta}$$

$$F_f = 2f\frac{G}{2\sin\theta} = \frac{f}{\sin\theta}G = f_v G$$

式中，$f_v = f/\sin\theta$，称为当量摩擦因数。很明显 $f_v > f$，说明槽面摩擦产生的摩擦力大于平面摩擦产生的摩擦力。

如果将图 4-6a 所示滑块作成图 4-6d 所示的圆柱形，其法向反力的总和为

$$N_{21} = kG$$

式中，$k$ 是与接触有关的系数，$k = 0 \sim \pi/2$。

总摩擦力 $F_f = kGf$，令 $f_v = kf$，$F_f = f_v G$，可知：

$$f_v = 1 \sim \frac{\pi}{2}$$

其值的选择与接触精度有关。

（2）转动副中的摩擦　轴承是转动副的典型代表，可分为承受径向力的轴承和承受轴向力的轴承。

1）径向轴承的摩擦。图 4-7 所示为考虑到运动副间隙的径向轴承。轴颈 1 在没有转动前，径向载荷 $G$ 与 $A$ 点处的法向反力 $N_{21}$ 平衡。

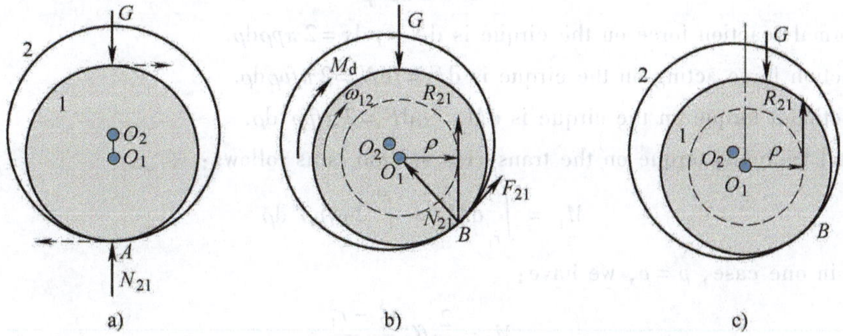


Fig. 4-7　Friction in a journal bearing（径向轴承中的摩擦）

在驱动力矩 $M_d$ 作用下，图 4-7b 所示的轴颈 1 由于受到接触点摩擦力的阻抗，接触点从 $A$ 点移动到 $B$ 点，摩擦力矩与驱动力矩平衡后开始转动。摩擦力 $F_{21}$ 与法向力 $N_{21}$ 的合力

ple which must be equal and opposite to the torque $M_d$.

Let the perpendicular distance from the center $O_1$ of the shaft to reaction force $R_{21}$ be $\rho$, and draw a circle with $\rho$ as radius; we call it as the friction circle of the journal.

From Fig. 4-7b, the friction moment is:

$$M_f = F_{21} r = R_{21} \rho = G\rho$$

As mentioned in Fig. 4-6d, the equivalent coefficient of friction may be used here. Thus:

$$F_{21} = f_v G$$

We have $\rho = f_v r$.

When shaft radius of the journal and the materials in contact are determined, the friction circle of the journal is constant.

If the clearance between the shaft and the bearing is greater, $f_v = f$, otherwise, $f_v = (1.27 - 1.57)f$.

The determination of the force $R_{21}$ is as follows.

The friction couple produced by the resultant force $R_{21}$ is opposite to the relative velocity $\omega_{12}$ and the resultant force is tangential to the friction circle.

The friction angle and friction circle are very useful concepts for force analysis.

2) Friction in thrust bearing. When a rotating shaft is subjected to an axial load, the thrust is taken by a collar bearing, or thrust bearing.

Fig. 4-8a shows a thrust bearing acted by an axial load $G$. Before calculating the friction torque, we have two kinds of assumptions, and each assumption leads to a different value of torque.

The first is that the intensity of the pressure on the bearing surface is constant, thus:

$$p = c$$

The second is that the product of the normal pressure $p$ and the corresponding radius $\rho$ must be constant, thus:

$$p\rho = c$$

This is a uniform wearing of the bearing surface.

Fig. 4-8b is a transverse of the bearing bottom, and if take a cirque within the section, its area is as:

$$ds = 2\pi\rho d\rho$$

The normal reaction force on the cirque is $dN = p ds = 2\pi p\rho d\rho$.

The friction force acting on the cirque is $dF = f dN = 2\pi f p\rho d\rho$.

The frictional torque on the cirque is $dM_f = \rho dF = 2\pi f p\rho^2 d\rho$.

The total frictional torque on the transverse section is as follows:

$$M_f = \int_{r_1}^{r_2} dM_f = \int_{r_1}^{r_2} 2\pi f p\rho^2 d\rho$$

If it is in one case, $p = c$, we have:

$$M_f = \frac{2}{3} f G \frac{r_2^3 - r_1^3}{r_2^2 - r_1^2}$$

If it is in second case, $p\rho = c$, then we have:

$$M_f = \frac{1}{2} f G (r_2 + r_1)$$

$R_{21}$为轴承 2 对轴颈 1 的总反力。总反力 $R_{21}$到轴心的距离为 $\rho$。

摩擦力矩为

$$M_f = F_{21} r = R_{21}\rho = G\rho$$

由于径向轴承为曲线状接触面，可引入当量摩擦因数 $f_v$，所以摩擦力与径向载荷之间的关系为：$F_{21} = f_v G$，将其代入上式，可求出总反力 $R_{21}$到轴心的距离 $\rho$ 为

$$\rho = f_v r$$

轴颈尺寸与材料确定以后，$\rho$ 为常量。

以 $\rho$ 为半径的圆称为摩擦圆，当 $\omega_{12} \neq 0$ 且匀速转动时，总反力 $R_{21}$相切于摩擦圆。

当量摩擦因数 $f_v$ 的选取遵循以下原则：①对于较大间隙的轴承，$f_v = f$；②对于较小间隙的轴承，未经跑合时 $f_v = 1.57f$，经过跑合时 $f_v = 1.27f$。

总反力 $R_{21}$方向的判别方法如下：轴承 2 给轴颈 1 的总反力 $R_{21}$对轴心力矩的方向与轴颈 1 相对于轴承 2 的相对角速度 $\omega_{12}$方向相反并相切于摩擦圆。

图 4-7c 所示的外力的合力 $G$ 作用在摩擦圆之内，若轴颈原来静止，则发生自锁；若原来运动，则减速到停止运动。

外力合力与摩擦圆相切，若轴颈原来静止，则发生自锁；若原来运动，则做等速运动。

外力的合力作用在摩擦圆之外，轴颈加速运动。

转动副的自锁条件可以描述为：外力的合力作用在摩擦圆之内，该转动副自锁。

2）推力轴承的摩擦。推力轴承是指作用力通过轴线的轴承。

图 4-8a 所示为推力轴承示意图，$G$ 为轴向载荷。未经跑合时，接触面压强 $p$ 为常数，$p = c$。经过跑合时，压强与半径的乘积为常数，即 $p\rho = c$。

在图 4-8b 所示的底平面半径 $\rho$ 处取微小圆环面积，其值为

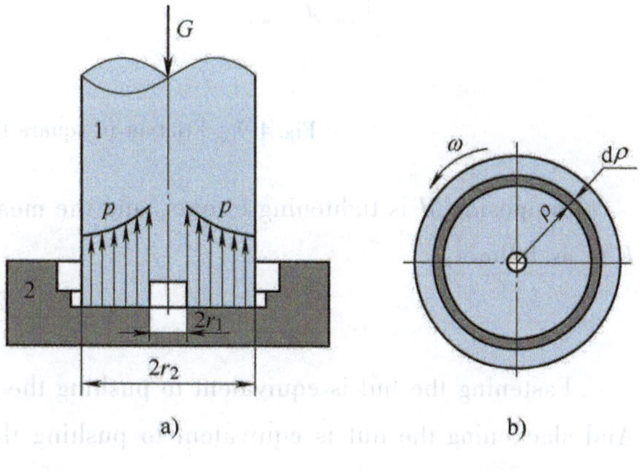


Fig. 4-8 Friction in a thrust bearing（推力轴承的摩擦）

$$ds = 2\pi\rho d\rho$$

小圆环面积上的正压力为

$$dN = p ds = 2\pi p\rho d\rho$$

小圆环面积上的摩擦力为

$$dF = f dN = 2\pi f p\rho d\rho$$

小圆环面积上的摩擦力矩为

$$dM_f = \rho dF = 2\pi f p\rho^2 d\rho$$

整个圆环接触面积上的摩擦力矩为

$$M_f = \int_{r_1}^{r_2} dM_f = \int_{r_1}^{r_2} 2\pi f p\rho^2 d\rho$$

未经跑合的推力轴承，$p = c$，且

$$M_f = \frac{2}{3} fG \frac{r_2^3 - r_1^3}{r_2^2 - r_1^2}$$

经过跑合的推力轴承，$p\rho = c$，且

The theorem of friction in thrust bearing can be used to design the frictional clutch, steam turbine and propeller of a ship, etc.

(3) Friction in helical pairs  According to the tooth shape, the screw can be divided into square threads and V-threads. A square-thread screw is used to transmit power, and V-threaded screw is used to fasten an element to another.

1) Friction in square thread. Fig. 4-9a shows a square-threaded screw; it may be thought of simply as an inclined plane or wedge wrapped around a cylinder. It will be clear that the motion of the nut relative to the thread is similar to that of a body sliding on an inclined plane of the angle $\alpha$ where $\tan\alpha = l/\pi d$. Tightening of the nut will be equivalent to pushing the body up the plane, and slackening will be equivalent to pushing the body down the plane.

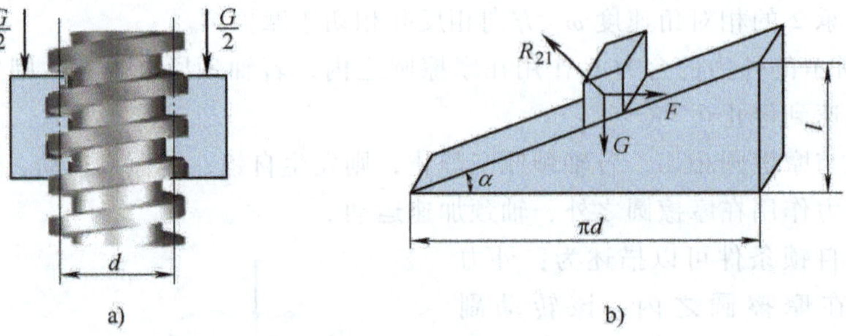


Fig. 4-9  Friction in square thread (矩形螺纹的摩擦)

Supposing $M$ is tightening torque, and the mean diameter of the thread is $d$, the horizontal force $F$ is as follows:

$$F = \frac{2M}{d}$$

Fastening the nut is equivalent to pushing the body up the plane; it is called as positive travel. And slackening the nut is equivalent to pushing the body down the plane; it is called as negative travel.

The relationship between $G$ and $F$ for tightening is given by:

$$F = G\tan(\alpha + \varphi)$$

If the negative travel is needed for self-locking, the inclined plane angle must be less than the frictional angle, that is:

$$\alpha \leqslant \varphi$$

2) Friction in V-thread. Fig. 4-10a shows a V-thread; its thread angle is $2\beta$, and the angle between the V-faces of the thread is $2\theta$. Fig. 4-10b shows a triangle which is the development of a helix of diameter $d$ and lead $l$.

Fastening the nut is equivalent to pushing the body up the oblique plane; it is called as positive travel. And slackening the nut is equivalent to pushing the body down the oblique plane; it is called as negative travel.

If the negative travel is needed for self-locking, the inclined plane angle must less than the equivalent frictional angle, that is:

$$M_f = \frac{1}{2}fG(r_2 + r_1)$$

推力轴承的摩擦原理是设计摩擦离合器的理论依据。

(3) 螺旋副中的摩擦 根据螺纹牙型可将螺纹分为矩形螺纹和三角形螺纹。

1) 矩形螺纹螺旋副中的摩擦。图 4-9a 所示为一矩形螺纹，将螺母 2 简化为图 4-9b 所示的滑块，承受轴向载荷 $G$，由于螺纹可以看成是斜面缠绕在圆柱体上形成的，故将矩形螺纹沿螺纹中径 $d$ 展开，该螺纹成为图 4-9b 所示的斜面，斜面底长为螺纹中径处的圆周长，高度为螺纹的导程 $l$。驱动力 $F$ 等于拧紧力矩 $M$ 除以螺纹半径 $d/2$，方向一般垂直于螺纹轴线。其值为

$$F = \frac{2M}{d}$$

拧紧螺母时，相当于滑块沿斜面等速上升，一般称为正行程；放松螺母时，相当于滑块等速沿斜面下降，称为反行程。其自锁性能完全与斜面摩擦相同。斜面的倾角 $\alpha$ 就是螺纹升角，控制升角的大小就可以控制螺旋副的自锁。

当要求螺旋副反行程自锁时，必须满足 $\alpha \leqslant \varphi$。

2) 三角形螺纹螺旋副中的摩擦。图 4-10a 所示的三角形螺纹中，牙型角为 $2\beta$，半牙型角为 $\beta$，槽角为 $2\theta$。将其螺纹展开，成为图 4-10b 所示的带半槽面的斜面，半牙型角 $\beta$ 与半槽角 $\theta$ 之和为 90°。斜面底长为螺纹中径处圆周长。

拧紧螺母时，相当于滑块沿斜槽面等速上升，为正行程；放松螺母时，相当于滑块沿斜槽面等速下降，为反行程。

当反行程要求螺纹自锁时，必须满足 $\alpha \leqslant \varphi_v$。这里的斜面倾角 $\alpha$ 就是螺纹升角，控制升角的大小就可以控制螺旋副的自锁条件。由于引入槽面摩擦的概念，当量摩擦因数 $f_v$ 的值可由槽面摩擦求出

$$f_v = \frac{f}{\sin\theta} = \frac{f}{\sin(90° - \beta)} = \frac{f}{\cos\beta}$$

当量摩擦角 $\varphi_v$ 的值为

$$\varphi_v = \arctan f_v$$

由于三角形螺纹的当量摩擦角较大，反行程容易发生自锁，故主要用于连接。而矩形螺纹则用于传动。

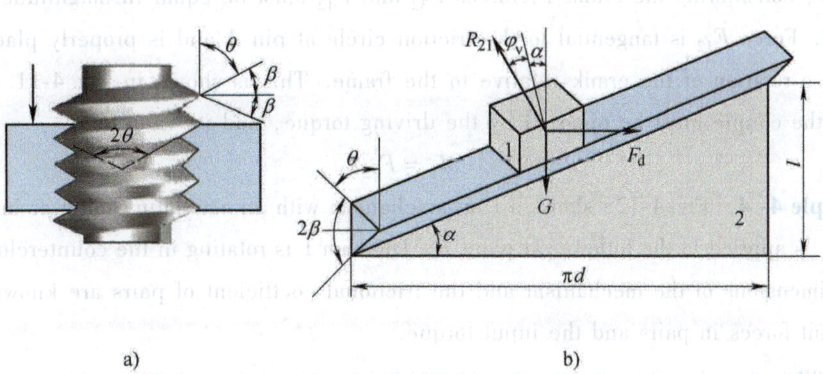


Fig. 4-10 Friction in V-thread（三角形螺纹的摩擦）

$$\alpha \leqslant \varphi_v$$

$$\varphi_v = \arctan f_v$$

According to Fig. 4-10b, the equivalent coefficient of friction is:

$$f_v = \frac{f}{\sin\theta} = \frac{f}{\cos\beta}$$

**2. Force Analysis Including Friction**

In static force analysis of a mechanism, the axial forces in links act along the longitudinal axes of the links. But if we consider the friction in the force analysis, the resultant forces on a journal are tangent to the friction circle. Similarly, in pin joined links, the line of constraint forces on a link is tangent to the friction circles at the two pins, and the forces must act in such a way as to oppose relative motion at joints.

**Example 4-3** Fig. 4-11 shows a slider-crank mechanism in which resistant force $F_r$ is applied to the piston 4. The crank 1 is rotating in the clockwise direction; the dimensions of links and the frictional coefficient of pairs are known. Determine the constraint forces in pairs and the input torque.

**Solution**

The friction circles are constructed for the three rotating pairs $A$, $B$, $C$, and the relative motions are as shown in Fig. 4-11.

First, consider the connecting rod 2. This link is a two-force link and it obviously is in compression for the loading. Therefore, the force $F_{23}$ and $F_{43}$ must be collinear, with force $F_{23}$ producing a clockwise moment about pin $B$ opposing the counterclockwise rotation of link 3 relative link 2, and force $F_{43}$ producing a clockwise moment about pin $C$ opposing the counterclockwise rotation of link 3 relative link 4. The line of action of the compressive forces is shown in Fig. 4-11.

Next, we can now proceed with the analysis of the piston. Resistant force $F_r$ is known completely and $F_{34}$ is opposed to $F_{43}$, so it is known also. Force $F_{14}$, exerted by the frame on the piston, must pass through the intersection of $F_r$ and $F_{34}$, and must act at the friction angle $(90° + \varphi)$ with respect to the relative velocity $v_{41}$. The angle $\varphi$ is measured as shown, according to a friction force to the left opposing the sliding of the block to the right. From this information, the force polygon can be constructed, yielding forces $F_{14}$ and $F_{34}$, and in turn, forces $F_{43}$, $F_{23}$ and $F_{32}$.

Finally, considering the crank 1, forces $F_{32}$ and $F_{12}$ must be equal in magnitude and opposite in direction. Force $F_{12}$ is tangential to the friction circle at pin $A$ and is properly placed to oppose the clockwise rotation of the crank relative to the frame. This is shown in Fig. 4-11. The moment created by the couple must be opposed by the driving torque, and its value is:

$$M_b = F_{12} h$$

**Example 4-4** Fig. 4-12a shows a cam mechanism with an oscillating follower in which resistant force $F_r$ is applied to the follower at point $F$. The cam 1 is rotating in the counterclockwise direction; the dimensions of the mechanism and the frictional coefficient of pairs are known. Determine the constraint forces in pairs and the input torque.

**Solution**

The friction circles are constructed for the two rotating pairs $A$, $C$, and the relative motions are

**2. 计入摩擦力的力分析**

当力分析过程需要考虑摩擦力时，仍依据力系的平衡条件，只是运动副的总反力方向发生了变化。移动副中的总反力与相对运动方向成（$90°+\varphi$）角，转动副中的总反力要相切于摩擦圆。

**例 4-3** 图 4-11 所示的曲柄滑块机构中，已知各构件的尺寸、曲柄的位置、作用在滑块 4 上的阻力 $F_r$ 以及各运动副中的摩擦因数 $f$，忽略各构件的质量和惯性力。在图 4-11 上标注出各运动副的反力以及加在曲柄上的平衡力矩 $M_b$。

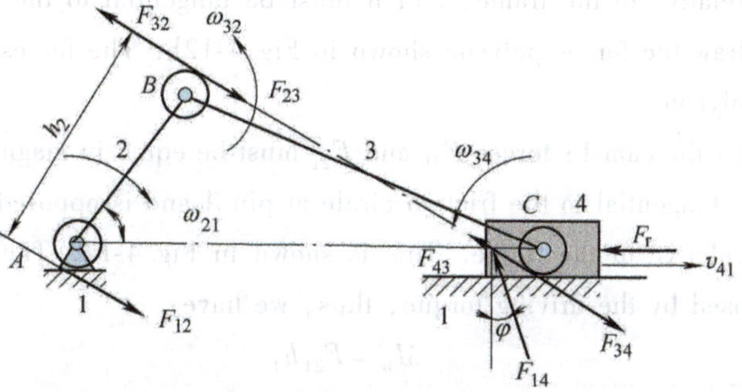


Fig. 4-11 Force analysis considering the friction in a slider-crank linkage
（考虑摩擦的曲柄滑块机构力分析）

**解**

1）根据轴径尺寸和摩擦因数，求出摩擦圆半径，摩擦圆如图 4-11 所示。

2）连杆 3 为受压的二力共线杆，根据连杆 3 相对曲柄 2 的相对运动方向 $\omega_{32}$ 判断曲柄 2 对连杆 3 的反力 $F_{23}$ 的方向；根据连杆 3 相对滑块 4 的相对运动方向 $\omega_{34}$ 判断滑块 4 对连杆 3 的反力 $F_{43}$ 的方向。两者在两摩擦圆的内公切线方向共线。

3）滑块 4 为三力汇交构件，根据滑块 4 相对机架 1 的运动方向 $v_{41}$，可知机架 1 对滑块 4 的反力 $F_{14}$ 与 $v_{41}$ 成（$90°+\varphi$）角。

4）曲柄 2 为分离体，连杆 3 对曲柄 2 的力 $F_{32}$ 的方向已求出，机架 1 对曲柄 2 的反作用力 $F_{12}$ 对轴心 $A$ 的矩与 $\omega_{21}$ 反向。$F_{32}$ 与 $F_{12}$ 大小相等，方向相反，形成力偶矩。

5）加在曲柄 2 上的平衡力矩 $M_b=F_{12}h$。

**例 4-4** 图 4-12a 所示的摆动从动件盘形凸轮机构中，已知凸轮机构的尺寸、轴径尺寸、运动副处的摩擦因数 $f$ 以及作用在从动件 $F$ 点处的阻力 $F_r$，在不计构件质量和惯性力时，求各运动副处的反作用力及作用在凸轮上的平衡力矩 $M_b$。

**解**

1）根据轴径尺寸和摩擦因数，画出转动副 $A$、$C$ 处的摩擦圆如图 4-12a 所示。

2）分析凸轮受力，凸轮 1 为二力构件，摆杆 2 对凸轮 1 的反力 $F_{21}$ 与凸轮 1 相对摆杆 2 的相对速度方向成（$90°+\varphi$）角。$F_{31}$ 对轴心 $A$ 的矩与 $\omega_{13}$ 反向，可判断凸轮 1 的受力如图 4-12a所示。

3）分析摆杆 2 的受力，摆杆 2 上作用有 $F_{12}$、$F_{32}$、$F_r$ 三个力，构成三力汇交的平衡力系。$F_{32}$ 对轴心 $C$ 的矩与 $\omega_{23}$ 反向，作图 4-12b 所示的力多边形，可求出 $F_{32}$、$F_{12}$，$F_{12}=-F_{21}$。

4）求平衡力矩，凸轮 1 中，$F_{21}=-F_{31}$，力臂为 $h_1$，则作用在凸轮 1 上的平衡力矩 $M_b=F_{21}h_1$，方向如图 4-12a 所示。

as shown in Fig. 4-12a.

First, consider the cam 1; it is obviously that forces $F_{31}$, $F_{21}$ must be equal in magnitude and opposite in direction, and produce a couple. The force $F_{21}$ must act at a frictional angle $\varphi$ with respect to the normal of the cam and the flat follower at contact point $B$, and it must have an angle $(90° + \varphi)$ with the relative velocity $v_{12}$. So we can find out the direction of force $F_{21}$.

Next, consider the follower 2 on which the three forces $F_r$, $F_{32}$ and $F_{12}$ act, and they must be intersected at one point $D$. The moment produced by force $F_{32}$ must oppose the clockwise rotation $\omega_{23}$ of the follower relative to the frame, and it must be tangential to the friction circle at pin $C$. Therefore, we can draw the forces polygon shown in Fig. 4-12b. The forces $F_{32}$ and $F_{12}$ can be determined from the polygon.

Finally, consider the cam 1; forces $F_{31}$ and $F_{21}$ must be equal in magnitude and opposite in direction. Force $F_{31}$ is tangential to the friction circle at pin $A$ and is opposed to the counterclockwise rotation of the cam relative to the frame. This is shown in Fig. 4-12. The moment created by the couple must be opposed by the driving torque, thus, we have:

$$M_b = F_{21} h_1$$

## 4.4 Friction and Design of Self-locking Mechanisms

### 1. Self-locking in Kinematic Pairs

There are two forces in a kinematic pair which is used to connect two links; the one is driving force and the other is friction force which resists the motion of the link. If the driving force acting in a pair is infinite, and the link is not movable, it is called as self-locking of a pair.

For a sliding pair, applied resultant force acts within the frictional angle of a body, the self-locking will occur in the sliding pair.

For a rotating pair, applied resultant force acts within the frictional circle of a body, the self-locking will occur in the turning pair.

The condition occurring self-locking is theoretical base for designing some self-locking mechanisms.

### 2. Self-locking Mechanisms

(1) Travel of a mechanism  Any mechanism has two travels; they are positive travel and negative travel.

1) Positive travel of a mechanism. A driving force acts on the driver $A$ shown in Fig. 4-13; the work is done by the resistant force acting on the driven link $B$. This travel is positive travel.

2) Negative travel of a mechanism. If take the resistant as a driving force and it acts on the link $B$, and the original driving link, such as $A$, becomes a driven link, this travel is negative travel.

Fig. 4-13 shows a helical mechanism. When the screw is rotating, the nut will be translated along its axis. This is positive travel of the mechanism. When we push the nut along its axis, the screw would rotate about its axis, this travel is negative travel.

(2) Self-locking mechanism  If the negative travel of a mechanism is in self-locking, it is called as self-locking mechanism. Self-locking mechanism can be used widely in mechanical engi-

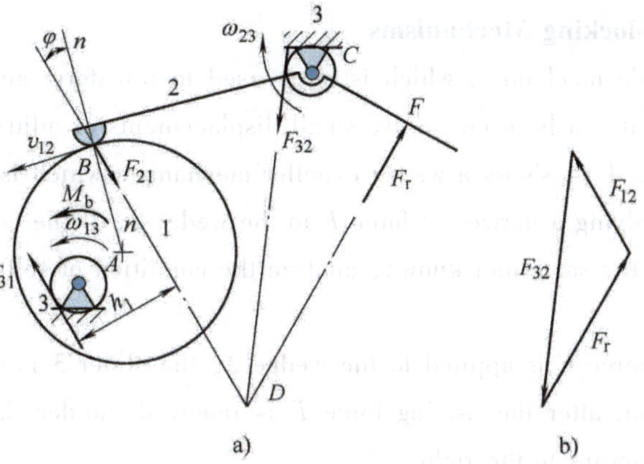


Fig. 4-12　Force analysis considering the friction in a cam mechanism

（考虑摩擦的凸轮机构的力分析）

## 4.4　摩擦与自锁机构的设计

**1. 运动副的自锁**

连接构件间的运动副中存在两种力，即使构件运动的驱动力和阻碍构件运动的摩擦力。如果驱动力无论多么大，都不能使构件运动，这种现象称为运动副的自锁。

对移动副而言，若外力合力作用在摩擦角之内，则移动副发生自锁；对于斜面移动副而言，经常用斜面倾角 $\alpha$ 与摩擦角 $\varphi$ 的关系判断自锁。滑块沿斜面上升时的自锁条件为 $\alpha>(90°-\varphi)$；滑块沿斜面下降时的自锁条件为 $\alpha\leqslant\varphi$。

对转动副而言，当外力合力作用在摩擦圆之内，则转动副发生自锁。

运动副的自锁条件是设计自锁机构的基础。

**2. 自锁机构**

（1）机构的行程　机构有正、反两个运动行程。

1）机构的正行程。当驱动力作用在图 4-13 所示的原动件 $A$ 上，从动件 $B$ 克服生产阻力 $F$ 做功时，一般称为正行程或工作行程。

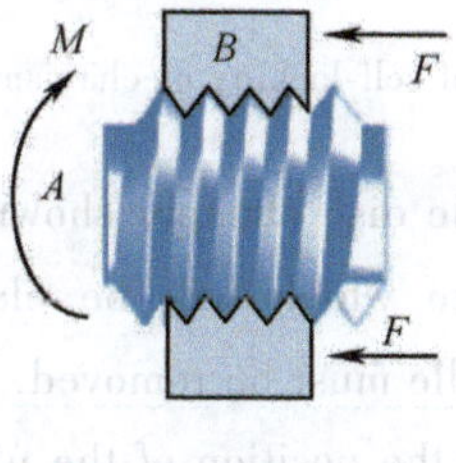


Fig. 4-13　Travel of mechanism

（机构的行程）

2）机构的反行程。当正行程的生产阻力为驱动力，作用在图 4-13 所示的从动件 $B$ 上，原动件 $A$ 为从动件时，称为机构的反行程。

neering.

**3. Design of Self-locking Mechanisms**

A wedge is a simple mechanism which is often used to transform an applied force into much large forces, and also, it can be used to give small displacements or adjustments to heavy loads.

**Example 4-5** Fig. 4-14 shows a wedge expeller mechanism which is used to compress a body 4 with loaded $F_r$ by applying a horizontal force $F$ to the wedge 2. If the coefficient of friction of the contact surfaces are all the same and known, analyze the condition of self-locking of the wedges.

**Solution**

When the driving force $F$ is applied to the wedge 2, the slider 3 is raised in the guides compressing the body 4, and after the driving force $F$ is removed, under the action of load $F_r$, the wedge 2 can not move toward to the right.

This mechanism consists of three sliding pairs formed by the frame 1, wedge 2, and slider 3.

Taking the wedge 2 as a body setting on the frame 1, the direction of loosing the wedge is toward to the right, and the direction of the reaction force is $F_{12}$ is inclined at an angle $(90° + \varphi)$ with respect to the relative velocity $v_{21}$. The force $F_{32}$ which the slider 3 applied to the wedge 2 is equivalent to the total resultant force, and if it acts within the frictional angle $\varphi$, the sliding pair will be in self-locking. This is shown in Fig. 4-14b, so we have:

$$\alpha - \varphi \leqslant \varphi$$

$$\alpha \leqslant 2\varphi$$

That is to say, $\alpha \leqslant 2\varphi$, the wedge is a self-locking mechanism.

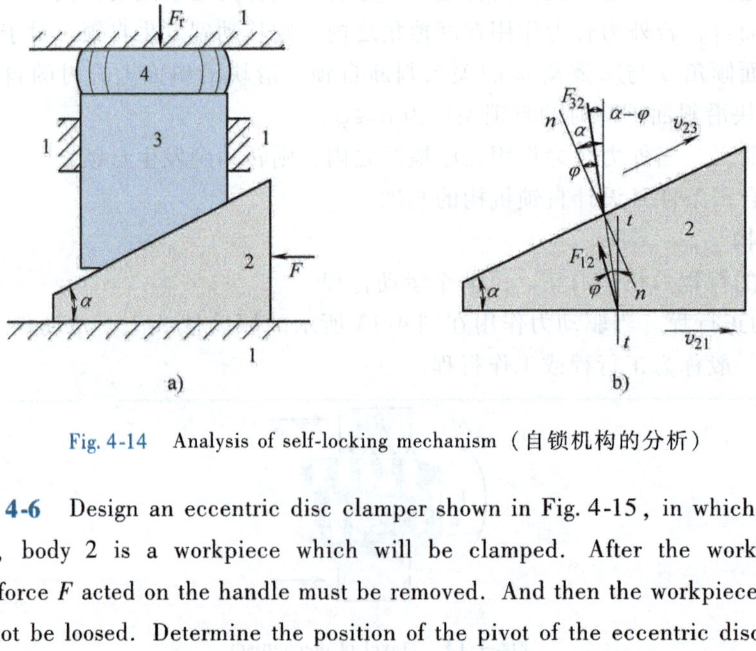


Fig. 4-14 Analysis of self-locking mechanism (自锁机构的分析)

**Example 4-6** Design an eccentric disc clamper shown in Fig. 4-15, in which the link 1 is an eccentric disc, body 2 is a workpiece which will be clamped. After the workpiece has been clamped, the force $F$ acted on the handle must be removed. And then the workpiece which has been clamped can not be loosed. Determine the position of the pivot of the eccentric disc.

**Solution**

The position of the pivot $O$ can be expressed by the distance $e$ and angle $\alpha$ shown in Fig. 4-15. $e$ is the distance between the centers of the disc and the pin; $\alpha$ is an intersection angle between the normal passing through the contact point and the diameter passing through the pivot center.

在图 4-13 所示的螺旋传动中，螺杆转动，螺母移动为正行程。反行程则为用力推动螺母，而使螺杆转动的过程。

（2）自锁机构　反行程发生自锁的机构，称为自锁机构。自锁机构在机械工程领域有广泛的应用。

**3. 自锁机构的分析与设计**

**例 4-5**　在图 4-14 所示的斜面压榨机中，设各接触平面之间的摩擦因数均为 $f$。若在滑块 2 上施加一定的力 $F$，可以将物体 4 压紧。$F_r$ 为被压紧的物体对滑块 3 的反作用力。当力 $F$ 撤去后，该机构在力 $F_r$ 的作用下应具有自锁性。试分析其自锁条件。

**解**　取图 4-14b 所示的滑块 2 为示力体，当力 $F$ 撤去后，滑块 2 可能松脱的运动方向分别为 $v_{21}$、$v_{23}$。若滑块自重忽略不计，构件 1 对滑块 2 的反力 $F_{12}$ 及构件 3 对滑块 2 的反力 $F_{32}$ 的判别方法为：$F_{12}$ 与 $v_{21}$ 成 $(90° + \varphi)$ 角，$F_{32}$ 与 $v_{23}$ 成 $(90° + \varphi)$ 角。$F_{32}$ 是使滑块 2 水平向右滑出的驱动力。当这个驱动力的作用线位于滑块 2 与构件 1 所形成的摩擦角之内时，构件 1、滑块 2 组成的移动副产生自锁现象。从图 4-14b 可以得出自锁条件为

$$\alpha - \varphi \leqslant \varphi$$

$$\alpha \leqslant \varphi$$

**例 4-6**　图 4-15 所示偏心圆盘夹紧机构中，1 为偏心圆盘，2 为待夹紧的工件，3 为夹具体。机构在驱动力 $F$ 的作用下夹紧工件，当力 $F$ 取消后，在总反力 $F_{21}$ 的作用下，工件不能自动松脱，求该机构的反行程必须满足的自锁条件。

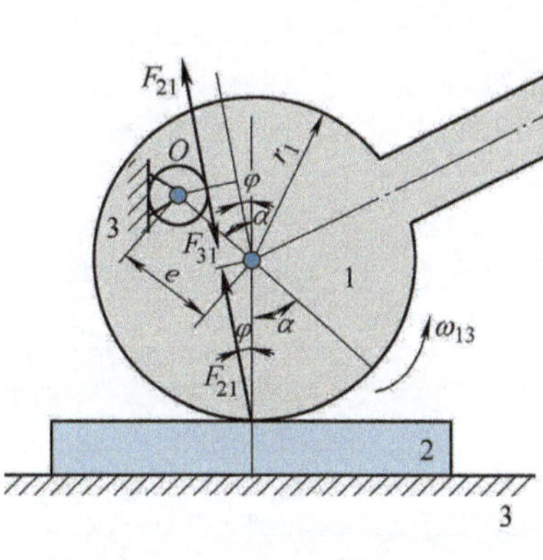


Fig. 4-15　Design of self-locking mechanism（自锁机构的设计）

**解**　该机构要能满足自锁，关键问题是确定偏心圆盘的转动中心 $O$ 点的位置。

设偏心圆盘的半径为 $r_1$，轴径 $O$ 处的摩擦因数为 $f_0$，偏心圆盘与工件的摩擦因数为 $f$，轴径圆心到偏心盘圆心的距离为 $e$，摩擦圆半径为 $\rho$，转轴中心 $O$ 和偏心盘圆心连线与工件接触点的法线之间夹角为 $\alpha$。

Suppose the radius of the disc is $r_1$, the coefficient of friction in the turning pair of the disc is $f_0$, the coefficient of friction between the disc and the workpiece is $f$.

When the force $F$ has been removed, the loose direction of the disc is counterclockwise, and the direction of angular velocity $\omega_{13}$ is shown in the figure. The resultant reaction $F_{21}$ will incline at an angle $\varphi$ to the vertical and opposing the relative motion $v_{12}$ (or $v_{13}$). The disc now is a two-force link, so the force $F_{21} = -F_{31}$. Therefore, the resultant reaction $F_{31}$ must be tangential or intersectional to the frictional circle at the right; its moment must be opposite to the relative velocity $\omega_{13}$.

To ensure the reliability of self-locking of the mechanism, the line of reaction of $F_{21}$ must be intercross with the friction circle at the right, thus we have:

$$e\sin(\alpha - \varphi) - r_1 \sin\varphi \leqslant \rho$$

$$e\sin(\alpha - \varphi) \leqslant r_1 \sin\varphi + \rho$$

Select the distance $e$ and angle $\alpha$ possibly, the self-locking mechanism can be designed by a designer.

如反行程能自锁，总反力 $F_{21}$ 与转轴中心 $O$ 处的摩擦圆相切于 $O$ 点右侧，才能保整 $F_{31}$ 对轴心的力矩与松脱方向 $\omega_{13}$ 相反。确定总反力 $F_{31}$ 后，即可初步确定 $O$ 点的位置应在 $F_{31}$ 的左侧。用公式表达可有下式

$$e\sin(\alpha-\varphi)-r_1\sin\varphi\leqslant\rho$$

从中解出

$$e\sin(\alpha-\varphi)\leqslant r_1\sin\varphi+\rho$$

式中，$r_1$、$\varphi$、$\rho$ 均为已知数据或可以求出的数据，选择适当的 $e$ 和 $\alpha$ 后，便可设计出该自锁机构。

# Chapter 5

## Synthesis of Planar Linkages

# 第5章

## 平面连杆机构及其设计

## 5.1 Characteristics and Types of Planar Linkages

A linkage is one of the mechanisms, but in a linkage all the pairs that are used to connect links are lower pairs, so the linkage is also called as lower pair mechanism. However, it is not necessary to distinguish a linkage from mechanisms. One of the most useful and common mechanisms is the four-bar linkage which consists of four rigid links connected by four lower pairs. A four-bar linkage is the most fundamental to the plane mechanism.

**1. The Characteristics of Planar Linkages**

1) A linkage has a simple structure and it is easy to manufacture, so it has a low cost.

2) The pressure in a lower pair with surface contact is lower, so the linkage has a large load capacity.

3) A linkage can achieve various motions by designing its dimensions of the links.

4) A linkage can transmit motion within a long distance.

5) It is difficult to perform precision motion; this is a disadvantage of linkages.

**2. Types of Planar Linkages**

Fig. 5-1a shows a linkage, in which all the links are connected with rotating pairs; it is called a four-bar linkage. In this linkage, the link 4 is the frame; the links connected to the frame are called side links, and if the side link 1 can rotate through a full revolution, it is called as a crank. If the side link 3 can not rotate a full revolution, such as oscillating through an angle $\psi$, it is called a rocker. If the turning pair can have a whole revolution, we often call it as a full turning pair, such as pairs $A$ and $B$. If the pair can not rotate a full revolution, we called it as an oscillating turning pair, such as pairs $C$ and $D$.

(1) Crank-rocker linkage   If a four-bar linkage shown in Fig. 5-1a is designed so that link 1 can rotate continuously while the link 3 only oscillates through an angle, it is called a crank-rocker linkage.

(2) Double-crank linkage   Both link 2 and link 4 in Fig. 5-1b can rotate continuously relative to the frame 1; this linkage is called a double-crank linkage.

(3) Double-rocker linkage   Both link 2 and link 4 in Fig. 5-1c can not rotate completely relative to the frame 3; this linkage is called a double-rocker linkage.

(4) Parallel crank four-bar linkage   If in a four-bar linkage, two opposite links are parallel and equal in length, as shown in Fig. 5-1d, cranks 2 and 4 always have the same angular velocity. This mechanism is called a parallel crank four-bar linkage. It is widely used, such as in the coupling of locomotive wheels, in pantograph, etc.

(5) Isosceles trapezium linkage   If two rockers of a double-rocker linkage are equal in length (see the Fig. 5-1e), this linkage is called an isosceles trapezium linkage; it can be used in the steering mechanism in automobiles.

(6) Slider-crank linkage   This linkage is shown in Fig. 5-2a, in which the side link 1 is a crank, and the other side link 3 is a slider. Turning pairs $A$ and $B$ are full turning pairs, and $C$ is oscillating turning pair.

## 5.1　平面连杆机构的特点与基本型式

各构件的运动平面重合或相互平行的连杆机构，称为平面连杆机构。四个构件组成的连杆机构结构最简单，应用最广泛。因此，平面四杆机构是平面连杆机构的基础。

**1. 平面连杆机构的特点**

1）平面连杆机构结构简单、易于制造、成本低廉。

2）连杆机构是低副连接的机构，故承载能力大。

3）通过适当地设计各杆件尺寸，连杆机构可实现运动规律与运动轨迹的多样化。

4）可进行远距离的传动。

5）连杆机构不宜应用在高速运转场合。

**2. 平面连杆机构的基本型式**

图 5-1a 所示机构中，各构件均以转动副相连接，称为铰链四杆机构。其中构件 4 为机架，连架杆能做整周转动的构件称为曲柄，如构件 1。连架杆中只能做往复摆动的构件称为摇杆，如构件 3。不与机架相连接的构件 2 称为连杆。其中，转动副 $A$、$B$ 能做 360° 的整周转动，称为整转副。转动副 $C$、$D$ 不能做 360° 的整周转动，称为摆转副。在铰链四杆机构中，转动副的运动范围与机架选择无关。

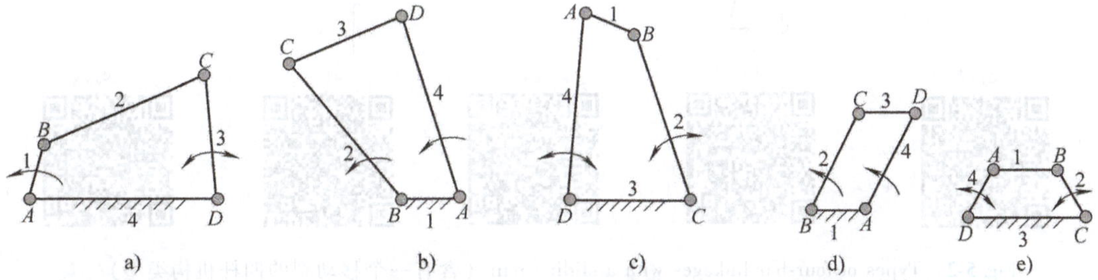


Fig. 5-1　Types of four-bar linkages（铰链四杆机构的类型）

（1）曲柄摇杆机构　若两个连架杆中一个为曲柄，另一个为摇杆，则此铰链四杆机构称为曲柄摇杆机构，如图 5-1a 所示。

（2）双曲柄机构　若将图 5-1a 所示曲柄摇杆机构中的曲柄 1 选为机架，转动副 $A$、$B$ 为整转副，则连架杆 2、4 均为曲柄，该机构称为双曲柄机构，如图 5-1b 所示。

（3）双摇杆机构　若将图 5-1a 所示曲柄摇杆机构中的摆杆 3 选为机架，转动副 $C$、$D$ 为摆转副，则连架杆 2、4 均为摇杆，该机构称为双摇杆机构，如图 5-1c 所示。

（4）平行四边形机构　图 5-1b 所示双曲柄机构中，如两曲柄平行且相等，则该机构演化为平行四边形机构，如图 5-1d 所示。

（5）等腰梯形机构　图 5-1c 所示双摇杆机构中，如两摇杆长度相等，则该机构演化为等腰梯形机构，如图 5-1e 所示。

(7) Rotating guide-bar linkage  If both the side link 2 and side link 4 can rotate completely, this linkage is called a rotating guide-bar linkage shown in Fig. 5-2b, in which the coupler becomes a slider in shape.

(8) Rocking-block linkage  The crank rotates completely while the block only oscillates about its pivot center $C$, this linkage is called rock-block linkage shown in Fig. 5-2c. It is widely used in hydraulic cylinder mechanism.

(9) Sliding guide-bar linkage  If the guide link 4 reciprocates along the axis of the fixed block, and the other side link 2 oscillates about the pivot $C$, the linkage is called a sliding guide-bar linkage shown in Fig. 5-2d. It can be used in some water pumps.

(10) Rocking guide-bar linkage  If the link 2 can rotates about pivot $B$ completely, and the link 4 oscillates about the pivot $A$ shown in Fig. 5-2e, this linkage is called a rocking guide-bar linkage. It is used in machine tools such as shapers.

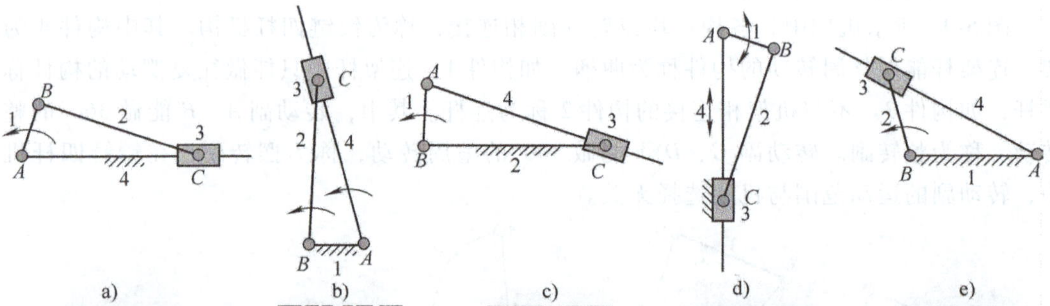


Fig. 5-2  Types of four-bar linkages with a sliding pair（含有一个移动副的四杆机构类型）

(11) Double-slider-crank linkage  In a four-bar linkage, if two side links 1 and 3 become sliders which are reciprocating along each axis of cross frame, and the two pairs of the same kind are adjacent, it is known as a double-slider-crank linkage shown in Fig. 5-3a. This linkage can draw ellipses.

(12) Double rotating block linkage  If two sliders are rotating completely about their pivots $A$ and $B$ relative to the frame 2, this linkage is called a double rotating block mechanism shown in Fig. 5-3b. It can also be called Oldham linkage, and it is used to connect two shafts having parallel misalignment.

(13) Sine linkage  This linkage can also be called a Scotch-yoke linkage, in which the crank 2 rotates about its pivot $A$ completely, and the link 4 reciprocates in the fixed link 1 to produce a simple harmonic motion. It is shown in Fig. 5-3c.

$$s = l_2 \sin\varphi$$

The Scotch-yoke linkage is used to convert the rotary motion into a sliding motion, such in testing machines has simple harmonic motion.

(14) Tangent linkage  In a tangent mechanism, the side link 2 oscillates about its pivot $A$, the other link 4 reciprocates along its axis, and the link 3 is a slider in shape. It is also called the

（6）曲柄滑块机构　图 5-2a 所示四杆机构中，一连架杆为曲柄，另一个连架杆为滑块，该机构称为曲柄滑块机构。其中，转动副 $A$、$B$ 为整转副，转动副 $C$ 为摆转副。

（7）转动导杆机构　若将图 5-2a 所示曲柄滑块机构中的曲柄 1 选为机架，转动副 $A$、$B$ 为整转副，则连架杆 2、4 均为曲柄，滑块 3 沿连架杆 4 移动，且随连架杆 4 转动，该机构称为转动导杆机构，如图 5-2b 所示。

（8）曲柄摇块机构　若将图 5-2a 所示机构中的构件 2 选为机架，转动副 $A$、$B$ 仍为整转副，连架杆 1 仍为曲柄，另一连架杆（滑块 3）只能绕 $C$ 点往复摆动，该机构称为曲柄摇块机构，如图 5-2c 所示。

（9）移动导杆机构　若将图 5-2a 所示机构中的滑块 3 选为机架，转动副 $A$、$B$ 仍为整转副，连架杆 4 只能沿滑块往复移动，该机构称为移动导杆机构，如图 5-2d 所示。

（10）摆动导杆机构　若将图 5-2b 所示转动导杆机构中的机架加长，使 $l_{BC} < l_{AB}$，转动副 $A$ 演化为摆转副，连架杆 4 往复摆动，则该机构称为摆动导杆机构，如图 5-2e 所示。

（11）双滑块机构　在含有两个移动副的四杆机构中，若将两个连架杆做成块状，且相对十字形机架做相对移动，则该机构称为双滑块机构，如图 5-3a 所示。

（12）双转块机构　若两个块状连架杆相对机架做定轴转动，则该机构称为双转块机构，如图 5-3b 所示。

（13）正弦机构　图 5-3c 中，曲柄 2 绕 $A$ 点转动时，通过滑块 3 驱动构件 4 做水平移动，其位移量 $s = l_2\sin\varphi$，与曲柄转角 $\varphi$ 成正弦函数关系，该机构称为正弦机构。

（14）正切机构　图 5-3d 中，构件 2 转动时，构件 4 竖直移动，其位移量 $s = a\tan\varphi$，该机构称为正切机构。

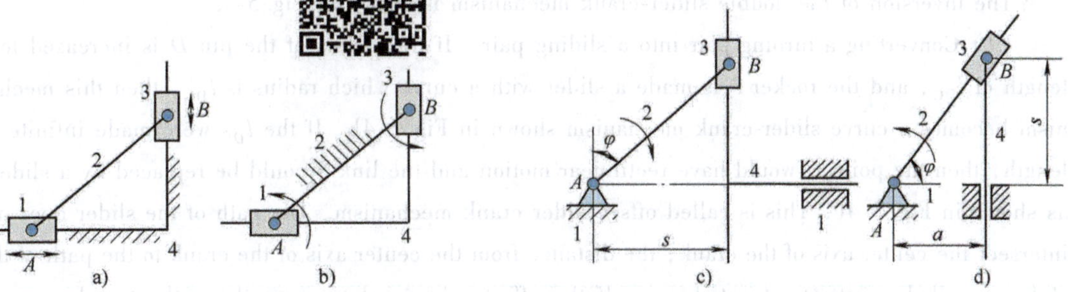


Fig. 5-3　Types of four-bar linkages with two sliding pairs
（含有两个移动副的四杆机构类型）

**3. 四杆机构的演化与变异**

四杆机构的类型多种多样，但它们之间存在密切的关系。下面介绍四杆机构的演化、变异方法。

（1）转换机架法　图 5-1a 所示机构为曲柄摇杆机构，若以曲柄 1 为机架，则得到图 5-1b所示的双曲柄机构；若以摇杆 3 为机架，则得到图 5-1c 所示的双摇杆机构。图 5-2a 所示机构为曲柄滑块机构；若以曲柄 1 为机架，则得到图 5-2b 所示的转动导杆机构；若以连杆 2 为机架，则得到图 5-2c 所示的曲柄摇块机构；若以滑块 3 为机架，则得到图 5-2d 所示的移动导杆机构。图 5-3 所示的双转块机构的演化也符合该方法。其基本原理是机构中各构

Rapson's slide linkage. The displacement of link 4 has a tangent motion.

$$s = a\tan\varphi$$

**3. Evolution and Mutation of Planar Linkages**

All kinds of four-bar linkages are relevant to each other inherently.

(1) Inversion of a four-bar linkage Different mechanisms can be obtained by fixing different links of a mechanism. This is known as inversion. The principle is that the relative motion between links of a four-bar linkage does not change in different inversion.

In a crank-rocker linkage shown in Fig. 5-1a, the link 1 is a crank, the link 2 is the coupler, the link 3 is the rocker, and the link 4 is the frame. The turning pairs $A$ and $B$ are full turning pairs, and pairs $C$ and $D$ are oscillating pairs.

If the link 1 is fixed instead of the link 4, this makes the link 2 and the link 4 cranks and rotate about their pivots $A$ and $B$ respectively. It is a double-crank mechanism shown in Fig. 5-1b.

If the link 3 is fixed instead of link 4, this makes link 2 and link 4 rockers and oscillate about their pivots $D$ and $C$ respectively. It is a double-rocker mechanism shown in Fig. 5-1c.

In a slider-crank mechanism shown in Fig. 5-2a, If the link 1 is fixed instead of the link 4, this makes link 2 and link 4 crank and rotate about their pivots $A$ and $B$ respectively. The link 3, a slider, becomes a coupler. It is a rotating guide-bar mechanism shown in Fig. 5-2b.

If the link 2 is fixed instead of the link 4, this makes the link 1 a crank, and the block oscillates about its pivots $C$. It is a rocking-block mechanism shown in Fig. 5-2c.

If the link 3 is fixed instead of link 4, the mechanism becomes a sliding guide-bar mechanism shown in Fig. 5-2d.

The inversion of the double slider-crank mechanism is shown in Fig. 5-3.

(2) Converting a turning pair into a sliding pair If the radius of the pin $D$ is increased to a length of $l_{DC}$, and the rocker 3 is made a slider with a curve which radius is $l_{DC}$, then this mechanism becomes a curve slider-crank mechanism shown in Fig. 5-4b. If the $l_{DC}$ were made infinite in length, then the point $C$ would have rectilinear motion and the link 3 could be replaced by a slider, as shown in Fig. 5-4c. This is called offset slider-crank mechanism. The path of the slider does not intersect the center axis of the crank; the distance from the center axis of the crank to the path of the slider is called an offset, denoted as $e$. If the offset $e$ is equal to zero, the path of slider passes through the crank center; it is called an inline slider-crank mechanism shown in Fig. 5-4d.

(3) Expansion of pin size in a turning pair If we increase the size of the crank pin $B$ shown in Fig. 5-5a until it is larger than the length of the crank, this enlarged crank pin is called an eccentric disk and can be used to replace the crank shown in Fig. 5-5a. The crank consists of a circular disk with center $B$, which is pivoted off-center at $A$ to the frame. The disk rotates inside the ring end of coupler 2. The motion of the eccentric mechanism is equivalent to that of a slider-crank mechanism having a crank length equal to $AB$ and an coupler of length $BC$.

**4. Applications of Planar Linkages**

Planar linkages have a lot of advantages, such as lower cost, simply structure, large load capacity, performing various motions, etc., so they are used widely in mechanical engineering.

(1) Applications of linkages consisted of turning pairs Fig. 5-6a is a jaw crusher. It consists

件相对运动与机架选择无关。

(2) 转动副向移动副的演化　图 5-4a 所示曲柄摇杆机构中，摇杆上 $C$ 点的运动轨迹是以 $D$ 为圆心、以$\overline{DC}$为半径的圆弧。将摇杆 $DC$ 作成图 5-4b 所示的块状构件，在以摇杆长度为半径的圆弧上滑动，则曲柄摇杆机构演化为曲柄曲线滑块机构，两者的运动完全等效。若曲线滑动导轨的曲率半径无穷大，则该曲线滑块机构演化为图 5-4c 所示的曲柄滑块机构。该曲柄滑块机构的导路方向线不通过曲柄转动中心，偏开的距离 $e$ 称为偏距。这种机构称为偏置曲柄滑块机构。如果偏置曲柄滑块机构的偏距 $e=0$，则导路的方向线通过曲柄的转动中心，这种曲柄滑块机构称为对心曲柄滑块机构，如图 5-4d 所示。转动副转化为移动副的过程说明了全转动副的四杆机构可以演化为含有移动副的四杆机构。

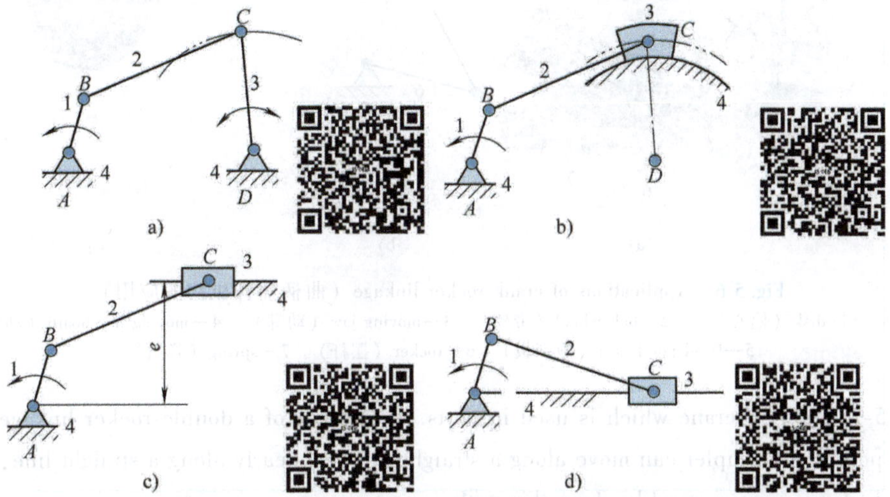


Fig. 5-4　Evolution from turning pair to sliding pair（转动副向移动副的演化）

(3) 转动副的销钉扩大　图 5-5a 所示的曲柄滑块机构中，当曲柄 $AB$ 的尺寸较小时，可将转动副 $B$ 的销钉扩大；当销钉 $B$ 的半径大于曲柄的长度时，该机构演化为图 5-4b 所示的偏心盘机构。该机构可增加曲柄 $AB$ 强度。

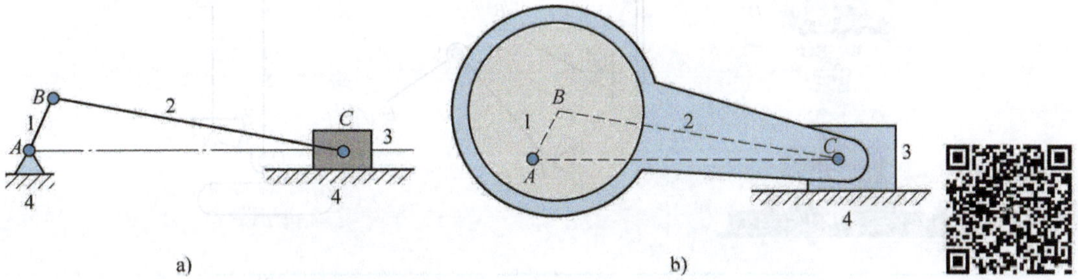


Fig. 5-5　Eccentric disk mechanism（偏心盘机构）

**4. 平面连杆机构的应用**

平面连杆机构的应用非常普遍，这里仅作些简单介绍。

(1) 全转动副四杆机构的应用　图 5-6 所示为曲柄摇杆机构的应用。其中，图 5-6a 所示为矿石破碎机，图 5-6b 所示为其机构简图。曲柄摇杆机构是该机器的主体机构。图 5-6c

of a crank-rocker linkage. The connecting rod is the moving jaw used to crush the rocks. Its kinematic diagram is shown in Fig. 5-6b. Fig. 5-6c shows a dough mixing machine; it consists of a crank-rocker linkage also. The coupler curves can be used to generate quite useful path motions for machine design problem, and in this mixer the coupler curve is like the mixing motion of hands.

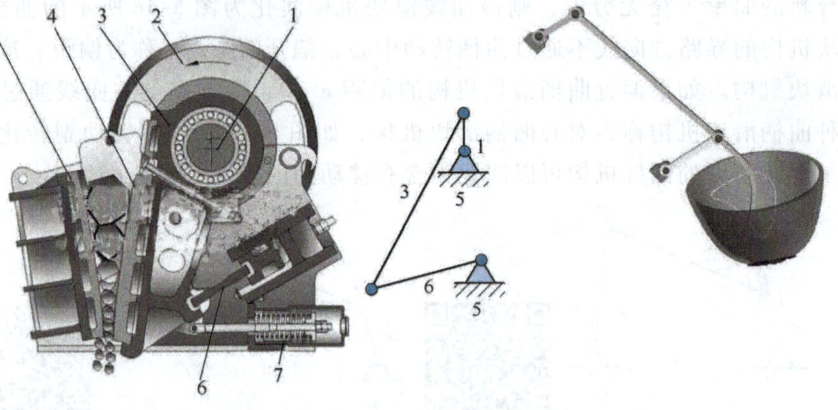


a) b) c)

Fig. 5-6 Applications of crank-rocker linkage（曲柄摇杆机构的应用）
1—eccentric disk（偏心盘） 2—belt wheel（带轮） 3—moving jaw（动鄂） 4—moving jaw board（动鄂板）
5—fixed jaw board（静鄂板） 6—rocker（摆杆） 7—spring（弹簧）

Fig. 5-7a shows a crane which is used in ports. It consists of a double-rocker linkage.

The point $E$ on coupler can move along a straight line, or nearly along a straight line, therefore the lifted body may move steadily along this path.

Fig. 5-7b shows a steering linkage of vehicles. Two equal rockers $AB$ and $CD$ are fixed to the stub axes to form two similar bell-crank levers. This linkage is known as a isosceles trapezium linkage.

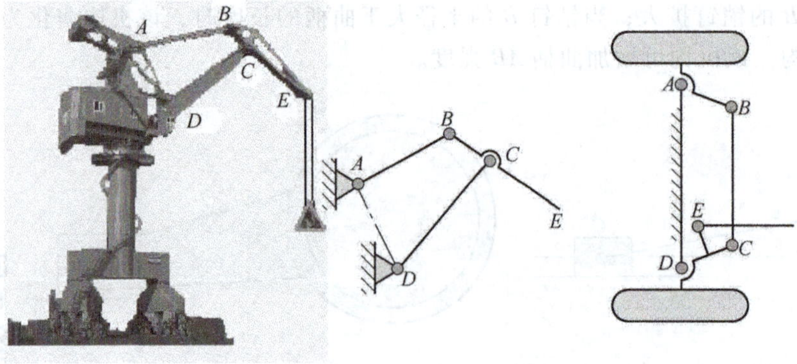


a) b)

Fig. 5-7 Applications of double-rocker linkage（双摇杆机构的应用）

Fig. 5-8a shows a inertia vibrating screen. It is a double-crank linkage, and is also called a drag-link linkage. If the driver $AC$ rotates counterclockwise with constant angular velocity, the slider makes a slow stroke and returns with a quick stroke. The inertia forces can cause vibration of the screen to perform separation of the materials.

所示为利用曲柄摇杆机构设计的和面机示意图。

图 5-7 所示为双摇杆机构的应用。图 5-7a 所示为鹤式起重机，当摇杆 $CD$ 摆动时，另一摇杆 $AB$ 随之摆动，使得悬挂在连杆 $E$ 点上的重物在近似的水平直线上运动，避免重物平移时因不必要的升降而消耗能量。图 5-7b 所示为汽车和拖拉机前轮转向机构，该机构为等腰梯形双摇杆机构。

图 5-8a 所示为双曲柄机构在惯性振动筛中的应用，图 5-8b 所示为机车动轮中平行四边形机构的应用，图 5-8c 所示为平行四边形机构在升降机中的应用。

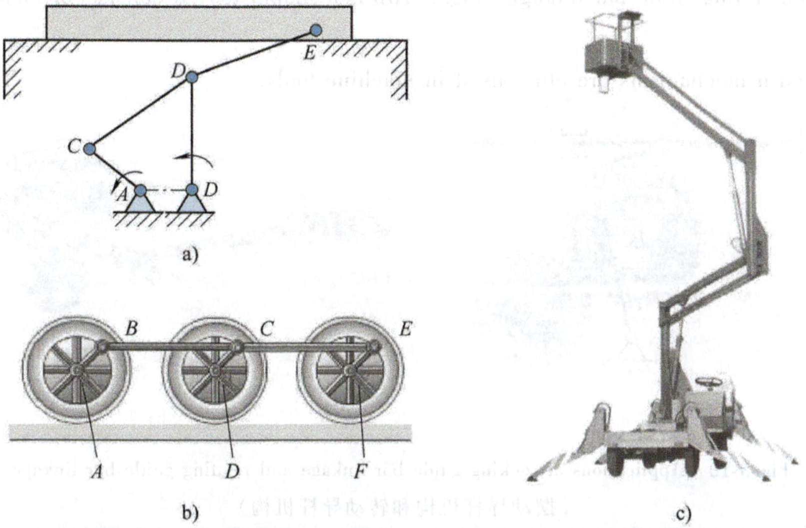


Fig. 5-8　Applications of double-crank linkage（双曲柄机构的应用）

（2）曲柄滑块机构的应用　图 5-9a 所示为曲柄滑块机构在多缸内燃机中的应用，图 5-9b所示为曲柄滑块机构在剪床中的应用。

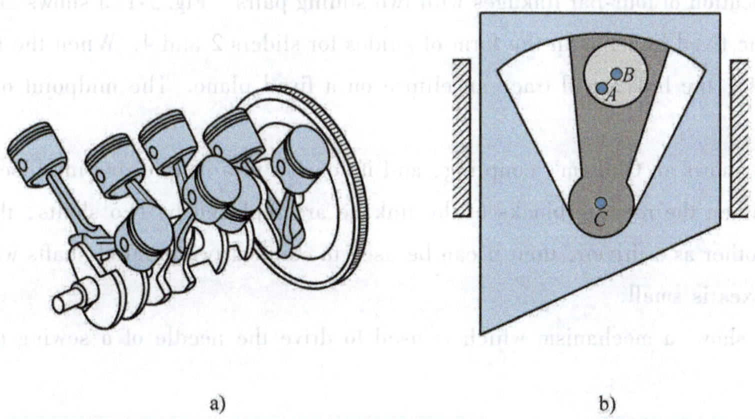


Fig. 5-9　Applications of slider-crank linkage（曲柄滑块机构的应用）

（3）导杆机构的应用　导杆机构包括转动导杆机构、摆动导杆机构、移动导杆机构。

图 5-10a 所示为摆动导杆机构在牛头刨床中的应用。图 5-10b 所示为转动导杆机构在小

Fig. 5-8b shows a parallel double-crank linkage used in the coupler of locomotive wheels, and Fig. 5-8c shows a hydraulic lift with two sets of parallel double-crank linkages.

(2) Applications of slider-crank linkage  Slider-crank linkages are used widely in various multi-cylinder engines. Fig. 5-9a is a V-engine with eight cylinders. This engine consists of eight slider-crank linkages.

Fig. 5-9b shows a shearing machine. It consists of one slider-crank linkage.

(3) Application of crank-slider quick-return linkage  Fig. 5-10a is a shaper; it consists of gear mechanism and sliding guide-bar linkage. Fig. 5-10b is a shaper too; it consists of rotating guide-bar linkage.

Quick return mechanisms are often used in machine tools.

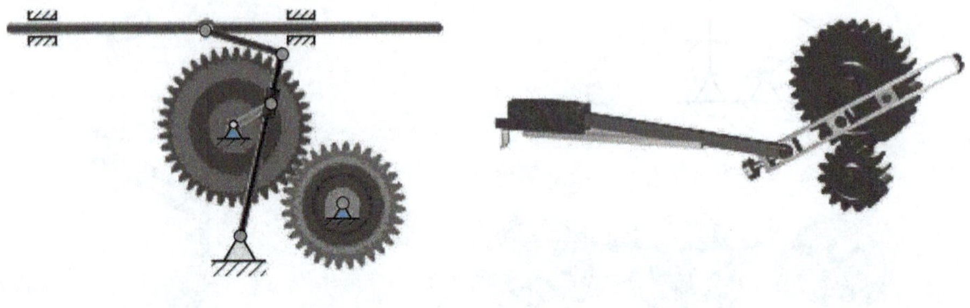

a)  b)

Fig. 5-10  Applications of rocking guide-bar linkage and rotating guide-bar linkage
(摆动导杆机构和转动导杆机构)

Fig. 5-11a shows a hand-pump, in which link 4 is made in the form of a cylinder and a plunger fixed to the link 3 reciprocates in it; it consists of sliding guide-bar linkage.

Fig. 5-11b shows a self loading truck, in which a rocking-block linkage is used to perform self loading work.

(4) Application of four-bar linkages with two sliding pairs  Fig. 5-12a shows an elliptical trammel in which the fixed link 1 is in the form of guides for sliders 2 and 4. When the two sliders move along their guide, the link 3 will trace an ellipse on a fixed plane. The midpoint of the link 3 will trace a circle.

Fig. 5-12b shows an Oldham's coupling, and it consists of a double rotating block linkage shown in Fig. 5-3b. When the rotating blocks of the linkage are replaced by two shafts, the one acts as a driver and the other as a driven, then it can be used to connect two parallel shafts when the distance between their axes is small.

Fig. 5-12c shows a mechanism which is used to drive the needle of a sewing machine; it is a sine linkage.

## 5.2 Fundamental Features of Four-Bar Linkages

There are more distinguishing features that are useful to know about the four-bar linkage. These features include Grashof criteria, quick-return characteristic, transmission angle and dead-point po-

型牛头刨床中的应用。

图 5-11a 所示为移动导杆机构在手摇水泵中的应用。当扳动手柄 1 时，活塞 4 便在筒体 3 内做往复移动，从而完成抽水和压水的工作。图 5-11b 所示为曲柄摇块机构在自动装卸卡车中的应用。

Fig. 5-11　Applications of sliding guide-bar linkage and rocking-block linkage
（移动导杆机构和曲柄摇块机构的应用）
1—handle（手柄）　2—link（连杆）　3—cylinder（筒体）　4—plunger（活塞）

（4）含有两个移动副的四杆机构的应用　图 5-12a 所示为双滑块机构在椭圆规中的应用实例。图 5-12b 所示为双转块机构在联轴器中的应用，该联轴器可连接两不共轴线的轴。图 5-12c 所示为正弦机构在缝纫机走线机构中的应用。

连杆机构在工程机械、纺织机械、印刷机械、食品机械、矿山机械等大量机械中都有广泛应用，这里仅作初步介绍。

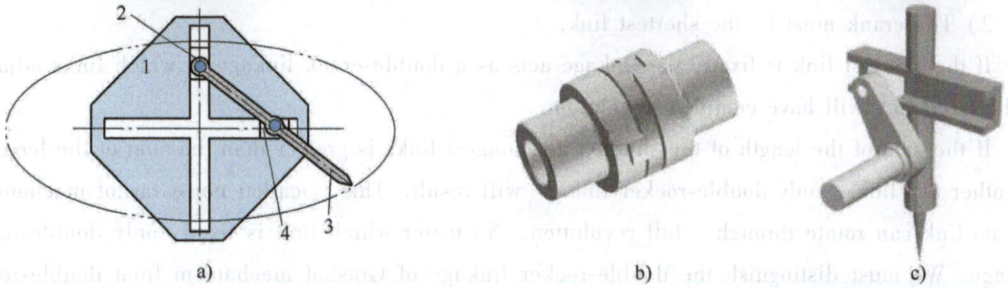


Fig. 5-12　Applications of four-bar linkages with two sliding pairs（含有两个移动副机构的应用）
1—fixed link（固定块）　2、4—slider（滑块）　3—coupler（连杆）

## 5.2　平面连杆机构的基本性质

了解四杆机构的基本特性是设计平面连杆机构的基础。

**1. 曲柄存在条件**

在铰链四杆机构中，欲使连架杆做整周转动而成为曲杆，各杆长度必须满足一定的条件，即所谓的曲柄存在条件。

sition.

**1. Grashof Criteria**

Fig. 5-13 shows a four-bar linkage *ABCD*, and if the link *AB* can rotate through a full revolution with an angular velocity $\omega$ clockwise, it must pass through the positions $AB_1$ and $AB_2$.

Let the lengths of the links of the linkage be $a$, $b$, $c$ and $d$ respectively.

From the triangle $\triangle B_1C_1D$, we have:

$$a+d\leqslant b+c \tag{5-1}$$

From the triangle $\triangle B_2C_2D$, we have:

$$b\leqslant(d-a)+c$$

$$c\leqslant(d-a)+b$$

Arranging the above inequalities, we have:

$$a+b\leqslant c+d \tag{5-2}$$

$$a+c\leqslant b+d \tag{5-3}$$

Adding the inequalities (5-1), (5-2), (5-3) respectively, then we get:

$$a\leqslant b \tag{5-4}$$

$$a\leqslant c \tag{5-5}$$

$$a\leqslant d \tag{5-6}$$

From the above inequalities, we have the following conclusions:

The link 1 which can rotate through a full revolution must be the shortest link; the sum of the length of shortest link and the longest link (there must be a longest link among $b$, $c$, $d$) can not be greater than the sum of the remaining two links.

The Grashof criteria can be stated as following:

1) A linkage, in which the sum of the length of the shortest and longest links is less than the sum of the length of the other two links, must have a crank.

2) The crank must be the shortest link.

If the shortest link is fixed, the linkage acts as a double-crank linkage in which links adjacent to the fixed link will have complete revolution.

If the sum of the length of the shortest and longest links is greater than the sum of the length of the other two links, only double-rocker linkage will result. This is called non-Grashof mechanism, and no link can rotate through a full revolution. No mater which link is fixed, only double-rocker linkage. We must distinguish the double-rocker linkage of Grashof mechanism from double-rocker linkage of non-Grashof mechanism.

**2. Quick-return Motion**

When the crank rotates through a full revolution with an angular velocity $\omega$ clockwise, and the rocker only oscillates through an angle $\psi$, the crank and coupler will form a straight line at each extreme position. Fig. 5-14 shows a crank-rocker linkage in the two extreme positions of the driven link.

The acute angle $\theta$ between the corresponding positions of the cranks when the rocker is at two extreme positions is called the angle corresponding the extreme positions.

In this particular case, the crank executes the angle $\varphi_1$ $(180°+\theta)$ while the rocker oscillates

图 5-13 所示的铰链四杆机构中，设构件 1 ~4 的长度分别为 $a$、$b$、$c$、$d$，并取 $a<d$。当构件 1 能绕点 $A$ 做整周转动时，构件 1 必须能通过与构件 4 共线的两位置 $AB_1$ 和 $AB_2$。

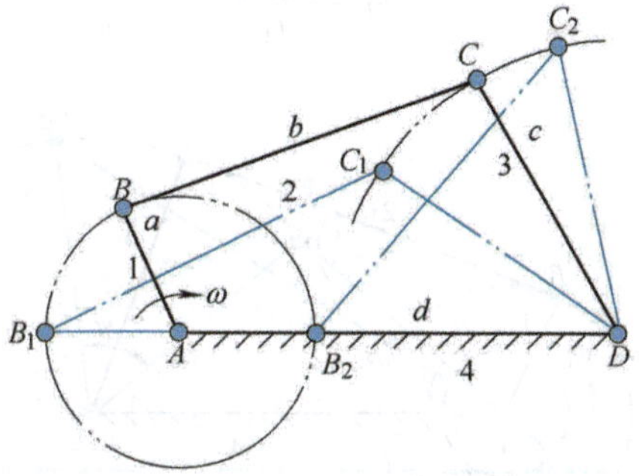


Fig. 5-13　Grashof law（曲柄存在条件）

当构件 1 转至 $AB_1$ 时，形成 $\triangle B_1C_1D$，根据三角形任意两边长度之和必大于第三边长度的几何关系并考虑到极限情况，得

$$a+d\leqslant b+c \tag{5-1}$$

当构件 1 转至 $AB_2$ 时，形成 $\triangle B_2C_2D$，同理可得

$$b\leqslant(d-a)+c \text{ 及 } c\leqslant(d-a)+b$$

整理后可写成

$$a+b\leqslant c+d \tag{5-2}$$

$$a+c\leqslant b+d \tag{5-3}$$

将式（5-1）~式（5-3）两两相加，化简后得

$$a\leqslant b \tag{5-4}$$

$$a\leqslant c \tag{5-5}$$

$$a\leqslant d \tag{5-6}$$

在铰链四杆机构中，要使构件 1 为曲柄，它必须是四杆中的最短杆，且最短杆与最长杆长度之和小于或等于其余两杆长度之和。考虑到更一般的情形，可将铰链四杆机构曲柄存在条件概括如下：

1）连架杆和机架中必有一个是最短杆。

2）最短杆与最长杆长度之和必小于或等于其余两杆长度之和。

当铰链四杆机构中最短杆与最长杆长度之和大于其余两杆长度之和时，不论以哪一构件为机架，都不存在曲柄，而只能是双摇杆机构。该双摇杆机构中不存在能作整周转动的运动副。前面提到的汽车前轮转向机构就是没有整转副的双摇杆机构。

**2. 急回特性**

图 5-14 所示曲柄摇杆机构中，设曲柄 $AB$ 为主动件，摇杆 $CD$ 为从动件。主动曲柄 $AB$ 以等角速度 $\omega$ 顺时针方向转动，当曲柄转至 $AB_1$ 位置与连杆 $B_1C_1$ 重叠共线时，摇杆 $CD$ 处于左极限位置 $C_1D$；而当曲柄转至 $AB_2$ 位置与连杆 $B_2C_2$ 拉伸共线时，从动摇杆处于右极限位置 $C_2D$。摇杆处于左、右两极限位置时，对应曲柄两位置所夹的锐角 $\theta$ 称为极位夹角。摇杆两极限位置间的夹角 $\psi$ 称为摇杆的摆角。

from $C_1$ to $C_2$, through an angle $\psi$ with time $t_1$, so the average speed of the forward stroke $v_1$ is:

$$v_1 = \frac{\overset{\frown}{C_1 C_2}}{t_1}$$

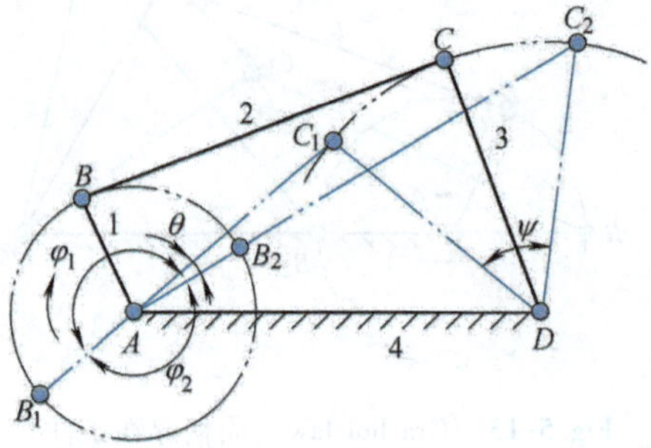


Fig. 5-14 Quick-return characteristics (急回特性)

If the crank executes the angle $\varphi_2(180° - \theta)$ while the rocker oscillates from $C_2$ to $C_1$, through an angle $\psi$ with time $t_2$, the average speed of the return stroke $v_2$ is:

$$v_2 = \frac{\overset{\frown}{C_2 C_1}}{t_2}$$

It will be observed that the crank angle $\varphi_1$, corresponding to the forward stroke $C_1C_2$ of the rocker, is larger than the return angle $\varphi_2$, corresponding to the return stroke $C_2C_1$.

The ratio of the average speed $v_2$ of the return stroke and $v_1$ of the forward stroke is called the coefficient of travel speed variation, and is denoted as $K$.

$$K = \frac{v_2}{v_1} = \frac{\dfrac{\overset{\frown}{C_2 C_1}}{t_2}}{\dfrac{\overset{\frown}{C_1 C_2}}{t_1}} = \frac{t_1}{t_2} = \frac{\varphi_1}{\varphi_2} = \frac{180° + \theta}{180° - \theta} \tag{5-7}$$

$$\theta = 180° \frac{K-1}{K+1} \tag{5-8}$$

To produce a quick-return motion, this ratio must be obviously greater than unity and as large as possible. The crank-rocker linkage, guider-bar linkage and offset slider-crank linkage are all quick-return mechanisms.

### 3. Pressure Angle and Transmission Angle

The pressure angle $\alpha$ is the acute angle between the direction of the static force transferred through the coupler and the absolute velocity of the output link at the connected point, such as at $C$. Since the coupler is a two-force link, the direction of the static force of the coupler is along the line of its pin joints.

The acute angle $\gamma$ between the coupler and output link is known as transmission angle shown in Fig. 5-15. Notice that in this case, $\alpha + \gamma = 90°$.

当曲柄从 $AB_1$ 位置转至 $AB_2$ 位置时，其对应转角 $\varphi_1=180°+\theta$，而当摇杆由 $C_1D$ 位置摆至 $C_2D$ 位置时，摆角为 $\psi$，设所需时间为 $t_1$，$C$ 点的平均速度为 $v_1$；当曲柄再继续从 $AB_2$ 位置转至 $AB_1$ 位置时，其对应转角 $\varphi_2=180°-\theta$，而摇杆则由 $C_2D$ 位置摆回 $C_1D$ 位置，摆角仍为 $\psi$，设所需时间为 $t_2$，$C$ 点的平均速度为 $v_2$。摇杆往复摆动的角度虽然相同，但是对应的曲柄转角不等，即 $\varphi_1>\varphi_2$，而曲柄又是等速转动，所以有 $t_1>t_2$，$v_2>v_1$。由此可见，当曲柄等速转动时，摇杆往复摆动的平均速度是不同的，摇杆的这种运动特性称为急回特性。通常用 $v_2$ 与 $v_1$ 的比值 $K$ 来衡量，$K$ 称为行程速度变化系数，即

$$K=\frac{v_2}{v_1}=\frac{\dfrac{\overset{\frown}{C_2C_1}}{t_2}}{\dfrac{\overset{\frown}{C_1C_2}}{t_1}}=\frac{t_1}{t_2}=\frac{\varphi_1}{\varphi_2}=\frac{180°+\theta}{180°-\theta}\tag{5-7}$$

当给定行程速度变化系数 $K$ 后，机构的极位夹角可由式（5-8）计算

$$\theta=180°\frac{K-1}{K+1}\tag{5-8}$$

平面连杆机构有无急回运动取决于极位夹角 $\theta$。只要极位夹角 $\theta$ 不为零，该机构就有急回特性，其行程速度变化系数 $K$ 可用式（5-7）计算。

四杆机构的这种急回特性，可以用来节省空回行程的时间，提高生产率。牛头刨床和摇摆式输送机都利用了这一特性。

**3. 机构的压力角与传动角**

压力角或传动角是判断连杆机构传力性能优劣的重要标志。图 5-15 所示曲柄摇杆机构中，若忽略各杆的质量和运动副中的摩擦，连杆 $BC$ 作用于从动摇杆 $CD$ 上的力 $F$ 是沿杆 $BC$ 方向的。把从动摇杆 $CD$ 所受力 $F$ 与力作用点 $C$ 的速度 $v$ 之间所夹的锐角 $\alpha$ 称为压力角。压力角 $\alpha$ 越小，传力性能越好。因此，压力角的大小可以作为判别一个连杆机构传力性能好坏的依据。

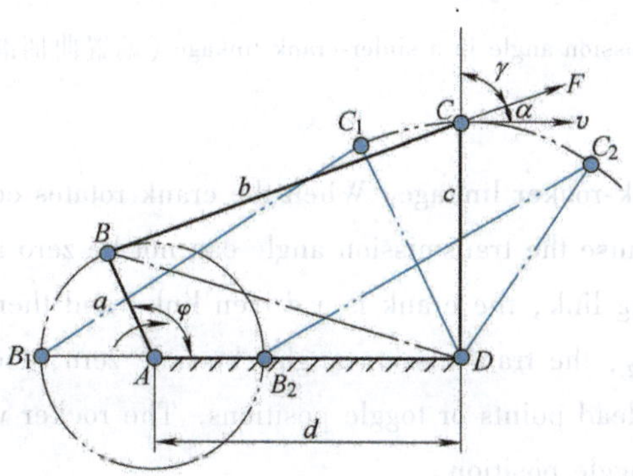


Fig. 5-15　Transmission angle and pressure angle（传动角和压力角）

由图 5-15 可知，$\alpha+\gamma=90°$，$\alpha$ 与 $\gamma$ 互为余角。通常用 $\gamma$ 值来衡量机构的传力性能。$\alpha$ 越小，则 $\gamma$ 越大，机构的传力性能越好，反之越差。当连杆 $BC$ 与摇杆 $CD$ 间的夹角为锐角时，该角即为传动角 $\gamma$；而当连杆 $BC$ 与摇杆 $CD$ 间的夹角为钝角时，传动角 $\gamma$ 则为其补角。在机构运动过程中，传动角的大小是随机构位置的改变而变化的。为了确保机构能正常工

During the motion of a mechanism, the transmission angle will of course change in value, and the minimum transmission angle will usually be kept more than 40° to promote smooth running and good force transmission.

Applying cosine law to triangles $\triangle ABD$ and $\triangle BCD$, we have:

$$\overline{BD}^2 = a^2 + d^2 - 2ad\cos\varphi$$

$$\overline{BD}^2 = b^2 + c^2 - 2bc\cos\gamma$$

Simultaneously solve the above equations:

$$\cos\gamma = \frac{b^2 + c^2 - a^2 - d^2 + 2ad\cos\varphi}{2bc} \tag{5-9}$$

The maximum value and minimum value of the transmission angle can be found by equation (5-9); when $\varphi = 0$, the transmission angle is minimum, shown $\angle B_2C_2D$. Another position of minimum transmission angle may be $\varphi = 180°$, shown $\angle B_1C_1D$ or its supplementary angle.

The transmission angle in a slider-crank linkage is shown in Fig. 5-16, and from this figure, we get:

$$\cos\gamma = \frac{a\sin\varphi + e}{b} \tag{5-10}$$

The maximum value and minimum value of the transmission angle can be found easily.

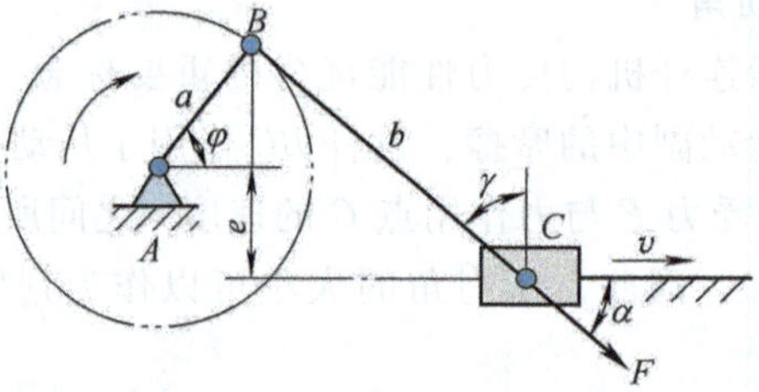


Fig. 5-16 Transmission angle in a slider-crank linkage (偏置曲柄滑块机构的传动角)

#### 4. Dead Point

Fig. 5-17 shows a crank-rocker linkage. When the crank rotates completely, there is no danger of the linkage locking, because the transmission angle can not be zero at any positions. However, if the rocker becomes a driving link, the crank is a driven link, and then the rocker moves to the extreme position $DC_1$ and $DC_2$, the transmission angles become zero, the linkage locking will occur. We call these positions as dead points or toggle positions. The rocker will then have to be driven to get the linkage out of the toggle position.

To avoid dead points, it is necessary to provide a flywheel to pass through these dangerous positions.

Fig. 5-18 shows a single-cylinder engine with four-stroke cycle. The flywheel can carry the piston through the engine's exhaust stroke, intake stroke and compression stroke. During this time, work must be done on the system.

作，应使一个运动循环中最小传动角 $\gamma_{min}>40°$，具体数值可根据传递功率的大小而定。传递功率大时，$\gamma_{min}$应取大些，如颚式破碎机、压力机等可取 $\gamma_{min}\geqslant 50°$。

铰链四杆机构的最小传动角可按以下关系求得。在△ABD 和△BCD 中分别有

$$\overline{BD}^2=a^2+d^2-2ad\cos\varphi$$

$$\overline{BD}^2=b^2+c^2-2bc\cos\gamma$$

联立两式求解，得

$$\cos\gamma=\frac{b^2+c^2-a^2-d^2+2ad\cos\varphi}{2bc} \tag{5-9}$$

由式（5-9）可知，$\gamma$ 仅取决于曲柄的转角 $\varphi$。当 $\varphi=0°$时，$\cos\varphi=1$，$\cos\gamma$ 为最大，$\gamma$ 最小。如图5-15所示位置 $AB_2C_2D$；当 $\varphi=180°$时，$\cos\varphi=-1$，$\cos\gamma$ 为最小，$\gamma$ 最大，如图 5-15 所示位置 $AB_1C_1D$。当 $\gamma$ 大于 90°时，取其补角为传动角即可。只要比较这两个位置的值，即可求得该机构的最小传动角 $\gamma_{min}$。

机构的最小传动角 $\gamma_{min}$ 可能发生在曲柄与机架两次共线位置之一处。进行连杆机构设计时，必须检验最小传动角是否满足要求。

偏置曲柄滑块机构的传动角如图 5-16 所示。最小传动角可用式（5-10）求出。

$$\cos\gamma=\frac{a\sin\varphi+e}{b} \tag{5-10}$$

**4. 机构的死点位置**

图 5-17 所示的曲柄摇杆机构中，若摇杆 CD 为主动件，则当摇杆在两极限位置 $C_1D$、$C_2D$ 时，连杆 BC 与从动曲柄 AB 将两次共线，出现 $\gamma=0°$的情况。该作用力对 A 点的力矩为零，故曲柄 AB 不会转动。机构的该位置称为死点位置。

就传动机构来说，存在死点是不利的，必须采取措施使机构能顺利通过死点位置。克服机构死点的常用方法有：

1）利用构件的惯性来通过死点位置。

2）利用机构的错位排列通过死点位置。

图 5-18 所示的单缸四冲程内燃机就是借助于飞轮的惯性通过曲柄滑块机构的死点位置。

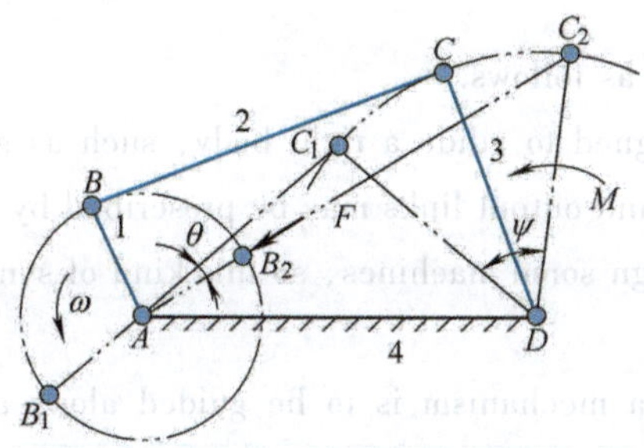


Fig. 5-17 Dead points（死点）

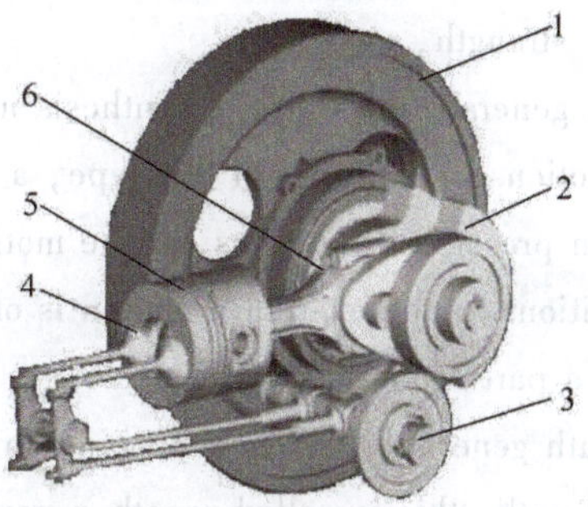


Fig. 5-18 Overcome dead points by using flywheel（利用飞轮克服死点）

1—flywheel（飞轮） 2—crank shaft（曲轴）
3—cam shaft（凸轮轴） 4—valve（气门）
5—piston（活塞） 6—connect link（连杆）

Another method to overcome the dead points is that two sets of mechanism are installed in a machine and one of them has different phase. Fig. 5-19 is a driving mechanism of the locomotive wheels; the phase angle of the cranks is 90°. If one of them is at the dead position, the other can be driven easily.

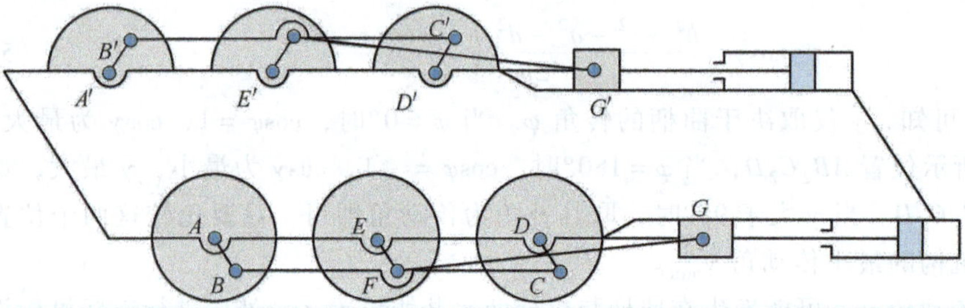


Fig. 5-19 Overcome dead points by using different phase (错位排列克服死点)

Sometimes, the dead point position is very useful for us, such as we can make use of the dead point of a mechanism to design some self-locking devices.

Fig. 5-20 shows a clamping device, in which after a small force $F$ applied on the coupler has been removed, the reaction force of the clamped workpiece to the link $AB$ can not drive the linkage to move at all. We can also design many devices using the dead point position.

## 5.3 Synthesis of Four-Bar Linkages

### 1. Introduction

Synthesis of a four-bar linkage necessitates determining the principal dimensions of various links that satisfy the required motion of the mechanism, and it is not used to deal with the structures, materials, strength, etc.

In general, the types of synthesis may be classified as follows.

Motion generation: In this type, a machine is designed to guide a rigid body, such as a coupler, in prescribed positions, or the motion of the input and output links may be prescribed by angular positions. A quick-return motion is often used to design some machines, so this kind of synthesis is also a part of motion synthesis.

Path generation: When a point on the coupler of a mechanism is to be guided along a prescribed path, this is called a path generation problem.

(1) Motion generation  Fig. 5-21a shows a overturning device used in a casting workshop. This device is designed to guide the sandbox which is fixed to the coupler in two positions $B_1C_1$ and $B_2C_2$. Fig. 5-21b shows a speed change device which is controlled by hand. It is designed to actualize the input and output links through the three positions.

图 5-19 所示的机车驱动轮联动机构中，采用机构错位排列，使两组机构的位置相互错开，可使机构顺利通过死点位置。

在工程中，也可利用机构的死点位置来实现一定的工作要求。图 5-20 所示的夹具，就是利用机构死点位置来夹紧工件的。在连杆 $BC$ 的手柄处施以压力 $F$ 之后，连杆 $BC$ 与连架杆 $CD$ 成一直线。撤去外力 $F$ 之后，在工件反弹力作用下，连架杆 $CD$ 处于死点位置。即使此反弹力很大，也不会使工件松脱。

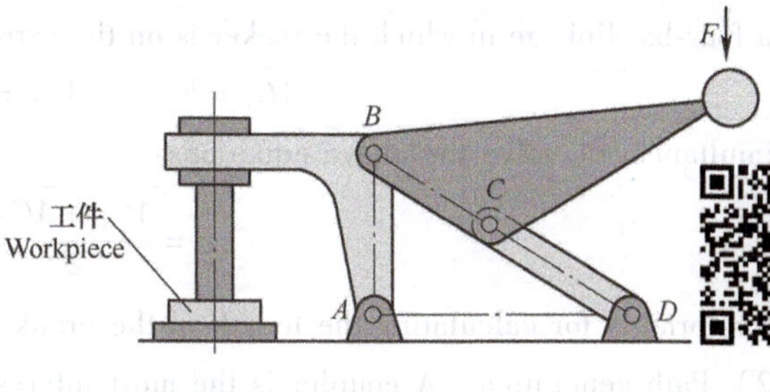


Fig. 5-20 Self-locking clamp（自锁夹具）

## 5.3 平面连杆机构的设计

### 1. 概述

平面连杆机构的设计是指找出机构运动简图的尺寸，不涉及构件的强度、材料、结构、工艺、公差、热处理以及运动副的具体结构等问题，这种设计又称为综合。

本书仅讨论平面四杆机构的设计。

平面四杆机构的设计可分为两大类：其一是按照给定的运动规律设计四杆机构；其二是按照给定的运动轨迹设计四杆机构。

（1）实现给定的运动规律 按照连杆的一系列位置设计四杆机构、按照连架杆的一系列位置设计四杆机构和按照行程速度变化系数设计四杆机构，是实现机构运动规律的基本途径。

图 5-21a 所示的铸造车间翻转台，是按照连杆的一系列位置设计四杆机构的示例。该机构是按照平台的两个位置 $B_1C_1$ 和 $B_2C_2$ 设计的。图 5-21b 所示车床变速机构是按照主动件和从动件的转角位置 $\varphi$、$\psi$ 之间的对应关系设计的。变速手柄位于 1、2、3 位置，换档齿轮位于 1、2、3 档。主动件和从动件的对应转角位置能实现一定的对应关系。

按照行程速度变化系数设计四杆机构时，实际上是按照连架杆的两个极限位置 $DC_1$、$DC_2$，摆角 $\psi$ 以及反映机构急回特性的极位夹角 $\theta$ 来设计四杆机构的。

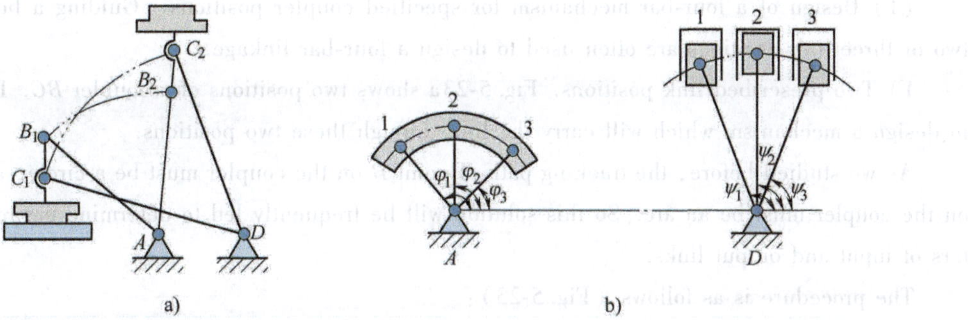


Fig. 5-21 Problems of synthesis of a four-bar linkage 1（四杆机构设计基本问题 1）

When we design a mechanism which has a quick-return motion, the extreme positions of the rocker and the angle between the two extreme positions must be determined by designers. Fig. 5-22a shows a four-bar linkage in which the rocker is on the extreme positions, and from this figure we have:

$$\overline{AC_1} = b + a,\ \overline{AC_2} = b - a$$

Simultaneously solve the above equations:

$$a = \frac{\overline{AC_1} - \overline{AC_2}}{2} \tag{5-11}$$

This formula for calculating the length of the crank is very useful to design a four-bar linkage.

(2) Path generation A coupler is the most interesting link in a mechanism. It is in complex motion, and each point on the coupler can have a path of high degree. The coupler can be extended infinitely in the plane, and there is, of course, an infinity of points on the coupler, each of which generate a different curve. Therefore, coupler curves can be used to generate quite useful path motions for machine design. Fig. 5-22b shows some coupler curves on the extended link plane in a four-bar linkage.

The synthesis of mechanism may be done by graphical method or by analytical method. The simplest and quickest method is graphical method, and it is usually necessary to use a computer when we use the analytical method. In this chapter, both graphical method and analytical method to design a four-bar linkage are being discussed.

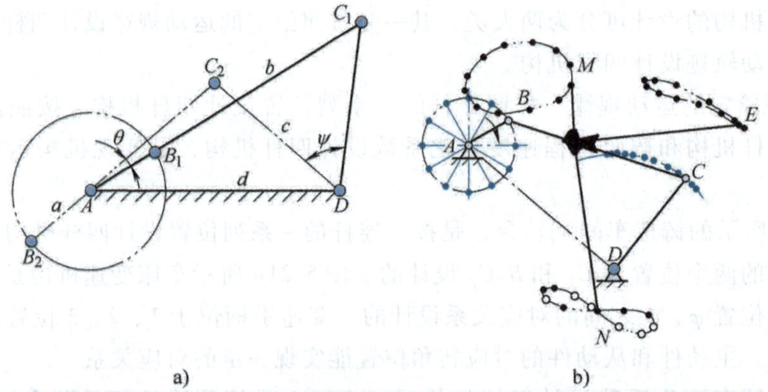


Fig. 5-22 Problems of synthesis of four-bar linkages 2 (四杆机构设计基本问题 2)

**2. Graphical Synthesis of Four-Bar Linkage**

(1) Design of a four-bar mechanism for specified coupler positions Guiding a body through two or three link positions are often used to design a four-bar linkage.

1) Two prescribed link positions. Fig. 5-23a shows two positions of a coupler $BC$. It is desired to design a mechanism which will carry the link through these two positions.

As we studied before, the tracking path of point $B$ on the coupler must be a circle, and point $C$ on the coupler must be an arc. So this solution will be frequently led to determine the rotating centers of input and output links.

The procedure is as follows (Fig. 5-23):

Draw the coupler $BC$ in its two desired positions $B_1C_1$ and $B_2C_2$ in the plane.

Construct the perpendicular bisectors $b_{12}$ of $B_1B_2$ and $c_{12}$ of $C_1C_2$, and then select any conven-

图 5-22a 中，设曲柄、连杆、摇杆和机架尺寸分别为 $a$、$b$、$c$、$d$，则有：$\overline{AC_1}=b+a$，$\overline{AC_2}=b-a$，联立求解得

$$a=\frac{\overline{AC_1}-\overline{AC_2}}{2} \tag{5-11}$$

求出曲柄长度后，其余尺寸可直接在图上用图解法求解。

(2) 实现给定的运动轨迹　连杆上各点能描绘出各种各样的高次曲线。图 5-22b 所示机构中，连杆上不同点描绘出不同曲线，称为连杆曲线。寻求能再现这些点位置的连杆机构则是实现按给定运动轨迹设计四杆机构的基本任务。

机构综合方法有图解法和解析法。图解法简单易行，但精度低、费时。解析法的设计精度高，但需要编制程序在计算机上运行。近年来解析法的应用越来越广泛，但是在机构尺寸的初步设计阶段，图解法也有其独特的优点。

**2. 图解设计法**

(1) 按照连杆的一系列位置设计四杆机构　通常情况下，会给定连杆的两个或三个位置，要求设计四杆机构。根据四杆机构的性质，连杆两端铰链点的运动轨迹均为圆曲线，所以设计的关键问题是分别找出两个连架杆的转动中心。

1）按照连杆的两个位置设计四杆机构。如图 5-23a 所示，设已知连杆 $BC$ 的长度和预定占据的两个位置 $B_1C_1$、$B_2C_2$，设计此四杆机构。

连杆上铰链中心 $B$ 点、$C$ 点的轨迹都是圆弧，两固定铰链的中心必分别位于 $B_1B_2$ 和 $C_1C_2$ 的垂直平分线 $b_{12}$ 和 $c_{12}$ 上。两固定铰链 $A$、$D$ 可分别在 $b_{12}$、$c_{12}$ 上适当选取，故此类四杆机构设计有无数解。此时，可根据其他要求确定 $A$、$D$ 的位置，然后连接 $AB_1$ 及 $C_1D$，如图 5-23b 所示四杆机构 $AB_1C_1D$ 是对应连杆位置 $B_1C_1$ 的机构运动简图。

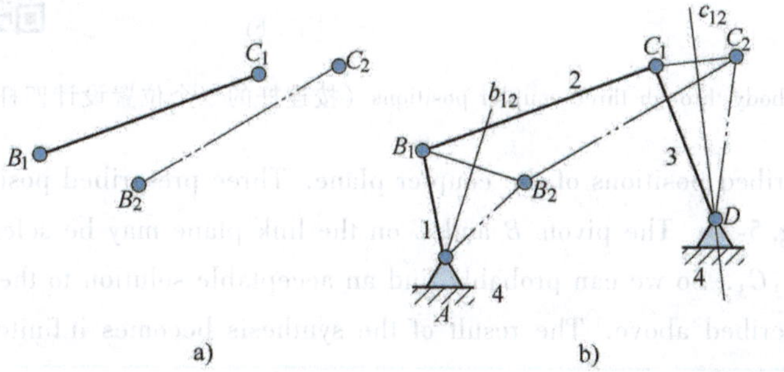


Fig. 5-23　Guiding a body through two coupler positions（按照连杆的两个位置设计四杆机构）

2）按照连杆的三个对应位置设计铰链四杆机构。如图 5-24a 所示，若要求连杆占据预定的三个位置 $B_1C_1$、$B_2C_2$、$B_3C_3$，则可用上述方法分别作出 $B_1B_2$ 和 $B_1B_3$ 的垂直平分线 $b_{12}$ 和 $b_{13}$，其交点即为转动副 $A$ 的位置；同理，分别作 $C_1C_2$ 和 $C_1C_3$ 的垂直平分线 $c_{12}$ 和 $c_{13}$，其交点即为转动副 $D$ 的位置。连接 $AB_1$ 及 $C_1D$，即得所求的四杆机构在位置 1 的简图，作图过程如图 5-24b 所示。

给定连杆的四个位置，可能有解，也可能无解，这取决于能否找到连架杆的转动中心。

3）按照连杆平面位置设计四杆机构。若给定图 5-25 所示的连杆平面的两个或三个位

ient points on each bisector as the fixed rotating centers $A$ and $D$.

Connect $A$ with $B_1$ or $B_2$, and call it link 1; connect $D$ with $C_1$ or $C_2$, and call it link 3.

The line $AD$ is the link 4, the frame.

Check the Grashof condition and the minimum transmission angle. If it doesn't satisfy the condition, repeat the above steps.

Calculate dimensions of all the links.

2) Three prescribed link positions. The procedure is similar to the above processes (Fig. 5-24). Draw the coupler $BC$ in its three desired positions $B_1C_1$, $B_2C_2$ and $B_3C_3$.

Construct the perpendicular bisectors $b_{12}$ of $B_1B_2$ and $b_{13}$ of $B_1B_3$; the intersecting point of $b_{12}$ and $b_{13}$ is the rotating center $A$. In the same way, we can get the point $D$.

Connect $A$ with $B_1$ or $B_2$ and $B_3$, and call it link 1; connect $D$ with $C_1$ or $C_2$ and $C_3$, and call it link 3.

The line $AD$ is the link 4, the frame.

Check the Grashof condition and the minimum transmission angle. If it doesn't satisfy the condition, modify the given conditions and repeat the above steps.

Calculate the dimensions of all the links.

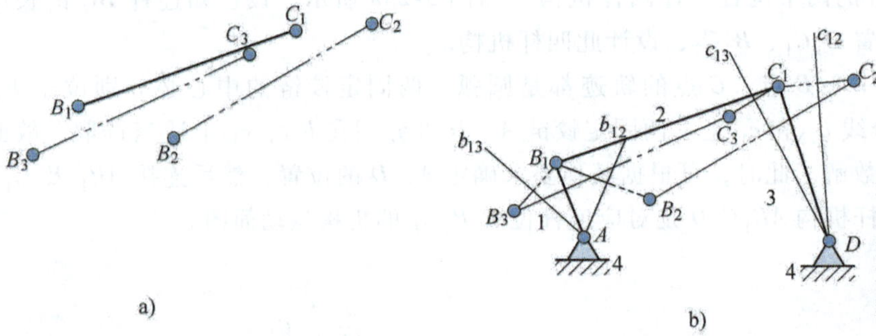


Fig. 5-24 Guiding a body through three coupler positions (按连杆的三个位置设计四杆机构)

3) Two or three prescribed positions of the coupler plane. Three prescribed positions of a coupler plane are shown in Fig. 5-25. The pivots $B$ and $C$ on the link plane may be selected arbitrary, such as $B_1C_1$, $B_2C_2$ and $B_3C_3$. So we can probably find an acceptable solution to the three-position problem by the method described above. The result of the synthesis becomes infinite, because the pivots $B$ and $C$ are varied selects.

(2) Design of a four-bar linkage for correlated angular positions of input and output links A common requirement of setting a mechanism is to coordinate the angular positions of a driven link with those of some driving link. For example, usually the angular positions of input and output links are to be coordinated, such as value $\varphi$ corresponding to value $\psi$, and the length of input link and the frame are known. The problem of this kind of synthesis can be solved by the inversion method which is based upon the notation of relative motion.

1) Inversion method. Fig. 5-26 shows a four-bar linkage; the motion of the input link $AB$ from the $\varphi_1$ to $\varphi_2$ causes a motion of the output link $DC$ from $\psi_1$ to $\psi_2$. Let us hold $DC_1$ stationary and hold the four-bar mechanism $AB_2C_2D$ to be a structure, then rotate it about the pivot $D$ through an

置，可在连杆平面中假设出 $BC$ 位置，按上述方法求解即可。由于每假设一组 $BC$ 就对应一组解，故此时有无穷多解。

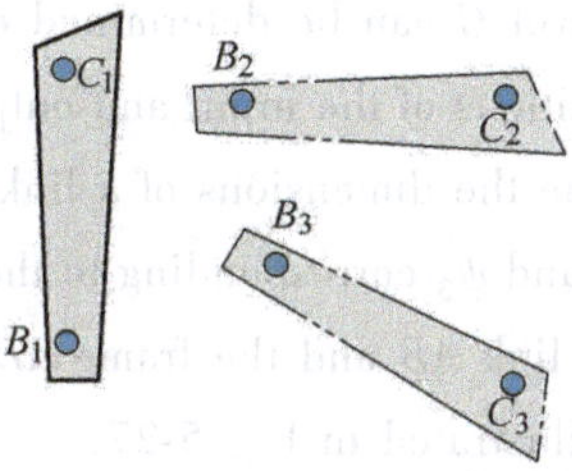


Fig. 5-25 Guiding a body through a number of link plane positions
（按照连杆平面位置设计四杆机构）

（2）按照连架杆的一系列对应位置设计四杆机构 通常情况下，会给定连架杆的两组或三组对应位置，而且机架和其中一个连架杆的尺寸是给定的，设计的关键问题是找出另一个连架杆与连杆的铰链点的位置。该类问题可转化为按照连杆的一系列位置设计四杆机构。

1）反转法的原理。在图5-26中，给出了四杆机构的两个位置 $AB_1C_1D$、$AB_2C_2D$，两连架杆的对应转角分别为 $\varphi_1$、$\varphi_2$ 和 $\psi_1$、$\psi_2$。设想将机构 $AB_2C_2D$ 整体刚化，并绕轴心 $D$ 转过 $(\psi_2-\psi_1)$ 角。构件 $DC_2$ 与 $DC_1$ 重合，$AB_2$ 运动到了 $A'B_2'$ 位置。经过这样的转换，可以认为此机构已转换为以 $C_1D$ 为机架，以 $AB_1$、$A'B_2'$ 为连杆位置的设计问题。

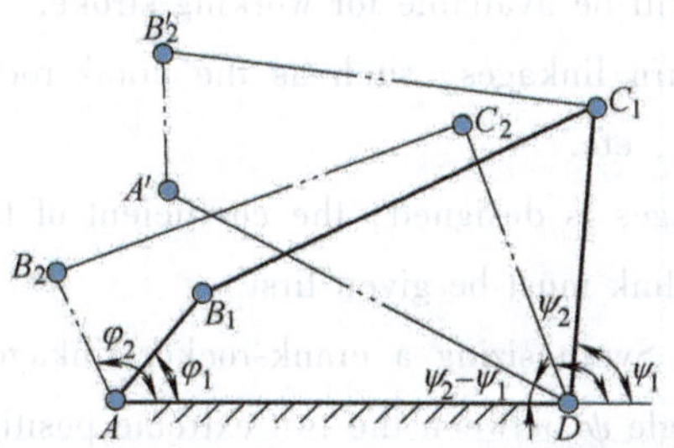


Fig. 5-26 Principle of inversion（反转法的原理）

2）按照两连架杆的三组对应位置设计四杆机构。如图5-27a所示，已知构件 $AB$ 和机架 $AD$ 的长度，要求在该机构的传动过程中，构件 $AB$ 和构件 $CD$ 上某一标线 $DE$ 能占据三组预定的对应位置 $AB_1$、$AB_2$、$AB_3$ 及 $DE_1$、$DE_2$、$DE_3$，三组对应位置的对应角度为 $\varphi_1$、$\varphi_2$、$\varphi_3$ 和 $\psi_1$、$\psi_2$、$\psi_3$。设计此四杆机构。

此类设计问题可以转化为以构件 $CD$ 为机架，以构件 $AB$ 为连杆的设计问题。设计过程如下（图5-27b）：

① 选适当的比例尺画出机构的三组对应位置。

② 以 $D$ 为圆心，任选半径画弧，交三个方向线于 $E_1$、$E_2$、$E_2$ 点。

angle $\psi_{12}=\psi_1-\psi_2$ clockwise, then we obtain the position $A'B_2'C_1D$ shown in Fig. 5-26. Therefore, the input link $AB$ is an assumed coupler and link $DC_1$ is an assumed frame. This problem becomes that of synthesizing a four-bar linkage for coupler position. Two positions of the coupler are $AB_1$ and $A'B_2$, and the frame is $DC_1$. The pivot $C$ can be determined easily.

2) Three prescribed angular positions of the input and output links. Fig. 5-27 illustrates a problem in which it is desired to determine the dimensions of a linkage in which the output link is to occupy three angular positions $\psi_1$, $\psi_2$ and $\psi_3$ corresponding to the three given positions $\varphi_1$, $\varphi_2$ and $\varphi_3$ of the input link. The lengths of the link $AB$ and the frame $AD$ are given by a designer.

The solution to the problem is illustrated in Fig. 5-27.

Draw the input link $AB$ in its three specified angular positions $\varphi_1$, $\varphi_2$ and $\varphi_3$ and locate a desired position for pivot $D$. Draw three rays $DE_1$, $DE_2$ and $DE_3$ with specified angles $\psi_1$, $\psi_2$ and $\psi_3$.

Draw a circular arc of convenient radius $DE$ about the pivot $D$, and it intersects the three rays at points $E_1$, $E_2$ and $E_3$.

Construct the three quadrilaterals $AB_1E_1D$, $AB_2E_2D$, $AB_3E_3D$, and suppose that they are structures. Rotate quadrilaterals $AB_2E_2D$, $AB_3E_3D$ about pivot $D$ through an angles ($\psi_1-\psi_2$), ($\psi_1-\psi_3$) counterclockwise respectively. The lines $DE_2$ and $DE_3$ are coincident with line $DE_1$ at this time. The three positions of the coupler are $AB_1$, $A_2B_2'$ and $A_3B'_3$. The $DE_1$ is the assumed frame.

Draw the perpendicular bisectors $b_{12'}$ of $B_1B_2'$ and $b_{2'3'}$ of $B_2'B_2'$. The intersecting point of $b_{12'}$ and $b_{2'3'}$ is the pin center $C_1$ of the driven link.

The complete mechanism in its first position is $AB_1C_1D$.

(3) Design of a four-bar linkage for quick-return motion Some external work is being done by the linkage on the forward stroke, and the return stroke needs to be accomplished as rapidly as possible so that a maximum time will be available for working stroke.

There are some quick-return linkages, such as the crank-rocker linkage, offset slider-crank linkages and guide-bar linkages, etc.

When a quick-return linkages is designed, the coefficient of travel speed variation $K$ and the limiting positions of the driven link must be given first.

1) Crank-rocker linkage. Synthesizing a crank-rocker linkage for specified value of $K$, the length of rocker $DC$ and the angle $\psi$ between the two extreme positions of the rocker are known.

The procedures are in Fig. 5-28.

Calculate the acute angle between the corresponding positions of the crank.

$$\theta=180°\frac{K-1}{K+1}$$

Locate the fixed pivot center $D$ of the rocker; draw the rocker $DC$ in both extreme positions separated by angle $\psi$ as given.

Connect the point $C_1$ with $C_2$; draw the perpendicular line through the point $C_2$ and construct a line through the point $C_1$ at the angle ($90°-\theta$) to $C_1C_2$. The intersection of the two lines is the point $P$.

$$\theta=\angle C_1PC_2$$

Construct a circumcircle of the right triangle $C_1PC_2$; the rotating center $A$ of the crank must be located on the circle which is called the pivot circle.

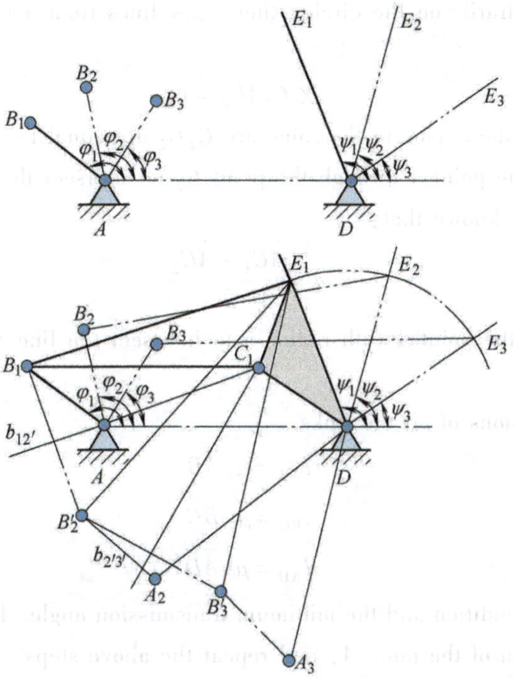


Fig. 5-27　Coordination of the positions of the input and output links

（按照两连架杆的三组对应位置设计四杆机构）

③ 连接四边形 $AB_1E_1D$、$AB_2E_2D$、$AB_3E_3D$，分别反转 $AB_2E_2D$、$AB_3E_3D$，使 $E_2D$、$E_3D$ 与 $E_1D$ 重合。此时，转换为以 $E_1D$ 为机架，以 $AB$、$A_2B'_2$，$A_3B'_3$ 为连杆三个位置的设计问题。

④ 作 $B_1B'_2$、$B'_2B'_3$ 的中垂线，交点 $C_1$ 即为所求，$AB_1C_1D$ 为机构的第一位置。

如果给出两连架杆的两组对应位置，则有无穷多解。如果其中一个连架杆演化为滑块，上述反转方法仍然适用。

（3）按行程速度变化系数设计四杆机构　设计具有急回特性的机构时，通常已知行程速度变化系数 $K$ 和其他条件，设计方法如下。

1）曲柄摇杆机构。已知摇杆的长度 $l_{CD}$、摇杆摆角 $\psi$ 和行程速度变化系数 $K$，设计曲柄摇杆机构。

设计的实质是确定固定铰链中心 $A$ 的位置，求出其他三个构件的尺寸 $l_{AB}$、$l_{BC}$ 和 $l_{AD}$。其设计步骤如下：

① 求解极位夹角 $\theta$，即

$$\theta = 180° \frac{K-1}{K+1}$$

② 任选一固定铰链点 $D$，选取长度比例尺 $\mu_l$，按摇杆长 $l_{CD}$ 和摆角 $\psi$ 作出摇杆的两个极限位置 $C_1D$ 和 $C_2D$，如图 5-28 所示。

③ 连接 $C_1C_2$，自 $C_2$ 作 $C_1C_2$ 的垂直线 $C_2P$。

④ 作 $\angle C_2C_1P = 90° - \theta$，直角三角形 $C_1PC_2$ 中，$\angle C_1PC_2 = \theta$。

⑤ 作直角三角形 $C_1PC_2$ 的外接圆，在圆弧 $NC_1$ 或 $MC_2$ 上任选一点 $A$ 作为曲柄 $AB$ 的转

Select a point $A$ arbitrarily on the circle, then draw lines from $A$ to $C_1$ and from $A$ to $C_2$, we have:

$$\angle C_1AC_2 = \theta$$

This is because that the angles in the same arc $C_1C_2$ are equal to each other.

Draw an arc about the point $A$ through the point $C_2$ to intersect the line $AC_1$ at point $E$. From previous section, we have known that:

$$a = \frac{\overline{AC_1} - \overline{AC_2}}{2}$$

Draw a circle about the point $A$ with radius a to intersect the line $AC_1$ at the point $B_1$ and the line $AC_2$ at point $B_2$.

Calculate the dimensions of all the links.

$$l_{AB} = \mu_l \overline{AB}$$

$$l_{BC} = \mu_l \overline{BC}$$

$$l_{AD} = \mu_l \overline{AD}$$

Check the Grashof condition and the minimum transmission angle. If it does not satisfy the condition, change the location of the point $A$, and repeat the above steps.

2) Slider-crank linkage. The specified values are the coefficient of travel speed variation of $K$, the stroke $s$ of the slider, and offset $e$.

The procedures are shown in Fig. 5-29.

Calculate the acute angle between the corresponding positions of the crank.

$$\theta = 180° \frac{K-1}{K+1}$$

Connect the point $C_1$ with $C_2$; draw the perpendicular line through the point $C_2$ and construct a line through the point $C_1$ at the angle $(90° - \theta)$ to $C_1C_2$; the intersection of the two lines is the point $P$.

$$\theta = \angle C_1PC_2$$

Construct a circumcircle of the right triangle $C_1PC_2$, and draw a line which is parallel to the path of the slider at a distance $e$ apart. The intersections with the pivot circle are fixed pivots of the crank. We can choose one of them to be the pivot $A$.

Draw lines from $A$ to $C_1$ and from $A$ to $C_2$, we have:

$$a = \frac{\overline{AC_1} - \overline{AC_2}}{2}$$

Draw a circle about the point $A$ with radius $a$ to intersect the line $AC_1$ at point $B_1$ and the line $AC_2$ at point $B_2$.

Calculate the dimensions of all the links.

$$l_{BC} = \mu_l \overline{BC}$$

3) Guide-bar mechanism. The prescribed values are the coefficient of travel speed variation of $K$, the length of the frame $l_{AC}$.

The procedures are shown in Fig. 5-30.

Calculate the angle $\theta$ as before.

动中心，并分别与 $C_1$、$C_2$ 相连，则 $\angle C_1AC_2 = \angle C_1PC_2 = \theta$。

⑥ 以 $A$ 为圆心，$\overline{AC_2}$ 为半径作圆弧，交 $AC_1$ 直线于 $E$ 点，则 $\overline{EC_1} = 2\,\overline{AB}$。然后，再以 $A$ 为圆心，以 $\overline{EC_1}/2$ 为半径作圆交 $C_1A$ 于 $B_1$ 点，交 $C_2A$ 延长线于 $B_2$ 点，则 $\overline{AB_1} = \overline{AB_2} = \overline{AB}$ 即为曲柄长，$\overline{B_1C_1} = \overline{B_2C_2} = \overline{BC}$ 为连杆长，$\overline{AD}$ 为机架。

故曲柄、连杆和机架的实际长度分别为

$$l_{AB} = \mu_l\,\overline{AB}$$

$$l_{BC} = \mu_l\,\overline{BC}$$

$$l_{AD} = \mu_l\,\overline{AD}$$

$A$ 点位置不同，机构传动角大小也不同。为了获得较好的传力性能，可按最小传动角或其他辅助条件来确定 $A$ 点位置。当 $K=1$ 时，$A$ 点在 $C_1C_2$ 的延长线上选取。

2）曲柄滑块机构。已知曲柄滑块机构的行程速度变化系数 $K$、行程 $H$ 和偏距 $e$，设计该曲柄滑块机构。

① 根据行程速度变化系数 $K$，计算出极位夹角 $\theta$。

② 如图 5-29 所示，作一直线 $C_1C_2$，$\overline{C_1C_2} = H$，由点 $C_2$ 作 $C_1C_2$ 的垂线 $C_2P$，再由点 $C_1$ 作一直线 $C_1P$ 与 $C_2C_1$ 成（$90° - \theta$）的夹角。

③ 作直角三角形 $C_1PC_2$ 的外接圆，然后作 $C_1C_2$ 的平行线，距离等于偏距 $e$，此直线与圆的交点即为曲柄 $AB$ 转动中心 $A$。连接 $AC_1$、$AC_2$，则 $\angle C_1AC_2 = \angle C_1PC_2 = \theta$。

④ $A$ 点确定后，根据机构在极限位置时曲柄与连杆共线的特点，即可求出曲柄的长度及连杆的长度。

如 $K=1$，$A$ 点在 $C_1C_2$ 的延长线上选取，该机构为对心曲柄滑块机构。

3）导杆机构。已知摆动导杆机构中机架的长度 $l_{AC}$，行程速度变化系数 $K$，设计该导杆机构。

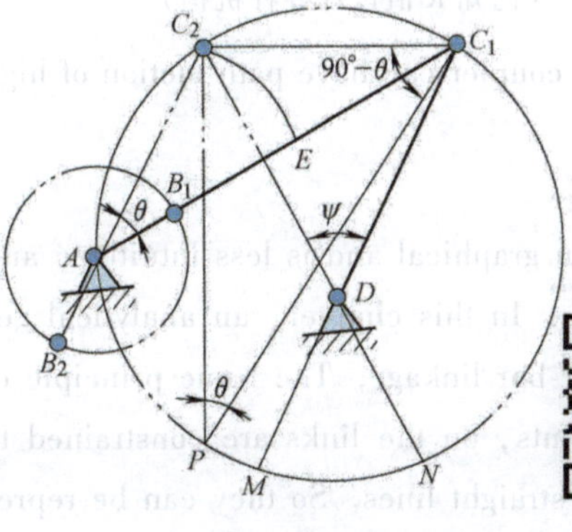


Fig. 5-28　Synthesis of a crank-rocker linkage for a given $K$（已知 $K$ 设计曲柄摇杆机构）

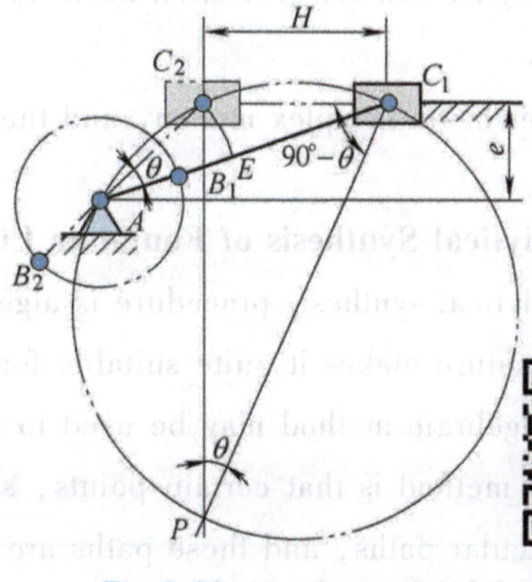


Fig. 5-29　Synthesis of a slider-crank linkage for a given $K$（已知 $K$ 值设计曲柄滑块机构）

由图 5-30 可知，导杆机构的极位夹角 $\theta$ 等于导杆的摆角 $\psi$，所需确定的尺寸是曲柄长度 $l_{AB}$。其设计步骤如下：

① 由行程速度变化系数 $K$ 计算极位夹角 $\theta$。

$$\theta = 180°\frac{K-1}{K+1}$$

Locate point $C$ arbitrarily and construct two positions $Cm$ and $Cn$ of the guide-bar separated by the angle $\theta$ shown in Fig. 5-30.

Bisect the angle $\angle mCn$ and locate the fixed pivot $A$, with $AC = l_{AC}$. Notice that $\angle mCn = \theta$.

Draw two lines from the point $A$, perpendicular to the lines $Cm$ and $Cn$ respectively; the intersections are $B_1$ and $B_2$ which are joins of the crank and the slider.

The length of the crank is as:

$$a = \mu_l l_{AB}$$

(4) Design of a four-bar linkage for specified tracing path Coupler curves can be capable of approximating straight lines and large circle arcs with remote centers, so they are quite useful path motions for machine design.

Several methods are given to produce a specified coupler path, such as coupler-point curve atlas method and experimental method.

There are 7300 coupler curves in the book written by John A. Hrones and George L. Nelson in 1951. The designer can select the dimension of the four-bar linkage from the atlas.

Fig. 5-31 shows a four-bar linkage, in which the point $F$ on the coupler can generate a close path. If the equation of the position at point $F$ can be changed, we may obtain a lot of coupler curves in which the designer can select desired curve to design a mechanism.

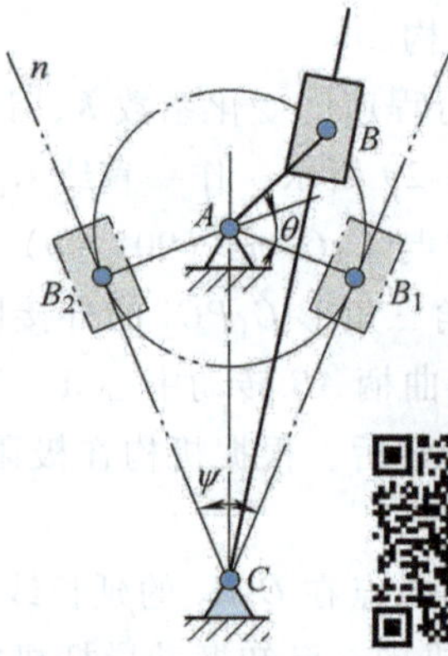

Fig. 5-30 Synthesis of a guide-bar linkage for a given $K$
(已知 $K$ 值设计导杆机构)

A coupler has a complex motion, and these points on the coupler can have path motion of high degree.

### 3. Analytical Synthesis of Four-bar Linkage

The analytical synthesis procedure is algebraic rather than graphical and is less intuitive, and its algebraic nature makes it quite suitable for computerization. In this chapter, an analytical geometry and algebraic method may be used to synthesizing four-bar linkage. The basic principle of the constraint method is that certain points, such as pivot points, on the links are constrained to move in particular paths, and these paths are often circles or straight lines. So they can be represented by simple algebraic equations, and this constraint method can be extended to more complex linkages.

(1) Design of a four-bar linkage for specified coupler positions The length of the coupler and coupler positions are given. Fig. 5-32 shows three positions of a coupler $BC$ in which the coordinates of the point $B$ and $C$ are shown in Fig. 5-32. It is desired to design a mechanism which will carry the link through these positions by using analytical method.

$$\theta = 180° \frac{K-1}{K+1}$$

② 选取适当的长度比例尺 $\mu_l$，任选固定铰链点 $C$，以夹角 $\psi$ 作出导杆两极限位置 $Cm$ 和 $Cn$。

③ 作摆角 $\psi$ 的平分线 $AC$，并在线上取 $\overline{AC} = l_{AC}/\mu_l$，得固定铰链点 $A$ 的位置。

④ 过 $A$ 点作导杆极限位置的垂线 $AB_1$（或 $AB_2$），即得曲柄长度 $l_{AB} = \mu_l \overline{AB_1}$。

（4）按照连杆曲线设计四杆机构　按照连杆曲线设计四杆机构时，经常采用图谱法。参照图谱中的曲线，可直接查阅出对应的连杆机构。利用连杆曲线方程绘制连杆曲线图谱也很方便。图 5-31 所示为连杆上 $F$ 点的运动轨迹。改变连杆上 $BE$ 与 $EF$ 的尺寸，可以生成许多不同的连杆曲线，再对照连杆曲线选择相关的连杆机构。

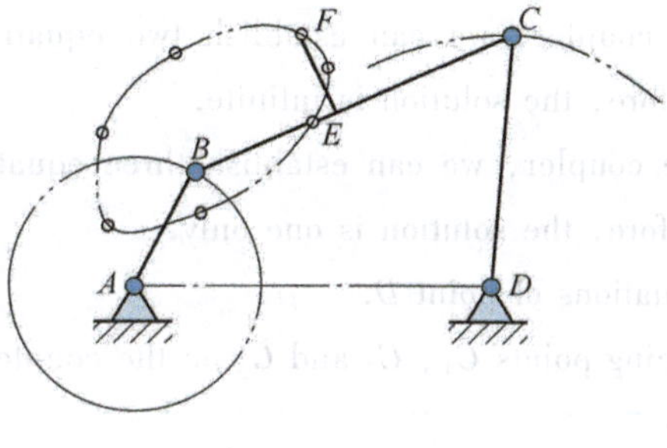


Fig. 5-31　Coupler curve（连杆曲线）

**3. 解析设计法**

用数学方法进行机构的尺寸综合，称为解析法。本书采用简单易学的几何代数法来进行机构的尺寸综合。

（1）按照连杆的对应位置设计四杆机构　给定连杆长度和连杆的一系列位置。通过设定连杆铰链点的坐标值给定连杆的两组或三组对应位置，求出两端圆心坐标后，可求解各构件的尺寸。图 5-32 给出了连杆对应位置的铰链 $B$、$C$ 点的坐标。设铰链 $B$ 点转动半径为 $a$，铰链 $C$ 点转动半径为 $c$。

由解析几何知识可写出各点的坐标方程。

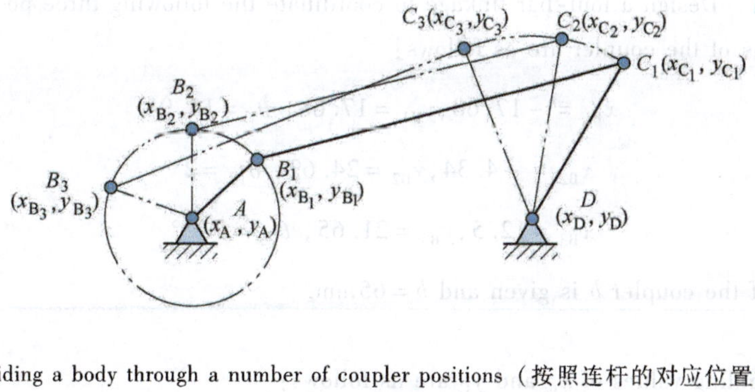


Fig. 5-32　Guiding a body through a number of coupler positions（按照连杆的对应位置设计四杆机构）

How we locate the fixing point $A$ and $D$ is the designing task.

As we know before, the tracing points $B_1$, $B_2$ and $B_3$ on the coupler must be on a circle of radius $a$ about fixing center $A$. We have the following equations:

$$(x_{B_1} - x_A)^2 + (y_{B_1} - y_A)^2 = a^2 \tag{5-12}$$

$$(x_{B_2} - x_A)^2 + (y_{B_2} - y_A)^2 = a^2 \tag{5-13}$$

$$(x_{B_3} - x_A)^2 + (y_{B_3} - y_A)^2 = a^2 \tag{5-14}$$

The general formula is:

$$(x_{B_i} - x_A)^2 + (y_{B_i} - y_A)^2 = a^2 \tag{5-15}$$

For two positions of the coupler, we can establish two equations, and there are three unknowns, $x_A$, $y_A$ and $a$, therefore, the solution is infinite.

For three positions of the coupler, we can establish three equations, and there are three unknowns, $x_A$, $y_A$ and $a$, therefore, the solution is one only.

We can formulate the equations of point $D$.

In the same way, the tracing points $C_1$, $C_2$ and $C_3$ on the coupler must be on a circle of radius $c$ about fixing center $D$, then:

$$(x_{C_1} - x_D)^2 + (y_{C_1} - y_D)^2 = c^2 \tag{5-16}$$

$$(x_{C_2} - x_D)^2 + (y_{C_2} - y_D)^2 = c^2 \tag{5-17}$$

$$(x_{C_3} - x_D)^2 + (y_{C_3} - y_D)^2 = c^2 \tag{5-18}$$

The general formula is

$$(x_{C_i} - x_D)^2 + (y_{C_i} - y_D)^2 = c^2 \tag{5-19}$$

If the value of $c$ is infinite, the linkage becomes a slide-crank mechanism.

The length of the frame $d$ can be calculated from the following equation

$$(x_D - x_A)^2 + (y_D - y_A)^2 = d^2$$

**Example 5-1** Design a four-bar linkage to coordinate the following three positions of the coupler. The positions of the coupler are as follows:

$$x_{B1} = -17.68, y_{B1} = 17.68, \theta_1 = 15.9°$$

$$x_{B2} = -4.34, y_{B2} = 24.62, \theta_{12} = 2°$$

$$x_{B3} = 12.5, y_{B3} = 21.65, \theta_{23} = 7.2°$$

The length of the coupler $b$ is given and $b = 65$mm.

**Solution**

The corresponding values of $x_C$ and $y_C$ are as follows:

铰链 $A$ 点坐标为

$$(x_{B_1}-x_A)^2+(y_{B_1}-y_A)^2=a^2 \tag{5-12}$$

$$(x_{B_2}-x_A)^2+(y_{B_2}-y_A)^2=a^2 \tag{5-13}$$

$$(x_{B_3}-x_A)^2+(y_{B_3}-y_A)^2=a^2 \tag{5-14}$$

写出通式

$$(x_{B_i}-x_A)^2+(y_{B_i}-y_A)^2=a^2 \tag{5-15}$$

给定连杆的两组位置，可列出式（5-12）、式（5-13），未知数有三个：$x_A$、$y_A$、$a$，此时有无数解。

给定连杆的三组位置，可列出式（5-12）、式（5-13）、式（5-14），未知数有三个，此时有唯一解。

同理可求解铰链 $D$ 点坐标为

$$(x_{C_1}-x_D)^2+(y_{C_1}-y_D)^2=c^2 \tag{5-16}$$

$$(x_{C_2}-x_D)^2+(y_{C_2}-y_D)^2=c^2 \tag{5-17}$$

$$(x_{C_3}-x_D)^2+(y_{C_3}-y_D)^2=c^2 \tag{5-18}$$

写出通式

$$(x_{C_i}-x_D)^2+(y_{C_i}-y_D)^2=c^2 \tag{5-19}$$

若 $c$ 值无穷大，则该机构演化为曲柄滑块机构。

机架长度 $d$ 可按下式计算

$$(x_D-x_A)^2+(y_D-y_A)^2=d^2$$

**例 5-1**　按连杆的三组对应位置设计一铰链四杆机构。连杆的长度为 65mm，三组对应位置（图 5-33）如下：

$$x_{B1}=-17.68, y_{B1}=17.68, \theta_1=15.9°$$

$$x_{B2}=-4.34, y_{B2}=24.62, \theta_{12}=2°$$

$$x_{B3}=12.5, y_{B3}=21.65, \theta_{23}=7.2°$$

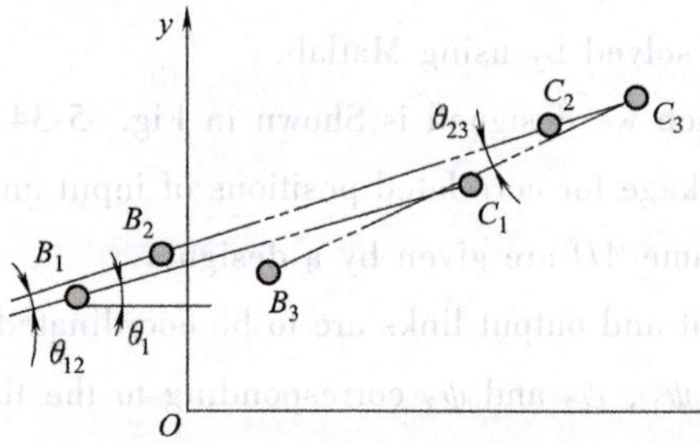


Fig. 5-33　Three positions of the coupler（连杆的三相对应位置）

**解**　根据给定条件，求出连杆在三个给定位置时 $C$ 点的坐标值 $x_C$ 和 $y_C$。

$$x_{C1}=x_{B1}+b\cos\theta_1$$

$$y_{C1}=x_{B1}+b\sin\theta_1$$

$$x_{C2}=x_{B2}+b\cos(\theta_1+\theta_{12})$$

$$y_{C2}=y_{B2}+b\sin(\theta_1+\theta_{12})$$

$$x_{C3}=x_{B3}+b\cos(\theta_1+\theta_{12}+\theta_{23})$$

$$x_{C1}=x_{B1}+b\cos\theta_1$$

$$y_{C1}=x_{B1}+b\sin\theta_1$$

$$x_{C2}=x_{B2}+b\cos(\theta_1+\theta_{12})$$

$$y_{C2}=y_{B2}+b\sin(\theta_1+\theta_{12})$$

$$x_{C3}=x_{B3}+b\cos(\theta_1+\theta_{12}+\theta_{23})$$

$$y_{C3}=y_{B3}+b\sin(\theta_1+\theta_{12}+\theta_{23})$$

The coordinates of Pivot $A$ and Point $B$ on the coupler are given as follows.

$$(x_{B1}-x_A)^2+(y_{B1}-y_A)^2=a^2$$

$$(x_{B2}-x_A)^2+(y_{B2}-y_A)^2=a^2$$

$$(x_{B3}-x_A)^2+(y_{B3}-y_A)^2=a^2$$

There are three unknowns, which are $x_A$, $y_A$, $a$. We can obtain the solution.

As the same, we can establish following three equations for link $CD$:

$$(x_{C1}-x_D)^2+(y_{C1}-y_D)^2=c^2$$

$$(x_{C2}-x_D)^2+(y_{C2}-y_D)^2=c^2$$

$$(x_{C3}-x_D)^2+(y_{C3}-y_D)^2=c^2$$

The length of the frame $d$ can be calculated from the following equation:

$$(x_D-x_A)^2+(y_D-y_A)^2=d^2$$

The above equations can be solved by using Matlab.

The four bar mechanism which we designed is Shown in Fig. 5-34.

(2) Design of a four-bar linkage for correlated positions of input and output links Usually the lengths of the link $AB$ and the frame $AD$ are given by a designer.

The angular positions of input and output links are to be coordinated, such as the output link is to occupy three angular positions $\psi_1$, $\psi_2$ and $\psi_3$ corresponding to the three given positions $\varphi_1$, $\varphi_2$ and $\varphi_3$ of the input link; this is shown in Fig. 5-35.

The constraint equation of point C can be written by using the equal length of the coupler $BC$.

$$(x_{B_i}-x_{C_i})^2+(y_{B_i}-y_{C_i})^2=b^2 \quad i=2,\ 3 \tag{5-20}$$

$$x_{B_i}=a\cos\varphi_i,\ y_{B_i}=a\sin\varphi_i$$

$$x_{C_i}=d+c\cos(\psi_i+\delta),\ y_{C_i}=c\sin(\psi_i+\delta)$$

Rearranging the above equations, we get the following general formula.

$$y_{C3}=y_{B3}+b\sin(\theta_1+\theta_{12}+\theta_{23})$$

根据定杆长条件，求解曲柄和摇杆转动中心 A、D 的坐标。

$$(x_{B1}-x_A)^2+(y_{B1}-y_A)^2=a^2$$
$$(x_{B2}-x_A)^2+(y_{B2}-y_A)^2=a^2$$
$$(x_{B3}-x_A)^2+(y_{B3}-y_A)^2=a^2$$
$$(x_{C1}-x_D)^2+(y_{C1}-y_D)^2=c^2$$
$$(x_{C2}-x_D)^2+(y_{C2}-y_D)^2=c^2$$
$$(x_{C3}-x_D)^2+(y_{C3}-y_D)^2=c^2$$

求解机架长度 $d$。

$$(x_D-x_A)^2+(y_D-y_A)^2=d^2$$

应用 Matlab 求解，结果如下：

$$x_A=0,y_A=0,a=25$$
$$x_D=80,y_D=0,c=50$$
$$d=80$$

所求四杆机构如图 5-34 所示。

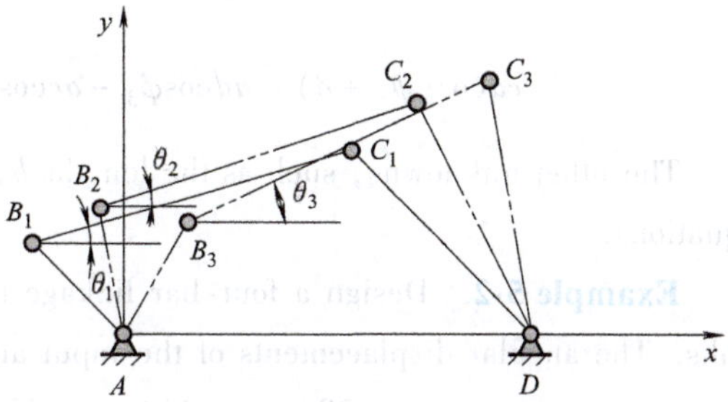


Fig. 5-34 Design of the four bar mechanism for given three coupler positions（按给定的三组对应位置设计铰链四杆机构）

（2）按照连架杆的对应位置设计四杆机构 已知连架杆的对应角位置 $(\varphi_1-\psi_1)$、$(\varphi_2-\psi_2)$、$(\varphi_3-\psi_3)$，其上点的坐标的对应位置也可以写出通式。图 5-35 给出了连架杆的两组和三组对应角位置，关键是求解 C 点坐标。根据连杆满足定杆长的条件，可写出机构在不同位置的通式为

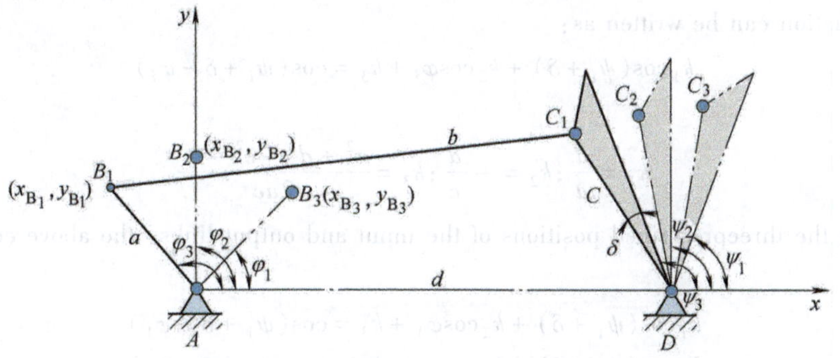


Fig. 5-35 Coordination of the positions of the input and output links（按照连架杆的对应位置设计四杆机构）

$$(x_{B_i}-x_{C_i})^2+(y_{B_i}-y_{C_i})^2=b^2 \qquad i=2,3 \tag{5-20}$$
$$x_{B_i}=a\cos\varphi_i,\ y_{B_i}=a\sin\varphi_i$$
$$x_{C_i}=d+c\cos(\psi_i+\delta),\ y_{C_i}=c\sin(\psi_i+\delta)$$

若 C 点在位置线左侧，则 $\delta$ 取正号；若 C 点在位置线右侧，则 $\delta$ 取负号。$x_{B_i}$、$x_{C_i}$、$y_{B_i}$、$y_{C_i}$分别为 $B_i$、$C_i$ 点的坐标。

整理后得

$$cd\cos(\psi_i+\delta)-ad\cos\varphi_i-ac\cos(\psi_i+\delta-\varphi_i)+\frac{a^2+d^2+c^2-b^2}{2}=0 \tag{5-21}$$

$$cd\cos(\psi_i+\delta)-ad\cos\varphi_i-ac\cos(\psi_i+\delta-\varphi_i)+\frac{a^2+d^2+c^2-b^2}{2}=0 \tag{5-21}$$

There are five unknowns, $a$, $b$, $c$, $d$ and $\delta$, so we can establish five equations, but it is very difficult to solve five equations.

Usually, two of the them, such as the length $a$ of the input and the length $d$ of the frame can be given by designers, then we have:

$$cd\cos(\psi_1+\delta)-ad\cos\varphi_1-ac\cos(\psi_1+\delta-\varphi_1)+\frac{a^2+d^2+c^2-b^2}{2}=0$$

$$cd\cos(\psi_2+\delta)-ad\cos\varphi_2-ac\cos(\psi_2+\delta-\varphi_2)+\frac{a^2+d^2+c^2-b^2}{2}=0$$

$$cd\cos(\psi_3+\delta)-ad\cos\varphi_3-ac\cos(\psi_3+\delta-\varphi_3)+\frac{a^2+d^2+c^2-b^2}{2}=0$$

The other unknowns, such as the lengths $b$, $c$ and the angle $\delta$ can be determined from the three equations.

**Example 5-2** Design a four-bar linkage to coordinate three positions of the input and output links. The angular displacements of the input and the output links are as follows:

$$\varphi_1=20°, \psi_1=35°; \varphi_2=35°, \psi_2=45°; \varphi_3=50°, \psi_3=60°$$

The angular $\delta$ is zero and the length of the frame $d$ is 100mm.

**Solution**

The equation (5-21) is divided throughout by $ac$ and rearranging it, we get the following formula.

$$\frac{d}{a}\cos(\psi_i+\delta)-\frac{d}{c}\cos\varphi_i+\frac{a^2+d^2+c^2-b^2}{2ac}=\cos(\psi_i+\delta-\varphi_i)$$

This equation can be written as:

$$k_1\cos(\psi_i+\delta)+k_2\cos\varphi_i+k_3=\cos(\psi_i+\delta-\varphi_i)$$

Where

$$k_1=\frac{d}{a}; k_2=-\frac{d}{c}; k_3=\frac{a^2+d^2+c^2-b^2}{2ac}$$

If giving the threeprescribed positions of the input and output links, the above equation can be written as:

$$k_1\cos(\psi_1+\delta)+k_2\cos\varphi_1+k_3=\cos(\psi_1+\delta-\varphi_1)$$
$$k_1\cos(\psi_2+\delta)+k_2\cos\varphi_2+k_3=\cos(\psi_2+\delta-\varphi_2)$$
$$k_1\cos(\psi_3+\delta)+k_2\cos\varphi_3+k_3=\cos(\psi_3+\delta-\varphi_3)$$

As we know, $\delta=0°$, the above equations can be written as following formulas:

$$k_1\cos\psi_1+k_2\cos\varphi_1+k_3=\cos(\psi_1-\varphi_1)$$

$$k_1\cos\psi_2+k_2\cos\varphi_2+k_3=\cos(\psi_2-\varphi_2)$$

$$k_1\cos\psi_3+k_2\cos\varphi_3+k_3=\cos(\psi_3-\varphi_3)$$

Knowing$k_1$, $k_2$, $k_3$, the values of a, b, c can be computed from the equation.

$k_1$, $k_2$, $k_3$ can be evaluated by Gaussian eliminated or by the Cramer rule

该方程未知数为 $a$、$b$、$c$、$d$、$\delta$，给定连架杆五组对应位置可列出五个方程求解。但方程过多时，求解较难。工程中，常按三组对应位置设计。此时，假定 $a$、$d$ 已知。

$$cd\cos(\psi_1+\delta)-ad\cos\varphi_1-ac\cos(\psi_1+\delta-\varphi_1)+\frac{a^2+d^2+c^2-b^2}{2}=0$$

$$cd\cos(\psi_2+\delta)-ad\cos\varphi_2-ac\cos(\psi_2+\delta-\varphi_2)+\frac{a^2+d^2+c^2-b^2}{2}=0$$

$$cd\cos(\psi_3+\delta)-ad\cos\varphi_3-ac\cos(\psi_3+\delta-\varphi_3)+\frac{a^2+d^2+c^2-b^2}{2}=0$$

上述三个方程可求解三个未知数。

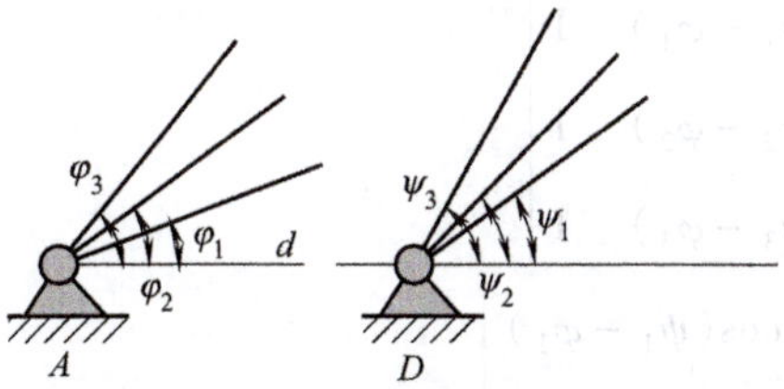


Fig. 5-36 Coordinate of three positions of the input and output links（连架杆的三组对应位置）

**例 5-2** 按给定连架杆的三组对应位置设计四杆机构，已知机架长 $d=100\text{mm}$，连架杆的三组对应角位置（图 5-36）分别为：$\varphi_1=20°$，$\psi_1=35°$；$\varphi_2=35°$，$\psi_2=45°$；$\varphi_3=50°$，$\psi_3=60°$

**解** 式（5-21）两边除以 $ac$，可得

$$\frac{d}{a}\cos(\psi_i+\delta)-\frac{d}{c}\cos\varphi_i+\frac{a^2+d^2+c^2-b^2}{2ac}=\cos(\psi_i+\delta-\varphi_i)$$

该方程可简化为

$$k_1\cos(\psi_i+\delta)+k_2\cos\varphi_i+k_3=\cos(\psi_i+\delta-\varphi_i)$$

其中：

$$k_1=\frac{d}{a},k_2=-\frac{d}{c},k_3=\frac{a^2+d^2+c^2-b^2}{2ac}$$

如果给定连架杆的三个对应位置，则有

$$k_1\cos(\psi_1+\delta)+k_2\cos\varphi_1+k_3=\cos(\psi_1+\delta-\varphi_1)$$
$$k_1\cos(\psi_2+\delta)+k_2\cos\varphi_2+k_3=\cos(\psi_2+\delta-\varphi_2)$$
$$k_1\cos(\psi_3+\delta)+k_2\cos\varphi_3+k_3=\cos(\psi_3+\delta-\varphi_3)$$

如假设 $\delta=0°$，该方程组可简化为

$$k_1\cos\psi_1+k_2\cos\varphi_1+k_3=\cos(\psi_1-\varphi_1)$$
$$k_1\cos\psi_2+k_2\cos\varphi_2+k_3=\cos(\psi_2-\varphi_2)$$
$$k_1\cos\psi_3+k_2\cos\varphi_3+k_3=\cos(\psi_3-\varphi_3)$$

$k_1$、$k_2$、$k_3$ 可用高斯消去法或克莱姆法则计算。

$$k_1=\frac{D_1}{D},k_2=\frac{D_2}{D},k_3=\frac{D_3}{D}$$

其中：

$$D=\begin{vmatrix}\cos\psi_1 & \cos\varphi_1 & 1\\ \cos\psi_2 & \cos\varphi_2 & 1\\ \cos\psi_3 & \cos\varphi_3 & 1\end{vmatrix}$$

$$D_1=\begin{vmatrix}\cos(\psi_1-\varphi_1) & \cos\varphi_1 & 1\\ \cos(\psi_2-\varphi_2) & \cos\varphi_2 & 1\\ \cos(\psi_3-\varphi_3) & \cos\varphi_3 & 1\end{vmatrix}$$

$$D=\begin{vmatrix}\cos\psi_1 & \cos\varphi_1 & 1\\ \cos\psi_2 & \cos\varphi_2 & 1\\ \cos\psi_3 & \cos\varphi_3 & 1\end{vmatrix}$$

$$D_1=\begin{vmatrix}\cos(\psi_1-\varphi_1) & \cos\varphi_1 & 1\\ \cos(\psi_2-\varphi_2) & \cos\varphi_2 & 1\\ \cos(\psi_3-\varphi_3) & \cos\varphi_3 & 1\end{vmatrix}$$

$$D_2=\begin{vmatrix}\cos\psi_1 & \cos(\psi_1-\varphi_1) & 1\\ \cos\psi_2 & \cos(\psi_2-\varphi_2) & 1\\ \cos\psi_3 & \cos(\psi_3-\varphi_3) & 1\end{vmatrix}$$

$$D_3=\begin{vmatrix}\cos\psi_1 & \cos\varphi_1 & \cos(\psi_1-\varphi_1)\\ \cos\psi_2 & \cos\varphi_2 & \cos(\psi_2-\varphi_2)\\ \cos\psi_3 & \cos\varphi_3 & \cos(\psi_3-\varphi_3)\end{vmatrix}$$

$k_1$, $k_2$, $k_3$ are giving by

$$k_1=\frac{D_1}{D},k_2=\frac{D_2}{D},k_3=\frac{D_3}{D}$$

Substituting the knowing values for the angles in the above formulas, we have

$$D=\begin{vmatrix}\cos 35^\circ & \cos 20^\circ & 1\\ \cos 45^\circ & \cos 35^\circ & 1\\ \cos 60^\circ & \cos 50^\circ & 1\end{vmatrix}=-0.005204$$

$$D_1=\begin{vmatrix}\cos(35^\circ-20^\circ) & \cos 20^\circ & 1\\ \cos(45^\circ-35^\circ) & \cos 35^\circ & 1\\ \cos(60^\circ-50^\circ) & \cos 50^\circ & 1\end{vmatrix}=-0.003333$$

$$D_2=\begin{vmatrix}\cos 35^\circ & \cos(35^\circ-20^\circ) & 1\\ \cos 45^\circ & \cos(45^\circ-35^\circ) & 1\\ \cos 60^\circ & \cos(60^\circ-50^\circ) & 1\end{vmatrix}=0.003911$$

$$D_3=\begin{vmatrix}\cos 35^\circ & \cos 20^\circ & \cos(35^\circ-20^\circ)\\ \cos 45^\circ & \cos 35^\circ & \cos(45^\circ-35^\circ)\\ \cos 60^\circ & \cos 50^\circ & \cos(60^\circ-50^\circ)\end{vmatrix}=-0.005974$$

Knowing $k_1$, $k_2$, $k_3$, the values of the link length can be computed as follows.

$d=100\text{mm}$

$$k_1=\frac{D_1}{D}=\frac{-0.003333}{-0.005204}=\frac{d}{a}=\frac{100}{a},a=156\text{mm}$$

$$k_2=\frac{D_2}{D}=\frac{0.003911}{-0.005204}=-\frac{d}{c}=-\frac{100}{c},c=133\text{mm}$$

$$D_2=\begin{vmatrix}\cos\psi_1 & \cos(\psi_1-\varphi_1) & 1\\ \cos\psi_2 & \cos(\psi_2-\varphi_2) & 1\\ \cos\psi_3 & \cos(\psi_3-\varphi_3) & 1\end{vmatrix}$$

$$D_3=\begin{vmatrix}\cos\psi_1 & \cos\varphi_1 & \cos(\psi_1-\varphi_1)\\ \cos\psi_2 & \cos\varphi_2 & \cos(\psi_2-\varphi_2)\\ \cos\psi_3 & \cos\varphi_3 & \cos(\psi_3-\varphi_3)\end{vmatrix}$$

代入对应的角度值，则有

$$D=\begin{vmatrix}\cos 35^\circ & \cos 20^\circ & 1\\ \cos 45^\circ & \cos 35^\circ & 1\\ \cos 60^\circ & \cos 50^\circ & 1\end{vmatrix}=-0.005204$$

$$D_1=\begin{vmatrix}\cos(35^\circ-20^\circ) & \cos 20^\circ & 1\\ \cos(45^\circ-35^\circ) & \cos 35^\circ & 1\\ \cos(60^\circ-50^\circ) & \cos 50^\circ & 1\end{vmatrix}=-0.003333$$

$$D_2=\begin{vmatrix}\cos 35^\circ & \cos(35^\circ-20^\circ) & 1\\ \cos 45^\circ & \cos(45^\circ-35^\circ) & 1\\ \cos 60^\circ & \cos(60^\circ-50^\circ) & 1\end{vmatrix}=0.003911$$

$$D_3=\begin{vmatrix}\cos 35^\circ & \cos 20^\circ & \cos(35^\circ-20^\circ)\\ \cos 45^\circ & \cos 35^\circ & \cos(45^\circ-35^\circ)\\ \cos 60^\circ & \cos 50^\circ & \cos(60^\circ-50^\circ)\end{vmatrix}=-0.005974$$

求出 $k_1$、$k_2$、$k_3$ 的值，即可求解各杆件尺寸。

$d=100\text{mm}$

$$k_1=\frac{D_1}{D}=\frac{-0.003333}{-0.005204}=\frac{d}{a}=\frac{100}{a},a=156\text{mm}$$

$$k_2=\frac{D_2}{D}=\frac{0.003911}{-0.005204}=-\frac{d}{c}=-\frac{100}{c},c=133\text{mm}$$

$$k_3=\frac{D_3}{D}=\frac{-0.005974}{-0.005204}=\frac{a^2+c^2+d^2-b^2}{2ac}=\frac{156^2+133^2+100^2-b^2}{2\times156\times133},b=66\text{mm}$$

机构运动简图如图 5-37 所示。

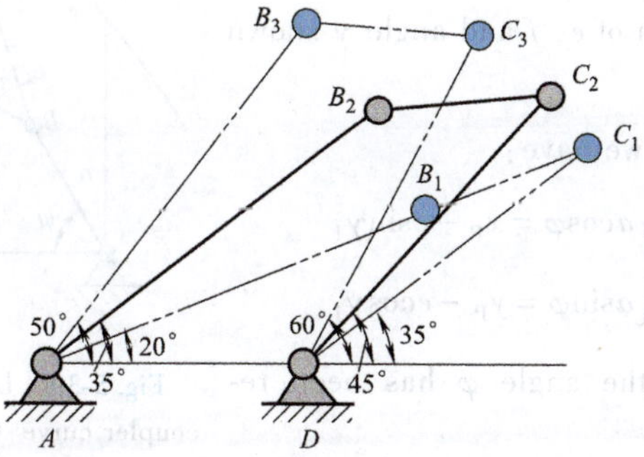


Fig. 5-37　Synthesize a four-bar linkage to coordinate three Positions of the input and output lin（按给定的三组对应位置设计四杆机构）

$$k_3=\frac{D_3}{D}=\frac{-0.005974}{-0.005204}=\frac{a^2+c^2+d^2-b^2}{2ac}=\frac{156^2+133^2+100^2-b^2}{2\times156\times133},b=66\text{mm}$$

The schematic diagram of the required mechanism is shown in Fig. 5-37.

(3) Design of a four-bar linkage for quick-return motion Fig. 5-38 shows a four-bar linkage in the extreme positions.

Where, in the $\triangle DC_1C_2$, we have:

$$\overline{C_1C_2}=2c\sin\frac{\psi}{2}$$

In the $\triangle AC_1C_2$, we have:

$$\overline{C_1C_2}^2=(b+a)^2+(b-a)^2-2(b+a)(b-a)\cos\theta$$

$$\theta=180°\frac{K-1}{K+1}$$

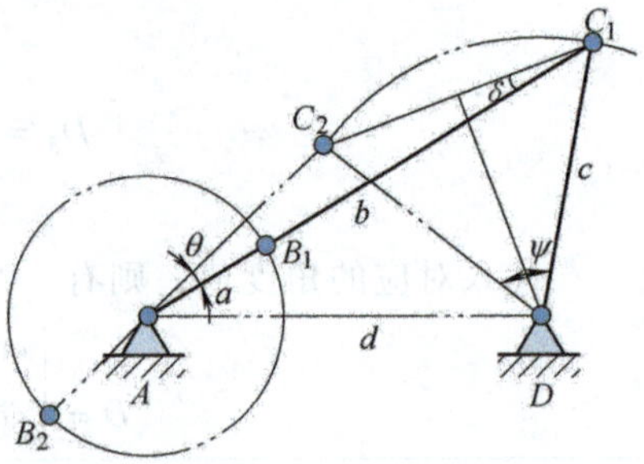

Fig. 5-38 Synthesis of a four-bar linkage for a given $K$（已知 $K$ 设计四杆机构）

Arranging the above equations, we have:

$$\left(2c\sin\frac{\psi}{2}\right)^2=(b+a)^2+(b-a)^2-2(b+a)(b-a)\cos\theta \tag{5-22}$$

Usually, the angle $\psi$ and length $c$ of output are known, and the angle $\theta$ can be calculated from $K$, so there are two unknowns $a$ and $b$. The solutions are infinite. We often select a length of input or coupler in design, then determine the another.

From the triangle $\triangle AC_1C_2$, we can obtain the angle $\delta$ by using the sine law.

$$\frac{\overline{C_1C_2}}{\sin\theta}=\frac{b-a}{\sin\delta},\sin\delta=\frac{(b-a)\sin\theta}{\overline{C_1C_2}}$$

From the triangle $\triangle AC_1D$, we can obtain the length of the frame d by using the cosine theorem.

$$d^2=c^2+(b+a)^2-2c(b+a)\cos\left(90°-\frac{\psi}{2}-\delta\right)=c^2+(b+a)^2-2c(b+a)\sin\left(\frac{\psi}{2}+\delta\right)$$

(4) Design of a four-bar linkage for coupler curve

A four-bar linkage $ABCD$ with a coupler point $P$ is shown in Fig. 5-39. The position of point $P$ on the coupler may be located by length of $e$, $f$ and angle $\gamma$ shown in Fig. 5-39.

From the vectors $ABP$, we have:

$$\begin{cases}x_P=a\cos\varphi+e\sin\gamma_1\\ y_P=a\sin\varphi+e\cos\gamma_1\end{cases}\quad\text{or}\quad\begin{cases}a\cos\varphi=x_P-e\sin\gamma_1\\ a\sin\varphi=y_P-e\cos\gamma_1\end{cases}$$

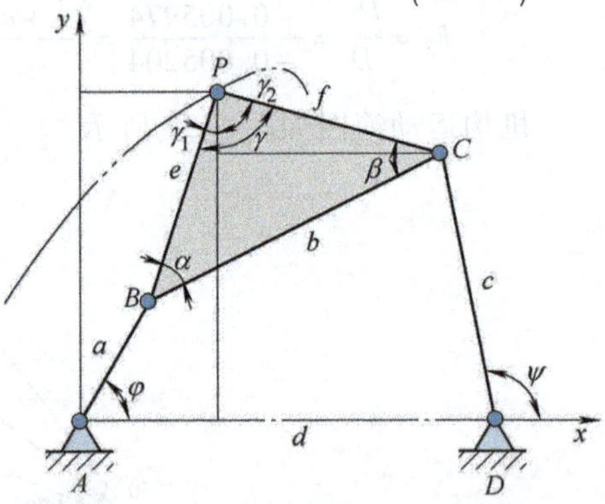

Fig. 5-39 Design of a four-bar linkage for coupler curve（按照连杆曲线设计四杆机构）

Squaring and adding, the angle $\varphi$ has been removed.

$$2e(x_P\sin\gamma_1+y_P\cos\gamma_1)=x_P^2+y_P^2+e^2-a^2$$

(3) 按照行程速度变化系数设计四杆机构　已知条件仍为摇杆长度 $c$、摆角 $\psi$、行程速度变化系数 $K$。求解曲柄 $a$、连杆 $b$ 和机架 $d$ 的尺寸。图 5-38 所示为待设计机构的极限位置。

$\triangle DC_1C_2$ 中，有

$$\overline{C_1C_2} = 2c\sin\frac{\psi}{2}$$

$\triangle AC_1C_2$ 中，有

$$\overline{C_1C_2}^2 = (b+a)^2 + (b-a)^2 - 2(b+a)(b-a)\cos\theta,\ \theta = 180^\circ\frac{K-1}{K+1}$$

$$\left(2c\sin\frac{\psi}{2}\right)^2 = (b+a)^2 + (b-a)^2 - 2(b+a)(b-a)\cos\theta \tag{5-22}$$

该方程有两个未知数，假定 $a$ 后，可求 $b$，故有无数解。求出 $a$、$b$ 后，可利用 $a$、$b$ 求解 $d$。

利用正弦定理，可从三角形$\triangle AC_1C_2$ 中求出角度 $\delta$ 的值。

$$\frac{\overline{C_1C_2}}{\sin\theta} = \frac{b-a}{\sin\delta},\ \sin\delta = \frac{(b-a)\sin\theta}{\overline{C_1C_2}}$$

利用余弦定理，可从$\triangle AC_1D$ 中求出机架尺寸。

$$d^2 = c^2 + (b+a)^2 - 2c(b+a)\cos\left(90^\circ - \frac{\psi}{2} - \delta\right) = c^2 + (b+a)^2 - 2c(b+a)\sin\left(\frac{\psi}{2} + \delta\right)$$

同样的道理，也可设计曲柄滑块机构。

(4) 按照连杆曲线设计四杆机构　给定连杆曲线设计四杆机构时，一般是给定连杆曲线上几个关键点的坐标，所设计的四杆机构能准确或近似通过所选的关键点。

如图 5-39 所示，已知 $P$ 点在连杆曲线上，求待设计的四杆机构的尺寸与描述 $P$ 点位置的 $e$、$f$、$\gamma$。

图 5-39 中，可从 $ABP$ 支路写出 $P$ 点坐标为

$$\begin{cases} x_P = a\cos\varphi + e\sin\gamma_1 \\ \quad\uparrow \\ y_P = a\sin\varphi + e\cos\gamma_1 \end{cases} \quad 或 \quad \begin{cases} a\cos\varphi = x_P - e\sin\gamma_1 \\ \quad\uparrow \\ a\sin\varphi = y_P - e\cos\gamma_1 \end{cases}$$

两边平方再相加，消去 $\varphi$，得

$$2e(x_P\sin\gamma_1 + y_P\cos\gamma_1) = x_P^2 + y_P^2 + e^2 - a^2$$

从 $ADCP$ 支路写出 $P$ 点坐标为

$$\begin{cases} x_P = d + c\cos\psi - f\sin\gamma_2 \\ \quad\uparrow \\ y_P = c\sin\psi + f\cos\gamma_2 \end{cases} \quad 或 \quad \begin{cases} c\cos\psi = x_P - d + f\sin\gamma_2 \\ \quad\uparrow \\ c\sin\psi = y_P - f\cos\gamma_2 \end{cases}$$

两边平方再相加，消去 $\psi$，得

$$2f[(x_P - d)\sin\gamma_2 - y_P\cos\gamma_2)] = (x_P - d)^2 + y_P^2 + f^2 - c^2$$

From the vectors $ADCP$, we have:

$$x_P = d + c\cos\psi - f\sin\gamma_2 \qquad c\cos\psi = x_P - d + f\sin\gamma_2$$

or

$$y_P = c\sin\psi + f\cos\gamma_2 \qquad c\sin\psi = y_P - f\cos\gamma_2$$

Squaring and adding, the angle $\psi$ has been removed.

$$2f[(x_P - d)\sin\gamma_2 - y_P\cos\gamma_2)] = (x_P - d)^2 + y_P^2 + f^2 - c^2$$

Because $\gamma = \gamma_1 + \gamma_2$, so

$$\gamma = \arccos\frac{e^2 + f^2 - b^2}{2ef}$$

Rearranging the above equations, we have:

$$U^2 + V^2 = W^2 \tag{5-23}$$

In which

$$\begin{aligned} U &= f[(x_P - d)\cos\gamma + y\sin\gamma](x_P^2 + y_P^2 + e^2 - a^2) - ex_P[(x_P - d)^2 + y_P^2 + f^2 - c^2] \\ V &= f[(x_P - d)\sin\gamma - y\cos\gamma](x_P^2 + y_P^2 + e^2 - a^2) + ey_P[(x_P - d)^2 + y_P^2 + f^2 - c^2] \\ W &= 2ef\sin\gamma[x_P(x_P - d) + y_P^2 - dy_P\cot\gamma] \end{aligned} \tag{5-24}$$

This is called algebraic curve equation which is very complex. In general, the more links, the higher the degree of curve generated. The four-bar slider-crank linkage has fourth-degree coupler curves; the four-bar linkage has sixth degree.

There are seven unknowns, lengths of $a$, $b$, $c$, $d$, $e$, $f$ and an angle $\gamma$. We can locate seven point coordinates, such as $P_i(x_{P_i}, y_{P_i})$, $i = 1, 2, 3, 4, 5, 6, 7$, then seven equations can be written. But it is very difficult to solve these seven equations. Usually, we only locate three points on the coupler to design a four-bar linkage.

$$\gamma=\gamma_1+\gamma_2\,,\ \gamma=\arccos\frac{e^2+f^2-b^2}{2ef}$$

整理后，得

$$U^2+V^2=W^2 \tag{5-23}$$

该连杆曲线方程为六阶方程。

$$\begin{cases}U=f[(x_P-d)\cos\gamma+y\sin\gamma](x_P^2+y_P^2+e^2-a^2)-ex_P[(x_P-d)^2+y_P^2+f^2-c^2]\\V=f[(x_P-d)\sin\gamma-y\cos\gamma](x_P^2+y_P^2+e^2-a^2)+ey_P[(x_P-d)^2+y_P^2+f^2-c^2]\\W=2ef\sin\gamma[x_P(x_P-d)+y_P^2-dy_P\cot\gamma]\end{cases} \tag{5-24}$$

该方程中的未知数为 $a$、$b$、$c$、$d$、$e$、$f$、$\gamma$，共计七个。从连杆曲线上取七个关键点坐标 $P_i(x_{P_i},\ y_{P_i})$，$i=1$，2，3，4，5，6，7，可写出七个连杆曲线方程，再求解七个未知数。由于方程复杂，导致求解难度大，一般满足曲线上三个点就可以了。

# Chapter 6

# Design of Cam Mechanisms

# 第6章

# 凸轮机构及其设计

## 6.1 Introduction

**1. Composition and Characteristics of Cam Mechanisms**

A cam mechanism consists of cam, follower and frame, and it belongs to the category of higher pair mechanism. The cam which has a curve shape may be rotating or reciprocating whereas the follower may be reciprocating or oscillating. A cam can transmit a desired motion to a follower by direct contact, and can produce complicated output motion which is difficult to achieve with other types of mechanism.

Fig. 6-1 shows two kinds of cam mechanisms, in which Fig. 6-1a shows a disk cam with a translating follower, and Fig. 6-1b shows a disk cam with an oscillating follower.

Cam mechanism is widely used in automatic machines, internal combustion engines, machine tools, printing machines, and so on. Compared to linkages, cams are easier designed to give a specific output motion. Cams also have a simple structure and reliable performance.

**2. Types of Cam Mechanisms**

Cam mechanisms can be classified in a number of ways: by the shape of the cam, by the shape of the follower, by the manner of constraint of the follower and by the general type of the motion imparted to the follower.

(1) Types of cams

1) Disc cam. Fig. 6-1 shows some disc cams, in which the radius of curvature is usually varied. They are also called radial cams or plate cams, and usually rotate at a constant velocity. These cams are very popular due to their simplicity and compactness.

Fig. 6-1 Cam mechanisms (盘形凸轮机构)

2) Translating cam. Fig. 6-2 shows some translating cams which usually have a translating motion; the follower can either translate or oscillate.

3) Spatial cam. Cylindrical cams are the common spatial cams. In Fig. 6-3a, a groove is cut on the surface of the cam, and a roller follower has a constrained oscillating motion. Fig. 6-3b shows an end cylindrical cam in which the end of the cylinder is the working surface.

Cylindrical cams are also called as barrel cams or drum cams.

Globoid cams and spherical cams are spatial cams too, but we do not deal with them.

## 6.1　凸轮机构概述

### 1. 凸轮机构的组成及其特点

凸轮机构是一种由凸轮、从动件和机架组成的高副机构。其中，凸轮是具有曲线轮廓形状的构件，一般做定轴转动。从动件可做往复移动或往复摆动。

图 6-1 所示为两种最常用的盘形凸轮机构。图 6-1a 所示为直动从动件盘形凸轮机构，当凸轮 1 绕轴 *O* 旋转时，推动从动件 2 沿机架 3 做往复直线移动。图 6-1b 所示为摆动从动件盘形凸轮机构，凸轮 1 转动时，摆杆绕铰链 *A* 做往复摆动。通常，凸轮为机构的主动件。

凸轮机构之所以能够得到广泛应用，主要有以下优点：

1）从动件可以实现复杂的运动规律。

2）结构简单、紧凑，能准确实现预期运动，运动特性好。

3）性能稳定，故障少，维护保养方便。

4）设计简单。

主要缺点是凸轮与从动件为高副接触，易于磨损。由于凸轮的轮廓曲线通常都比较复杂，因而加工比较困难。

### 2. 凸轮机构的分类

根据以下几种情况对凸轮机构进行分类。

（1）按凸轮的形状分类

1）盘形凸轮。呈盘状、具有变化的向径，且做定轴转动的凸轮。图 6-1 所示的凸轮都属于盘形凸轮。

2）移动凸轮。做往复直线移动的凸轮。图 6-2 所示为移动凸轮机构。

3）圆柱凸轮。圆柱凸轮的圆柱面上开有曲线凹槽（图 6-3a），或者端面上具有曲线状轮廓（图 6-3b）。由于圆柱凸轮与从动件的运动不在同一平面内，因此，它属于空间凸轮机构。

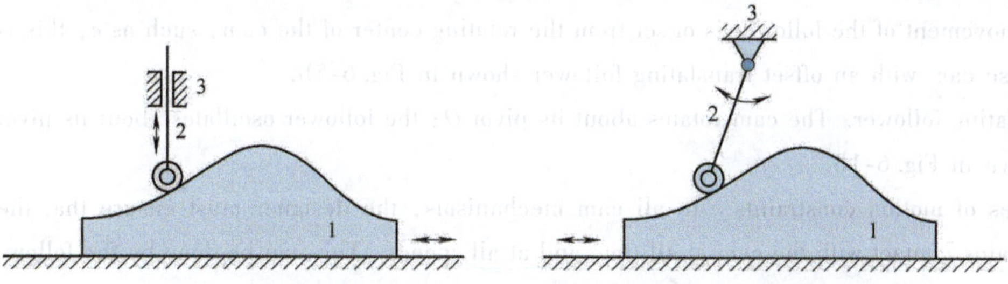


Fig. 6-2　Translating cam mechanisms（移动凸轮机构）

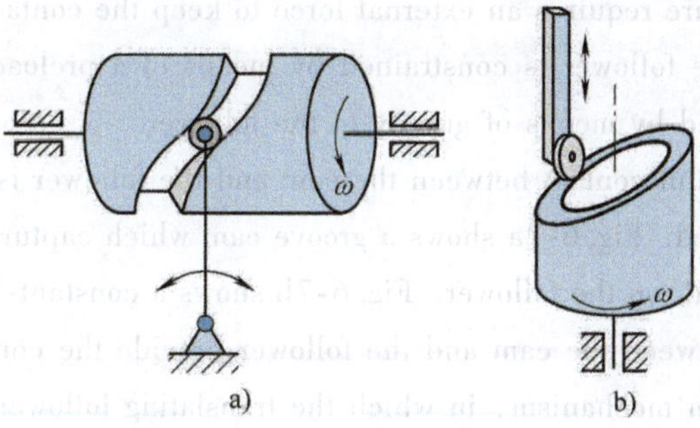


Fig. 6-3　Spatial cam mechanisms（空间凸轮机构）

(2) Types of follower shapes  Fig. 6-4 shows some different types of the followers.

1) Knife-edge follower. Fig. 6-4a shows a translating knife-edge follower; Fig. 6-4e shows an oscillating knife-edge follower. They are simple in construction; however, they can produce a great wear of the surface at the contact point, so their use is limited.

2) Roller follower. Fig. 6-4b shows a translating roller follower; Fig. 6-4f shows an oscillating roller follower which has a cylindrical roller free to rotate about its pin joint. The follower has a pure rolling contact with the cam surface at low speeds, and at high speeds, some sliding occurs also.

3) Flat-faced follower. Fig. 6-4c shows a translating flat-faced follower; Fig. 6-4g shows an oscillating flat-faced follower. Flat-faced follower can be packaged smaller than roller follower for some cam designs and is often favored for that reason as well as cost for automotive valve trains.

4) Spherical-faced follower. Fig. 6-4d shows a translating spherical-faced follower; Fig. 6-4h shows an oscillating spherical-faced follower. They are also called mushroom follower. The advantage is that it does not pose the problem of jamming.

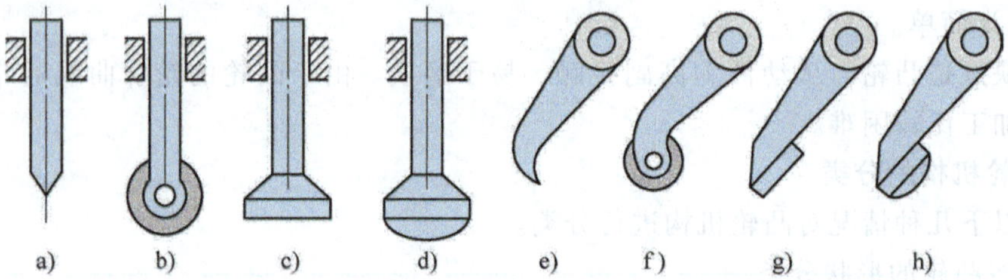


Fig. 6-4 Follower types (从动件的分类)

(3) Types of follower motion

1) Reciprocating follower. The cam rotates about its pivot; the follower reciprocates or translates in the guides, as shown in Fig. 6-1a. If the line of movement of the follower passes through the rotating center of the cam, this is called the disc cam with a radial follower shown in Fig. 6-5a, and if the line of movement of the follower is offset from the rotating center of the cam, such as $e$, this is known as a disc cam with an offset translating follower shown in Fig. 6-5b.

2) Oscillating follower. The cam rotates about its pivot $O$; the follower oscillates about its pivot $A$ too, as shown in Fig. 6-1b.

(4) Types of motion constraints  In all cam mechanisms, the designer must ensure that the follower maintains contact with the cam at all time and at all speeds. This can be done by the following methods.

1) Force closure. Force closure requires an external force to keep the contact between the cam and the follower. In Fig. 6-6a, the follower is constrained by means of a preloaded spring, and in Fig. 6-6b the follower is constrained by means of gravity of the follower.

2) Form closure. The constraint contact between the cam and the follower is maintained by geometry; no external force is required. Fig. 6-7a shows a groove cam which captures a single follower in the groove and both push and pull on the follower. Fig. 6-7b shows a constant-breadth cam mechanism where two contact points between the cam and the follower provide the constraints. Fig. 6-7c shows a conjugate yoke radical cam mechanism, in which the translating follower has two rollers on diametrically opposite side of the cam. Fig. 6-7d shows a conjugate cam mechanism, in which there

（2）按从动件的形状分类　图 6-4 所示为常用的从动件。

1）尖底从动件。图 6-4a、e 所示为尖底从动件。尖底从动件的结构非常简单，其尖底能与任意复杂形状的凸轮轮廓保持接触。但其尖底处容易磨损，一般适用于传力较小和速度较低的场合。

2）滚子从动件。图 6-4b、f 所示为滚子从动件。滚子的存在使得凸轮与从动件之间的滑动摩擦转化为滚动摩擦，减小了凸轮机构的磨损，因而可以传递较大的动力，在工程中应用最为广泛。

3）平底从动件。图 6-4c、g 所示为平底从动件。平底从动件的优点是受力平稳，传动效率高。平底从动件常用于高速场合。其缺点是要求相应的凸轮轮廓曲线必须全部外凸。

4）曲底从动件。图 6-4d、h 所示为曲底从动件。曲底从动件端部为一曲面，兼有尖底与平底从动件的优点，在生产实际中的应用也较多。

（3）按从动件的运动形式分类　按从动件的运动方式可分为直动从动件（图 6-4a ~ d）和摆动从动件（图 6-4e ~ h）。对于直动从动件凸轮机构，当从动件的导路中心线通过凸轮的回转中心时，称为对心直动从动件凸轮机构，反之则称为偏置从动件凸轮机构，偏置的距离称为偏距，常用 $e$ 表示。例如：图 6-5a 所示为对心直动尖底从动件盘形凸轮机构，而图 6-5b 所示为偏置直动尖底从动件盘形凸轮机构，偏距为 $e$。

（4）按凸轮与从动件维持高副接触的方式分类　凸轮与从动件必须永远保持高副接触。把凸轮与从动件维持高副接触的方式称为封闭方式或锁合方式，工程中主要靠外力或特殊的几何形状来保证两者的接触。

1）力封闭方式。利用弹簧力、从动件本身的重力或其他外力来保证凸轮与从动件始终保持接触。图 6-6a 所示凸轮机构是利用弹簧的回复力来保持高副接触的，而图 6-6b 所示凸轮机构则是利用从动件的重力来保持高副接触的。

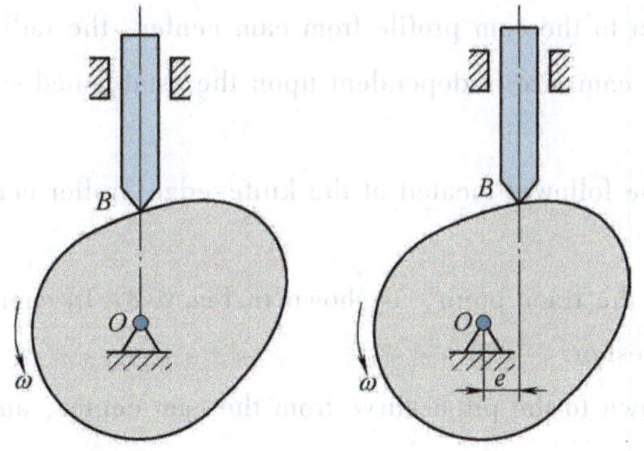


Fig. 6-5　Disc cam with radial follower
（直动从动件盘形凸轮机构）

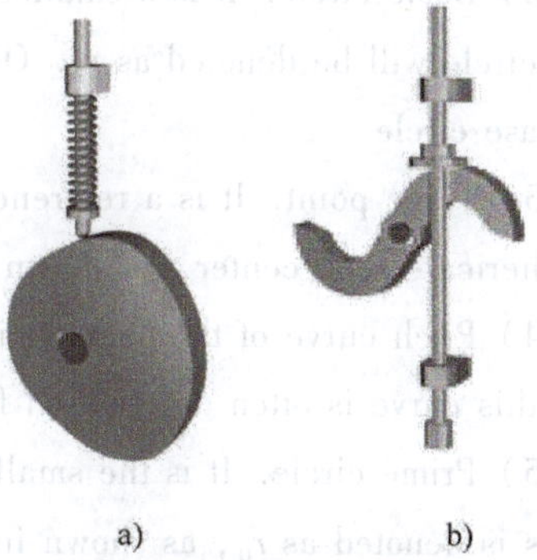


Fig. 6-6　Force-closed cam mechanism
（力封闭凸轮机构）

2）形封闭方式。依靠凸轮和从动件的特殊几何形状来维持凸轮机构的高副接触。图 6-7a 所示的端面凸轮机构是利用凸轮端面上的沟槽和放于槽中的滚子使凸轮与从动件保持接触，这类凸轮又称为端面凸轮。图 6-7b 所示的凸轮机构中，凸轮与从动件的两个高副接触点之间的距离处处相等，且等于从动件的槽宽，凸轮和从动件始终保持接触。这种凸轮机构称为等宽凸轮机构。图 6-7c 所示的凸轮机构中，两滚子中心距离与对应凸轮径向距离处处

are two cams fixed on a common shaft. They are mathematical conjugates of one another. Two roller followers, attached to a common arm, are each pushed in opposite direction by the conjugate cams.

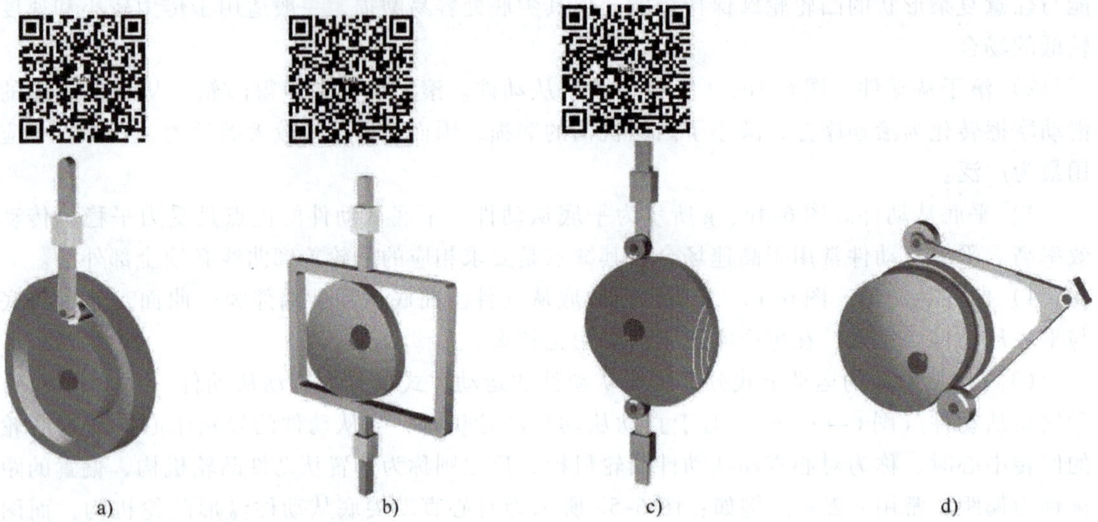


Fig. 6-7 Form-closed cam mechanisms (形封闭式凸轮机构)

### 3. Cam Terminology

Let us define the terms necessary for the investigation of the cam shape and actions.

Fig. 6-8 shows a disc cam mechanism with a translating knife-edge follower. As the cam rotates about its axis through one revolution, the follower is made to execute a series of events such as rises, dwells and returns.

1) Cam profile. It is the actual working surface of the cam, as shown in Fig. 6-8.

2) Basic circle. It is a smallest circle drawn to the cam profile from cam center; the radius of base circle will be denoted as $r_b$. Obviously, the cam size is dependent upon the established size of the base circle.

3) Trace point. It is a reference point on the follower located at the knife-edge, roller center, or spherical-faced center, as shown in Fig. 6-8.

4) Pitch curve of the cam. It is the path of the trace point, as shown in Fig. 6-8. In cam lay-out, this curve is often determined first in cam design.

5) Prime circle. It is the smallest circle drawn to the pitch curve from the cam center, and its radius is denoted as $r_0$, as shown in Fig. 6-8.

6) Pressure angle. It is an angle between the normal to the pitch curve at a contact point and the direction of the follower motion, as shown in Fig. 6-8. It varies in magnitude at all instants of the follower motion. This angle is important in cam design since it represents the steepness of the cam profile.

7) Rise travel. As the cam rotates, the follower moves apart from the cam center. This process is called the rise travel.

8) Return travel. As the cam rotates, the follower moves toward the cam center. This process is called the return travel.

9) Stroke. It is the great distance through which the follower moves. It is the total rise of the

相等，保证从动件上的两个滚子同时与凸轮接触，这种凸轮机构称为等径凸轮机构。图 6-7d 所示凸轮机构中，安装在同一轴上的两个凸轮与摆杆上的两个滚子同时保持接触，一个凸轮推动摆杆做正行程运动，而另一个凸轮推动摆杆做反行程运动。设计出其中一个凸轮的轮廓曲线后，另一个凸轮的轮廓曲线可根据共轭条件求出，故称之为共轭凸轮。

形封闭式凸轮机构，需要有较高的加工精度才能满足准确的形封闭条件。

**3. 凸轮机构的名词术语**

1）凸轮实际廓线。图 6-8 所示凸轮机构中，与从动件直接接触的凸轮廓线称为实际廓线。

2）实际廓线基圆。以凸轮的回转中心为圆心，以凸轮实际轮廓的最小向径为半径所作的圆，称为凸轮的实际廓线基圆，其半径用 $r_b$ 表示，如图 6-8 所示。

3）轨迹点。从动件上的尖点、滚子的圆心点的运动位置称为从动件上的轨迹点。

4）理论廓线。由一系列轨迹点形成的曲线，称为凸轮节曲线或理论廓线。

5）理论廓线基圆。以凸轮理论轮廓的最小向径为半径所作的圆，称为凸轮的理论廓线基圆，其半径用 $r_0$ 表示。

6）压力角。凸轮和从动件接触点的公法线与从动件运动方向所夹的锐角称为压力角。压力角是凸轮设计的重要参数。

7）推程。从动件从距凸轮回转中心的最近点向最远点运动的过程称为推程。

8）回程。从动件从距凸轮回转中心的最远点向最近点运动的过程称为回路。

9）行程。从动件从距凸轮回转中心的最近点运动到最远点所通过的距离，或从最远点回到最近点所通过的距离称为行程，是指从动件的最大运动距离，常用 $h$ 来表示。

10）推程运动角。从动件从距凸轮回转中心的最近点运动到最远点时，对应凸轮所转过的角度称为推程运动角，用 $\Phi$ 表示。

11）回程运动角。从动件从距凸轮回转中心的最远点运动到最近点时，对应凸轮所转过的角度称为回程运动角，用 $\Phi'$ 表示。

12）远休止角。从动件在距凸轮回转中心的最远点静止不动时，对应凸轮所转过的角度称为远休止角，用 $\Phi_s$ 表示。

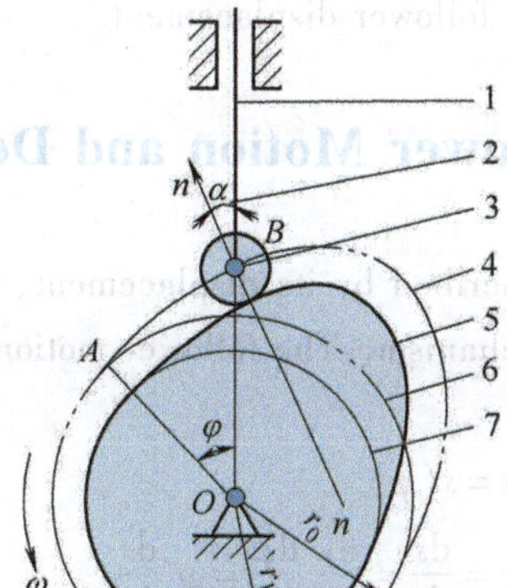


Fig. 6-8　Nomenclatures of cam mechanism（凸轮机构名词术语）

1—follower（从动件）　2—pressure angle（压力角）　3—trace point（轨迹点）

4—pitch curve（理论廓线）　5—cam profile（凸轮实际廓线）

6—prime circle（理论廓线基圆）　7—base circle（实际廓线基圆）

follower, denoted as $h$.

10) Angle of ascent. When the follower moves from the initial position to the terminal position, the angle through which the cam rotates is called the angle of ascent, denoted as $\Phi$.

11) Angle of descent. When the follower moves from the terminal position to the initial position, the angle through which the cam rotates is called the angle of descent, denoted as $\Phi'$.

12) Angle dwelt at the highest position. It is an angle through which the cam rotates while the follower remains stationary at the highest position, and it is denoted as $\Phi_s$.

13) Angle dwelt at the lowest position. It is an angle through which the cam rotates while the follower remains stationary at the lowest position, and it is denoted as $\Phi_s'$.

14) Cam angle. It is an angular displacement of the cam measured from the initial position of the follower on the prime circle to the intersection point of the follower axis or oscillating arc and the prime circle.

15) Follower displacement. It is the distance that the follower moves, and it is measured from the lowest position of the follower on the prime circle, no matter which travel it is. It is denoted as $s$. Fig. 6-9 shows the cam angle and follower displacement of the cam mechanism. In Fig. 6-9c, the cam angle is measured from the initial position $OA$ on the prime circle, and the displacement is $s$.

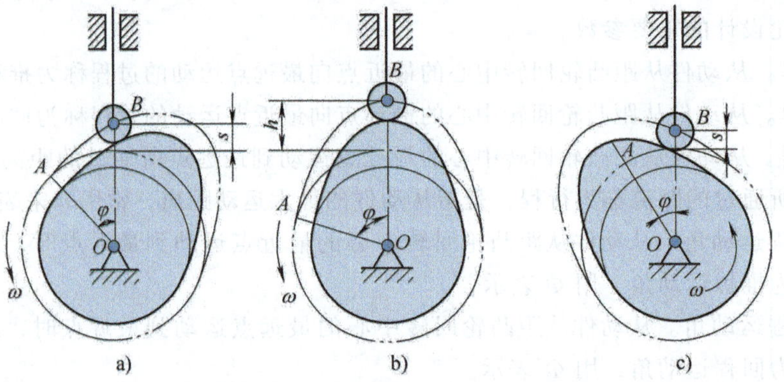


Fig. 6-9 Cam angle and follower displacement (凸轮转角与从动件的位移)

Fig. 6-10 shows the displacement diagram of the cam mechanism. It can describe the relationship between the cam angle and the follower displacement.

## 6.2 Basic Types of Follower Motion and Design

The follower motion can be described by its displacement, velocity and acceleration, and it is the foundation of designing cam mechanism. The follower motions are given by the mathematical relationship as follows.

$$
\begin{aligned}
s &= s(\varphi) \\
v &= \frac{\mathrm{d}s}{\mathrm{d}t} = \frac{\mathrm{d}s}{\mathrm{d}\varphi}\frac{\mathrm{d}\varphi}{\mathrm{d}t} = \omega\frac{\mathrm{d}s}{\mathrm{d}\varphi} \\
a &= \frac{\mathrm{d}^2 s}{\mathrm{d}t^2} = \frac{\mathrm{d}v}{\mathrm{d}t} = \frac{\mathrm{d}v}{\mathrm{d}\varphi}\frac{\mathrm{d}\varphi}{\mathrm{d}t} = \omega^2\frac{\mathrm{d}^2 s}{\mathrm{d}\varphi^2}
\end{aligned}
\tag{6-1}
$$

Where $s$ is the displacement of the follower, $\varphi$ is the angle of the cam, $v$ is the velocity of the follower, $a$ is the acceleration of the follower, $t$ is the time for cam to rotate through angle $\varphi$, and $\omega$ is

13）近休止角。从动件在距凸轮回转中心的最近点静止不动时，对应凸轮所转过的角度称为近休止角，用 $\Phi_s'$ 表示。

14）凸轮转角。凸轮绕自身轴线转过的角度，称为凸轮转角，用 $\varphi$ 表示。一般情况下，凸轮转角从行程的起始点在基圆上开始度量，其值等于行程起点和从动件的导路中心线与基圆的交点所组成的圆弧对应的基圆圆心角。

15）从动件的位移。凸轮转过转角 $\varphi$ 时，从动件所运动的距离称为从动件的位移。位移 $s$ 从距凸轮回转中心的最近点开始度量。图 6-9a 所示为推程阶段的凸轮转角与对应的从动件位移，图 6-9c 所示为回程阶段的凸轮转角与对应的从动件位移，图 6-9b 所示为从动件位于推程终止点位置的凸轮转角示意图。

对于摆动从动件，其位移为角位移。

图 6-10 所示为图 6-9 所示凸轮机构的运动循环图。显然，在一个运动循环中，推程运动角、远休止角、回程运动角和近休止角之间应该满足以下关系：

$$\Phi + \Phi_s + \Phi' + \Phi_s' = 360°$$

在设计凸轮机构时，凸轮的 $\Phi$、$\Phi'$、$\Phi_s$ 和 $\Phi_s'$ 应根据实际的工作要求选择，如果没有远休止和近休止过程，则其远休止角和近休止角均等于零。

通过该运动循环图也可以理解凸轮转角与从动件位移的度量原则。

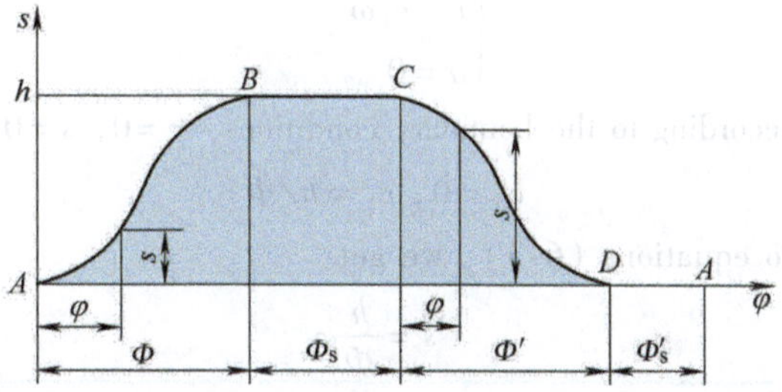


Fig. 6-10 Motion of the follower（凸轮机构运动循环图）

## 6.2 从动件的运动规律及其设计

从动件的运动规律是指从动件的位移 $s$、速度 $v$、加速度 $a$ 与凸轮转角 $\varphi$ 或时间 $t$ 之间的函数关系，可以用方程表示，也可以用线图表示。

从动件运动规律的一般方程表达式为：$s = s(\varphi)$，$v = v(\varphi)$，$a = a(\varphi)$。对应的曲线分别称为从动件的位移曲线、速度曲线和加速度曲线，统称为从动件的运动规律线图。

凸轮一般为主动件，且做匀速回转运动。设凸轮的角速度为 $\omega$，则从动件的位移、速度和加速度与凸轮转角之间的关系为

$$\begin{aligned} s &= s(\varphi) \\ v &= \frac{\mathrm{d}s}{\mathrm{d}t} = \frac{\mathrm{d}s}{\mathrm{d}\varphi}\frac{\mathrm{d}\varphi}{\mathrm{d}t} = \omega\frac{\mathrm{d}s}{\mathrm{d}\varphi} \\ a &= \frac{\mathrm{d}^2 s}{\mathrm{d}t^2} = \frac{\mathrm{d}v}{\mathrm{d}t} = \frac{\mathrm{d}v}{\mathrm{d}\varphi}\frac{\mathrm{d}\varphi}{\mathrm{d}t} = \omega^2\frac{\mathrm{d}^2 s}{\mathrm{d}\varphi^2} \end{aligned} \tag{6-1}$$

the angular velocity of the cam.

**1. Basic Types of Follower Motions**

Before a cam contour can be determined, it is necessary to select the follower motion. There are many basic motions of follower. We only discuss a few of them.

(1) Polynomial motions The displacement equation of the simple polynomial motion is of the form: $s = c\varphi^n$. The general form of the polynomial motion can be written as follows:

$$\begin{cases} s = c_0 + c_1\varphi + c_2\varphi^2 + c_3\varphi^3 + \cdots + c_n\varphi^n \\ v = \omega(c_1 + 2c_2\varphi + 3c_3\varphi^2 + \cdots + nc_n\varphi^{n-1}) \\ a = \omega^2(2c_2 + 6c_3\varphi + \cdots + n(n-1)c_n\varphi^{n-2}) \end{cases} \tag{6-2}$$

Where $n$ is any number, and $c_0$, $c_1$, $c_2$, ..., $c_n$ are constant coefficients. These constant coefficients are the unknowns to be determined in the polynomial equation, and they can be decided from the boundary conditions of the follower motion.

1) Constant velocity motion. This curve of the polynomial family is the simplest of all, when $n$ is equal to one. It is called a first-degree polynomial.

The motion equations for the rise travel are given by:

$$\begin{cases} s = c_0 + c_1\varphi \\ v = c_1\omega \\ a = 0 \end{cases} \tag{6-3}$$

When $\varphi \in [0, \Phi]$, according to the boundary conditions, $\varphi = 0$, $s = 0$; $\varphi = \Phi$, $s = h$, we have:

$$c_0 = 0, \ c_1 = h/\Phi$$

Substituting them into equations (6-3), we get:

$$\begin{aligned} s &= \frac{h}{\Phi}\varphi \\ v &= \frac{h}{\Phi}\omega \\ a &= 0 \end{aligned} \tag{6-4}$$

When the follower is in the return travel, that is $\varphi \in [0, \Phi']$. According to the boundary conditions, $\varphi = 0$, $s = h$, $\varphi = \Phi'$, $s = 0$, we have:

$$c_0 = h, \ c_1 = -h/\Phi'$$

Substituting $c_0 = h$, $c_1 = -h/\Phi'$ into equations (6-3), we get:

$$\begin{aligned} s &= h - \frac{h}{\Phi'}\varphi \\ v &= -\frac{h}{\Phi'}\omega \\ a &= 0 \end{aligned} \tag{6-5}$$

The displacement diagram, velocity diagram and acceleration diagram are shown in Fig. 6-11. We see that it has a straight-line displacement curve at a constant slope, and the velocity is constant; the acceleration is zero. Though acceleration is zero during the rise or return of the follower, it is infinite at the beginning and end of the motion as there are abrupt changes in velocity at these points. A theoretically infinite acceleration can transmit a high shock throughout the follower. So this curve is not suitable for a dwell-rise-dwell cam, only applies at very low speed. But this curve has

**1. 从动件的基本运动规律**

从动件运动规律有多种，这里仅介绍几种最基本的运动规律。

（1）多项式类运动规律　多项式类运动规律的一般形式为

$$
\begin{aligned}
s &= c_0 + c_1\varphi + c_2\varphi^2 + c_3\varphi^3 + \cdots + c_n\varphi^n \\
v &= \omega(c_1 + 2c_2\varphi + 3c_3\varphi^2 + \cdots + nc_n\varphi^{n-1}) \\
a &= \omega^2(2c_2 + 6c_3\varphi + \cdots + n(n-1)c_n\varphi^{n-2})
\end{aligned}
\tag{6-2}
$$

式中，$c_0$、$c_1$、$c_2$、…、$c_n$ 为待定系数，可根据工作要求的边界条件确定。

本书介绍几种最简单的多项式运动规律。

1）一次多项式运动规律。在多项式运动规律中，如果 $n=1$，则有

$$
\begin{aligned}
s &= c_0 + c_1\varphi \\
v &= c_1\omega \\
a &= 0
\end{aligned}
\tag{6-3}
$$

推程阶段，$\varphi\in[0, \Phi]$，根据边界条件：$\varphi=0$ 时 $s=0$，$\varphi=\Phi$ 时 $s=h$，可解出待定常数：$c_0=0$，$c_1=h/\Phi$。将 $c_0$、$c_1$ 代入式（6-3）中并整理，即可得到从动件在推程阶段的运动方程。

$$
\begin{aligned}
s &= \frac{h}{\Phi}\varphi \\
v &= \frac{h}{\Phi}\omega \\
a &= 0
\end{aligned}
\tag{6-4}
$$

回程阶段，$\varphi\in[0, \Phi']$，根据边界条件：$\varphi=0$ 时 $s=h$，$\varphi=\Phi'$时 $s=0$，可解出待定常数：$c_0=h$，$c_1=-h/\Phi'$。将 $c_0$、$c_1$ 代入式（6-3）中并整理，即可得到从动件在回程阶段的运动方程为

$$
\begin{aligned}
s &= h - \frac{h}{\Phi'}\varphi \\
v &= -\frac{h}{\Phi'}\omega \\
a &= 0
\end{aligned}
\tag{6-5}
$$

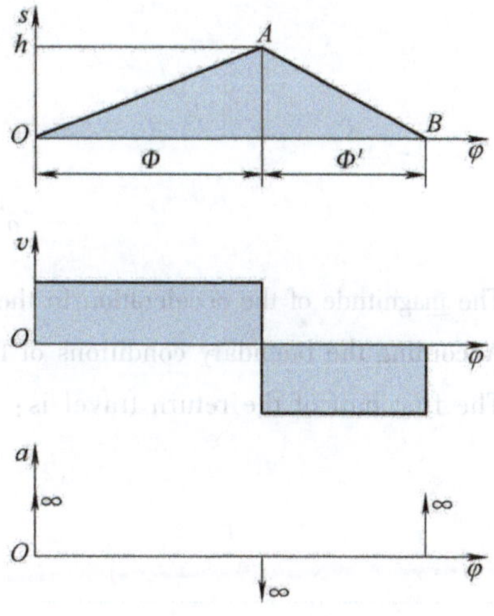

Fig. 6-11　Constant-velocity curve
（等速运动规律）

当 $n=1$ 时，从动件做等速运动，因此，一次多项式运动规律也称为等速运动规律，其位移为凸轮转角的一次函数，位移曲线为一条斜直线，位移、速度、加速度相对于凸轮转角的变化规律线图如图 6-11 所示。在行程的起点、中点与终点（$O$、$A$、$B$）处，由于速度发生突变，加速度在理论上趋于无穷大，从而导致在从动件上产生非常大的惯性力冲击，这种冲击称为刚性冲击。所以，等速运动规律常用于从

frequently been employed in combination with other curves.

2) Constant acceleration and deceleration motion. Another polynomial motion is the parabolic motion. When $n$ is equal to two, it is called a second-degree polynomial; this curve of polynomial motion has the property of constant positive acceleration and negative acceleration.

The motion equations for the rise travel are given by:

$$\begin{cases} s = c_0 + c_1\varphi + c_2\varphi^2 \\ v = \omega(c_1 + 2c_2\varphi) \\ a = 2c_2\omega^2 \end{cases} \tag{6-6}$$

It is acceleration in the first half travel of the follower, whereas it is deceleration during the last half travel, and the acceleration is the same and constant in the two halves. The displacement curve is found to be parabolic.

When $\varphi \in [0, \Phi/2]$, according to the boundary conditions, $\varphi = 0$, $s = 0$, $v = 0$; $\varphi = \Phi/2$, $s = h/2$, the constant coefficient can be solved.

$$c_0 = 0,\ c_1 = 0,\ c_2 = 2h/\Phi^2$$

Substituting them into equations (6-6), we get:

$$\begin{aligned} s &= \frac{2h}{\Phi^2}\varphi^2 \\ v &= \frac{4h\omega}{\Phi^2}\varphi \\ a &= \frac{4h\omega^2}{\Phi^2} \end{aligned} \tag{6-7}$$

In this half travel, the acceleration is a constant. Its magnitude is:

$$a = \frac{4h\omega^2}{\Phi^2}$$

When $\varphi \in [\Phi/2, \Phi]$, according to the boundary conditions, $\varphi = \Phi/2$, $s = h/2$, $v = 2h\omega/\Phi$, $\varphi = \Phi$, $s = h$, $v = 0$, substituting them into equations (6-6), we get:

$$c_0 = -h,\ c_1 = 4h/\Phi,\ c_2 = -2h/\Phi^2$$

Rearranging the above equations, we have:

$$\begin{aligned} s &= h - \frac{2h}{\Phi^2}(\Phi - \varphi)^2 \\ v &= \frac{4h\omega}{\Phi^2}(\Phi - \varphi) \\ a &= -\frac{4h\omega^2}{\Phi^2} \end{aligned} \tag{6-8}$$

The magnitude of the acceleration in the last half travel is a constant too, but its direction is opposite. According the boundary conditions of the return travel, we can establish the following equations. The first half of the return travel is:

$$\begin{aligned} s &= h - \frac{2h}{\Phi'^2}\varphi^2 \\ v &= -\frac{4h\omega}{\Phi'^2}\varphi \\ a &= -\frac{4h\omega^2}{\Phi'^2} \end{aligned} \tag{6-9}$$

动件具有等速运动要求、从动件的质量不大或低速场合。

2）二次多项式运动规律。多项式运动规律中，令 $n=2$，则有

$$
\begin{aligned}
s &= c_0 + c_1\varphi + c_2\varphi^2 \\
v &= \omega(c_1 + 2c_2\varphi) \\
a &= 2c_2\omega^2
\end{aligned}
\tag{6-6}
$$

推程前半阶段，$\varphi \in [0, \Phi/2]$，则根据边界条件：$\varphi=0$，$s=0$，$v=0$；$\varphi=\Phi/2$，$s=h/2$，即可解出待定常数：$c_0=0$，$c_1=0$，$c_2=2h/\Phi^2$。将 $c_0$、$c_1$、$c_2$ 代入式（6-6）并整理，即可得到从动件在推程前半阶段的运动方程为

$$
\begin{aligned}
s &= \frac{2h}{\Phi^2}\varphi^2 \\
v &= \frac{4h\omega}{\Phi^2}\varphi \\
a &= \frac{4h\omega^2}{\Phi^2}
\end{aligned}
\tag{6-7}
$$

在前半推程中，从动件的加速度 $a=4h\omega^2/\Phi^2=$ 常数，因此，从动件做等加速运动。

在推程后半阶段，$\varphi \in [\Phi/2, \Phi]$，则根据边界条件：$\varphi=\Phi/2$，$s=h/2$，$v=2h\omega/\Phi$；$\varphi=\Phi$，$s=h$，$v=0$，即可解出待定常数：$c_0=-h$，$c_1=4h/\Phi$，$c_2=-2h/\Phi^2$。将 $c_0$、$c_1$、$c_2$ 代入式（6-6）并整理，即可得到从动件在推程后半阶段的运动方程为

$$
\begin{aligned}
s &= h - \frac{2h}{\Phi^2}(\Phi-\varphi)^2 \\
v &= \frac{4h\omega}{\Phi^2}(\Phi-\varphi) \\
a &= -\frac{4h\omega^2}{\Phi^2}
\end{aligned}
\tag{6-8}
$$

在该阶段，从动件加速度 $a=-4h\omega^2/\Phi^2$ 为一负常数，因此，从动件做等减速运动。

根据从动件在回程阶段的边界条件，同理可求出从动件在回程阶段的运动方程，即式（6-9）和式（6-10）。

回程的等加速阶段：

$$
\begin{aligned}
s &= h - \frac{2h}{\Phi'^2}\varphi^2 \\
v &= -\frac{4h\omega}{\Phi'^2}\varphi \\
a &= -\frac{4h\omega^2}{\Phi'^2}
\end{aligned}
\tag{6-9}
$$

The last half of the return travel is:

$$s = \frac{2h}{\Phi'^2}(\Phi' - \varphi)^2$$

$$v = -\frac{4h\omega}{\Phi'^2}(\Phi' - \varphi) \tag{6-10}$$

$$a = \frac{4h\omega^2}{\Phi'^2}$$

Fig. 6-12 shows the displacement diagram, velocity diagram and acceleration diagram.

It can be observed from the Fig. 6-12 that there are abrupt changes in the acceleration at the beginning, midway and the end of the follower motion; thus this motion of the follower is adopted only up to moderate speeds.

3) The 3-4-5 polynomial motion. Rewriting the equation (6-2), and letting $n=5$, we get the fifth-degree polynomial motion equations.

$$s = c_0 + c_1\varphi + c_2\varphi^2 + c_3\varphi^3 + c_4\varphi^4 + c_5\varphi^5$$

$$v = \omega(c_1 + 2c_2\varphi + 3c_3\varphi^2 + 4c_4\varphi^3 + 5c_5\varphi^4) \tag{6-11}$$

$$a = \omega^2(2c_2 + 6c_3\varphi + 12c_4\varphi^2 + 20c_5\varphi^3)$$

When $\varphi \in [0, \Phi]$, according to the boundary conditions $\varphi=0$, $s=0$, $v=0$, $a=0$; $\varphi=\Phi$, $s=h$, $v=0$, $a=0$, substituting them into equations (6-11), we get:

$$c_0 = c_1 = c_2 = 0,\ c_3 = 10h/\Phi^3,\ c_4 = -15h/\Phi^4,\ c_5 = 6h/\Phi^5$$

Rearrang the above equations, we have:

$$s = h\left(\frac{10}{\Phi^3}\varphi^3 - \frac{15}{\Phi^4}\varphi^4 + \frac{6}{\Phi^5}\varphi^5\right)$$

$$v = h\omega\left(\frac{30}{\Phi^3}\varphi^2 - \frac{60}{\Phi^4}\varphi^3 + \frac{30}{\Phi^5}\varphi^4\right) \tag{6-12}$$

$$a = h\omega^2\left(\frac{60}{\Phi^3}\varphi - \frac{180}{\Phi^4}\varphi^2 + \frac{120}{\Phi^5}\varphi^3\right)$$

In the same way, we can obtain the equations for return travel of the follower.

$$s = h - h\left(\frac{10}{\Phi'^3}\varphi^3 - \frac{15}{\Phi'^4}\varphi^4 + \frac{6}{\Phi'^5}\varphi^5\right)$$

$$v = -h\omega\left(\frac{30}{\Phi'^3}\varphi^2 - \frac{60}{\Phi'^4}\varphi^3 + \frac{30}{\Phi'^5}\varphi^4\right) \tag{6-13}$$

$$a = -h\omega^2\left(\frac{60}{\Phi'^3}\varphi - \frac{180}{\Phi'^4}\varphi^2 + \frac{120}{\Phi'^5}\varphi^3\right)$$

Fig. 6-13 shows the result diagrams for a 3-4-5 polynomial motion. Note that the acceleration is continuous, but it is difficult to manufacture the cam with sufficient accuracy to match the required displacement. Usually it is good for high speed.

回程的等减速阶段：

$$s=\frac{2h}{\Phi'^2}(\Phi'-\varphi)^2$$

$$v=-\frac{4h\omega}{\Phi'^2}(\Phi'-\varphi) \tag{6-10}$$

$$a=\frac{4h\omega^2}{\Phi'^2}$$

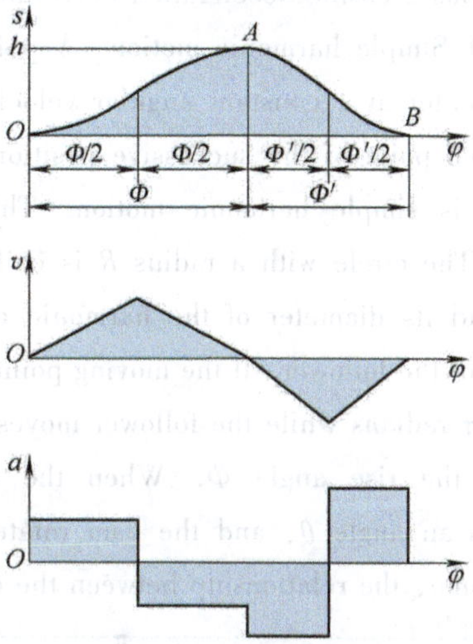


Fig. 6-12 Constant acceleration and deceleration curve（等加速等减速运动规律）

当 $n=2$ 时，从动件按等加速、等减速运动规律运动，因此，二次多项式运动规律也称为等加速等减速运动规律，其位移为凸轮转角的二次函数，位移曲线为抛物线。从动件的运动线图如图 6-12 所示。

由加速度线图可以看出，在行程的起点、中点和终点处，其加速度发生突变，因而在从动件上产生的惯性力也发生突变，会引起凸轮机构的冲击。由于加速度的突变为一有限值，所引起的惯性力突变也是有限值，对凸轮机构的冲击也是有限的，因此，这种冲击称为柔性冲击。

3）五次多项式运动规律。在多项式类运动规律的一般形式中，令 $n=5$，此时从动件的运动规律为

$$\begin{cases} s=c_0+c_1\varphi+c_2\varphi^2+c_3\varphi^3+c_4\varphi^4+c_5\varphi^5 \\ v=\omega(c_1+2c_2\varphi+3c_3\varphi^2+4c_4\varphi^3+5c_5\varphi^4) \\ a=\omega^2(2c_2+6c_3\varphi+12c_4\varphi^2+20c_5\varphi^3) \end{cases} \tag{6-11}$$

在推程阶段，$\varphi\in[0,\Phi]$，根据边界条件：$\varphi=0$，$s=0$，$v=0$，$a=0$；$\varphi=\Phi$，$s=h$，$v=0$，$a=0$，可解出待定常数：$c_0=c_1=c_2=0$，$c_3=10h/\Phi^3$，$c_4=-15h/\Phi^4$，$c_5=6h/\Phi^5$。将 $c_0$、$c_1$、$c_2$、$c_3$、$c_4$、$c_5$ 代入式（6-11）中并整理，即可得到从动件在推程阶段的运动方程为

$$s=h\left(\frac{10}{\Phi^3}\varphi^3-\frac{15}{\Phi^4}\varphi^4+\frac{6}{\Phi^5}\varphi^5\right)$$

$$v=h\omega\left(\frac{30}{\Phi^3}\varphi^2-\frac{60}{\Phi^4}\varphi^3+\frac{30}{\Phi^5}\varphi^4\right) \tag{6-12}$$

$$a=h\omega^2\left(\frac{60}{\Phi^3}\varphi-\frac{180}{\Phi^4}\varphi^2+\frac{120}{\Phi^5}\varphi^3\right)$$

同理可得从动件在回程阶段的运动方程为

$$s=h-h\left(\frac{10}{\Phi'^3}\varphi^3-\frac{15}{\Phi'^4}\varphi^4+\frac{6}{\Phi'^5}\varphi^5\right)$$

$$v=-h\omega\left(\frac{30}{\Phi'^3}\varphi^2-\frac{60}{\Phi'^4}\varphi^3+\frac{30}{\Phi'^5}\varphi^4\right) \tag{6-13}$$

$$a=-h\omega^2\left(\frac{60}{\Phi'^3}\varphi-\frac{180}{\Phi'^4}\varphi^2+\frac{120}{\Phi'^5}\varphi^3\right)$$

(2) Trigonometric motions  The motions of trigonometric form are simple harmonic motion which has a cosine acceleration curve and cycloid motion which has a sine acceleration curve.

1) Simple harmonic motion. A point $M$ rotates about a fixed center at a constant angular velocity $\omega$. The motion of projected point of the successive positions on the vertical diameter is simple harmonic motion. This is shown in Fig. 6-14. The circle with a radius $R$ is called the harmonic circle, and its diameter of the harmonic circle is equal to the stroke of the follower. If the moving point $M$ passes through an angle $\pi$ radians while the follower moves a stroke $h$, the cam rotates the rise angle $\Phi$. When the moving point passes through an angle $\theta$, and the cam rotates an angle $\varphi$ at the same time, the relationship between the $\theta$ and $\varphi$ is as follows:

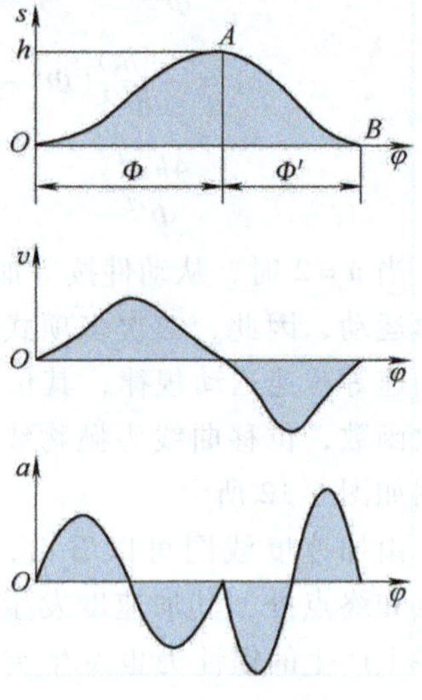

Fig. 6-13 3-4-5polynomial curve
(五次多项式运动规律)

$$\theta = \frac{\pi}{\Phi}\varphi$$

The equation of simple harmonic motion for the rise travel is:

$$s = R - R\cos\theta = \frac{h}{2} - \frac{h}{2}\cos\left(\frac{\pi}{\Phi}\varphi\right) \tag{6-14a}$$

The second and third derivatives are as follows:

$$v = \frac{\pi h\omega}{2\Phi}\sin\left(\frac{\pi}{\Phi}\varphi\right) \tag{6-14b}$$

$$a = \frac{\pi^2 h\omega^2}{2\Phi^2}\cos\left(\frac{\pi}{\Phi}\varphi\right) \tag{6-14c}$$

In the same way, we can obtain the equations for the return travel.

$$\begin{aligned} s &= \frac{h}{2} + \frac{h}{2}\cos\left(\frac{\pi}{\Phi'}\varphi\right) \\ v &= -\frac{\pi h\omega}{2\Phi'}\sin\left(\frac{\pi}{\Phi'}\varphi\right) \\ a &= -\frac{\pi^2 h\omega^2}{2\Phi'^2}\cos\left(\frac{\pi}{\Phi'}\varphi\right) \end{aligned} \tag{6-15}$$

The diagram of simple harmonic motion is in Fig. 6-15.

There are abrupt changes in the acceleration at the beginning and at the end of the follower motion. The flexing shock will occur. This is undesirable for high speed cam since it will result in noise, vibration and wear.

2) Cycloid motion. In Fig. 6-16, a cycloid is the locus of a point on a circle which is rolled on a straight line. This line is the $s$ axis; the circumference of the circle is made equal to the total rise $h$ of the follower. When the circle makes one revolution, a point $M$ on the circumference will move a distance equal to the total rise $h$, thus:

$h = 2\pi R$; the radius of the circle is:

从动件按照五次多项式运动规律运动时的位移、速度和加速度对凸轮转角的变化规律线图如图 6-13 所示。五次多项式运动规律的加速度曲线是连续曲线，因此，既不存在刚性冲击，也不存在柔性冲击，运动平稳性好，适用于高速凸轮机构。

（2）三角函数类运动规律　三角函数类运动规律是指从动件的加速度按余弦规律或正弦规律变化，也称为从动件位移按简谐运动和摆线运动变化。

1）余弦运动规律。如图 6-14 所示，当动点 $M$ 从 $O$ 点开始做逆时针方向的圆周运动时，$M$ 点在坐标轴 $s$ 上投影的变化规律为简谐运动。取动点 $M$ 在坐标轴 $s$ 上投影的变化为从动件的运动规律，并设从动件的行程 $h$ 等于圆周的直径 $2R$，则当点 $M$ 由 $O$ 点开始转过 180°时，从动件到达推程的最高点，即 $h=2R$。设点 $M$ 转过角 $\theta$ 时，从动件的位移为 $s$，凸轮转角为 $\varphi$，则点 $M$ 转过的角度 $\theta$ 与凸轮转角 $\varphi$ 之间的关系为

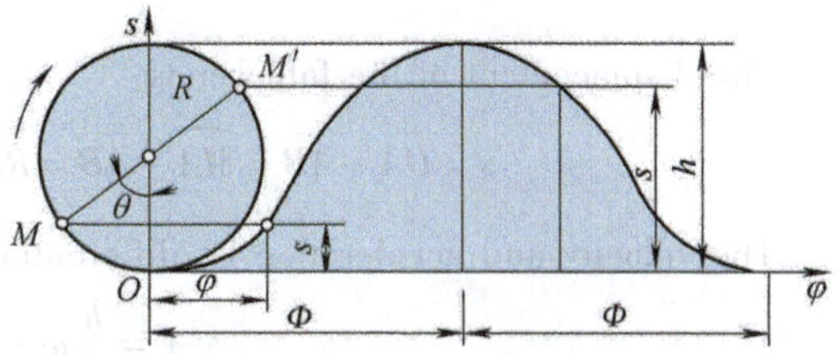


Fig. 6-14　Simple harmonic motion（简谐运动）

$$\theta=\frac{\pi}{\Phi}\varphi$$

根据图 6-14 中的几何关系，从动件在推程阶段的位移方程为

$$s=R-R\cos\theta=\frac{h}{2}-\frac{h}{2}\cos\left(\frac{\pi}{\Phi}\varphi\right) \tag{6-14a}$$

上式分别对时间求一阶、二阶导数并整理，即可得到从动件在推程阶段的速度和加速度方程为

$$v=\frac{\pi h\omega}{2\Phi}\sin\left(\frac{\pi}{\Phi}\varphi\right) \tag{6-14b}$$

$$a=\frac{\pi^2 h\omega^2}{2\Phi^2}\cos\left(\frac{\pi}{\Phi}\varphi\right) \tag{6-14c}$$

同理可得，从动件在回程阶段的运动方程为

$$\left.\begin{aligned} s&=\frac{h}{2}+\frac{h}{2}\cos\left(\frac{\pi}{\Phi'}\varphi\right)\\ v&=-\frac{\pi h\omega}{2\Phi'}\sin\left(\frac{\pi}{\Phi'}\varphi\right)\\ a&=-\frac{\pi^2 h\omega^2}{2\Phi'^2}\cos\left(\frac{\pi}{\Phi'}\varphi\right)\end{aligned}\right\} \tag{6-15}$$

当 $\Phi=\Phi'$ 时，从动件的位移、速度与加速度相对于凸轮转角的变化规律线图如图 6-15 所示。简谐运动规律的特征是从动件的加速度按余弦规律变化，因此，简谐运动规律也称为余弦加速度运动规律。

由加速度线图还可以看出，当从动件以余弦加速度运动规律运动时，在行程的起点和终点处存在有限突变，故会产生柔性冲击。但是，在无休止角的“升→降→升”类型的凸轮机构中，加速度曲线变成连续曲线，避免了柔性冲击的产生。

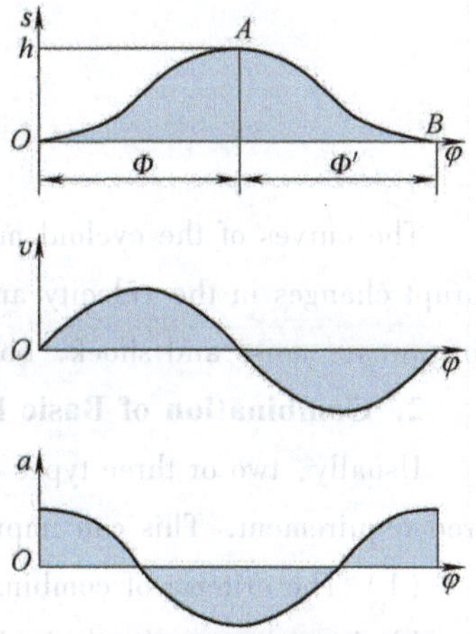


Fig. 6-15　Cosine acceleration curve（余弦运动规律）

$$R = \frac{h}{2\pi}$$

If the circle rolls an angle $\theta$, the follower moves a distance $s$, and the cam rotates an angle $\varphi$. And when the circle rolls a revolution, the follower moves the total rise $h$, and the cam rotates the rise angle $\Phi$; we have:

$$\theta = \frac{2\pi}{\Phi}\varphi$$

The displacement of the follower is:

$$s = \overline{OA} - \overline{AB} = \widehat{MA} - \overline{AB} = R\theta - R\sin\theta = \frac{h}{\Phi}\varphi - \frac{h}{2\pi}\sin\left(\frac{2\pi}{\Phi}\varphi\right) \tag{6-16a}$$

The velocity and acceleration by differentiating are:

$$v = \frac{h}{\Phi}\omega - \frac{h\omega}{\Phi}\cos\left(\frac{2\pi}{\Phi}\varphi\right) \tag{6-16b}$$

$$a = \frac{2\pi h\omega^2}{\Phi^2}\sin\left(\frac{2\pi}{\Phi}\varphi\right) \tag{6-16c}$$

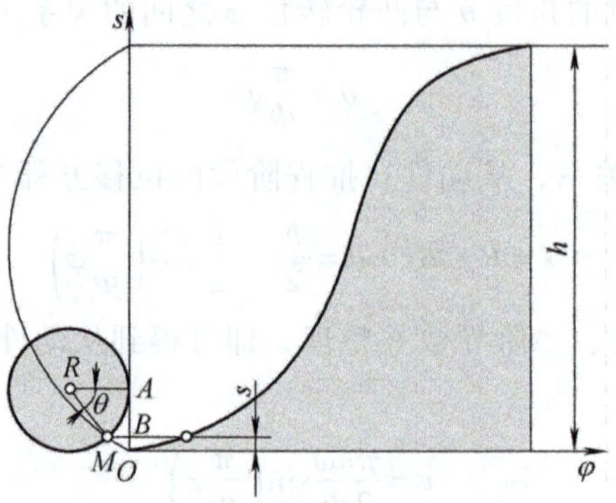


Fig. 6-16 Cycloid motion (摆线运动)

In the same way, we can obtain the equations of return travel of the follower.

$$\begin{aligned} s &= h - \frac{h}{\Phi'}\varphi + \frac{h}{2\pi}\sin\left(\frac{2\pi}{\Phi'}\varphi\right) \\ v &= -\left[\frac{h}{\Phi'}\omega - \frac{h\omega}{\Phi'}\cos\left(\frac{2\pi}{\Phi'}\varphi\right)\right] \\ a &= -\frac{2\pi h\omega^2}{\Phi'^2}\sin\left(\frac{2\pi}{\Phi'}\varphi\right) \end{aligned} \tag{6-17}$$

The curves of the cycloid motion are shown in Fig. 6-17. It is observed that there are no any abrupt changes in the velocity and acceleration at the stage of the motion. It has the lowest vibration, wear, noise and shock. So, it is the most ideal for high speed motion.

**2. Combination of Basic Follower Motions**

Usually, two or three types of basic follower motions can be combined together to satisfy the desired requirement. This can improve the dynamic performance of the cam mechanism.

(1) The criteria of combination of the follower motion

1) According to the desired function of the follower, a basic curve has been determined first, and then another basic curve which would like to minimize the inertia force and kinetic energy of the follower has been selected.

2）正弦运动规律。图 6-16 所示，当半径为 $R$ 的圆沿坐标轴线 $s$ 做纯滚动时，圆上一点 $M$ 在 $s$ 轴上投影的变化为摆线运动规律。取该圆滚动一周时点 $M$ 沿 $s$ 轴上升的距离为从动件的行程 $h$，则有 $h=2\pi R$，此时凸轮转过了推程运动角 $\Phi$。设滚圆转过 $\theta$ 角，对应从动件的位移为 $s$，凸轮转角为 $\varphi$，则滚圆转角 $\theta$ 与凸轮转角 $\varphi$ 之间的关系为

$$\theta=\frac{2\pi}{\Phi}\varphi$$

因此，根据图 6-16 中的几何关系，从动件在推程阶段的位移方程为

$$s=OA-AB=\overset{\frown}{MA}-AB=R\theta-R\sin\theta=\frac{h}{\Phi}\varphi-\frac{h}{2\pi}\sin\left(\frac{2\pi}{\Phi}\varphi\right) \tag{6-16a}$$

对式（6-16a）分别求时间的一阶、二阶导数并整理，即可得到从动件在推程阶段的速度和加速度方程为

$$v=\frac{h}{\Phi}\omega-\frac{h\omega}{\Phi}\cos\left(\frac{2\pi}{\Phi}\varphi\right) \tag{6-16b}$$

$$a=\frac{2\pi h\omega^2}{\Phi^2}\sin\left(\frac{2\pi}{\Phi}\varphi\right) \tag{6-16c}$$

同理可得，从动件在回程阶段的运动方程为

$$\left.\begin{aligned}s&=h-\frac{h}{\Phi'}\varphi+\frac{h}{2\pi}\sin\left(\frac{2\pi}{\Phi'}\varphi\right)\\ v&=-\left[\frac{h}{\Phi'}\omega-\frac{h\omega}{\Phi'}\cos\left(\frac{2\pi}{\Phi'}\varphi\right)\right]\\ a&=-\frac{2\pi h\omega^2}{\Phi'^2}\sin\left(\frac{2\pi}{\Phi'}\varphi\right)\end{aligned}\right\} \tag{6-17}$$

从动件的运动线图如图 6-17 所示。摆线运动规律的特征是从动件的加速度按照正弦规律变化，因此，摆线运动规律也称为正弦加速度运动规律。其速度和加速度均无突变，故在运动中不会产生冲击，适用于高速场合。

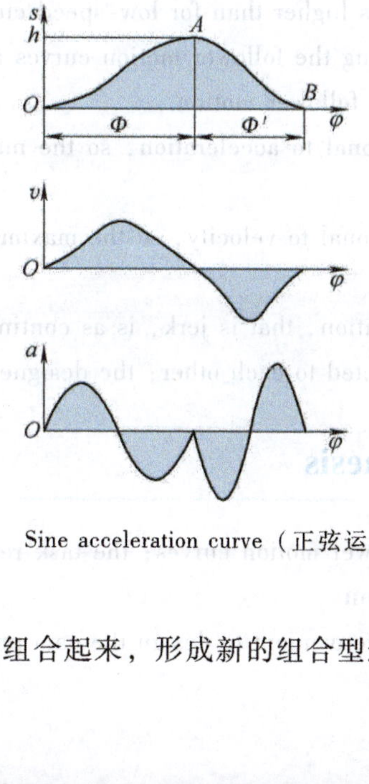


Fig. 6-17　Sine acceleration curve（正弦运动规律）

**2. 组合型运动规律**

将几种不同的基本运动规律组合起来，形成新的组合型运动规律，可以改善凸轮机构的

2) The combined curve must have excellent boundary conditions at the beginning point and the end point of the travel.

3) The displacement curve, velocity curve, acceleration curve or jerk curve must be continuous at the connecting points between the two curves to satisfy a smooth surface of the cam. This can reduce or eliminate the vibration and noise.

4) It is useful to improve the dynamic features.

(2) Examples of combined motion curve  As we know in Fig. 6-11, the accelerations at the beginning point and the end point of the follower motion are infinite. A modified curve, such as cycloid curve, polynomial curve, can be evolved, in which the accelerations are reduced to a finite values. This can be done by rounding the sharp corners of the displacement curve so that the velocity changes are gradual at the beginning point and the end point of the follower motion.

When we use half-cycloid match constant velocity curve at the beginning and the end of the displacement curve, the modified constant velocity curve can be obtained shown in Fig. 6-18.

We will have to choose a value of $\beta$ for the interval of this half-cycloid which is appropriate to the problem. Usually, the value of $\beta$ may be about $\Phi/8$-$\Phi/6$.

In this case, the magnitude of the follower velocity may be changed a little; sometimes this error can be neglected.

**3. Selection of the Basic Motion of the Follower**

The constant velocity curve is poor at any speed due to the sudden contour bump, but the modified constant velocity curve may be used in specific occasions.

The constant acceleration curve and deceleration curve should be limited to moderate speed or lower speed, but the modified trapezoidal acceleration curve can be used widely.

The simple harmonic curve can give reasonable cam-follower action at moderate speed, but the sudden change in acceleration at the beginning and end of action in dwell-rise-dwell curves precludes high-speed application.

The cycloid curve and polynomial curve are the best of all for most machines. However, the necessary accuracy of fabrication is higher than for low-speed curves.

Therefore, the criteria selecting the follower motion curves are as follows:

1) It must satisfy the desired follower motion.

2) Dynamic force is proportional to acceleration, so the maximum acceleration of the follower must be as small as possible.

3) Kinetic energy is proportional to velocity, so the maximum velocity of the follower must be as small as possible.

4) The derivation of acceleration, that is jerk, is as continuous as possible.

Some qualifications are restricted to each other; the designer should consider them impartially.

## 6.3 Cam Profile Synthesis

We are aware of several follower motion curves; the task remains to generate the cam profile.

**1. The Principle of Inversion**

The principle of inversion is convenient to design the cam profile, in which the cam is imagined

运动和动力特性。

(1) 运动规律的组合原则

1) 选择一种基本运动规律作为主体运动规律，然后用其他运动规律与之组合。

2) 在行程的起点和终点处，有较好的边界条件。

3) 在运动规律的连接点处，应满足位移、速度、加速度甚至是更高一阶导数的连续条件，以减小或避免冲击。

4) 各段运动规律要有较好的动力特性。

(2) 组合型运动规律举例　要求从动件做等速运动，行程的起点和终点处避免任何形式的冲击。因此，以等速运动规律为主体，在行程的起点和终点处可用正弦运动规律或五次多项式运动规律来组合。图 6-18 所示为等速运动规律与正弦运动规律的组合，对应凸轮转角 $\beta=\Phi/8\sim\Phi/6$。改进后，直线的斜率略有变化，其速度也发生一些变化，但对运动影响不大。

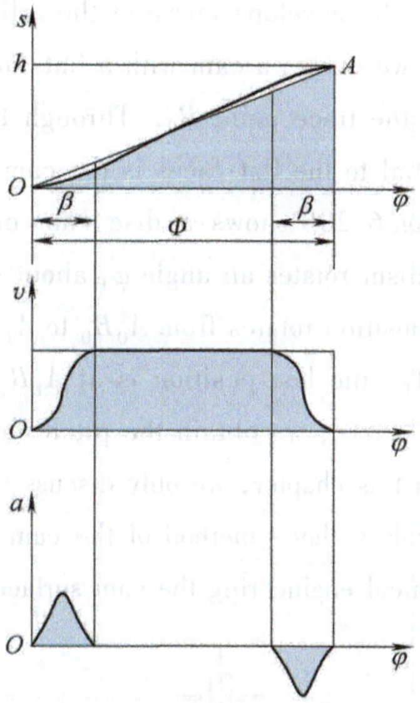


Fig. 6-18　Modified constant velocity curve（改进的等速运动规律）

**3. 从动件运动规律的选择与设计原则**

选择与设计从动件的运动规律是凸轮机构设计的一项重要内容。在进行运动规律的选择与设计时，不但要考虑凸轮机构的工作要求，还要考虑凸轮机构的工作速度和载荷的大小、从动件系统的质量、动力特性以及加工制造等因素。具体地讲，主要需要注意以下几点：

1) 从动件的最大速度 $v_{max}$ 应尽量小。$v_{max}$ 越大，则最大动量 $mv_{max}$ 越大，过大的动量会导致凸轮机构引起极大的冲击力。

2) 从动件的最大加速度 $a_{max}$ 应尽量小，且无突变。$a_{max}$ 越大，机构的惯性力就越大。特别是对于高速凸轮，应该限制最大加速度 $a_{max}$。

3) 从动件的最大跃度 $j_{max}$ 应尽量小。跃度是加速度的一阶导数，它反映了惯性力的变化率，直接影响着机构的振动和运动平稳性，因此希望越小越好。

总之，在选择与设计从动件的运动规律时，一般都希望 $v_{max}$、$a_{max}$ 和 $j_{max}$ 值尽可能小，但因为这些值之间是互相制约的，往往是此抑彼长。一般需要根据实际工作要求，分清主次来选择理想的运动规律。必要时，可对从动件运动规律的 $v_{max}$、$a_{max}$ 和 $j_{max}$ 进行优化计算。

## 6.3　凸轮轮廓曲线的设计

凸轮轮廓曲线设计方法有作图法和解析法，他们都以相对运动原理为基础。

**1. 凸轮机构的相对运动原理**

如图 6-19a 所示，在直动尖底从动件盘形凸轮机构中，当凸轮以等角速度 $\omega$ 做逆时针方向转动时，从动件做往复直线移动。设想给整个凸轮机构加上一个绕凸轮回转中心 $O$ 的反

to be stationary, while the follower rotates about the cam center at the cam's angular velocity in the opposite direction. The locus formed by the follower trace point which has the absolute angular motion about the cam center and the relative translating motion along the follower guide is the cam profile.

The fundamental basis for all cam synthesis is that the cam profile is developed by fixing the cam and moving the follower around the cam to its respective relative positions. This tenet must be remembered and the synthesis procedures will be easily understood and retained.

Fig. 6-19a shows inverse procedures of a cam with the knife-edge follower. We can obtain the cam profile. In Fig. 6-19b, the trace point is the roller center. We can only obtain the cam's pitch curve. The envelope curve of the roller is the actual profile of the cam.

If we design a cam with a flat-faced follower, the intersection of the follower guide and the flat-face is the trace point $B_0$. Through these points, such as $B_1$, $B_2$, draw the flat-faces; the curve tangential to the flat-faces is the cam actual profile. This is shown in Fig. 6-20a.

Fig. 6-20b shows a disc cam mechanism with an oscillating roller follower. When the cam mechanism rotates an angle $\varphi_1$ about cam center at an angular velocity in opposite direction, the follower position rotates from $A_0B_0$ to $A_1B_1'$, and then the follower $A_1B_1'$ oscillates an angle $\psi_1$ about its pivot $A_1$; the last position is at $A_1B_1$. Connecting the trace points $B_0$, $B_1$, ..., $B_n$, $B_0$ with a smooth curve, we obtain the pitch curve. The envelope curve of the rollers is the cam profile.

In this chapter, we only discuss the principle of inversion in cam design, and do not deal with the graphical synthesis method of the cam mechanism. The graphical synthesis is obsolete method, and in mechanical engineering the cam surface is manufactured by some numerical control machine tools.

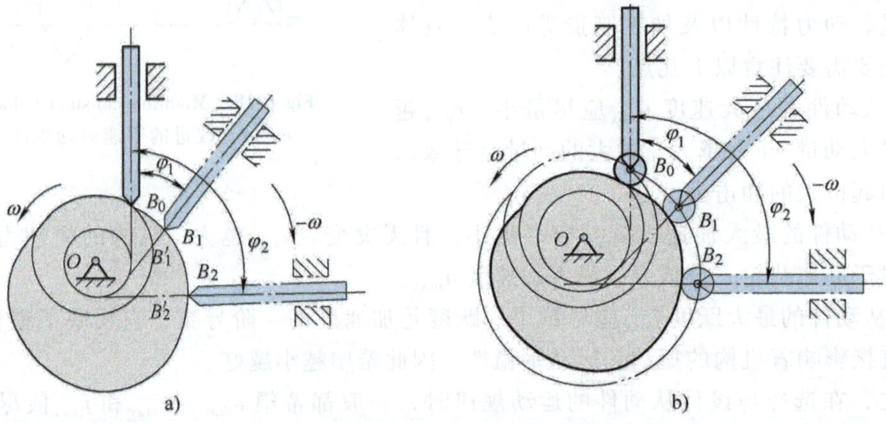


Fig. 6-19 Principle of inversion 1 (反转原理 1)

**2. Synthesis of Cam Profiles**

As cam rotates at a high speed, the graphical method often becomes inadequate for establishing accurate cam dimensions. Small errors in the profile are found to increase the dynamic loads and vibratory amplitudes considerably. Thus, a more precise determination of profiles is required. The analytical method can realize the precise synthesis, and this method is desirable to be able to determine the coordinates of points on the cam surface as well as the corresponding coordinates of the center of the milling cutter.

(1) Disc cam with translating roller follower  Fig. 6-21a shows the disc cam with an offset translating roller follower, and the following data relative to the cam synthesis are known. They are radius

向转动，使反转角速度等于凸轮的角速度，即反转角速度为 $-\omega$。此时，凸轮将静止不动，而从动件一方面随导路绕 $O$ 点以角速度 $-\omega$ 转动，分别占据 $B_1'$、$B_2'$，同时又沿其导路方向做相对移动，分别占据 $B_1$、$B_2$ 等位置。因此，从动件尖底导路的反转和从动件相对导路移动的复合运动轨迹，便形成了凸轮的轮廓曲线，这就是凸轮机构的相对运动原理，也称反转法原理。

图 6-19b 所示为直动滚子从动件盘形凸轮机构的反转示意图，把滚子中心看作尖底从动件的尖顶，仍按图 6-19a 所示的反转过程，此时产生的凸轮廓线称为理论廓线。以理论廓线上的各点为圆心，以滚子半径画圆，包络线为凸轮的实际廓线。

设计直动平底从动件盘形凸轮时。以平底与导杆交点作为尖底从动件的尖点，仍按上述方法反转，过各尖点作平底线。其包络线为凸轮的实际廓线，图 6-20a 所示为反转过程。平底从动件的假想尖点反转轨迹曲线不能称为理论廓线。

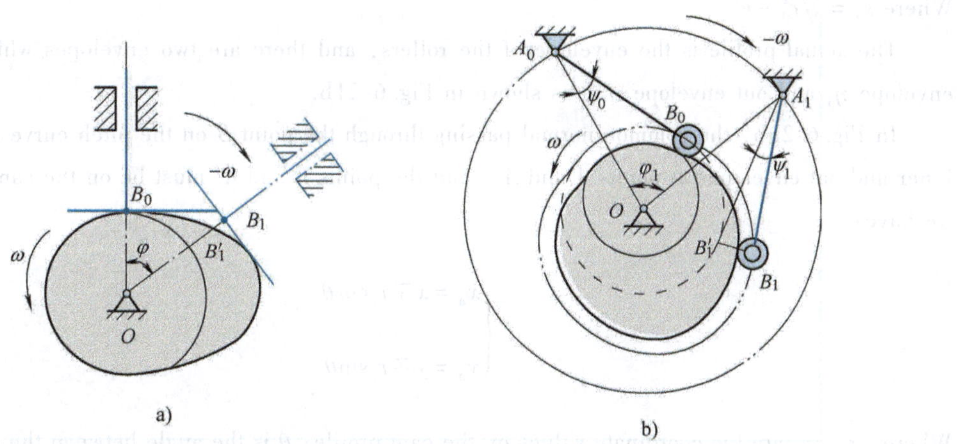


Fig. 6-20　Principle of inversion 2（反转原理 2）

同理，对图 6-20b 所示的摆动滚子从动件盘形凸轮机构施加角速度 $-\omega$ 的反转后，凸轮静止不动。从动件由初始位置 $A_0B_0$ 反转 $\varphi_1$ 角后到达 $A_1B_1'$，再绕 $A_1$ 点摆动 $\psi_1$ 角到达 $A_1B_1$。同样，可作出从动件由初始位置 $A_0B_0$ 反转到的其他位置。因此，将 $B_0$、$B_1$、$B_2$ 等点光滑连接，即可得到凸轮的理论轮廓曲线，其包络线为实际廓线。

**2. 凸轮廓线的设计**

用解析法进行凸轮廓线设计的主要任务是建立凸轮廓线方程。

（1）直动滚子从动件盘形凸轮廓线的设计　建立原点 $O$ 位于凸轮转动中心的直角坐标系 $Oxy$，如图 6-21a 所示。设初始位置时滚子中心 $B_0$ 点为凸轮推程段理论廓线的起始点。凸轮机构反转 $\varphi$ 角后，点 $B_0$ 到达点 $B$。此时，从动件的位移 $s=\overline{B'B}$。从动件上点 $B$ 的运动可以看作是由 $B_0$ 点先绕 $O$ 点反转 $\varphi$ 角到达凸轮基圆上的 $B'$点，然后，$B'$点再沿导路移动位移 $s$ 到达 $B$ 点。

设凸轮机构的偏距为 $e$，基圆半径为 $r_0$。由图 6-21a 所示几何关系可求出 $B$ 点坐标为

of prime circle $r_0$, offset of the follower axis $e$, radius of the roller $r_r$, and follower motion $s = s(\varphi)$.

The origin of the Cartesian $Oxy$ coordinate system is placed at the cam's center of rotation, and the $y$ axis is defined parallel to the follower guide. Suppose the first position of the follower is at the point $B_0$ which is the initial point of the rise travel on the cam's pitch curve, and when the mechanism rotates through an angle $\varphi$ about the cam's center in opposite direction, the follower position arrives the point $B$. The distance $\overline{B'B}$ represents the follower displacement $s$.

The coordinate of the point $B$ can be written easily from the Fig. 6-21a.

$$\begin{cases} x = (s_0 + s)\sin\varphi + e\cos\varphi \\ y = (s_0 + s)\cos\varphi - e\sin\varphi \end{cases} \tag{6-18}$$

These equations can determine the coordinate values of the pitch curve.

Where $s_0 = \sqrt{r_0^2 - e^2}$

The actual profile is the envelope of the rollers, and there are two envelopes which are inner envelope $\eta_1$ and out envelope $\eta_2$, as shown in Fig. 6-21b.

In Fig. 6-21b, the common normal passing through the point $B$ on the pitch curve intersects the inner and out envelopes at points $A$ and $A'$, and the points $A$ and $A'$ must be on the cam profiles, so we have:

$$\begin{cases} x_a = x \mp r_r\cos\theta \\ y_a = y \mp r_r\sin\theta \end{cases} \tag{6-19}$$

Where, $x_a$, $y_a$ are the coordinate values on the cam profile; $\theta$ is the angle between the common normal passing through the trace point $B$ on the pitch curve and the $x$ axis.

The upper signs are used to determine the profile $\eta_1$ and lower signs are used to determine the profile $\eta_2$.

From the advanced mathematics, we know that the slopes of normal and tangent passing through an arbitrary point on a curve are equal to the negative reciprocal each other, so we get:

$$\tan\theta = \frac{\sin\theta}{\cos\theta} = -\frac{dx}{dy} = \frac{\dfrac{dx}{d\varphi}}{-\dfrac{dy}{d\varphi}} \tag{6-20}$$

We differentiate the equations (6-18), yielding:

$$\begin{aligned} \frac{dx}{d\varphi} &= (s_0 + s)\cos\varphi + \frac{ds}{d\varphi}\sin\varphi - e\sin\varphi \\ \frac{dy}{d\varphi} &= -(s_0 + s)\sin\varphi + \frac{ds}{d\varphi}\cos\varphi - e\cos\varphi \end{aligned} \tag{6-21}$$

$$\begin{cases} x = (s_0 + s)\sin\varphi + e\cos\varphi \\ y = (s_0 + s)\cos\varphi - e\sin\varphi \end{cases} \tag{6-18}$$

式（6-18）为凸轮的理论廓线方程。式中，$s_0 = \sqrt{r_0^2 - e^2}$。

凸轮的实际廓线是圆心位于理论廓线上的一系列滚子圆族的包络线（图 6-21b），而且，滚子圆族的包络线应该有两条，分别对应于外凸轮和内凸轮的实际廓线。

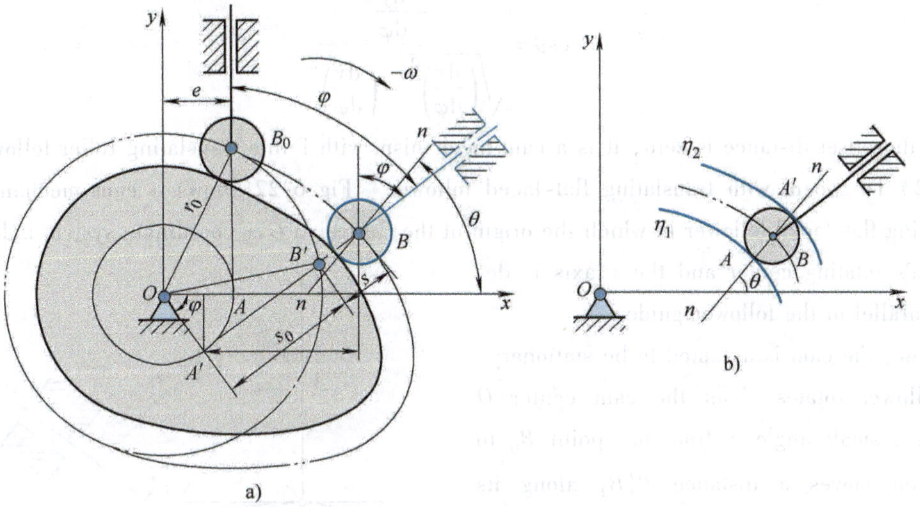


Fig. 6-21　Cam synthesis with translating roller follower
（直动滚子从动件盘形凸轮的轮廓曲线设计）

设过凸轮理论廓线上 $B$ 点的公法线与滚子圆族的包络线交于 $A$、$A'$点，则 $A$、$A'$点也是凸轮实际廓线上的点。设 $A$ 或 $A'$点的坐标为（$x_a$，$y_a$），则凸轮的实际廓线方程为

$$\begin{cases} x_a = x \mp r_r\cos\theta \\ y_a = y \mp r_r\sin\theta \end{cases} \tag{6-19}$$

式中，$r_r$ 为滚子半径；$\theta$ 为公法线与 $x$ 轴的夹角；$x$、$y$ 为滚子圆心位于理论廓线上的坐标；“－”号用于求解外凸轮的实际廓线 $\eta_1$；“+”号用于计算内凸轮的实际廓线 $\eta_2$。

利用高等数学的知识，曲线上任意一点法线的斜率与该点切线斜率互为负倒数，所以有

$$\tan\theta = \frac{\sin\theta}{\cos\theta} = -\frac{dx}{dy} = \frac{\frac{dx}{d\varphi}}{-\frac{dy}{d\varphi}} \tag{6-20}$$

对式（6-18）求导可得

$$\begin{aligned} \frac{dx}{d\varphi} &= (s_0 + s)\cos\varphi + \frac{ds}{d\varphi}\sin\varphi - e\sin\varphi \\ \frac{dy}{d\varphi} &= -(s_0 + s)\sin\varphi + \frac{ds}{d\varphi}\cos\varphi - e\cos\varphi \end{aligned} \tag{6-21}$$

整理后可得

Rearranging the above equations, we get:

$$\sin\theta = \frac{\dfrac{dx}{d\varphi}}{\sqrt{\left(\dfrac{dx}{d\varphi}\right)^2 + \left(\dfrac{dy}{d\varphi}\right)^2}}$$

$$\cos\theta = \frac{-\dfrac{dy}{d\varphi}}{\sqrt{\left(\dfrac{dx}{d\varphi}\right)^2 + \left(\dfrac{dy}{d\varphi}\right)^2}} \tag{6-22}$$

If the offset distance is zero, it is a cam mechanism with inline translating roller follower.

(2) Disc cam with translating flat-faced follower Fig. 6-22 shows a cam mechanism with translating flat-faced follower in which the origin of the Cartesian $Oxy$ coordinate system is located at the cam's rotating center and the $y$ axis is defined parallel to the follower guide.

Now, the cam is assumed to be stationary. The follower rotates about the cam center $O$ through a small angle $\varphi$ from the point $B_0$ to $B_1'$, then moves a distance $\overline{B_1'B_1}$ along its guide. The contact point between the cam profile and the flat-face is at point $B$. The displacement of the follower is $s = B_1'B_1$. The instant center of the cam and follower is located at point $P$, and we have:

$$\overline{OP} = \overline{BB_1} = \frac{v}{\omega} = \frac{ds}{d\varphi}$$

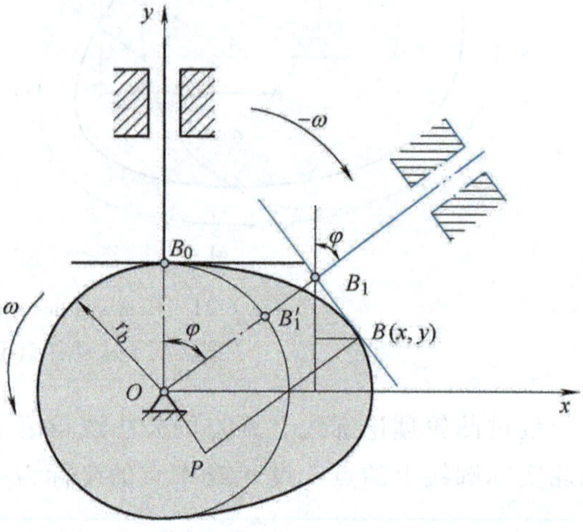

Fig. 6-22 Cam synthesis with translating flat-faced follower（直动平底从动件盘形凸轮的廓线设计）

The coordinates of point $B$ on the cam profile are as follows:

$$\begin{cases} x = (r_b + s)\sin\varphi + \dfrac{ds}{d\varphi}\cos\varphi \\ y = (r_b + s)\cos\varphi - \dfrac{ds}{d\varphi}\sin\varphi \end{cases} \tag{6-23}$$

(3) Disc cam with oscillating roller follower Fig. 6-23 shows a cam mechanism having an oscillating roller follower, in which the origin of the Cartesian $Oxy$ coordinate system is located at the cam rotating center $O$ and the $y$ axis passes through the oscillating pivot of the follower. The data relative to the cam synthesis are known. They are radius of prime circle $r_0$, length of the follower $l$, radius of the roller $r_r$, the distance between the cam center and oscillating center of the follower $a$, and follower motion $s = s(\varphi)$.

Now, the cam is assumed to be stationary. The follower rotates about the cam center $O$ through a small angle $\varphi$ from position $A_0B_0$ to $AB'$, then oscillates an angle $\psi$ about its center $A$. The last position is at $AB$.

$$\sin\theta=\frac{\dfrac{\mathrm{d}x}{\mathrm{d}\varphi}}{\sqrt{\left(\dfrac{\mathrm{d}x}{\mathrm{d}\varphi}\right)^2+\left(\dfrac{\mathrm{d}y}{\mathrm{d}\varphi}\right)^2}}$$

$$\cos\theta=\frac{-\dfrac{\mathrm{d}y}{\mathrm{d}\varphi}}{\sqrt{\left(\dfrac{\mathrm{d}x}{\mathrm{d}\varphi}\right)^2+\left(\dfrac{\mathrm{d}y}{\mathrm{d}\varphi}\right)^2}} \tag{6-22}$$

若设计对心直动滚子从动件盘形凸轮机构，则上述公式中令 $e=0$ 即可。

（2）直动平底从动件盘形凸轮廓线的设计　建立原点 $O$ 位于凸轮的回转中心的直角坐标系 $Oxy$，如图 6-22 所示。当从动件在初始位置时，从动件的平底相切于行程的起始点 $B_0$。当凸轮机构反转 $\varphi$ 角后，从动件导路线与平底的交点到达 $B_1$ 点，凸轮与从动件平底的切点从 $B_0$ 到达 $B$ 点。此时，从动件的位移 $s=\overline{B_1'B_1}$。

从图 6-22 上可以看出，在 $B$ 点接触时的瞬心为 $P$。

$$\overline{OP}=\overline{BB_1}=\frac{v}{\omega}=\frac{\mathrm{d}s}{\mathrm{d}\varphi}$$

$B$ 的坐标为

$$x=(r_b+s)\sin\varphi+\frac{\mathrm{d}s}{\mathrm{d}\varphi}\cos\varphi$$
$$y=(r_b+s)\cos\varphi-\frac{\mathrm{d}s}{\mathrm{d}\varphi}\sin\varphi \tag{6-23}$$

式（6-23）即为直动平底从动件盘形凸轮的实际廓线方程。

（3）摆动滚子从动件盘形凸轮廓线的设计　建立图 6-23 所示的直角坐标系 $Oxy$，原点 $O$ 位于凸轮的回转中心。当从动件在初始位置时，从动件位于行程的起始位置 $A_0B_0$。当凸轮机构反转 $\varphi$ 角后，从动件 $A_0B_0$ 运动到 $AB$ 位置。此时，从动件的角位移 $\psi=\angle B'AB$。从动件的运动可以看成是 $A_0B_0$ 先绕 $O$ 点反转 $\varphi$ 角到达 $AB'$ 位置，$AB'$ 再绕 $A$ 点摆动 $\psi$ 角到达 $AB$ 位置。设机架的长度为 $a$，摆杆的长度为 $l$，$B$ 点的坐标为 $(x, y)$，由图示几何关系，可求出 $B$ 点的坐标为

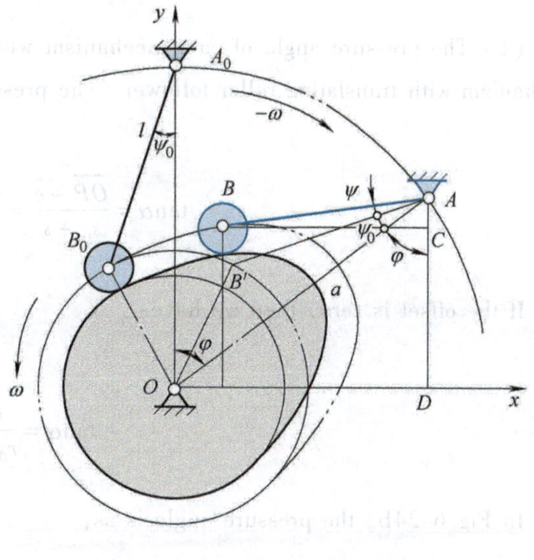


Fig. 6-23　Cam synthesis with oscillating roller follower
（摆动滚子从动件盘形凸轮的廓线设计）

The equations of the pitch curve can be obtained easily from Fig. 6-23.

$$\begin{cases} x = a\sin\varphi - l\sin(\varphi + \psi_0 + \psi) \\ y = a\cos\varphi - l\cos(\varphi + \psi_0 + \psi) \end{cases} \tag{6-24}$$

Where $\psi_0$ is the angle between the initial position of the follower and $y$ axis, and its value is as:

$$\psi_0 = \arccos\frac{a^2 + l^2 - r_0^2}{2al}$$

The equations of the cam profile are the same as the equations (6-19).

In the same way, the equations of the cam profile with an oscillating flat-faced follower can also be determined by this method.

## 6.4 Sizes of Cam Mechanisms

The follower motion is necessary to design a cam profile, but some sizes of the cam mechanism, such as the radius of the prime circle, radius of the roller or length of the flat-face, the offset distance of the follower, are very important to determine the cam sizes. The pressure angle, curvature, undercutting and practical factors of the follower location are necessary to design a cam mechanism.

**1. Pressure Angle of Cam Mechanism**

As previously stated, the pressure angle is the angle between the normal to the pitch curve at a contact point and the direction of the follower motion. It is the distribution of forces in follower. The maximum pressure angle establishes the cam size, torque, loads, wear life, and other pertinent factors.

(1) The pressure angle of cam mechanism with translating follower Fig. 6-24a shows a cam mechanism with translating roller follower. The pressure angle $\alpha$ is:

$$\tan\alpha = \frac{\overline{OP} - e}{s_0 + s} = \frac{\frac{ds}{d\varphi} - e}{\sqrt{r_0^2 - e^2} + s} \tag{6-25}$$

If the offset is zero, then we have:

$$\tan\alpha = \frac{\frac{ds}{d\varphi}}{r_0 + s} \tag{6-26}$$

In Fig. 6-24b, the pressure angle is as:

$$\alpha = 90° - \gamma$$

$$\begin{cases} x = a\sin\varphi - l\sin(\varphi + \psi_0 + \psi) \\ y = a\cos\varphi - l\cos(\varphi + \psi_0 + \psi) \end{cases} \tag{6-24}$$

式中，$\psi_0$ 为摆杆的初始位置角，且有

$$\psi_0 = \arccos\frac{a^2 + l^2 - r_0^2}{2al}$$

式（6-24）即为凸轮的理论廓线方程，实际廓线方程求法同前。

平底摆动从动件盘形凸轮机构廓线也可按上述方法设计。

## 6.4　凸轮机构基本尺寸的设计

设计凸轮的轮廓曲线时，不仅要求从动件能够实现预期的运动规律，还应该保证凸轮机构具有合理的结构尺寸和良好的运动、力学性能。因此，基圆半径、偏距和滚子半径、压力角等基本尺寸和参数的选择也是凸轮机构设计的重要内容。

**1. 凸轮机构的压力角**

凸轮机构的压力角是指不计摩擦时，凸轮与从动件在某瞬时接触点处的公法线方向与从动件运动方向之间所夹的锐角，常用 $\alpha$ 表示。压力角是衡量凸轮机构受力情况好坏的一个重要参数。

（1）直动从动件凸轮机构的压力角　图 6-24 所示为直动滚子从动件盘形凸轮机构，接触点 $B$ 处的压力角为 $\alpha$。$P$ 点为从动件与凸轮的瞬心。压力角 $\alpha$ 可从几何关系中找出。

$$\tan\alpha = \frac{\overline{OP} - e}{s_0 + s} = \frac{\dfrac{ds}{d\varphi} - e}{\sqrt{r_0^2 - e^2} + s} \tag{6-25}$$

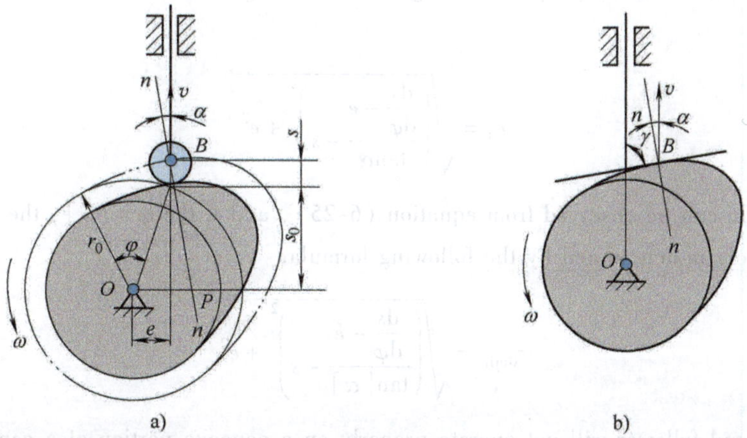


Fig. 6-24　Pressure angle of radial cam with translating follower（直动从动件盘形凸轮机构的压力角）

正确选择从动件的偏置方向有利于减小机构的压力角。此外，压力角还与凸轮的基圆半径和偏距等参数有关。

当偏距 $e=0$ 时，代入式（6-25），即可得到对心直动从动件盘形凸轮机构的压力角计算公式：

$$\tan\alpha = \frac{\dfrac{ds}{d\varphi}}{r_0 + s} \tag{6-26}$$

Where $\gamma$ is the angle between the guide axis and the flat-face. If $\gamma = 90°$, then $\alpha = 0°$, all the transmitted force goes into motion of the follower and none goes into slip velocity.

(2) The pressure angle of cam mechanism with oscillating follower  Fig. 6-25a shows a cam mechanism with oscillating roller follower. The pressure angle is described in this figure also, and the pressure angle of cam mechanism with oscillating flat-faced follower is described in Fig. 6-25b.

(3) Allowable pressure angle  The allowable pressure angles are shown in Tab. 6-1.

**Tab. 6-1 Allowable pressure angles of cam mechanisms** (凸轮机构的许用压力角)

| Type of closure 封闭形式 | Type of follower motion 从动件的运动方式 | Rise travel 推程 | Return travel 回程 |
|---|---|---|---|
| Force closure 力封闭 | Translating follower 直动从动件 | $[\alpha] = 25° \sim 35°$ | $[\alpha'] = 70° \sim 80°$ |
| | Oscillating follower 摆动从动件 | $[\alpha] = 35° \sim 45°$ | $[\alpha'] = 70° \sim 80°$ |
| Form closure 形封闭 | Translating follower 直动从动件 | $[\alpha] = 25° \sim 35°$ | $[\alpha'] = [\alpha]$ |
| | Oscillating follower 摆动从动件 | $[\alpha] = 35° \sim 45°$ | $[\alpha'] = [\alpha]$ |

If the maximum pressure angle is greater than the allowable pressure angle, this can increase the follower sliding or pivot friction to undesirable level and may tend to jam a translating follower in its guide.

Therefore, the maximum pressure angle must be less than the allowable pressure angle, that is:

$$\alpha_{\max} < [\alpha]$$

**2. Sizes of Cam Mechanism**

(1) Radius of prime circle  The shape of the cam in general is controlled by pressure angle. As the cam becomes smaller, the pressure angle becomes larger. This can be shown in the following formula.

$$r_0 = \sqrt{\left(\frac{\dfrac{ds}{d\varphi} - e}{\tan\alpha} - s\right)^2 + e^2} \tag{6-27}$$

This formula can be observed from equation (6-25), and if the $\alpha = [\alpha]$, the minimum radius of the prime circle is determined by the following formula.

$$r_{0\min} = \sqrt{\left(\frac{\dfrac{ds}{d\varphi} - e}{\tan[\alpha]} - s\right)^2 + e^2} \tag{6-28}$$

The flat-faced follower will not operate properly on a concave portion of a cam profile, therefore, the cam should be designed so that the follower inequality is maintained at all points on the profile. That is to say, the radius of curvature of the cam profile must be greater than zero.

$$\rho > 0$$

The radius of curvature is a mathematical property of a function, and its value is:

$$\rho = \frac{(1 + y'^2)^{3/2}}{y''} \tag{6-29}$$

对于图 6-24b 所示的直动平底从动件盘形凸轮机构，根据图中的几何关系，其压力角为

$$\alpha = 90° - \gamma$$

式中，$\gamma$ 为从动件的平底与导路中心线的夹角。显然，平底直动从动件凸轮机构的压力角为常数，机构的受力方向不变，运转平稳性好。如果从动件的平底与导路中心轴线之间的夹角 $\gamma = 90°$，则压力角 $\alpha = 0°$。

（2）摆动从动件凸轮机构的压力角　图 6-25 为摆动从动件盘形凸轮机构的压力角示意图，其中，图 6-25a 为滚子从动件的压力角示意图，图 6-25b 为平底从动件的压力角示意图。

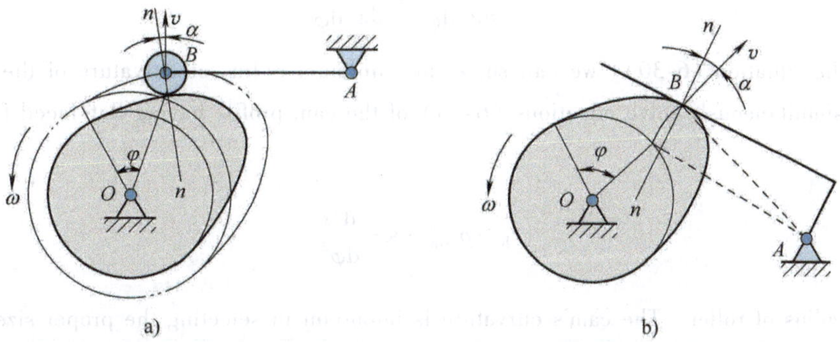


Fig. 6-25　Pressure angle of radial cam with oscillating follower（摆动从动件盘形凸轮机构的压力角示意图）

（3）凸轮机构的许用压力角　凸轮机构的压力角与基圆半径、偏距和滚子半径等基本尺寸有直接的关系，这些参数之间往往互相制约。增大凸轮的基圆半径可以获得较小的压力角，但凸轮尺寸增大；反之，减小凸轮的基圆半径，可以获得较为紧凑的结构，但同时又使凸轮机构的压力角增大。压力角过大会降低机械效率，因此，必须对凸轮机构的最大压力角加以限制，使其小于许用压力角，即 $\alpha_{max} < [\alpha]$。凸轮机构的许用压力角见表 6-1，供设计人员参考。

**2. 凸轮机构基本尺寸的设计**

（1）基圆半径的设计　对于直动滚子从动件盘形凸轮，可根据式（6-25）求解出凸轮的基圆半径为

$$r_0 = \sqrt{\left(\frac{\frac{ds}{d\varphi} - e}{\tan\alpha} - s\right)^2 + e^2} \tag{6-27}$$

显然，压力角 $\alpha$ 越大，基圆半径越小，机构的结构越紧凑。在其他参数不变的情况下，当 $\alpha = [\alpha]$ 时，可以使凸轮机构在满足压力角条件的同时，获得紧凑的结构。此时，最小基圆半径为

$$r_{0\min} = \sqrt{\left(\frac{\frac{ds}{d\varphi} - e}{\tan[\alpha]} - s\right)^2 + e^2} \tag{6-28}$$

对于直动平底从动件盘形凸轮，凸轮廓线上各点的曲率半径 $\rho > 0$。曲率半径的计算公式为

$$\rho = \frac{(1 + y'^2)^{3/2}}{y''} \tag{6-29}$$

Where $y' = \frac{dy}{dx} = \frac{\frac{dy}{d\varphi}}{\frac{dx}{d\varphi}}$

A derivation for radius of curvature can be found in any calculus text. The equation for the radius of curvature of the cam pitch curve can be written as follows:

$$\rho = \frac{\left[\left(\frac{dx}{d\varphi}\right)^2 + \left(\frac{dy}{d\varphi}\right)^2\right]^{3/2}}{\frac{dx}{d\varphi}\frac{d^2y}{d\varphi^2} - \frac{dy}{d\varphi}\frac{d^2x}{d\varphi^2}} \tag{6-30}$$

From the equation (6-30), we can solve the minimum radius of curvature of the cam profile $\rho_{min}$, then simultaneously solve equations (6-23) of the cam profile having flat-faced follower. We get:

$$r_b > \rho_{min} - s - \frac{d^2s}{d\varphi^2} \tag{6-31}$$

(2) Radius of roller The cam's curvature is important in selecting the proper size of roller or length of the flat-face of the follower.

In this expression, the $\rho$, $\rho_a$, $r_r$ are the radius of curvature of the cam pitch curve, radius of curvature of the cam profile curve and the radius of the roller.

Fig. 6-26a shows a convex cam having a roller follower, and from this figure we have:

$$\rho_a = \rho - r_r$$

If the radius of the roller is increased to the radius of cam pitch curve, that is $r_r = \rho$, then $\rho_a = 0$. This is shown in Fig. 6-26b. The cutter will create a sharp point, or cusp, on the cam profile. This cam will not run very well at speeds, and serious wears will occur at this point.

If the radius of the roller is greater than that of the cam, then $\rho_a < 0$. This is shown in Fig. 6-26c. The result is that the cam will be undercut and become a pointed cam. This cam no longer has the same displacement function so the cam should be carefully designed. Thus, to have a minimum radius of curvature of the cam profile, the radius of the roller must be less than that of the minimum radius of curvature of the cam pitch curve. That is:

$$r_r \leqslant 0.8\rho_{min}$$

The undercutting phenomenon exists only in the cam having convex surface. It is not possible for a concave surface of the cam. Fig. 6-26d shows a concave cam having a roller follower. Obviously we have:

$$\rho_a = \rho + r_r$$

No matter how large the radius of roller is, the radius of curvature of the cam profile is greater than zero.

From the viewpoint of strength, the radius of roller must satisfy the following condition:

$$r_r \geqslant (0.1 \sim 0.5) r_0$$

式中，$y'=\frac{dy}{dx}=\frac{\frac{dy}{d\varphi}}{\frac{dx}{d\varphi}}$，代入式（6-29）并整理得

$$\rho=\frac{\left[\left(\frac{dx}{d\varphi}\right)^2+\left(\frac{dy}{d\varphi}\right)^2\right]^{3/2}}{\frac{dx}{d\varphi}\frac{d^2y}{d\varphi^2}-\frac{dy}{d\varphi}\frac{d^2x}{d\varphi^2}} \tag{6-30}$$

令 $\rho>\rho_{min}$，代入平底从动件盘形凸轮的廓线方程，可得

$$r_b>\rho_{min}-s-\frac{d^2s}{d\varphi^2} \tag{6-31}$$

（2）滚子半径的设计　在设计滚子尺寸时，必须保证滚子同时满足运动特性要求和强度要求。

图 6-26 所示为外凸廓线中的滚子圆族的包络情况。设理论廓线上某点的曲率半径为 $\rho$，实际廓线在对应点的曲率半径为 $\rho_a$，滚子半径为 $r_r$，根据几何关系有：$\rho_a=\rho-r_r$。

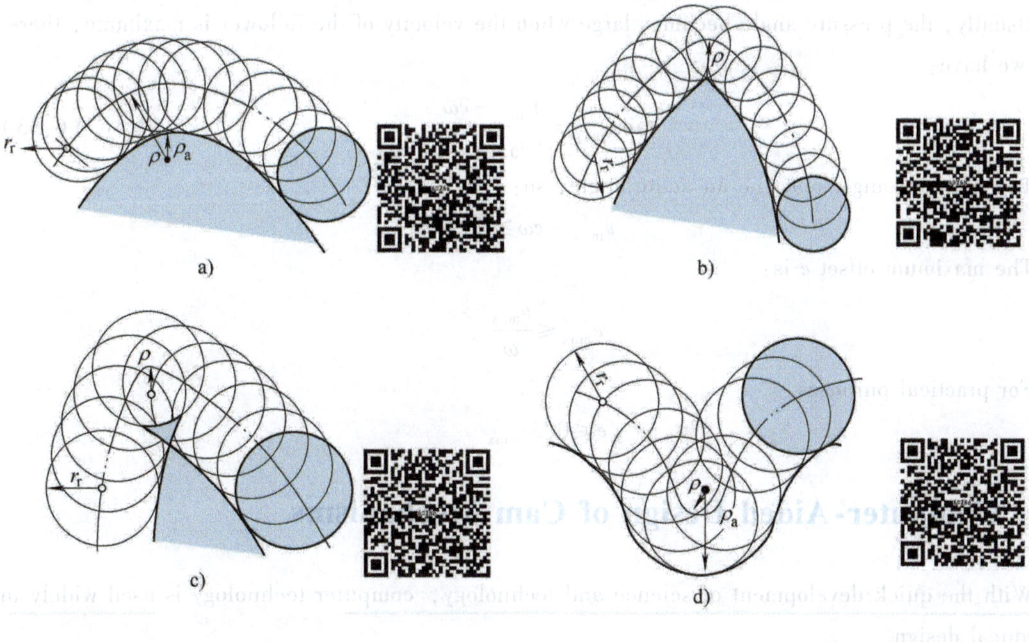


Fig. 6-26　Relationship between the roller and the cam pitch curve（凸轮滚子尺寸与廓线的关系）

图 6-26a 中，$\rho-r_r>0$，图 6-26b 中，$\rho-r_r=0$，实际廓线最小曲率半径为零，表明在该位置出现尖点，运动过程中容易磨损；图 6-26c 中，$\rho-r_r<0$，实际廓线曲率半径为负值，说明在包络加工过程中，图中交叉的阴影部分将被切掉，从而导致机构的运动发生失真。因此，为了避免产生这种现象，要对滚子的半径加以限制。通常情况下，应保证

$$r_r\leqslant 0.8\rho_{min}$$

对于图 6-26d 所示的内凹廓线中滚子圆族的包络情况，由于 $\rho_a=\rho+r_r$，不会出现运动失真问题。

从强度要求考虑，滚子半径应满足

$$r_r\geqslant(0.1\sim0.5)r_0$$

(3) The length of the flat-face of the follower  When the cam rotates about its center, the flat-faced follower must maintain contact with the cam surface at all times so that the flat-face has sufficient length.

The length of the flat-face of the follower can be determined in Fig. 6-27.

$$l = 2\,\overline{OP}_{max} + \Delta l = 2\left(\frac{ds}{d\varphi}\right)_{max} + \Delta l$$

Where $l$ is the length of the flat-faced follower; $P$ is the instantaneous center of the cam and the follower; $\Delta l$ is an additional length for safety, usually $\Delta l = 5 \sim 7$mm.

(4) Offset  The direction and magnitude of the offset can influence the pressure angle of cam mechanisms. To reduce the pressure angle, the location of the offset can be determined by observing.

The magnitude of the offset is calculated in the following formula:

$$\tan\alpha = \frac{\dfrac{ds}{d\varphi} - e}{\sqrt{r_0^2 - e^2} + s} = \frac{\dfrac{v}{\omega} - e}{s_0 + s} = \frac{v - e\omega}{(s_0 + s)\,\omega} \tag{6-32}$$

Usually, the pressure angle becomes large when the velocity of the follower is maximum, therefore, we have:

$$\tan\alpha_{max} = \frac{v_{max} - e\omega}{(s_0 + s)\,\omega} \tag{6-33}$$

The pressure angle must be an acute angle, so:

$$v_{max} - e\omega \geqslant 0$$

The maximum offset $e$ is:

$$e_{max} \leqslant \frac{v_{max}}{\omega}$$

For practical purposes,

$$e \leqslant 0.5e_{max}$$

## 6.5 Computer-Aided Design of Cam Mechanisms

With the quick development of science and technology, computer technology is used widely in mechanical design.

In this case, analytical expressions for displacement, velocity, acceleration and the equations of the cam profile in terms of the general parameters are derived. A digital computer facilitates the calculation work.

Fig. 6-28 shows a flow chart of computer-aided design of cam mechanism.

In the given motion of follower, the total rise, angle of ascent, angle of descent, angle of dwell, and the type of follower motion must be designed first.

The given data include the radius of prime circle of the cam, radius of roller, offset, allowable pressure angle, etc.

The analysis of the calculating result includes the cam sizes, the maximum pressure angle, jerk, maximum velocity and acceleration, etc.

（3）平底长度的设计　如图 6-27 所示，在平底从动件盘形凸轮机构运动过程中，应能保证从动件的平底在任意时刻均与凸轮接触，因此，平底的长度 $l$ 应满足以下条件：

$$l = 2\,\overline{OP}_{\max} + \Delta l = 2\left(\frac{\mathrm{d}s}{\mathrm{d}\varphi}\right)_{\max} + \Delta l$$

式中，$\Delta l$ 为附加长度，由具体的结构而定，一般取 $\Delta l = 5 \sim 7\text{mm}$。

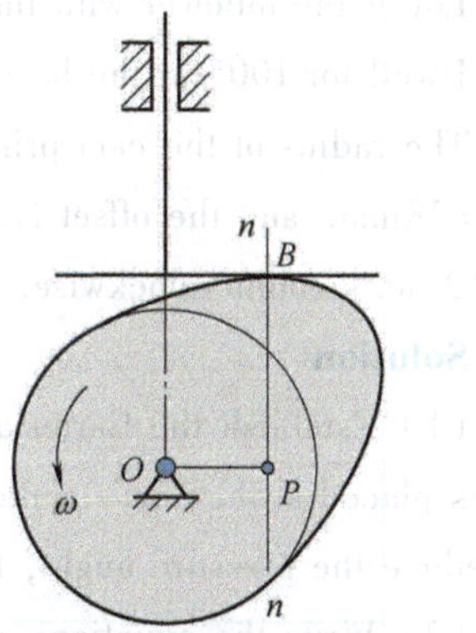


Fig. 6-27　Length of the flat-faced follower（平底从动件的长度）

（4）偏距的设计　从动件的偏置方向可直接影响凸轮机构压力角的大小，因此，在选择从动件的偏置方向时需要遵循的原则是：尽可能减小凸轮机构在推程阶段的压力角。其偏置的距离推导如下：

$$\tan\alpha = \frac{\dfrac{\mathrm{d}s}{\mathrm{d}\varphi} - e}{\sqrt{r_0^2 - e^2} + s} = \frac{\dfrac{v}{\omega} - e}{s_0 + s} = \frac{v - e\omega}{(s_0 + s)\omega} \tag{6-32}$$

一般情况下，从动件运动速度的最大值发生在凸轮机构压力角最大的位置，则式（6-32）可改写为

$$\tan\alpha_{\max} = \frac{v_{\max} - e\omega}{(s_0 + s)\omega} \tag{6-33}$$

由于压力角为锐角，故有 $v_{\max} - e\omega \geqslant 0$。

由式（6-33）可知，增大偏距，有利于减小凸轮机构的压力角，但偏距的增加也有限度，其最大值应满足

$$e_{\max} \leqslant \frac{v_{\max}}{\omega}$$

因此，当设计偏置式凸轮机构时，其从动件偏置方向的确定原则是：从动件应置于使该凸轮机构的压力角减小的位置。

综上所述，在进行凸轮机构基本尺寸的设计时，由于各参数之间有时是互相制约的，因此，应综合考虑各种因素，使其综合性能指标满足设计要求。

## 6.5　计算机辅助凸轮设计

随着计算机技术和精密数控加工技术的发展，凸轮加工已经摆脱了图解设计法，解析法应用日渐广泛。图 6-28 所示框图为解析法的具体应用实例，也是凸轮机构计算机辅助设计的基本过程。

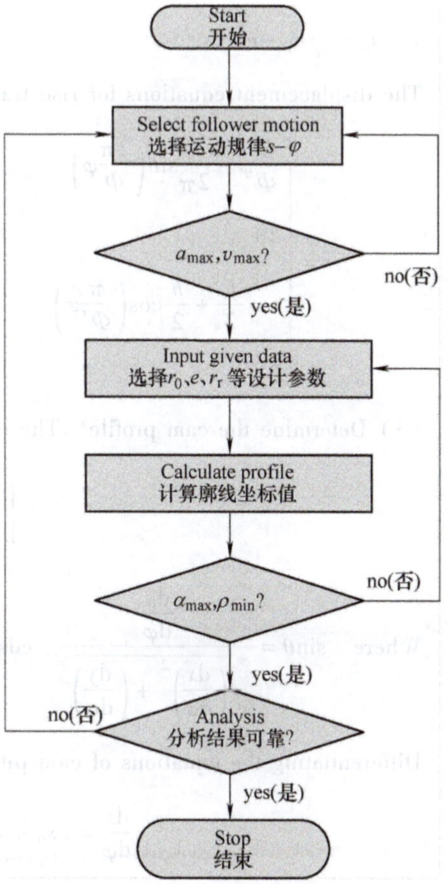


Fig. 6-28　Cam profile design procedure（凸轮廓线设计过程）

**Example 6-1**: Design a disc cam mechanism with translating offset roller follower shown in Fig. 6-29. The follower motions are:

Rise the follower with the cycloid motion for 135°;

Dwell for 45° at the highest position of the follower;

Lower the follower with the simple harmonic motion for 80°;

Dwell for 100° at the lowest position of the follower;

The radius of the cam prime circle is $r_0 = 65\text{mm}$; the radius of the roller is $r_r = 12\text{mm}$, the lift is $h = 28\text{mm}$. and the offset is $e = 12\text{mm}$. The cam rotates about its center with an angular velocity $\omega = 12\text{rad/s}$ counterclockwise.

**Solution**

(1) Establish the Cartesian $Oxy$ coordinate system shown in Fig. 6-29 The origin of the system is placed at the cam's center of rotation, and the $y$ axis is defined parallel to the follower guide. To reduce the pressure angle, the follower is on the right of the cam center.

(2) Write the equations of the pitch curve of the cam

$$\begin{cases} x = (s_0 + s)\sin\varphi + e\cos\varphi \\ y = (s_0 + s)\cos\varphi - e\sin\varphi \end{cases}$$

Where $s_0 = \sqrt{r_0^2 - e^2}$.

The displacement equations for rise travel, dwell and return travel are as follows:

$$s = \begin{cases} \dfrac{h}{\Phi}\varphi - \dfrac{h}{2\pi}\sin\left(\dfrac{2\pi}{\Phi}\varphi\right) & \varphi \in [0°,\ 135°] & \text{(rise travel)} \\ 28 & \varphi \in (135°,\ 180°) & \text{(dwell)} \\ s = \dfrac{h}{2} + \dfrac{h}{2}\cos\left(\dfrac{\pi}{\Phi'}\varphi\right) & \varphi \in [180°,\ 260°] & \text{(return travel)} \\ 0 & \varphi \in (260°,\ 360°) & \text{(dwell)} \end{cases}$$

(3) Determine the cam profile The equations of the actual profile of the cam are as follows:

$$\begin{cases} x_a = x - r_r\cos\theta \\ y_a = y - r_r\sin\theta \end{cases}$$

Where $\sin\theta = \dfrac{\dfrac{dx}{d\varphi}}{\sqrt{\left(\dfrac{dx}{d\varphi}\right)^2 + \left(\dfrac{dy}{d\varphi}\right)^2}}$, $\cos\theta = \dfrac{-\dfrac{dy}{d\varphi}}{\sqrt{\left(\dfrac{dx}{d\varphi}\right)^2 + \left(\dfrac{dy}{d\varphi}\right)^2}}$

Differentiating the equations of cam pitch curve with respect to $\varphi$, we have:

$$\frac{dx}{d\varphi} = (s_0 + s)\cos\varphi + \frac{ds}{d\varphi}\sin\varphi - e\sin\varphi$$

$$\frac{dy}{d\varphi} = -(s_0 + s)\sin\varphi + \frac{ds}{d\varphi}\cos\varphi - e\cos\varphi$$

**例 6-1**　设计一偏置直动滚子从动件盘形凸轮机构。已知从动件的行程 $h=28\text{mm}$，在推程阶段和回程阶段分别以摆线运动规律和简谐运动规律运动，且有：$\Phi=135°$，$\Phi_s=45°$，$\Phi'=80°$，$\Phi'_s=100°$；凸轮的角速度 $\omega=12\text{rad/s}$，以逆时针方向匀速转动，基圆半径 $r_0=65\text{mm}$；凸轮机构的偏距 $e=12\text{mm}$，滚子半径 $r_r=12\text{mm}$。试用解析法设计此凸轮机构的凸轮廓线。

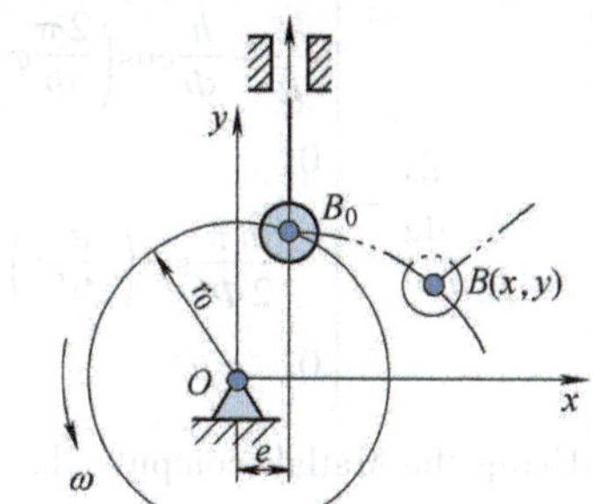

Fig. 6-29　Cam profile design
（凸轮轮廓曲线的设计）

**解**

（1）选择从动件的偏置方向　建立图 6-29 所示的直角坐标系 $Oxy$，为减小压力角，从动件置于凸轮转动中心右侧。

（2）求解凸轮的理论廓线　凸轮的理论廓线方程为

$$\begin{cases} x=(s_0+s)\sin\varphi+e\cos\varphi \\ y=(s_0+s)\cos\varphi-e\sin\varphi \end{cases}$$

式中，$s_0=\sqrt{r_0^2-e^2}$。

从动件在不同阶段的位移方程分别为

$$s=\begin{cases} \dfrac{h}{\Phi}\varphi-\dfrac{h}{2\pi}\sin\left(\dfrac{2\pi}{\Phi}\varphi\right) & \varphi\in[0°,\ 135°] \quad（推程阶段）\\ 28 & \varphi\in(135°,\ 180°) \quad（远休止阶段）\\ s=\dfrac{h}{2}+\dfrac{h}{2}\cos\left(\dfrac{\pi}{\Phi'}\varphi\right) & \varphi\in[180°,\ 260°] \quad（回程阶段）\\ 0 & \varphi\in(260°,\ 360°) \quad（近休止阶段）\end{cases}$$

（3）求解凸轮的实际廓线　设该凸轮机构中的凸轮构件是外凸轮，凸轮的实际廓线方程为

$$\begin{cases} x_a=x-r_r\cos\theta \\ y_a=y-r_r\sin\theta \end{cases}$$

式中

$$\sin\theta=\frac{\dfrac{dx}{d\varphi}}{\sqrt{\left(\dfrac{dx}{d\varphi}\right)^2+\left(\dfrac{dy}{d\varphi}\right)^2}},\ \cos\theta=\frac{-\dfrac{dy}{d\varphi}}{\sqrt{\left(\dfrac{dx}{d\varphi}\right)^2+\left(\dfrac{dy}{d\varphi}\right)^2}}$$

而

$$\frac{dx}{d\varphi}=(s_0+s)\cos\varphi+\frac{ds}{d\varphi}\sin\varphi-e\sin\varphi$$

$$\frac{dy}{d\varphi}=-(s_0+s)\sin\varphi+\frac{ds}{d\varphi}\cos\varphi-e\cos\varphi$$

The rate of change of the displacement with respect to the cam angle $\varphi$ is as:

$$\frac{ds}{d\varphi}=\begin{cases}\frac{h}{\Phi}-\frac{h}{\Phi}\cos\left(\frac{2\pi}{\Phi}\varphi\right) & \varphi\in[0^\circ,\ 135^\circ] \quad \text{(rise travel)}\\ 0 & \varphi\in(135^\circ,\ 180^\circ) \quad \text{(dwell)}\\ -\frac{h\pi}{2\Phi'}\sin\left(\frac{\pi}{\Phi'}\varphi\right) & \varphi\in[180^\circ,\ 260^\circ] \quad \text{(return travel)}\\ 0 & \varphi\in(260^\circ,\ 360^\circ) \quad \text{(dwell)}\end{cases}$$

Using the Matlab computer language program, the calculating results of the pitch curve and profile can be obtained by a computer. This is shown in Tab. 6-2. The pitch curve and profile of the cam can be drawn by a computer also, as shown in Fig. 6-30.

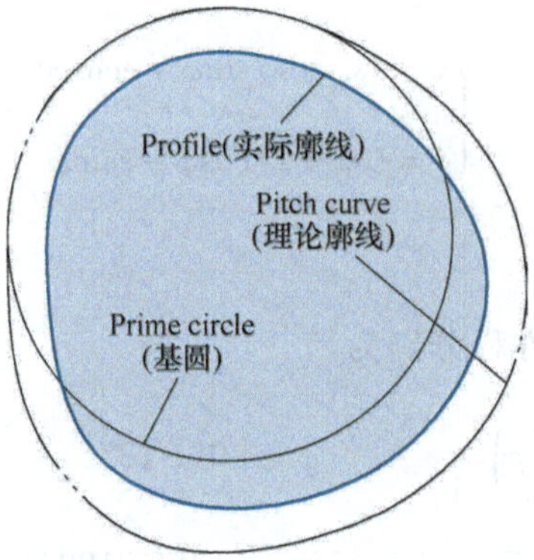


Fig. 6-30 Cam pitch curve and profile (凸轮的理论廓线和实际廓线)

**Tab. 6-2 Result of example 6-1** (例 6-1 的计算结果)

| Cam angle $\varphi$ 凸轮转角 | Pitch curve 理论廓线坐标 | | Cam profile 实际廓线坐标 | |
|---|---|---|---|---|
| | $x$ | $y$ | $x_a$ | $y_a$ |
| 0° | 12.000 | 63.883 | 9.785 | 52.089 |
| 10° | 22.924 | 60.901 | 18.912 | 49.592 |
| 20° | 33.322 | 56.465 | 27.989 | 45.715 |
| 30° | 43.250 | 50.912 | 36.909 | 40.724 |
| 40° | 52.844 | 44.309 | 45.585 | 34.753 |
| 50° | 62.112 | 36.453 | 53.845 | 27.754 |
| 60° | 70.781 | 27.009 | 61.377 | 19.555 |
| 70° | 78.263 | 15.715 | 67.71 | 10.004 |
| 80° | 83.748 | 2.582 | 72.257 | -0.876 |
| 90° | 86.409 | -12 | 74.436 | -12.807 |
| 100° | 85.635 | -27.285 | 73.804 | -25.277 |
| 110° | 81.210 | -42.328 | 70.167 | -37.631 |
| 120° | 73.359 | -56.21 | 63.62 | -49.2 |
| 130° | 62.666 | -68.248 | 54.519 | -59.436 |

（续）

| Cam angle $\varphi$ 凸轮转角 | Pitch curve 理论廓线坐标 | | Cam profile 实际廓线坐标 | |
|---|---|---|---|---|
| | $x$ | $y$ | $x_a$ | $y_a$ |
| 140° | 49.869 | -78.1 | 43.41 | -67.986 |
| 150° | 35.549 | -85.573 | 30.945 | -74.491 |
| 160° | 20.149 | -90.446 | 17.54 | -78.733 |
| 170° | 4.138 | -92.571 | 3.602 | -80.583 |
| 180° | -12.000 | -91.883 | -10.446 | -79.984 |
| 190° | -27.588 | -87.354 | -22.548 | -76.463 |
| 200° | -41.300 | -78.384 | -33.375 | -69.373 |
| 210° | -52.012 | -66.088 | -42.031 | -59.426 |
| 220° | -59.255 | -51.948 | -48.038 | -47.684 |
| 230° | -63.271 | -37.426 | -51.47 | -35.247 |
| 240° | -64.875 | -23.599 | -52.894 | -22.92 |
| 250° | -65.136 | -10.937 | -53.136 | -11.005 |
| 260° | -64.996 | 0.725 | -52.997 | 0.591 |
| 270° | -63.883 | 12 | -52.089 | 9.785 |
| 280° | -60.828 | 22.911 | -49.599 | 18.681 |
| 290° | -55.926 | 33.125 | -45.601 | 27.01 |
| 300° | -49.324 | 42.334 | -40.218 | 34.518 |
| 310° | -41.224 | 50.256 | -33.613 | 40.978 |
| 320° | -31.870 | 56.65 | -25.987 | 46.192 |
| 330° | -21.549 | 61.324 | -17.571 | 50.003 |
| 340° | -10.573 | 64.134 | -8.621 | 52.294 |
| 350° | 0.725 | 64.996 | 0.591 | 52.997 |
| 360° | 12.000 | 63.883 | 9.785 | 52.089 |

同样，由于位移 $s$ 与从动件所处的运动阶段有关，所以有

$$\frac{ds}{d\varphi}=\begin{cases}\dfrac{h}{\Phi}-\dfrac{h}{\Phi}\cos\left(\dfrac{2\pi}{\Phi}\varphi\right) & \varphi\in[0°,\ 135°] & （推程阶段）\\ 0 & \varphi\in(135°,\ 180°) & （远休止阶段）\\ -\dfrac{h\pi}{2\Phi'}\sin\left(\dfrac{\pi}{\Phi'}\varphi\right) & \varphi\in[180°,\ 260°] & （回程阶段）\\ 0 & \varphi\in(260°,\ 360°) & （近休止阶段）\end{cases}$$

代入已知条件，并利用 Matlab 软件求解，得到凸轮理论廓线和实际廓线的坐标值见表 6-2。分别将凸轮理论廓线和实际廓线上的所有坐标点光滑连接，即可得到该凸轮的理论廓线和实际廓线，如图 6-30 所示。

实际加工时，在行程起始点附近的凸轮转角分度不超过 1°，其他位置的凸轮转角小于 3°。可将位移方程直接写入数控机床或输入凸轮廓线上的坐标值。

# Chapter 7

## Design of Cear Mechanisms

# 第7章

## 齿轮机构及其设计

## 7.1 Classification of Gear Mechanisms

A gear may be thought of as a friction wheel with teeth cut around the circumference, thus gears are used to transmit motion from one shaft to another, and this process is accomplished by successively engaging teeth.

Although gears had been used as early as the 2nd century B. C., it was not until the invention of method for milling and for generating tooth profile in the latter half the 19th century that gear could be fabricated with sufficient accuracy to permit their use at high speed and heavy loads.

Gears can be conveniently classified according to the relative positions of their shaft axes. There are parallel shaft gears, intersecting shaft gears and skew shaft gears.

**1. Parallel Shaft Gears**

The parallel shaft gears are also called planar gear mechanisms.

(1) Spur gears The teeth are straight and parallel to the gear axis. This gear is spur gear. If the gears have external teeth on the outer surface of the cylinders, they rotate in the opposite direction, as shown in Fig. 7-1a. If the teeth are formed on the inner surface of an annulus ring, this gear is an internal gear. An internal gear can mesh with a external pinion only and they rotate in the same direction, as shown in Fig. 7-1b. When a spur gear is made of infinite diameter so that the pitch surface is a plane, this gear is called a rack. The pinion and rack combination converts rotary motion into translational motion or vice-versa, as shown in Fig. 7-1c.

a) b) c)

Fig. 7-1 Spur gears (直齿圆柱齿轮)

(2) Helical gears Like spur gears, helical gears can also be used to connect parallel shafts, but their teeth are not parallel to their shafts axes, each being helical in shape. Two meshing gears have the same helix angle, but have teeth of opposite hands, as shown in Fig. 7-2. Helical gears can also be classified as external contact, internal contact, helical pinion and rack.

The helical gears can be used at higher velocities than spur gears and have greater lord-carrying capacity. Helical gears have the disadvantage of having end thrust as there is a force component along the gear axis.

(3) Herringbone gears Herringbone gears are equivalent to a pair of helical gears with opposite helix angles mounted side by side, as shown in Fig. 7-3. The axial thrust forces of the two rows of teeth cancel each other. So they can be used at high speeds with less noise and vibrations.

## 7.1　齿轮机构的分类

齿轮是周向均布轮齿的轮子，是重要的机械传动机构，应用非常广泛，可以用来传递空间两任意轴之间的运动和动力。

**1. 平面齿轮机构**

用于传递两平行轴间运动和动力的齿轮机构称为平面齿轮机构。

（1）直齿圆柱齿轮机构　图 7-1 所示为直齿圆柱齿轮机构，各轮齿方向与齿轮的轴线平行。图 7-1a 所示为外啮合直齿圆柱齿轮机构；图 7-1b 所示为内啮合直齿圆柱齿轮机构；图 7-1c 所示为齿轮齿条机构，其中齿条可看成直径为无穷大的齿轮的一部分，齿轮做回转运动，而齿条做直线移动。

（2）斜齿圆柱齿轮机构　图 7-2 所示为斜齿圆柱齿轮机构，轮齿方向与其轴线方向有一倾斜角，称为斜齿圆柱齿轮的螺旋角。

（3）人字齿轮机构　图 7-3 所示为人字齿轮机构，其齿形如人字，可看成由两个螺旋方向相反的斜齿轮构成。

Fig. 7-2　Helical gears（斜齿圆柱齿轮）

Fig. 7-3　Herringbone gears（人字齿轮）

**2. 空间齿轮机构**

用于传递相交轴或交错轴间运动和动力的齿轮机构称为空间齿轮机构。

（1）锥齿轮机构　锥齿轮的轮齿分布在圆锥体的表面上，两齿轮的轴线相交。图7-4a 所示为直齿锥齿轮，图 7-4b 所示为斜齿锥齿轮，图 7-4c 所示为曲齿锥齿轮。直齿锥齿轮制造较为简单，应用广泛；斜齿锥齿轮的轮齿倾斜于圆锥母线，制造困难，应用较少；曲齿锥齿轮的轮齿为曲线形，传动平稳，适用于高速、重载传动，但制造成本较高。

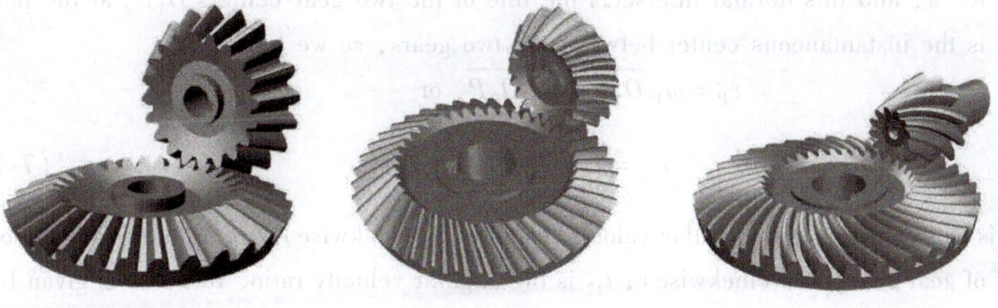

a)　　b)　　c)

Fig. 7-4　Bevel gears（锥齿轮）

### 2. Spatial Gears

When motion is to be transmitted between shafts whose axes intersect or skew by gears, the gears are called spatial gears.

(1) Intersecting shaft gears  If the teeth are formed on the cone surface, this gear is called a bevel gear. The motion is transmitted between two intersecting shafts; usually the bevel gears are made for a shaft angle of 90°.

If the teeth are straight and coincident with their cone elements, this gear is called straight-teeth bevel gear shown in Fig. 7-4a. Usually they are used to connect shafts at right angles which run at low speeds.

When the teeth of bevel gear are inclined at an angle to the face of bevel, this gear is known as helical bevel gear or spiral bevel gear, as shown in Fig. 7-4b. They are smooth in action than straight bevel gears.

The curved teeth bevel gear has teeth that are curved, but with a zero degree spiral angle, as shown in Fig. 7-4c. They are quieter in action than the other bevel gears.

(2) Skew shaft gears  To transmit motion between the skew shafts, the spiral gears or crossed helical gears, worm gears and Hypoid gears are often used in machines.

The spiral gears or crossed helical gears are limited to high load, and their shafts can be set at any angle, as shown in Fig. 7-5.

(3) Worm gears  Fig. 7-6 shows worm gears in which a few teeth of the smaller gear are wrapped around its circumference a number of times and form screw threads, and the larger wheel has concave shape in its diameter direction so that to envelope the portion of the smaller gear. The smaller gear is called worm; the larger gear is called worm gear.

The worm gears have a large transmission ratio, but wear will occur seriously.

Fig. 7-5 Crossed helical gears (交错轴斜齿轮)

## 7.2 Fundamental Law of Gearing

### 1. Law of Gearing

During meshing of gears, the profile of a tooth on the gear 1 is meshing at the point $K$ with another profile on that gear 2 to produce a rotary motion. Fig. 7-7 shows two contacting profiles with common normal $n$—$n$, and this normal intersects the line of the two gear centers $O_1O_2$ at the point $P$. The point $P$ is the instantaneous center between the two gears, so we have:

$$v_P = \omega_1 \overline{O_1P} = \omega_2 \overline{O_2P}, \text{ or}$$

$$i_{12} = \frac{\omega_1}{\omega_2} = \frac{\overline{O_2P}}{\overline{O_1P}} \tag{7-1}$$

Where $\omega_1$ is the instantaneous angular velocity of gear 1 (clockwise); $\omega_2$ is the instantaneous angular velocity of gear 2 (counterclockwise); $i_{12}$ is the angular velocity ratio. Its value is given by:

$$i_{12} = \frac{\omega_1}{\omega_2}$$

（2）交错轴斜齿轮机构　图 7-5 所示为交错轴斜齿圆柱齿轮机构，其中的每一个齿轮都是斜齿圆柱齿轮。

（3）蜗杆传动机构　蜗杆传动机构通常用于两垂直交错轴之间的传动，如图 7-6 所示。蜗杆传动可获得大传动比，应用广泛。

Fig. 7-6　Worm gears（蜗轮蜗杆）

在众多齿轮机构中，直齿圆柱齿轮是最简单、应用最广泛的一种，也是本课程重点学习的内容。

## 7.2　齿廓啮合基本定律

### 1. 齿廓啮合的基本定律

齿轮传动是靠主动齿轮的齿廓推动从动齿轮的齿廓来实现的，图 7-7 所示齿轮机构中，$O_1$、$O_2$ 分别为两齿轮的转动中心。主动轮 1 与从动轮 2 的角速度比值称为瞬时传动比，用 $i_{12}$ 表示。$i_{12}=\omega_1/\omega_2$，$\omega_1$ 和 $\omega_2$ 分别为两齿轮的角速度。

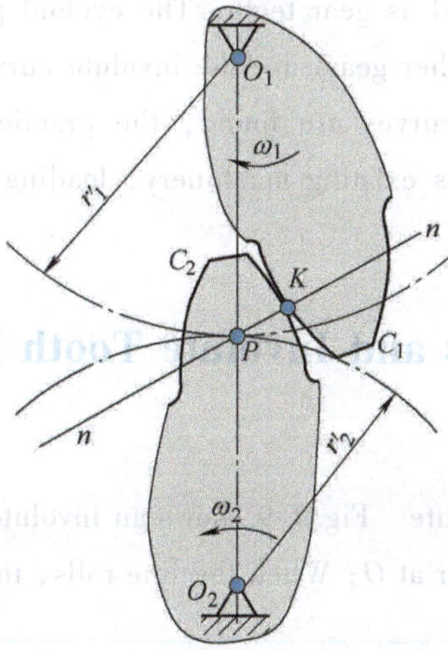


Fig. 7-7　Fundamental law of gearing（齿廓啮合的基本定律）

瞬时传动比与齿廓的形状有关。齿廓啮合基本定律揭示了齿廓曲线与两轮传动比之间的关系。主动齿轮 1 的齿廓 $C_1$ 与从动齿轮 2 的齿廓 $C_2$ 在 $K$ 点接触，过 $K$ 点作两齿廓的公法线 $n—n$，此公法线与连心线交于 $P$ 点，$P$ 点为两齿廓的速度瞬心。根据瞬心的概念可得

In gearing, the point $P$ is called the pitch point.

This equation is frequently used to define the law of gearing. It states that the angular velocity ratio of the two gears with direct contact is inversely proportional to the ratio of the two segments cut the line of centers by the common normal at the contact point. That is to say, the common normal passing through the contact point of the two profiles should always pass through a fixed point $P$, and it divides the center line in the inverse ratio of angular velocity of two gears.

If the pitch point is fixed, the ratio is a constant, and the locus of the point $P$ on the gear 1 is called the pitch circle 1 in which the radius is $r_1'$ and rotating center is at $O_1$. The locus of the point $P$ on the gear 2 is called the pitch circle 2 in which the radius is $r_2'$ and rotating center is $O_2$. The two pitch circles are always tangent and roll without slipping.

If the pitch point varies for all phase of the gearing, the ratio is not a constant. This type of gear is called noncircular gears. Fig. 7-8 shows a pair of noncircular gears.

**2. Conjugate Profiles**

When all the common normal lines for every instantaneous point of contact can pass through the pitch point, this is the fundamental law of gearing; any meshing profiles which satisfy the law of gearing can be called the conjugate profiles.

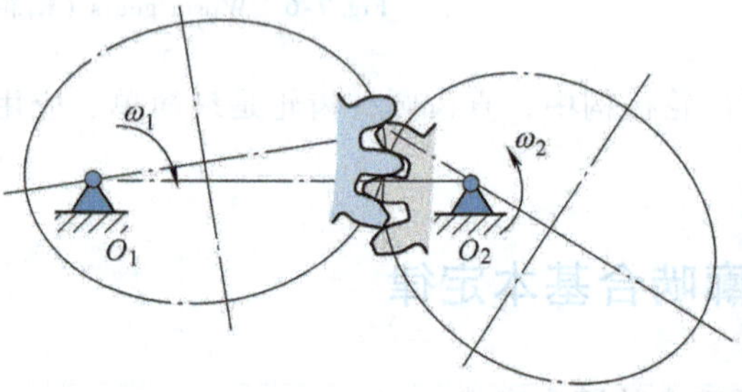

Fig. 7-8 Noncircular gears (非圆齿轮)

In order to maintain the fundamental law of gearing to be true, the gear tooth profiles on meshing gears must be conjugate of one another. There are infinite numbers of possible conjugate pairs that can be used, but only a few curves have been practically applied as gear teeth. The cycloid profile is still used in watches and clocks as a tooth form, but most other gears use the involute curve for their shape.

Even though new conjugate curves are found, the practical problems of reproducing these curves must be considered, such as existing machinery, loading capacity, and the changes in the distance centers, etc.

## 7.3 Involute Properties and Involute Tooth Profiles

**1. Involute**

(1) Development of the involute Fig. 7-9 shows an involute generated by a line rolling on the circumference of a circle with center at $O$; When the line rolls, the path generated by the point $K$ on the line is the involute curve.

The circle is known as the base circle with radius $r_b$; the line rolling on the base circle is called generating line, and it is always normal to the involute and tangent to the base circle. The point $A$ is the start point of the involute curve on the base circle, and the central angle $\angle AOK$ corresponding to the involute curve segment $AK$ is called the unfolding angle of the involute.

When one involute profile pushes against another at the point $K$, the force at any instant is a-

$$v_P=\omega_1\ \overline{O_1P}=\omega_2\ \overline{O_2P}$$

故两轮的瞬时传动比为

$$i_{12}=\frac{\omega_1}{\omega_2}=\frac{\overline{O_2P}}{\overline{O_1P}} \tag{7-1}$$

一对齿轮的瞬时传动比等于两齿廓接触点处的公法线分连心线$O_1O_2$所成两段线段长度的反比。这一结论称为齿廓啮合的基本定律。

由式（7-1）可知，两轮的瞬时传动比与瞬心 $P$ 的位置有关，而瞬心 $P$ 的位置与齿廓曲线的形状有关。在齿轮传动机构中，又把瞬心 $P$ 称为节点。

若两齿轮的瞬时传动比为常数，则 $P$ 必为定点，此时节点 $P$ 随齿轮 1 的运动轨迹为以 $O_1$ 为圆心、以$\overline{O_1P}$为半径的圆。同理，节点 $P$ 在齿轮 2 的运动平面上的轨迹为以 $O_2$ 为圆心、以$\overline{O_2P}$为半径的圆。这两个圆分别称为齿轮 1 和齿轮 2 的节圆，其半径分别用 $r_1'$ 和 $r_2'$ 表示，这种齿轮机构称为圆形齿轮机构。两轮在节点 $P$ 处的相对速度等于零，说明一对齿轮的啮合传动相当于两齿轮节圆的纯滚动。

如果节点 $P$ 的位置是变动的，则为变传动比齿轮机构。这时节点在两个齿轮的运动平面上的轨迹为非圆曲线，称为节线，这种齿轮机构称为非圆齿轮机构，图 7-8 所示的非圆齿轮机构即为变传动比齿轮机构的示例。

**2. 共轭齿廓及齿廓曲线的选择**

（1）共轭齿廓　能满足齿廓啮合定律的一对齿廓称为共轭齿廓。共轭齿廓的齿廓曲线称为共轭曲线。一对共轭齿廓上相互啮合的点称为共轭点。共轭点的集合就是共轭曲线。

（2）齿廓曲线的选择　给出一个齿轮的齿廓曲线，可根据齿廓啮合的基本定律求出与之共轭的另一个齿轮的齿廓曲线。因此，可以作为共轭齿廓的曲线是很多的。实际中选择齿廓曲线时除了满足给定传动比的要求外，还应考虑设计、制造、测量、安装、互换性和强度等方面的问题。渐开线齿廓能够较为全面地满足上述几方面的要求，因此渐开线是定传动比齿轮传动中最常用的齿廓曲线，此外摆线和圆弧曲线也有应用。

## 7.3　渐开线齿廓及其啮合特点

**1. 渐开线的形成、特性及渐开线方程**

（1）渐开线的形成　如图 7-9 所示，当一直线 $L$ 沿半径为 $r_b$ 的圆的圆周作纯滚动时，直线 $L$ 上任意一点 $K$ 的轨迹 $AK$ 称为该圆的渐开线，简称渐开线；这个圆称为渐开线的基圆，其半径用 $r_b$ 表示；直线 $L$ 称为渐开线的发生线；$A$ 为渐开线在基圆上的起始点；角 $\theta_K$（$\angle AOK$）称为渐开线 $AK$ 段的展角。

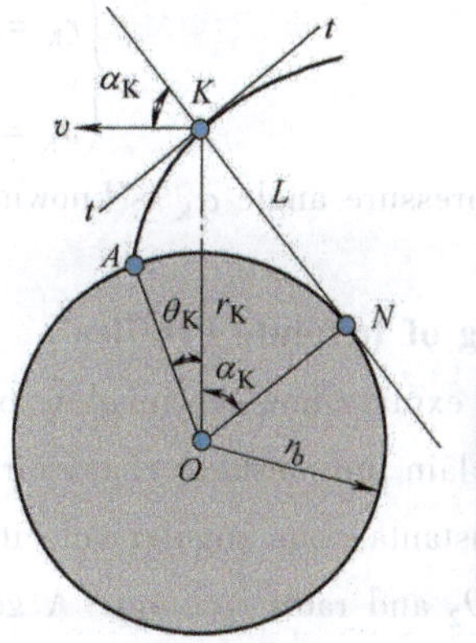


Fig. 7-9　Development of the involute（渐开线的形成）

（2）渐开线的特性

1）发生线沿基圆做纯滚动，因此发生线沿基圆滚过的长度$\overline{KN}$等于基圆被滚过的弧长$\overset{\frown}{AN}$，即

long the common normal, representing the direction of force action, such as $KN$. The angle between the line of the force action and the velocity direction at point $K$ of the gear is called the pressure angle of the involute at point $K$. It is often expressed by the angle $\angle NOK$.

(2) Involute properties The important properties possessed by the involute curve are as follows.

1) The length of the generating line is equal to the arc length which the generating line rolls without slipping on the base circle, that is:

$$\overline{KN} = \widehat{AN}$$

2) The generating line is always tangent to the base circle, and it is always a normal of the involute at a point $K$.

3) The length $\overline{NK}$ is the curvature radius of the involute at the point $K$. the point $N$ is the center of the curvature of the involute at the point $K$. Conversely, the farther the point $K$ is from the base circle, the greater the radius of the curvature is. The radius of curvature at the point $A$ on the involute is zero, so there is no involute within the base circle.

4) The shape of the involute depends upon the radius of the base circle. The smaller the radius of the base circle is, the steeper the involute is. If the radius of the base circle is infinite, the involute curve becomes a straight line. This is shown in Fig. 7-10.

(3) Equations of involute curve If we take the line $OA$ as the polar axis of the polar coordinate, the length of $OK$ is polar radius, and the $\angle AOK$ is the polar angle. The equations of the involute curve can be found easily from Fig. 7-9.

$$r_K = \frac{r_b}{\cos\alpha_K}$$

$$\tan\alpha_K = \frac{\overline{NK}}{\overline{ON}} = \frac{\widehat{AN}}{r_b} = \frac{r_b(\alpha_K + \theta_K)}{r_b} = \alpha_K + \theta_K$$

Let $\theta_K = \tan\alpha_K - \alpha_K$

The polar angle $\theta_K$ is the function of the pressure angle $\alpha_K$ at the point $K$, then:

$$\theta_K = \mathrm{inv}\alpha_K = \tan\alpha_K - \alpha_K$$

Rearranging the above equations, we have the equations of involute in polar coordinate.

$$\begin{cases} r_K = \dfrac{r_b}{\cos\alpha_K} \\ \theta_K = \mathrm{inv}\alpha_K = \tan\alpha_K - \alpha_K \end{cases} \tag{7-2}$$

When the pressure angle $\alpha_K$ is known, the polar angle $\theta_K$ can be determined by using table or by calculating.

**2. Meshing of Involute Profiles**

Let us now explain how the involute profile satisfies the requirement for transmission of uniform motion, and explain the meshing characteristics.

(1) The instantaneous angular velocity ratio is constant Fig. 7-11 shows two base circles with centers at $O_1$, $O_2$ and radii $r_{b1}$, $r_{b2}$. A generating line $N_1N_2$ is tangent to the two base circles at point $N_1$, $N_2$. When the generating line rolls on the two base circles without slipping, the points $K$ and $K'$ on the line generate the involutes $C_1$, $C_1'$ on the base circle 1 and $C_2$, $C_2'$ on the base circle

$$\overline{KN}=\widehat{AN}$$

2）渐开线上任意一点的法线必是基圆的切线。如图 7-9 所示，当发生线 $L$ 沿基圆纯滚动时，$N$ 为速度瞬心，因此发生线在 $K$ 点的速度方向与渐开线在该点的切线方向重合，故发生线 $L$ 就是渐开线在 $K$ 点的法线，也是基圆的切线。

3）发生线与基圆的切点 $N$ 是渐开线在 $K$ 点的曲率中心，线段 $KN$ 是渐开线在 $K$ 点的曲率半径。显然，离基圆越远，曲率半径越大。渐开线在基圆上点 $A$ 处的曲率半径为零，基圆内没有渐开线，点 $A$ 称为渐开线的起始点。

4）渐开线的形状取决于基圆的大小。如图 7-10 所示，在展角相同的情况下，基圆半径越小，渐开线的曲率半径越小，渐开线越弯曲；基圆半径越大，渐开线的曲率半径越大，渐开线越平直；当基圆半径为无穷大时，渐开线将变成垂直于 $N_3K$ 的一条直线，直线是特殊的渐开线。

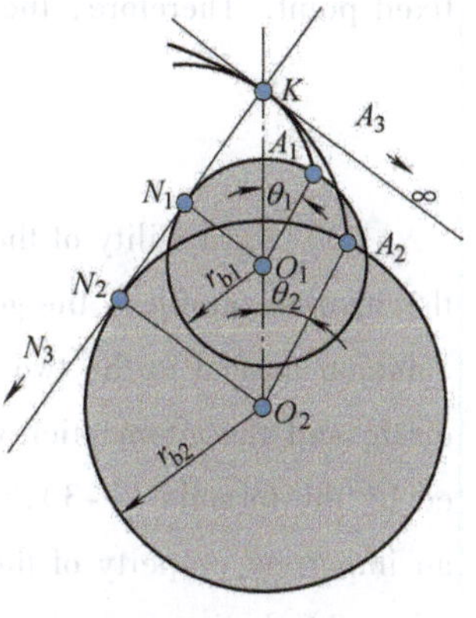

Fig. 7-10　Shape of involutes and radius of base circles
（渐开线的形状与基圆半径）

（3）渐开线方程　在研究渐开线齿轮啮合传动和几何尺寸计算时，要用到渐开线方程式及渐开线函数。下面就根据渐开线的形成原理来进行推导。

如图 7-9 所示，以 $OA$ 为极坐标轴，渐开线上的任意一点 $K$ 的位置可用向径 $r_K$ 和展角 $\theta_K$ 来确定。$K$ 点受力沿法线 $NK$ 方向，与该点速度方向（垂直于直线 $OK$）所夹的锐角称为渐开线在 $K$ 点的压力角，用 $\alpha_K$ 表示。

由图 7-9 所示的几何关系可得渐开线上任意点 $K$ 的向径 $r_K$、压力角 $\alpha_K$、基圆半径 $r_b$ 之间的关系为

$$r_K=\frac{r_b}{\cos\alpha_K}$$

又

$$\tan\alpha_K=\frac{\overline{NK}}{\overline{ON}}=\frac{\widehat{AN}}{r_b}=\frac{r_b(\alpha_K+\theta_K)}{r_b}=\alpha_K+\theta_K$$

故

$$\theta_K=\tan\alpha_K-\alpha_K$$

上式说明，展角 $\theta_K$ 是压力角 $\alpha_K$ 的函数，工程上常用 $\mathrm{inv}\alpha_K$ 表示 $\theta_K$，并称其为渐开线函数，即

$$\theta_K=\mathrm{inv}\alpha_K=\tan\alpha_K-\alpha_K$$

综上所述，渐开线的极坐标方程为

$$\begin{cases} r_K=\dfrac{r_b}{\cos\alpha_K} \\ \theta_K=\mathrm{inv}\alpha_K=\tan\alpha_K-\alpha_K \end{cases} \tag{7-2}$$

**2. 渐开线齿廓啮合传动的特点**

一对渐开线齿廓进行啮合传动时，有如下特点。

（1）瞬时传动比恒定不变　图 7-11 所示为一对渐开线齿廓啮合示意图。过啮合点 $K$ 作

2 respectively, as shown in Fig. 7-11. The profiles $C_1$, $C_2$ and $C'_1$, $C'_2$ are conjugate curves. The point $P$ is the pitch point, and also the instantaneous center of the gears. The radii of the pitch circles are $r'_1$, $r'_2$ respectively.

No matter where the profiles are in contact, such as at the point $K$, the common normal is always the common tangent $N_1N_2$ of the two base circles, so the common normal passing through any contact point always occupies the same position. That is to say, the point $P$ is does not move; it is a fixed point. Therefore, the involute curves satisfy the law of gearing, the gear ratio is:

$$i_{12}=\frac{\omega_1}{\omega_2}=\frac{\overline{O_2P}}{\overline{O_1P}}=\frac{r'_2}{r'_1}=\frac{r_{b2}}{r_{b1}}=c \tag{7-3}$$

(2) Separability of the center distance Any change in center distance will have no effect upon the involute profiles; the gear ratio can not be varied. If the involute profiles are still in contact, the common normal to the two profiles at the contact point will be the new common tangent to the base circle and the intersection with the center line as the new pitch point. The gear ratio is still calculated by the formula (7-3). Altering the center distance without destroying the tooth profiles action is an important property of the involute gears.

(3) Stationary action force The line of action in case of involute teeth is along the common normal at the contact point, and the common normal is the common tangent of the base circles. This shows that the angle of obliquity, which is the angle between the action line and the common tangent to pitch circles, remains constant. This is also an important advantage of involute gears.

## 7.4 Nomenclatures of Standard Spur Gear and Gear Sizes

Two involutes with opposite direction can represent a gear tooth, and many teeth can form a gear.

**1. Gear Teeth Nomenclatures**

The terminology of gear teeth is illustrated in Fig. 7-12 where most of the following definitions are shown.

(1) Addendum circle It is a circle passing through the topes of the teeth. The radius and diameter are denoted as $r_a$ and $d_a$.

(2) Dedendum circle It is a circle passing through the roots of the teeth. The radius and diameter are denoted as $r_f$ and $d_f$.

(3) Reference circle It is a datum circle in gear design and measurement. The radius and diameter are denoted as $r$ and $d$. Note that the subscript will be omitted when expressing the dimensions of a gear, such as $r$, $d$, $p$, etc.

(4) Base circle It is a circle generating the involute curves. The radius and diameter are denoted as $r_b$ and $d_b$.

(5) Tooth thickness, space width and circular pitch The tooth thickness is the thickness of the tooth measured along the circumference, such as $s_K$. The space width is the space between the adjacent teeth measured along the circumference, such as $e_K$. The circular pitch is the distance from

两齿廓的公法线，必与两齿轮的基圆相切且为其内公切线，$N_1$、$N_2$ 为切点。同一个方向上的内公切线只有一条，因此它与连心线的交点只有一个，即节点 $P$ 为定点，两轮的传动比 $i_{12}$ 为常数，即

$$i_{12}=\frac{\omega_1}{\omega_2}=\frac{\overline{O_2P}}{\overline{O_1P}}=\frac{r'_2}{r'_1}=\frac{r_{b2}}{r_{b1}}=常数 \qquad (7\text{-}3)$$

(2) 渐开线齿廓传动中心距具有可分性　一对渐开线齿廓齿轮啮合，其传动比 $i_{12}$ 恒等于两轮基圆半径的反比。齿轮加工完后，其基圆半径就已确定，如两轮的实际安装中心距 $a'$ 发生变化，其传动比不变。这种中心距改变而传动比不变的性质称为渐开线齿轮传动中心距的可分性。

(3) 轮齿受力方向不变　如图 7-11 所示，当不计齿面间的摩擦力时，齿面间的作用力始终沿啮合点的公法线方向，即作用力方向始终保持不变。这是渐开线齿廓的重要特性之一，对于齿轮传动的平稳性十分有利。

一对渐开线齿廓无论在何处啮合，其啮合点只能在 $N_1N_2$ 线上，即 $N_1N_2$ 为啮合点的轨迹，故 $N_1N_2$ 又称为啮合线。啮合线 $N_1N_2$ 与两轮节圆公切线 $t—t$ 所夹的锐角 $\alpha'$ 称为啮合角，它等于渐开线在节圆上的压力角。

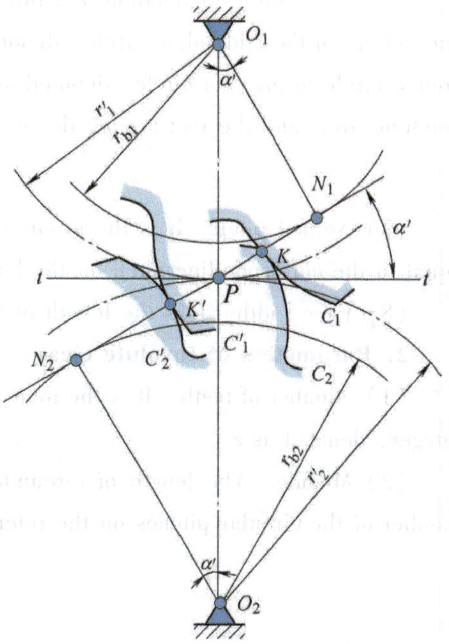

Fig. 7-11　Gearing of involute profiles
(渐开线齿廓的啮合)

## 7.4 渐开线标准直齿圆柱齿轮的基本参数和几何尺寸

### 1. 渐开线齿轮各部分的名称

图 7-12 所示为一渐开线直齿外圆柱齿轮的一部分，各部分名称如下：

(1) 齿顶圆　通过各轮齿顶部的圆，其半径和直径分别用 $r_a$ 和 $d_a$ 表示。

(2) 齿根圆　通过各齿槽底部的圆，其半径和直径分别用 $r_f$ 和 $d_f$ 表示。

(3) 分度圆　在齿顶圆和齿根圆之间规定的一个参考圆，此圆作为计算齿轮各部分几何尺寸的基准，其半径和直径分别用 $r$ 和 $d$ 表示。

(4) 基圆　生成轮齿渐开线齿廓的圆，其半径和直径分别用 $r_b$ 和 $d_b$ 表示。

(5) 齿厚、齿槽宽、齿距　在半径为 $r_K$ 的任意圆周上，一个轮齿两侧齿廓间的弧长称为该圆上的齿厚，用 $s_K$ 表示，分度圆上的齿厚用 $s$ 表示；一个齿槽两侧齿廓间的弧长称为该圆上的齿槽宽，用 $e_K$ 表示，分度圆上的齿槽宽用 $e$ 表示；相邻两齿的同向齿廓之间的弧长称为这个圆上的齿距，用 $p_K$ 表示，分度圆上的齿距用 $p$ 表示。显然，在同一圆周上，齿距等于齿厚与齿槽宽之和，即

$$p_K=s_K+e_K$$

(6) 齿顶高、齿根高、全齿高　轮齿在分度圆至齿顶圆沿半径方向的高度称为齿顶高，用 $h_a$ 表示；由分度圆至齿根圆沿半径方向的高度称为齿根高，用 $h_f$ 表示；由齿根圆至齿顶

a point on one tooth to the corresponding point on the adjacent tooth measured on the same circumference, such as $p_K$, and $p_K = s_K + e_K$. Obviously, the tooth thickness, space width and circular pitch are different on the different circumferences.

(6) Addendum, dedendum and full depth  The addendum is the radial height from the reference circle to the addendum circle, denoted as $h_a$. The dedendum is the radial height from the reference circle to the root circle, denoted as $h_f$. The full depth is the radial distance between the addendum circle and the root circle, denoted as $h$. Obviously:

$$h = h_a + h_f$$

(7) Normal pitch  It is the circular pitch measured along their normal, denoted as $p_b$. It is equal to the corresponding pitch on the base circle.

(8) Face width  It is the length of the tooth parallel to the gear axis, denoted as $B$.

**2. Parameters of Involute Gear**

(1) Number of teeth  It is the total number of the teeth the gear possesses, and it is always an integer, denoted as $z$.

(2) Module  The length of circumference of the reference circle is equal to the sum of the number of the circular pitches on the reference circle, so we have:

$$\pi d = pz$$

$$d = \frac{p}{\pi} z$$

As the expression for $d$ involves $\pi$, an indeterminate number, to obtain a more accurate size of the diameter of the gear, the influence of the irrational number $\pi$ must be dispelled.

Let $m = p/\pi$, then $d = mz$

The ratio of the circular pitch to $\pi$ is called the module of gear, denoted as $m$. The unit of the module is millimeter. Unfortunately, metric gears are not interchangeable with U. S. gears, despite both being involute tooth form, as their standards for tooth size are different.

Fig. 7-13 shows the actual sizes of 20° pressure angle, the number of teeth of 16, and module of 4, 2 and 1mm, we can see that the larger the module is, the stronger the teeth of the gear is.

The modules have been standardized (see the Tab. 7-1).

When we design a gear mechanism, the first series of modules must be selected first, and then the second series of the modules. The modules in brackets should be avoided if possible, because the relevant cutter is very scarcity. The geometrical sizes of gears will base on the module.

(3) Pressure angle  The definition of pressure angle has been demonstrated and here we emphasize that there are different angles on the different circumferences of the gear. In Fig. 7-12, the pressure angle $\alpha$ may be described as:

$$\cos\alpha = \frac{r_b}{r}$$

The relationship between the radius of base circle and reference circle can be given by:

$$r_b = r\cos\alpha = \frac{mz}{2}\cos\alpha \tag{7-4}$$

圆沿半径方向的高度称为全齿高，用 $h$ 表示。显然

$$h = h_a + h_f$$

（7）法向齿距　相邻两齿同向齿廓沿公法线方向所量得的距离称为齿轮的法向齿距。根据渐开线的性质，法向齿距等于基圆齿距，都用 $p_b$ 表示。

（8）齿宽　齿轮的轴向长度，用 $B$ 表示。

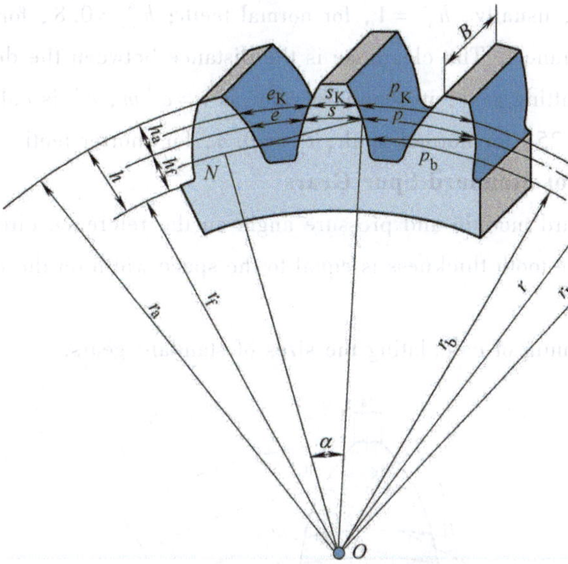


Fig. 7-12　Spur gear nomenclatures（渐开线直齿圆柱齿轮名词术语）

**2. 渐开线齿轮的基本参数**

（1）齿数 $z$　圆周上分布的轮齿数目，用 $z$ 表示，$z$ 为整数。

（2）模数 $m$　设齿轮的分度圆周长等于 $\pi d$，也等于齿距之和，因此有 $\pi d = pz$。分度圆直径为

$$d = \frac{p}{\pi} z$$

为了便于设计、计算、制造和检验，人为规定 $p/\pi = m$，且设定为标准值，称 $m$ 为齿轮分度圆模数，简称模数，单位为 mm。模数 $m$ 已经标准化，设计时必须按国家标准所规定的标准模数系列值选取。圆柱齿轮的标准模数系列见表 7-1。模数 $m$ 是齿轮的一个基本参数。在其他参数不变的情况下，模数不同，齿轮的尺寸也不同。

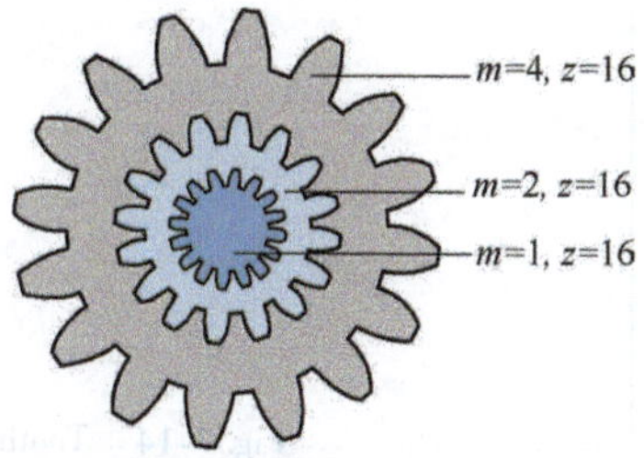


Fig. 7-13　Gear size with same number of teeth and different module of teeth（同齿数、不同模数齿轮尺寸）

图 7-13 给出齿数相同、模数不同的齿轮对比图，由图可见，模数越大，轮齿越大。

**Tab. 7-1　Modules of involute cylindrical gears**（圆柱齿轮标准模数系列）（单位：mm）

| | |
|---|---|
| 第一系列<br>First series | 0.1　0.12　0.15　0.2　0.25　0.3　0.4　0.5　0.6　0.8　1　1.25　1.5　2　2.5<br>3　4　5　6　8　10　12　16　20　25　32　40　50 |
| 第二系列<br>Second series | 0.35　0.7　0.9　1.75　2.25　2.75　(3.25)　3.5　(3.75)　4.5　5.5　(6.5)<br>7　9　(11)　14　18　22　28　(30)　36　45 |

（3）压力角 $\alpha$　齿轮齿廓上各点的压力角不同，通常所说的压力角是指齿轮分度圆上的

The pressure angles of gears are standardized at a few values; the standard values are 14.5°, 20°, and 25°, with 20° being the most commonly used.

Therefore, we can give a definition of the reference circle: It is a circle on which the module and pressure angle have been standardized.

(4) Coefficient of addendum  A standard value of the addendum is $h_a = h_a^* m$. $h_a^*$ is called the coefficient of addendum, usually, $h_a^* = 1$, for normal teeth; $h_a^* = 0.8$, for shorter teeth.

(5) Coefficient of clearance  The clearance is the distance between the dedendum of one gear and the addendum of the matting gear, its standard value is $c = c^* m$; $c^*$ is called the coefficient of clearance. Usually, $c^* = 0.25$, for normal teeth; $c^* = 0.3$, for shorter teeth.

**3. Geometrical Sizes of Standard Spur Gears**

If a gear has the standard module and pressure angle on the reference circle, standard addendum and dedendum, also the tooth thickness is equal to the space width on the reference circle, it is called the standard gear.

Tab. 7-2 shows the formula of calculating the sizes of standard gears.

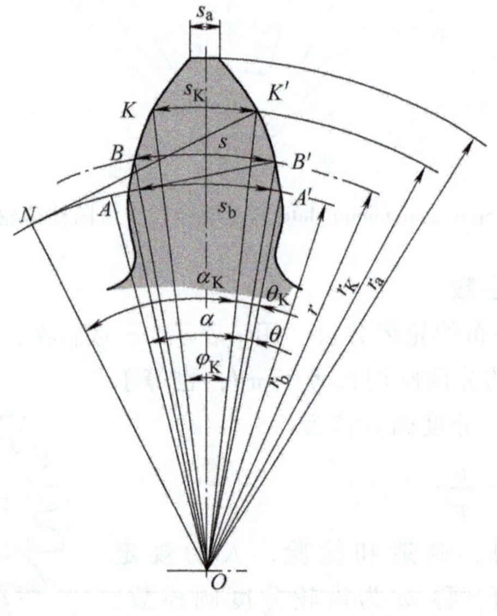


Fig. 7-14  Tooth thickness along an arbitrary circle (任意圆齿厚)

**4. Tooth Thickness Along an Arbitrary Circle**

In gear design and manufacturing, the tooth thickness on the arbitrary circle, such as tooth thickness on the gear top, sometimes may be necessary.

Fig. 7-14 shows a gear tooth that has thickness $s_K$ at radical location $r_K$. We have:

$$\angle KOK' = \angle BOB' - 2\angle BOK = \frac{s}{r} - 2(\theta_K - \theta)$$

Rearranging the above equation, we get:

$$s_K = r_K \frac{s}{r} - 2r_K(\theta_K - \theta) = s\frac{r_K}{r} - 2r_K(\mathrm{inv}\alpha_K - \mathrm{inv}\alpha) \tag{7-5}$$

Where $r$ is the radius of the reference circle; $s$ is the tooth thickness of the reference circle; $\alpha$ is the pressure angle of the reference circle; $\theta$ is the unfolding angle of the reference circle at point $B'$; $\theta_K$

压力角。由图 7-12 可知，齿轮的分度圆压力角 $\alpha$、基圆半径 $r_b$ 和分度圆半径 $r$ 之间的关系为

$$r_b = r\cos\alpha = \frac{mz}{2}\cos\alpha \tag{7-4}$$

模数已经标准化，齿数为整数。国家标准中规定分度圆压力角标准值为 $\alpha=20°$。在某些情况下也可采用 $\alpha=14.5°$、15°或 22.5°。此时可给分度圆作完整的定义：齿轮上具有标准模数、标准压力角的圆称为分度圆。

（4）齿顶高系数 $h_a^*$　齿轮的齿顶高 $h_a = h_a^* m$，$h_a^*$ 称为齿顶高系数。国家标准规定：正常齿 $h_a^*=1$，短齿 $h_a^*=0.8$。

（5）顶隙与顶隙系数 $c^*$　一个齿轮的齿顶圆和另一个齿轮的齿根圆之间的径向距离，称为顶隙，用 $c$ 表示。$c=c^* m$，$c^*$ 称为顶隙系数。国家标准规定：正常齿 $c^*=0.25$，短齿 $c^*=0.3$。

**3. 渐开线标准直齿圆柱齿轮几何尺寸计算**

具有标准模数 $m$、标准压力角 $\alpha$、标准齿顶高系数 $h_a^*$、标准顶隙系数 $c^*$，并且分度圆上的齿厚 $s$ 等于分度圆上的齿槽宽 $e$ 的齿轮称为标准齿轮。已知齿轮的基本参数，由表 7-2 即可计算出渐开线标准直齿圆柱齿轮各部分的几何尺寸。

**Tab. 7-2　Formula for standard spur gear**（渐开线标准直齿圆柱齿轮传动几何尺寸计算公式）

| 名称 Names | 符号 Symbol | 计算公式 Formula |
|---|---|---|
| 分度圆直径 Diameter of reference circle | $d$ | $d_1 = mz_1$，$d_2 = mz_2$ |
| 基圆直径 Diameter of base circle | $d_b$ | $d_{b1} = d_1\cos\alpha$，$d_{b2} = d_2\cos\alpha$ |
| 齿顶高 Addendum | $h_a$ | $h_a = h_a^* m$ |
| 齿根高 Dedendum | $h_f$ | $h_f = (h_a^* + c^*)m$ |
| 全齿高 Whole depth | $h$ | $h = h_a + h_f = (2h_a^* + c^*)m$ |
| 齿顶圆直径 Diameter of addendum circle | $d_a$ | $d_{a1} = d_1 + 2h_a$，$d_{a2} = d_2 + 2h_a$ |
| 齿根圆直径 Diameter of dedendum circle | $d_f$ | $d_{f1} = d_1 - 2h_f$，$d_{f2} = d_2 - 2h_f$ |
| 齿距 Circular pitch | $p$ | $p = \pi m$ |
| 齿厚 Tooth thickness | $s$ | $s = \dfrac{\pi m}{2}$ |
| 齿槽宽 Space width | $e$ | $e = \dfrac{\pi m}{2}$ |
| 基圆齿距 Base pitch | $p_b$ | $p_b = p\cos\alpha$ |
| 顶隙 Clearance | $c$ | $c = c^* m$ |
| 中心距 Center distance | $a$ | $a = \dfrac{m(z_1+z_2)}{2}$ |

**4. 任意圆弧齿厚**

在设计、加工和检验齿轮时，有时需要知道某一圆周上的齿厚。如为了确定齿轮啮合时的齿侧间隙，须确定节圆上的齿厚；为检测齿顶强度，需算出齿顶圆上的齿厚。因此，有必要推导出齿轮任意半径 $r_K$ 的圆周上的齿厚 $s_K$ 的计算公式。

图 7-14 所示为外齿轮的一个齿，$r$、$s$、$\alpha$ 和 $\theta$ 分别为分度圆的半径、齿厚、压力角和展角。由于

is the unfolding angle of the arbitrary circle at point $K'$; $\mathrm{inv}\alpha_K$ is the involute function of the pressure angle $\alpha_K$; $\mathrm{inv}\alpha$ is the involute function of the pressure angle $\alpha$.

The pressure angle at point $K$ can be determined by following formula:

$$\alpha_K = \arccos\left(\frac{r_b}{r_K}\right)$$

**5. Terminology for Internal Gear**

An internal gear has its teeth cut on the inside of the rim rather than on the outside, and it has concave tooth profiles, while the tooth profiles of the external gear are convex. Fig. 7-15 shows a typical internal gear, and the followings are different from the external gear.

1) The addendum circle is on the inside of the reference circle rather than on the outside, so:

$$d_a = d - 2h_a = (z - 2h_a^*)m$$

2) The dedendum circle is on the outside of the reference circle rather than on the inside, so:

$$d_f = d + 2h_f = (z + 2h_a^* + 2c^*)m$$

3) The center distance is less, and the value is:

$$a = \frac{m(z_2 - z_1)}{2}$$

The others sizes of internal gear are the same with external gear, such as $d = mz$, $d_b = d\cos\alpha$, $h_a = h_a^* m$, $h_f = (h_a^* + c^*)m$, etc.

Internal gears can be used widely in the gear trains design.

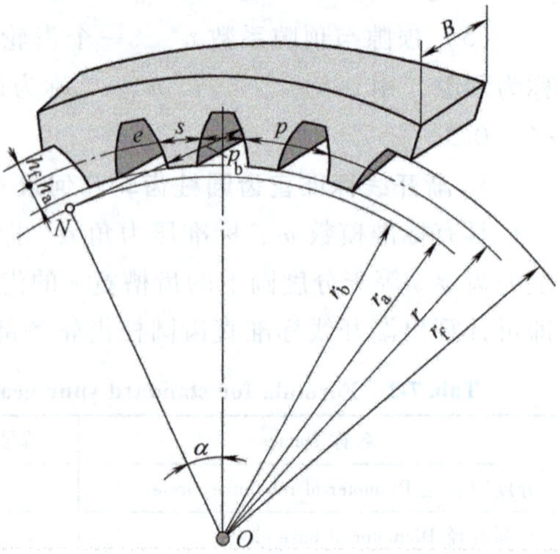

Fig. 7-15 Terminology for internal gear（渐开线内齿圆柱齿轮术语）

**6. Terminology for a Rack**

A rack is a portion of a gear having an infinite base diameter, thus its reference circle, addendum circle, dedendum circle are all straight lines. The involute profile of the rack becomes a straight line and is perpendicular to the line of action (see Fig. 7-16).

The characteristics of a rack are as follows.

1) The profiles are all skew lines, and they are parallel on the same side of the teeth. The pressure angles at different point on the profile are all the same and they are equal to the nominal pressure angle of 20°.

2) The pitch remains unchangeable on the reference line, addendum line, and so on. Its value is $p = \pi m$; the base pitch is $p_b = \pi m\cos\alpha$.

3) The addendum and dedendum are the same with the external gear.

## 7.5 Meshing Drive of Standard Spur Gears

**1. Conditions of Correctly Meshing for Involute Gears**

Gears transmit motion by means of successively engaging teeth, but not both gears are to be meshed together correctly. Fig. 7-17 shows a pair of meshing gears in which all the contact points be-

$$\angle KOK' = \angle BOB' - 2\angle BOK = \frac{s}{r} - 2(\theta_K - \theta)$$

则任意半径 $r_K$ 的圆周上的齿厚 $s_K$ 为

$$s_K = r_K \frac{s}{r} - 2r_K(\theta_K - \theta) = s\frac{r_K}{r} - 2r_K(\mathrm{inv}\alpha_K - \mathrm{inv}\alpha) \tag{7-5}$$

式中，$\alpha_K = \arccos(r_b/r_K)$ 为在任意半径 $r_K$ 上的渐开线齿廓压力角。

**5. 内齿轮的特点**

轮齿分布在圆柱体的内表面上，称为内齿轮，图 7-15 所示为一内齿圆柱齿轮的一部分。内齿轮的齿槽相当于外齿轮的轮齿，内齿轮的轮齿相当于外齿轮的齿槽；内齿轮的齿顶圆在内，齿根圆在外，即齿顶圆半径小于齿根圆半径；为保证内齿轮齿顶以外为渐开线，内齿轮的齿顶圆应大于基圆。

内齿轮的主要几何尺寸计算公式如下：

分度圆直径：$d = mz$，　基圆直径：$d_b = d\cos\alpha$

齿顶高：　$h_a = h_a^* m$，齿根高：　$h_f = (h_a^* + c^*)m$

齿顶圆直径：$d_a = d - 2h_a = (z - 2h_a^*)m$

齿根圆直径：$d_f = d + 2h_f = (z + 2h_a^* + 2c^*)m$

中心距：　$a = \dfrac{m(z_2 - z_1)}{2}$

**6. 齿条的结构及其特点**

当标准齿轮的齿数为无穷多时，其分度圆、齿顶圆、齿根圆分别演变为分度线、齿顶线、齿根线，且相互平行，此时基圆半径为无穷大，渐开线演变为一条直线，齿轮则演变为图 7-16 所示的做直线移动的齿条。

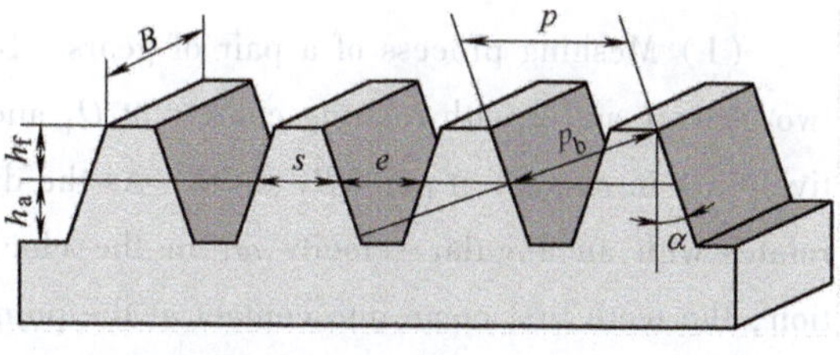


Fig. 7-16　Terminology for a rack（齿条术语）

齿条有如下特点：

1）齿条齿廓为斜直线，齿廓上各点的压力角均为标准值，且等于齿条齿廓的齿形角。

2）在平行于齿条齿顶线的各条直线上，齿条的齿距均相等，齿距 $p = \pi m$，其法向齿距 $p_b = \pi m\cos\alpha$；与齿顶线平行且其上齿厚等于齿槽宽的直线称为齿条分度线，它是计算齿条尺寸的基准线。

3）分度线至齿顶线的高度为齿顶高，$h_a = h_a^* m$；分度线至齿根线的高度为齿根高，$h_f = (h_a^* + c^*)m$。

## 7.5　渐开线直齿圆柱齿轮机构的啮合传动

**1. 渐开线齿轮的正确啮合条件**

齿轮传动是靠主动轮齿依次拨动从动轮齿的啮合来实现的，如图 7-17 所示。要使啮合正确进行，应保证处于啮合线上的各对轮齿都处于啮合状态，即前一对轮齿在啮合线 $N_1N_2$ 上的 $K$ 点啮合，后一对轮齿应在啮合线 $N_1N_2$ 上的 $K'$ 点啮合。线段 $KK'$ 是齿轮 1 和齿轮 2 的法向齿距。为保证两齿轮正确啮合，其条件是两轮的法向齿距必相等。

根据渐开线的性质，齿轮的法向齿距等于基圆齿距，因此，正确啮合条件为

tween the two gears with the involute profiles must lie on the line of action, so that the pitch point remain fixed.

From the properties of involute, we have known that:

$\overline{KK'} = \overset{\frown}{K_1K_1'} = p_{b1}$ and $\overline{KK'} = \overset{\frown}{K_2K_2'} = p_{b2}$

Where $\overline{KK'}$ is the normal distance between the corresponding sides of the two adjacent teeth. It is called the normal pitch; $\overset{\frown}{K_1K_1'}$ is defined as the distance from a point on one tooth to the corresponding point on the next tooth measured along the base circle. It is called the base pitch, denoted as $p_{b1}$; $\overset{\frown}{K_2K_2'}$ is a base pitch of another gear, denoted as $p_{b2}$.

Rearranging the above equations, we have:

$$p_{b1} = p_{b2}$$

$$p_{b1} = \pi m_1 \cos\alpha_1, \quad p_{b2} = \pi m_2 \cos\alpha_2$$

$$\pi m_1 \cos\alpha_1 = \pi m_2 \cos\alpha_2$$

The module and pressure angle have been standardized already. This equation can be satisfied by using $m_1 = m_2 = m$ and $\alpha_1 = \alpha_2 = \alpha$. Therefore, the conditions of correctly engaging for a pair of gears are that the module and pressure angle of the two meshing gears must be the same respectively.

**2. Conditions of Continuous Transmission of Gears**

(1) Meshing process of a pair of gears In Fig. 7-18, two gears 1 and 2 with rotating centers at $O_1$ and $O_2$ respectively are in contact at point $B_2$ and $K$. As the driving gear 1 rotates with an angular velocity $\omega_1$ in the clockwise direction, the teeth first come into contact at the point $B_2$, where the addendum circle of the driven gear cuts the line of action. The contact follows the line of action through point $P$, and ceases at point $B_1$, where the addendum circle of the driving gear cuts the line of action. The line $B_2B_1$ is called the actual path of the contact point, and the line $N_1N_2$ is the theoretical path of contact point. This is because that there is no involute within the base circle, so the maximum addendum circle can not pass through the point $N$.

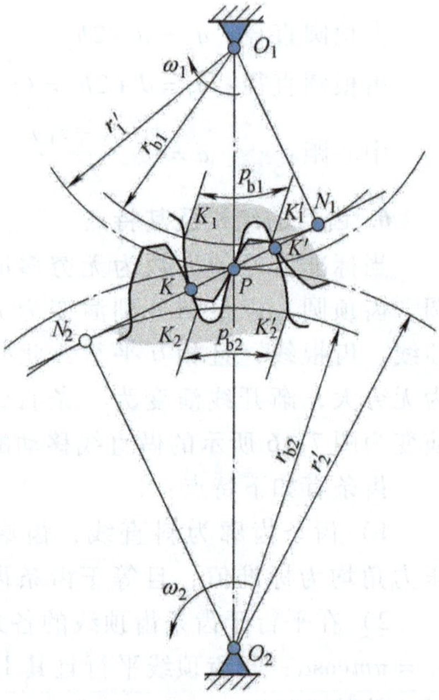

Fig. 7-17 Meshing of teeth (齿轮啮合)

The initial contact point is on the bottom of the driving gear and on the tip of the driven gear. The final contact point is on the tip of the driving gear and on the bottom of the driven gear.

(2) Conditions of continuous transmission of gears As we stated earlier, all the points of contact between two gear teeth with involute profiles lie on the pressure line. The initial contact occurs where the addendum circle of the driven gear intersects the pressure line. The final contact occurs at the point where the addendum circle of the driving gear intersects the pressure line.

Fig. 7-18 shows that the later pair of teeth is just coming into contact at the initial point $B_2$ and the previous pair of teeth is in contact at the point $K$, and the contact will not yet have reached final

$$\overline{KK'} = p_{b1} = p_{b2}$$

$p_{b1} = \pi m_1 \cos\alpha_1$，$p_{b2} = \pi m_2 \cos\alpha_2$，于是有

$$\pi m_1 \cos\alpha_1 = \pi m_2 \cos\alpha_2$$

由于齿轮的模数和压力角都已经标准化了，满足上式的条件为

$$m_1 = m_2 = m$$

$$\alpha_1 = \alpha_2 = \alpha$$

故一对渐开线直齿圆柱齿轮传动的正确啮合条件为两轮的模数和压力角分别相等，且为标准值。

**2. 渐开线齿轮的连续传动条件**

（1）轮齿的啮合过程　图 7-18 所示的齿轮机构中，齿轮 1 为主动轮，以角速度 $\omega_1$ 顺时针方向转动。齿轮 2 为从动轮，以角速度 $\omega_2$ 逆时针方向转动。当主动齿轮 1 根部渐开线与从动齿轮 2 的顶部渐开线在啮合线 $N_1N_2$ 上的点 $B_2$ 接触时，这对轮齿开始进入啮合状态，称点 $B_2$ 为啮合开始点。随着传动的进行，两齿轮齿廓的啮合点沿啮合线向左下方移动，直到主动齿轮 1 的齿顶与从动齿轮 2 的齿根在啮合线上的点 $B_1$ 处接触时，这对轮齿即将脱离啮合，称点 $B_1$ 为啮合终止点。因此，线段$B_1B_2$是啮合点实际所走过的轨迹，称为实际啮合线。显然点 $B_1$、$B_2$ 分别为齿轮 1、2 的齿顶圆与 $N_1N_2$ 的交点。如果增大两齿轮的齿顶圆半径，$B_1$、$B_2$ 将逐渐接近 $N_2$、$N_1$，但由于基圆内没有渐开线，因此它们永远也不会超过 $N_2$、$N_1$，线段$N_1N_2$是理论上最长的啮合线，称为理论啮合线。

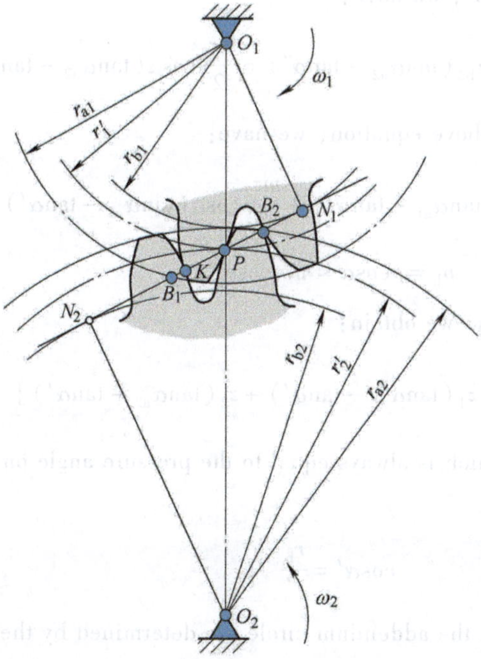


Fig. 7-18　Meshing process of teeth（轮齿的啮合过程）

（2）连续传动条件　图 7-18 所示的轮齿啮合过程中，要使齿轮传动连续进行，应使前一对轮齿在点 $B_1$ 退出啮合之前，后一对轮齿就已经从点 $B_2$ 进入啮合。为此，要求实际啮合

point $B_1$. Thus, for a short time there will be two pairs of teeth in contact, the continuous transmission is satisfied.

Therefore, the condition of continuous transmission of a pair of meshing gears is defined that the contact length must be greater than the base pitch of the gears, that is:

$$B_1B_2 \geqslant p_b$$

Where the normal pitch $B_2K$ is equal to the base pitch $p_b$ of the gear.

The ratio of the contact length to the base pitch of the gear is defined as contact ratio, denoted as $\varepsilon_a$.

$$\varepsilon_a = \frac{\overline{B_1B_2}}{p_b} \geqslant 1$$

Usually, $\varepsilon_a > 1.2$

(3) Value of contact ratio

1) Contact ratio for external spur gears. Fig. 7-19 shows a pair of meshing gears having involute teeth. The length $B_2B_1$ can be calculated from the following relationship.

$$\overline{B_2B_1} = \overline{PB_1} + \overline{PB_2}$$

In right $\triangle N_1O_1B_1$ and $\triangle N_1O_1P$, we have:

$$\overline{PB_1} = \overline{N_1B_1} - \overline{N_1P} = r_{b1}(\tan\alpha_{a1} - \tan\alpha') = \frac{mz_1}{2}\cos\alpha(\tan\alpha_{a1} - \tan\alpha')$$

In right $\triangle N_2O_2B_2$ and $\triangle N_2O_2P$, we have:

$$\overline{PB_2} = \overline{N_2B_2} - \overline{N_2P} = r_{b2}(\tan\alpha_{a2} - \tan\alpha') = \frac{mz_2}{2}\cos\alpha(\tan\alpha_{a2} - \tan\alpha')$$

Substituting $\overline{PB_1}$ and $\overline{PB_2}$ into above equation, we have:

$$\overline{B_2B_1} = \frac{mz_1}{2}\cos\alpha(\tan\alpha_{a1} - \tan\alpha') + \frac{mz_2}{2}\cos\alpha(\tan\alpha_{a2} - \tan\alpha')$$

$$p_b = p\cos\alpha = \pi m\cos\alpha$$

Rearranging the above equations, we obtain:

$$\varepsilon_a = \frac{\overline{B_2B_1}}{p_b} = \frac{1}{2\pi}[z_1(\tan\alpha_{a1} - \tan\alpha') + z_2(\tan\alpha_{a2} - \tan\alpha')] \tag{7-6}$$

Where $\alpha'$ is the angle of obliquity which is always equal to the pressure angle on the pitch circle. It can be determined as follows:

$$\cos\alpha' = \frac{r_b}{r'}$$

The pressure angles $\alpha_{a1}$, $\alpha_{a2}$ on the addendum circle are determined by the following formula:

$$\alpha_{a1} = \arccos\frac{r_{b1}}{r_{a1}}, \quad \alpha_{a2} = \arccos\frac{r_{b2}}{r_{a2}}$$

$z_1$ and $z_2$ are the number of teeth of the gear 1 and the gear 2.

$r_{b1}$ and $r_{b2}$ are radii of the base circle 1 and the base circle 2, $r_{a1}$ and $r_{a2}$ are radii of addendum circle 1 and addendum circle 2, $r_1'$ and $r_2'$ are radii of pitch circle 1 and pitch circle 2 respectively.

线段 $B_1B_2$ 的长度大于或等于轮齿的法向齿距$\overline{B_2K}$，$\overline{B_2K}=p_b$，即$\overline{B_1B_2}\geqslant p_b$。

将实际啮合线段 $B_1B_2$ 的长度与法向齿距 $p_b$ 的比值称为齿轮传动的重合度，用 $\varepsilon_a$ 表示。因此，齿轮连续传动的条件为

$$\varepsilon_a=\frac{\overline{B_1B_2}}{p_b}\geqslant 1$$

在实际应用中，$\varepsilon_a>1.2$。

(3) 重合度的计算公式

1) 外啮合直齿圆柱齿轮传动的重合度。在图 7-19 所示的齿轮传动过程中，实际啮合线长度计算过程如下：

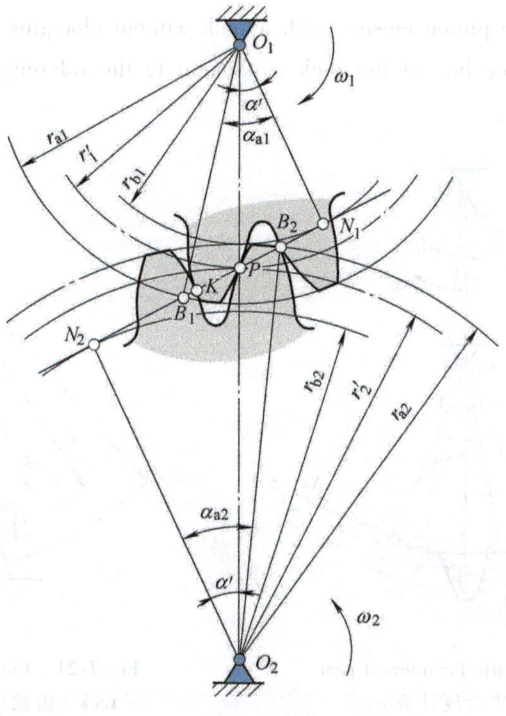


Fig. 7-19　Contact ratio for external gears（外啮合齿轮重合度计算）

$$\overline{B_2B_1}=\overline{PB_1}+\overline{PB_2}$$

$$\overline{PB_1}=\overline{N_1B_1}-\overline{N_1P}=r_{b1}(\tan\alpha_{a1}-\tan\alpha')=\frac{mz_1}{2}\cos\alpha(\tan\alpha_{a1}-\tan\alpha')$$

$$\overline{PB_2}=\overline{N_2B_2}-\overline{N_2P}=r_{b2}(\tan\alpha_{a2}-\tan\alpha')=\frac{mz_2}{2}\cos\alpha(\tan\alpha_{a2}-\tan\alpha')$$

$$\overline{B_2B_1}=\frac{mz_1}{2}\cos\alpha(\tan\alpha_{a1}-\tan\alpha')+\frac{mz_2}{2}\cos\alpha(\tan\alpha_{a2}-\tan\alpha')$$

$$p_b=p\cos\alpha=\pi m\cos\alpha$$

所以，一对外啮合直齿圆柱齿轮的重合度的计算公式为

$$\varepsilon_a=\frac{\overline{B_2B_1}}{p_b}=\frac{1}{2\pi}[z_1(\tan\alpha_{a1}-\tan\alpha')+z_2(\tan\alpha_{a2}-\tan\alpha')]\tag{7-6}$$

式中，$\alpha'$为啮合角，也就是节圆压力角；$\alpha_{a1}$、$\alpha_{a2}$分别为齿轮 1、2 的齿顶圆压力角，$\alpha_{a1}=$

2) Contact ratio for internal spur gears. The contact ratio for internal gear is illustrated in Fig. 7-20. In the same method, we can obtain the following formulas.

$$\overline{PB_1} = \overline{N_1B_1} - \overline{N_1P} = r_{b1}(\tan\alpha_{a1} - \tan\alpha') = \frac{mz_1}{2}\cos\alpha(\tan\alpha_{a1} - \tan\alpha')$$

$$\overline{PB_2} = \overline{N_2P} - \overline{N_2B_2} = r_{b2}(\tan\alpha' - \tan\alpha_{a2}) = -\frac{mz_2}{2}\cos\alpha(\tan\alpha_{a2} - \tan\alpha')$$

$$\varepsilon_a = \frac{\overline{B_2B_1}}{p_b} = \frac{1}{2\pi}[z_1(\tan\alpha_{a1} - \tan\alpha') - z_2(\tan\alpha_{a2} - \tan\alpha')] \tag{7-7}$$

3) Contact ratio for a pinion and a rack. The contact ratio for a pinion and a rack is illustrated in Fig. 7-21. Where $PB_1$ is as the same as before and $PB_2$ depends upon the relative position of the pinion and the rack. If the pinion meshes with a rack without changing the distance of the center, that is to say, the reference line of the rack is tangent to the reference circle of the pinion, the length of $PB_2$ is as follows:

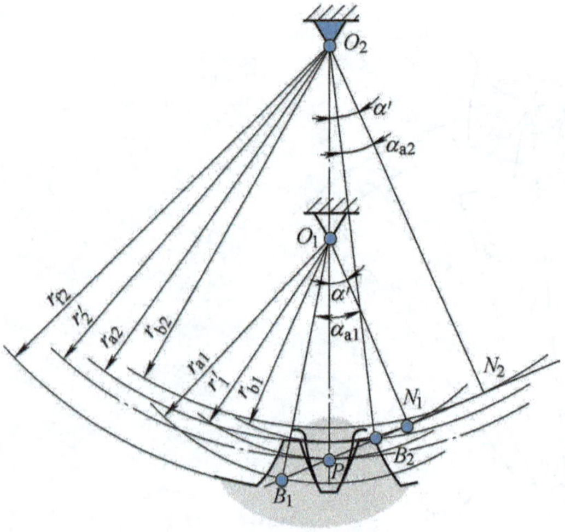


Fig. 7-20 Contact ratio for internal gear (内啮合齿轮重合度计算)

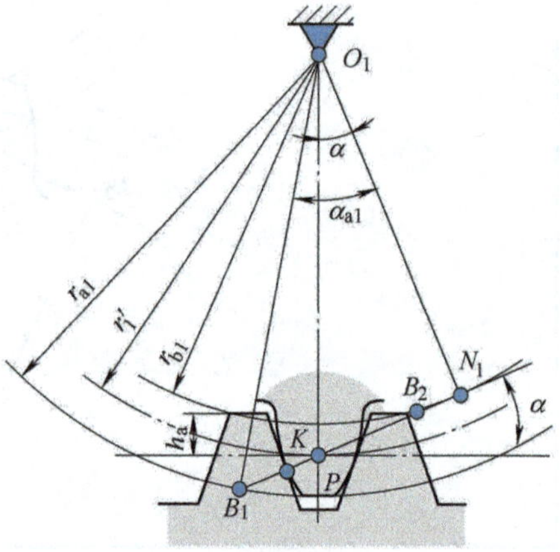


Fig. 7-21 Contact ratio for a pinion and a rack (齿轮齿条啮合的重合度计算)

$$\overline{PB_2} = \frac{h_a^* m}{\sin\alpha}$$

$$\overline{B_2B_1} = \overline{PB_1} + \overline{PB_2} = (\overline{N_1B_1} - \overline{N_1P}) + \frac{h_a}{\sin\alpha}$$

$$= \frac{mz_1}{2}\cos\alpha(\tan\alpha_{a1} - \tan\alpha) + \frac{h_a^* m}{\sin\alpha}$$

Arranging the above equations, we obtain:

$$\varepsilon_a = \frac{\overline{B_2B_1}}{p_b} = \frac{1}{2\pi}\left[z_1(\tan\alpha_{a1} - \tan\alpha) + \frac{4h^*}{\sin 2\alpha}\right] \tag{7-8}$$

Supposing the number of teeth of the two gears is infinite, the maximum value of the contact ratio can be obtained.

$$\varepsilon_{a\max} = \frac{2h_a^* m}{\pi m \sin\alpha\cos\alpha} = \frac{4h_a^*}{\pi\sin 2\alpha}$$

$\arccos\dfrac{r_{b1}}{r_{a1}}$，$\alpha_{a2}=\arccos\dfrac{r_{b2}}{r_{a2}}$。

2）内啮合直齿轮传动的重合度。图7-20所示为一对内啮合直齿圆柱齿轮传动，其重合度计算公式可参考上述过程推导。即

$$PB_1=N_1B_1-N_1P=r_{b1}(\tan\alpha_{a1}-\tan\alpha')=\frac{mz_1}{2}\cos\alpha(\tan\alpha_{a1}-\tan\alpha')$$

$$\overline{PB_2}=\overline{N_2P}-\overline{N_2B_2}=r_{b2}(\tan\alpha'-\tan\alpha_{a2})=-\frac{mz_2}{2}\cos\alpha(\tan\alpha_{a2}-\tan\alpha')$$

因此，内啮合直齿轮传动重合度的计算公式为

$$\varepsilon_a=\frac{\overline{B_2B_1}}{p_b}=\frac{1}{2\pi}[z_1(\tan\alpha_{a1}-\tan\alpha')-z_2(\tan\alpha_{a2}-\tan\alpha')] \tag{7-7}$$

3）齿轮与齿条啮合的重合度。当齿轮2的齿数增大到无穷多时即为齿轮与齿条啮合，图7-21所示的实际啮合线长度为

$$\overline{B_2B_1}=\overline{PB_1}+\overline{PB_2}=(\overline{N_1B_1}-\overline{N_1P})+\frac{h_a}{\sin\alpha}=\frac{mz_1}{2}\cos\alpha(\tan\alpha_{a1}-\tan\alpha)+\frac{h_a^* m}{\sin\alpha}$$

重合度为

$$\varepsilon_a=\frac{B_2B_1}{p_b}=\frac{1}{2\pi}\left[z_1(\tan\alpha_{a1}-\tan\alpha)+\frac{4h^*}{\sin 2\alpha}\right] \tag{7-8}$$

重合度反映了啮合线上同时参与啮合的轮齿对数的平均值，重合度越大，同时参与啮合的轮齿对数越多，传动越平稳。由式（7-6）~式（7-8）可知，重合度与齿轮的模数无关，增加齿轮的齿数 $z$ 及增大齿顶高系数 $h_a^*$ 均可使实际啮合线$B_1B_2$加长，从而使重合度 $\varepsilon_a$ 增大。假想当两齿轮的齿数 $z_1$、$z_2$ 都趋于无穷大时，重合度也趋于最大值 $\varepsilon_{amax}$，这时

$$\varepsilon_{amax}=\frac{2h_a^* m}{\pi m\sin\alpha\cos\alpha}=\frac{4h_a^*}{\pi\sin 2\alpha}$$

当 $h_a^*=1$，$\alpha=20°$时，$\varepsilon_{amax}=1.981$。

如果 $\varepsilon_a=1$，只有在 $B_1$ 和 $B_2$ 两点接触的瞬间，才有两对轮齿同时啮合，其余时间内只有一对轮齿啮合；如果 $\varepsilon_a=2$，只有瞬间才有三对轮齿啮合，其余时间为两对轮齿啮合。如果 $\varepsilon_a$ 不是整数，如图7-22所示，$\varepsilon_a=1.2$，表明在啮合线上 $B_2K'$和 $B_1K$（长度均为 $0.2p_b$）两段范围内，有两对轮齿同时啮合，该区域称为双齿啮合区；在节点 $P$ 附近的 $KK'$（长度为 $0.6p_b$）段内，只有一对轮齿啮合，该区域称为单齿啮合区。

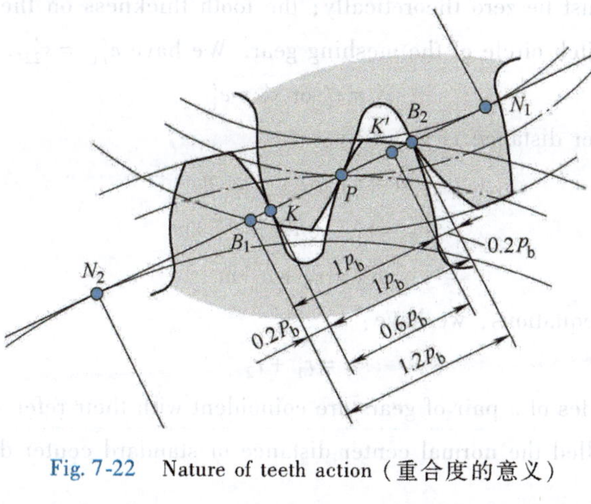


Fig. 7-22 Nature of teeth action（重合度的意义）

When $h_a^* = 1$, $\alpha = 20°$, then $\varepsilon_{amax} = 1.981$

The contact ratio means the average number of pairs of teeth which are in contact, and usually is not an integer. If the ratio is 1.2, as shown in Fig. 7-22, it does not mean that there are 1.2 pairs of teeth in contact. It means that there are alternately one pair and two pairs of teeth in contact, or one pair of teeth is always in contact, and two pairs of gears are in contact 20 percent of times. From Fig. 7-22, we know that there are two pairs of teeth in contact on the segments of $B_2K'$ and $KB_1$, and on the segment $KK'$ only one pair of teeth is in contact.

If a contact ratio is one, it means that one pair of teeth is in contact at all times, and if the contact ratio were less than one, there would be an interval during which no teeth would be in contact.

The larger the contact ratio is, the more quietly the gears will operate, and in practice, a contact ratio from 1.2 to 1.6 has been recommended.

**3. Relative Slide Between Contact Teeth**

From Fig. 7-22 we know that the working profile of the gear 1 is from the point $B_2$ to its tooth top, and the working profile of the gear 2 is from the point $B_1$ to its tooth top. When the pair of teeth contacts at an arbitrary point, such as the point $K'$, the slide will occur between the teeth surfaces along their tangent direction. This is because that the velocity of the point $K'$ is not the same on the two gears.

The approach phase of the action is period between initial contact point and pitch point. During the approach phase, contact is sliding down the face of the gear tooth toward the pitch circle. At the pitch point there is no slide, and the action is pure rolling.

**4. Center Distance of Gears**

The center distance is a distance between the centers of the two gears. When the shafts of a pair of standard gears are mounted, the actual center distance $a'$ is always equal to the sum of the radii of the pitch circles.

$$a' = r_1' + r_2'$$

When the shafts of a pair of gears are mounted correctly, the following conditions must be satisfied.

1) The radical clearance between the addendum circle of a gear and the dedendum circle of the meshing gear must be the standard value, that is $c = c^* m$; see the Fig. 7-23.

2) The backlash must be zero theoretically; the tooth thickness on the pitch circle is equal to the space width on the pitch circle of the meshing gear. We have $e'_{11'} = s'_{22'}$. This can be written as:

$$s_1' = e_2' \text{ or } s_2' = e_1'$$

Therefore, the center distance is given by:

$$a = r_{f1} + r_{a2} + c^* m$$

$$r_{f1} = r_1 - (h_a^* + c^*) m$$

$$r_{a2} = r_2 + h_a^* m$$

Rearranging above equations, we have:

$$a = r_1 + r_2$$

When the pitch circles of a pair of gears are coincident with their reference circles respectively, the center distance is called the normal center distance or standard center distance.

**3. 齿廓啮合的相对滑动**

由图 7-22 可以看出，齿轮 1 的齿根部点 $B_2$ 到齿顶的齿廓和齿轮 2 的齿根部点 $B_1$ 到齿顶的齿廓是实际接触的齿廓，称为工作齿廓。在节点 $P$ 处啮合时，两齿轮啮合点具有相等的速度；在非节点处啮合时，如在点 $K'$处啮合，两齿轮在点 $K'$处的速度不相等。这说明两齿廓在其公切线方向存在相对滑动速度。相对滑动速度越大，磨损越严重。采用运动分析方法可求解各啮合点的相对滑动速度。齿根部渐开线部分磨损较大，小齿轮根部的磨损最大。

**4. 齿轮传动的中心距及标准齿轮的安装**

（1）齿轮传动的中心距　两齿轮转动中心之间的距离，称为齿轮传动的中心距 $a$。两齿轮实际安装后的中心距，也称实际中心距。实际中心距 $a'$恒等于两齿轮节圆半径之和，即

$$a' = r_1' + r_2'$$

节圆与分度圆重合时的中心距称为标准中距，其值为

$$a = r_1 + r_2$$

（2）齿侧间隙　一齿轮的节圆齿槽宽与另一齿轮的节圆齿厚的差值，称为齿侧间隙。研究啮合原理时忽略齿侧间隙，认为齿轮是无侧隙啮合，而齿侧间隙靠设计公差保证。为保证无齿侧间隙啮合，一个齿轮节圆上的齿厚$s_1'$应等于另一个齿轮节圆上的齿槽宽$e_2'$，图 7-23 所示的无侧隙啮合中，$e_{11'}' = s_{22'}'$。

$$s_1' = e_2' \text{或} s_2' = e_1'$$

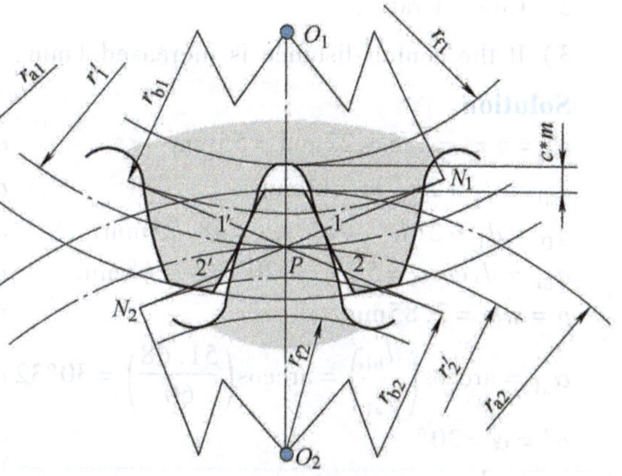

Fig. 7-23　Normal center distance（无侧隙啮合的径向间隙）

（3）径向间隙 $c$　一齿轮的齿根圆和另一齿轮的齿顶圆之间的径向距离的差值，称为径向间隙，如图 7-23 所示。为避免一齿轮的齿顶与另一齿轮的齿槽底相接触，并能有一定的空隙存储润滑油，径向间隙 $c = c^* m$，$c^*$ 为径向间隙系数。

（4）齿轮的标准安装　一对标准齿轮按标准中心距安装，做无侧隙啮合并具有标准顶隙。

**5. 齿轮和齿条传动**

齿轮与齿条啮合时，啮合线与齿轮的基圆相切且垂直于齿条的齿廓。

当齿轮与齿条标准安装时，齿轮的分度圆与节圆重合，齿条分度线与节线重合，啮合角 $\alpha'$等于齿轮分度圆压力角 $\alpha$。如将图 7-24 所示的齿条位置向远离齿轮圆心方向移动一段距离 $xm$，由于齿条同向齿廓上各点法向方向相同，因此啮合线不变，故节点 $P$ 不变，齿轮的分度圆仍然与节圆重合，但齿条的分度线与节线不再重合，而是相距一段距离 $xm$。

齿轮与齿条啮合传动时，无论是标准安装，还是非标准安装，齿轮分度圆永远与节圆重合。但只有在标准安装时，齿条的分度线才与节线重合。

**例 7-1**　已知一对标准安装的外啮合标准直齿圆柱齿轮的参数 $z_1 = 22$，$z_2 = 33$，$\alpha = 20°$，$m = 2.5\text{mm}$，$h_a^* = 1$，$c^* = 0.25$，求这对齿轮的主要尺寸和重合度。若两轮的中心距分开 1mm，重合度又为多少？

**解**

$d_1 = mz_1 = 2.5 \times 22\text{mm} = 55\text{mm}$　　　$d_2 = mz_2 = 2.5 \times 33\text{mm} = 82.5\text{mm}$

**5. Pinion and Rack**

Fig. 7-24 shows an involute rack in mesh with the pinion. If the pinion and the rack are mounted normally, the pitch line of the rack is tangent to the reference circle. When the rack moves an offset $xm$ outwards, the angle of obliquity is equal to the pressure angle, and the new pitch line is still tangent to the reference circle. The pitch point $P$ is not varied, and the pitch circle of the pinion is always coincident with its reference circle. This is very important for manufacturing gears with generating methods.

Where, $x$ is called coefficient of offset, and $m$ is the module of the gear.

**Example 7-1** Two involute gears in mesh have a module of 2.5mm and a pressure angle of 20°. The numbers of teeth are $z_1 = 22$, $z_2 = 33$. The coefficient of addendum is $h_a^* = 1$, and $c^* = 0.25$. Find the followings:

1) Sizes of the two gears;

2) Contact ratio;

3) If the center distance is increased 1mm, find the contact ratio.

**Solution**

$d_1 = mz_1 = 2.5 \times 22\text{mm} = 55\text{mm}$ $\quad d_2 = mz_2 = 2.5 \times 33\text{mm} = 82.5\text{mm}$

$d_{a1} = d_1 + 2h_a^* m = 60\text{mm}$ $\quad d_{a2} = d_2 + 2h_a^* m = 87.5\text{mm}$

$d_{f1} = d_1 - 2(h_a^* + c^*)m = 48.75\text{mm}$ $\quad d_{f2} = d_2 - 2(h_a^* + c^*)m = 76.25\text{mm}$

$d_{b1} = d_1 \cos\alpha = 55 \times \cos 20° = 51.68\text{mm}$ $\quad d_{b2} = d_2 \cos\alpha = 82.5 \times \cos 20° = 77.52\text{mm}$

$p = \pi m = 7.85\text{mm}$ $\quad p_b = p\cos\alpha = \pi m \cos 20° = 7.38\text{mm}$

$$\alpha_{a1} = \arccos\left(\frac{d_{b1}}{d_{a1}}\right) = \arccos\left(\frac{51.68}{60}\right) = 30°32' \alpha_{a2} = \arccos\left(\frac{d_{b2}}{d_{a2}}\right) = \arccos\left(\frac{77.52}{87.5}\right) = 27°38'$$

$\alpha' = \alpha = 20°$

$$\varepsilon_a = \frac{1}{2\pi}[z_1(\tan\alpha_{a1} - \tan\alpha') + z_2(\tan\alpha_{a2} - \tan\alpha')]$$

$$= \frac{1}{2\pi}[22 \times (\tan 30°32' - \tan 20°) + 33 \times (\tan 27°38' - \tan 20°)] = 1.629$$

$a = r_1 + r_2 = 27.5 + 41.25 = 68.75\text{mm}$

If the center distance increases 1mm, then:

$a' = a + 1 = 69.75\text{mm}$

From the following formulas, we can obtain the angle of obliquity.

$a'\cos\alpha' = a\cos\alpha$

$$\alpha' = \arccos\left(\frac{a}{a'}\right)\cos\alpha = \arccos\left(\frac{68.75}{69.75} \times 0.9397\right) = 22°9'$$

The contact ratio is then:

$$\varepsilon_a = \frac{1}{2\pi}[z_1(\tan\alpha_{a1} - \tan\alpha') + z_2(\tan\alpha_{a2} - \tan\alpha')]$$

$$= \frac{1}{2\pi}[22 \times (\tan 30°32' - \tan 22°9') + 33 \times (\tan 27°38' - \tan 22°9')] = 1.252$$

## 7.6 Forming and Undercutting of Gear Teeth

**1. Gear Teeth Forming**

There are many methods of forming the teeth of gears, such as shell molding, investment cast-

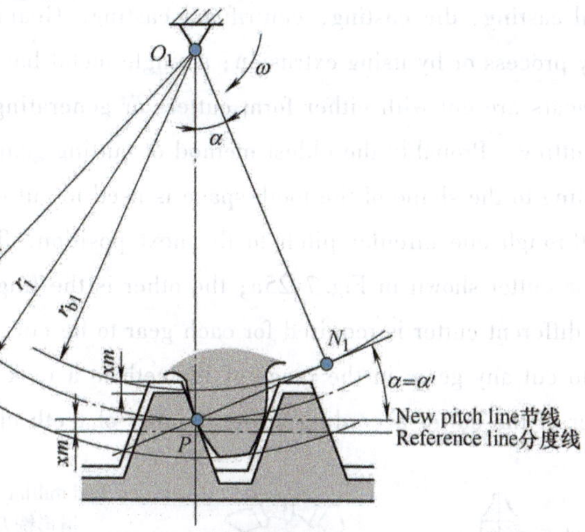


Fig. 7-24 Meshing of a pinion and a rack（齿轮与齿条啮合）

$d_{a1} = d_1 + 2h_a^* m = 60\text{mm}$ $\qquad d_{a2} = d_2 + 2h_a^* m = 87.5\text{mm}$

$d_{f1} = d_1 - 2(h_a^* + c^*)m = 48.75\text{mm}$ $\qquad d_{f2} = d_2 - 2(h_a^* + c^*)m = 76.25\text{mm}$

$d_{b1} = d_1\cos\alpha = 55\text{mm} \times \cos 20° = 51.68\text{mm}$ $\qquad d_{b2} = d_2\cos\alpha = 82.5 \times \cos 20° = 77.52\text{mm}$

$p = \pi m = 7.85\text{mm}$ $\qquad p_b = p\cos\alpha = \pi m\cos 20° = 7.38\text{mm}$

$\alpha_{a1} = \arccos\dfrac{d_{b1}}{d_{a1}} = \arccos\dfrac{51.68}{60} = 30°32'$ $\qquad \alpha_{a2} = \arccos\dfrac{d_{b2}}{d_{a2}} = \arccos\dfrac{77.52}{87.5} = 27°38'$

$\alpha' = \alpha = 20°$

$$\varepsilon_a = \frac{1}{2\pi}[z_1(\tan\alpha_{a1} - \tan\alpha') + z_2(\tan\alpha_{a2} - \tan\alpha')]$$

$$= \frac{1}{2\pi}[22 \times (\tan 30°32' - \tan 20°) + 33 \times (\tan 27°38' - \tan 20°)] = 1.629$$

标准中心距 $a = r_1 + r_2 = (27.5 + 41.25)\text{mm} = 68.75\text{mm}$

增大 1mm 时，中心距 $a' = a + 1\text{mm} = 69.75\text{mm}$

由 $a'\cos\alpha' = a\cos\alpha$，求得啮合角为

$$\alpha' = \arccos\frac{a\cos\alpha}{a'} = \arccos\frac{68.75 \times 0.9397}{69.75} = 22°9'$$

由此可得

$$\varepsilon_a = \frac{1}{2\pi}[z_1(\tan\alpha_{a1} - \tan\alpha') + z_2(\tan\alpha_{a2} - \tan\alpha')]$$

$$= \frac{1}{2\pi}[22 \times (\tan 30°32' - \tan 22°9') + 33 \times (\tan 27°38' - \tan 22°9')] = 1.252$$

## 7.6 渐开线圆柱齿轮的加工及其根切现象

### 1. 渐开线齿轮轮齿的加工

齿轮的加工方法很多，有铸造法、热压法、冲压法、粉末冶金法和切削法。最常用的是

ing, permanent-mold casting, die casting, centrifugal casting. Gears can also be formed by using the power-metallurgy process or by using extrusion; a single metal bar can be formed and then sliced into gears. Usually gears are cut with either form cutters or generating cutters.

(1) Forming Cutting  Probably the oldest method of cutting gear teeth is milling. A form milling cutter corresponding to the shape of the tooth space is used to cut one tooth space at a time, after the gear is indexed through one circular pitch to the next position. There are two kinds of milling cutter: one is the disc cutter shown in Fig. 7-25a; the other is the finger cutter shown in Fig. 7-25b. With this method, a different cutter is required for each gear to be cut, but it is impossible. The eight cutters can be used to cut any gears in the range of 12 teeth to a rack with reasonable accuracy. We can select the number of the cutter according to the number of teeth of the gear which will be cut.

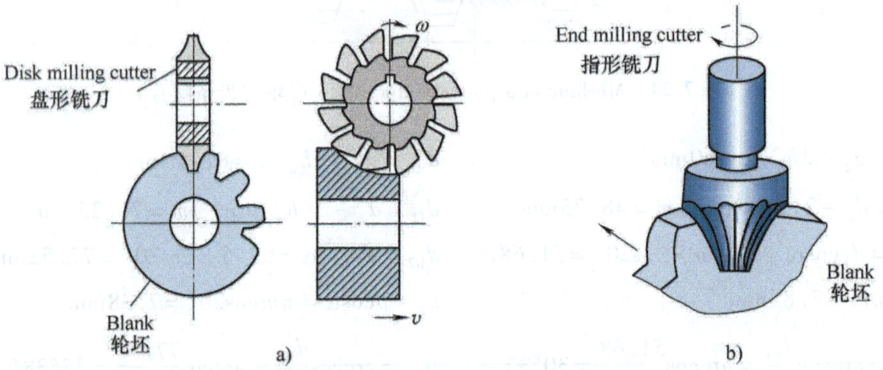


Fig. 7-25 Forming cutting (仿形加工)

(2) Generating Cutting  In generating, a tool having a shape different from the tooth profile is moved relative to the gear blank to obtain the proper tooth shape. The most common methods of generating gear teeth are shaping method and hobbing method.

1) Shaping teeth. Shaping is a highly favored method of generating teeth of gear. The cutting tool used in the shaping method is either a rack cutter or a pinion cutter. Fig. 7-26 shows shaping teeth with a pinion cutter; Fig. 7-27 shows shaping teeth with a rack cutter.

The reciprocating cutter is first fed into the blank until the pitch circle is tangent. Then, after each cutting stroke, the gear blank and the cutter roll slightly on their pitch circles. When the blank and cutter have rolled a distance equal to the circular pitch, the cutter is returned to the starting point and the process is continued until all the teeth have been cut.

The rack-shaped cutter generates at a velocity $v_c = r\omega$, as like as a rack meshing with a gear. For the pinion cutter, the generating motion with a pinion cutter must satisfy the following condition.

$$i_{12} = \frac{\omega_1}{\omega_2} = \frac{z_g}{z_c}$$

Where $\omega_1$ is the angular velocity of the pinion cutter; $\omega_2$ is the angular velocity of the gear to be cut; $z_g$ is the number of teeth to be cut; $z_c$ is the number of teeth of the pinion cutter.

2) Hobbing teeth. Fig. 7-28 illustrates the generating process with a hob.

A hob is a cylindrical cutter with one or more helical threads quite like a screw-thread tap, and has straight sides like a rack. The hob and the blank are rotate continuously at the proper angular

切削法，从加工原理上可将切削法分为仿形法和展成法两大类。

（1）仿形法　仿形法利用刀具的轴面齿形与所切制的渐开线齿轮的齿槽形状相同的特点，在轮坯上直接加工出齿轮的轮齿。常用刀具有盘形铣刀和指形铣刀两种：图 7-25a 所示为用盘形铣刀加工，切齿时刀具绕自身轴线转动，同时轮坯沿自身轴线移动；每铣完一个齿槽后，轮坯退回原处，利用分度机构将齿轮轮坯旋转 $360°/z$，之后再铣下一个齿槽，直至铣出全部轮齿；图 7-25b 所示为用指形铣刀加工。仿形法加工齿轮方法简单，在普通铣床上即可进行，但精度低，目前已经很少使用该方法加工齿轮。

（2）展成法　展成法是利用互相啮合的两个齿轮的齿廓曲线互为包络线的原理加工齿轮轮齿的。展成法切齿时，分为插齿法和滚齿法。插齿法所用刀具有齿轮插刀和齿条插刀，滚齿法所用刀具为齿轮滚刀。

1）齿轮插刀插制齿轮。图 7-26 所示齿轮插刀是带有切削刃的外齿轮。其模数和压力角与被切制齿轮相同。插齿机床的传动系统使插齿刀和轮坯按传动比 $i_{12} = \omega_1/\omega_2 = z_{被加工齿轮}/z_{刀具}$ 传动，此运动称为展成运动。为切出齿槽，刀具还需沿轮坯轴线方向做往复运动，称为切削运动。另外，为切出齿高，刀具还有沿轮坯径向的进给运动及插刀每次回程时轮坯沿径向的让刀运动。

2）齿条插刀插制齿轮。图 7-27 所示齿条插刀是带有切削刃的齿条。加工时，机床的传动系统使齿条插刀的移动速度 $v_刀$ 与被加工齿轮的分度圆线速度相等，即 $v_刀 = r\omega$。

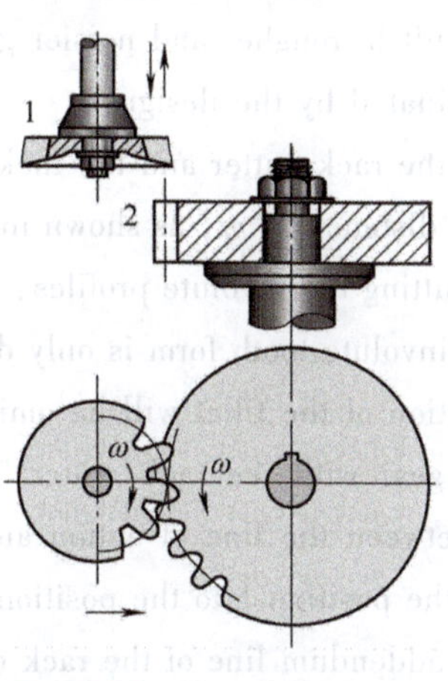

Fig. 7-26　Shaping teeth with a pinion cutter（齿轮插刀）

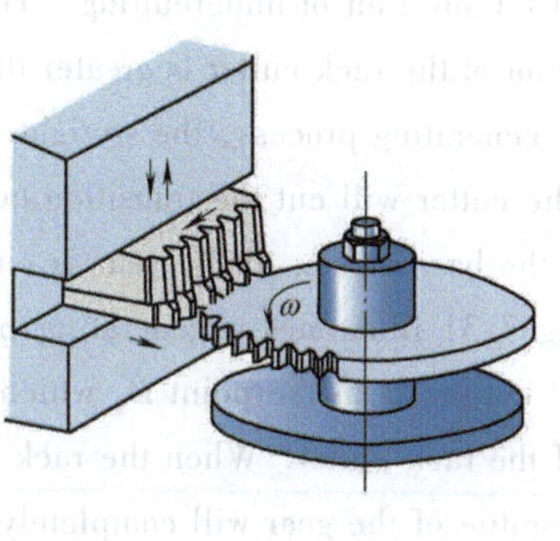

Fig. 7-27　Shaping teeth with a rack cutter（齿条插刀）

3）滚齿加工。插齿加工存在不连续的缺点，为了克服这个缺点可以采用齿轮滚刀加工，如图 7-28 所示。滚刀的外形类似一个螺杆，它的轴向剖面齿形与齿条插刀的齿形类似。当滚刀转动时，相当于直线齿廓的齿条连续不断地移动，从而包络出待加工的齿廓。此外，为了切制出具有一定宽度的齿轮，滚刀在转动的同时，还需沿轮坯轴线方向做进给运动。

用滚齿刀加工齿轮时，能连续切削，故生产率高，适用于大批量生产。

**2. 渐开线齿廓的根切**

用展成法加工齿轮时，刀具顶部可能把被加工齿轮的齿根部渐开线齿廓切去一部分，这种现象称为根切现象，如图 7-29 所示。发生根切的齿轮会削弱轮齿的抗弯强度，使实际啮合线缩短，重合度降低，影响传动的平稳性。因此，在设计齿轮时应避免发生根切。

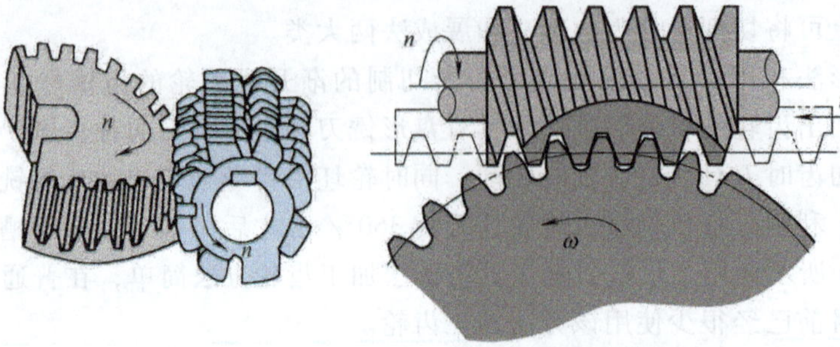

Fig. 7-28 Hobbing teeth (滚齿加工)

velocity ratio, and the hob is then fed slowly across the face of the blank from one end of teeth to the other. All teeth have been cut.

The cutting and rolling action is continued until the cutting process has been finished.

Some cutting processes, such as grinding, lapping, shaving and burnishing are often used in final finishing processes when accurate tooth profiles have been desired.

**2. Undercutting**

When gear teeth are produced by a generating process, the top of cutting tool removes the portion of the involute profile near the root teeth. This is called undercutting. The undercutting weakens the tooth by removing material at its root shown in Fig. 7-29. Severe undercutting will promote early tooth failure. It may also reduce the length of contact and result in rougher and noisier gear action. In machine design, the undercutting must be avoided or eliminated by the designer.

(1) Causation of undercutting The difference between the rack cutter and the rack is that the addendum of the rack cutter is greater than that of the rack, a distance $c^* m$, as shown in Fig. 7-30.

In generating process, the straight edge of the cutter is cutting the involute profiles, and the fillet of the cutter will cut the transition curve of the gear. The involute tooth form is only defined outside of the base circle. If the gear is cut with a hob, the portion of the fillet will be omitted.

Fig. 7-31 illustrates a generating process of a standard gear with the rack cutter. The initial point of cutting is at the point $B_1$ which is the intersection between the line of action and the right edge of the rack cutter. When the rack cutter is moved from the position 1 to the position 2, the involute profile of the gear will completely be cut. Because the addendum line of the rack exceeds the extreme point $N$, the generating process can not be stopped, such as at the position 3, obviously, the involute profile of the root tooth on the left edge of the cutter will be cut away.

Supposing the rack moves a distance $s_{23}$ along its pitch line from the position 2 to the position 3, the gear to be cut will rotates an angle $\varphi$, so the normal distance of the rack profiles between the position 2 and 3 is as follows:

$$\overline{K_2K_3} = r\varphi\cos\alpha = r_b\varphi$$

The corresponding arc length on base circle is as:

$\overset{\frown}{NN'} = r_b\varphi$ and $\overline{K_2K_3} = \overset{\frown}{NN'}$

In the same length of the arc and segment, the chord length of the arc must be less than the straight distance; the point $N'$ must be on the left of the cutter edge. That is to say, when the addendum line of rack cutter excesses the extreme point, the undercutting will occur.

（1）根切原因　齿条插刀在齿条的基础上，使齿顶增加高度为 $c^* m$ 的圆角部分，图 7-30所示为标准齿条插刀的齿廓形状。在展成加工标准齿轮的过程中，齿条插刀齿侧直刃切制出齿廓的渐开线部分，齿顶圆角刃切制出齿轮根部的介于渐开线与齿根圆弧间的过渡曲线。

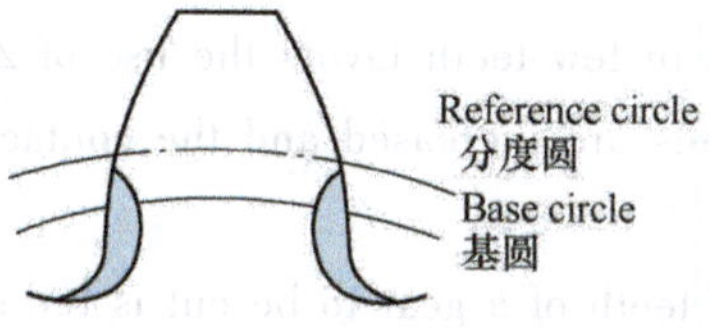


Fig. 7-29　Undercutting（根切现象）

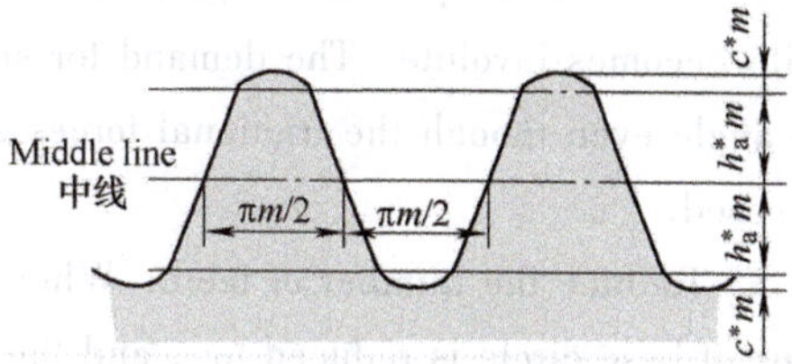


Fig. 7-30　Rack cutter profile
（齿条插刀的齿廓）

加工标准齿轮时，齿条插刀的中线与齿轮毛坯的分度圆相切，节点为 $P$，如图 7-31 所示。当切削刃处于位置 1 时，右切削刃与被切制齿轮在 $B_1$ 点啮合，开始加工轮坯上的渐开线齿廓，当切削刃处于位置 2 时，达到啮合极限点 $N$，加工出轮齿由基圆至齿顶圆间的渐开线齿廓。设刀具移动距离 $s_{23}=r\varphi$，到达位置 3。因刀具的中线与轮坯的分度圆做纯滚动，轮坯转过角度 $\varphi$，刀具法线移动距离为

$$\overline{K_2K_3}=r\varphi\cos\alpha=r_b\varphi$$

此时，轮坯上的点 $N$ 转过的弧长为

$$\overset{\frown}{NN'}=r_b\varphi$$

因此

$$\overline{K_2K_3}=\overset{\frown}{NN'}$$

由于 $\overline{K_2K_3}$ 是直线距离，而 $\overset{\frown}{NN'}$ 为弧长，故点 $N'$ 必位于齿廓的左侧。刀具的齿顶必定切入轮坯的齿根，基圆内的齿廓和基圆外的渐开线齿廓会被切去一部分，从而发生根切。

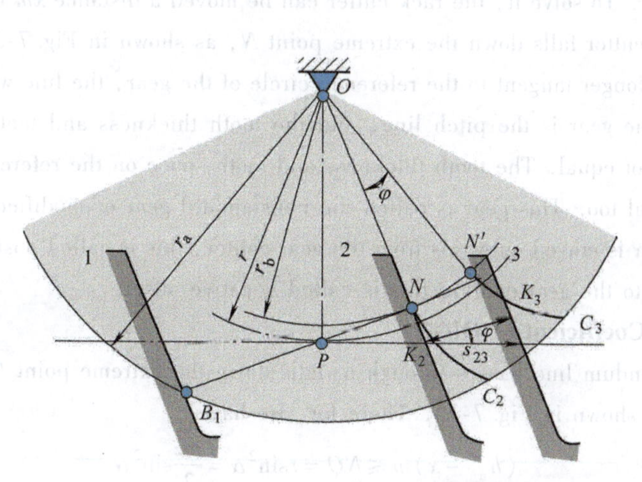


Fig. 7-31　Undercutting process（根切的形成）

(2) Minimum number of teeth to avoid undercutting The methods to eliminate undercutting are as follows:

1) Reduce the height of the tooth of the cutter. These gears to be cut are called short-tooth gears; they have a small contact ratio.

2) Increase the pressure angle. This results in a smaller base circle, so that more of the tooth profile becomes involute. The demand for smaller pinion with few teeth favors the use of 25° pressure angle even though the frictional forces and bearing loads are increased and the contact ratio is decreased.

3) Reduce the number of teeth. When the number of teeth of a gear to be cut is reduced, the radius of base circle is reduced too, and the extreme point $N$ is likely to fall on the right of the point $B_2$ where the addendum line of the rack cutter intersects the line of action. See the Fig. 7-32.

From this figure, we have:

$$h_a^* m \leqslant \overline{NQ} = r\sin^2\alpha = \frac{mz}{2}\sin^2\alpha$$

Solving it, we can obtain:

$$z \geqslant \frac{2h_a^*}{\sin^2\alpha}$$

The minimum number of teeth to avoid undercutting is:

$$z_{\min} = \frac{2h_a^*}{\sin^2\alpha} \tag{7-9}$$

When $h_a^* = 1$, $\alpha = 20°$, then $z_{\min} = 17$.

## 7.7 Nonstandard Spur Gears

### 1. Concept of Nonstandard Gears

In generating a standard gear with a rack cutter, if the addendum line of the rack cutter excesses the extreme point $N$ where the addendum line of the cutter intersects the line of action, the undercutting will occur. To solve it, the rack cutter can be moved a distance $xm$ outwards until the addendum line of the cutter falls down the extreme point $N$, as shown in Fig. 7-33. In this case, the reference line is no longer tangent to the reference circle of the gear, the line which is tangent to the reference circle of the gear is the pitch line, and the tooth thickness and tooth space on the pitch line of the rack is not equal. The tooth thickness and tooth space on the reference circle of the gear to be cut is not equal too. This gear is called the nonstandard gear or modified gear.

When the cutter is moved outwards from the gear center, this is called positive offset, and when it is moved towards to the gear center, this is called negative offset.

### 2. Minimum Coefficient of Offset

When the addendum line passes through or falls down the extreme point $N$, there is no undercutting to occur, as shown in Fig. 7-33. There for, we have:

$$(h_a^* - x)m \leqslant \overline{NQ} = r\sin^2\alpha = \frac{mz}{2}\sin^2\alpha$$

$$h_a^* - x \leqslant \frac{z}{2}\sin^2\alpha$$

(2) 避免根切的最少齿数 用展成法加工标准齿轮时，刀具的齿顶线如超过了啮合极限点 $N$，就会出现根切。因此，要避免发生根切，必须使刀具的齿顶线不超过啮合极限点 $N$，如图7-32所示。即刀具的齿顶线到中线的距离 $h_a^* m$ 应小于或等于啮合极限点 $N$ 到中线距离 $\overline{NQ}$，即

$$h_a^* m \leqslant \overline{NQ} = r\sin^2\alpha = \frac{mz}{2}\sin^2\alpha$$

$$z \geqslant \frac{2h_a^*}{\sin^2\alpha}$$

加工标准齿轮不出现根切的最少齿数为

$$z_{\min} = \frac{2h_a^*}{\sin^2\alpha} \tag{7-9}$$

当 $h_a^* = 1$、$\alpha = 20°$时，$z_{\min} = 17$。

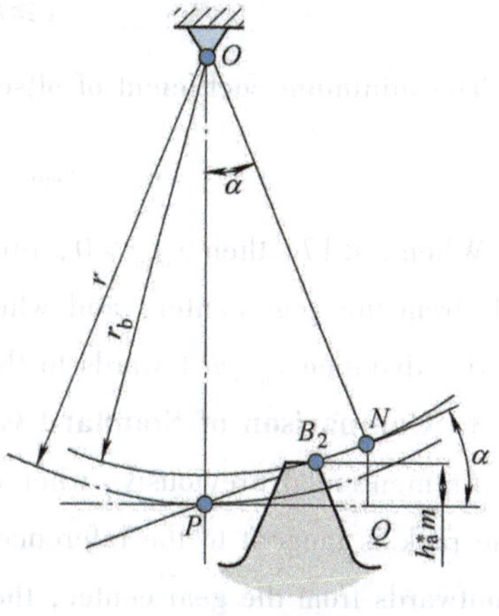


Fig. 7-32 Minimum number of teeth to avoid undercutting (避免根切的最少齿数)

## 7.7 变位齿轮概述

### 1. 变位齿轮的概念

加工标准齿轮时，当刀具齿顶线超过啮合极限点 $N$ 时将发生根切。若刀具向远离轮坯中心方向移动一段距离 $xm$，使刀具齿顶线落在点 $N$ 之下，则可避免发生根切。由于这时刀具的节线与中线不再重合，而是分离了 $xm$，故加工出的齿轮在分度圆上的齿厚与齿槽宽不相等，这种齿轮称为变位齿轮，$x$ 称为变位系数。通常，刀具由标准安装位置远离轮坯中心时，$x$ 为正值，称为正变位，加工出的齿轮称为正变位齿轮；如果被切制的齿轮齿数比较多，为了满足齿轮传动的某些要求，也可将刀具由标准安装位置移向轮坯中心 $xm$，此时 $x$ 为负值，称为负变位，加工出的齿轮称为负变位齿轮。

### 2. 最小变位系数

当刀具的齿顶线刚好通过轮坯与刀具的啮合极限点 $N$ 时，齿轮便完全没有根切。如图7-33所示。不发生根切的条件为

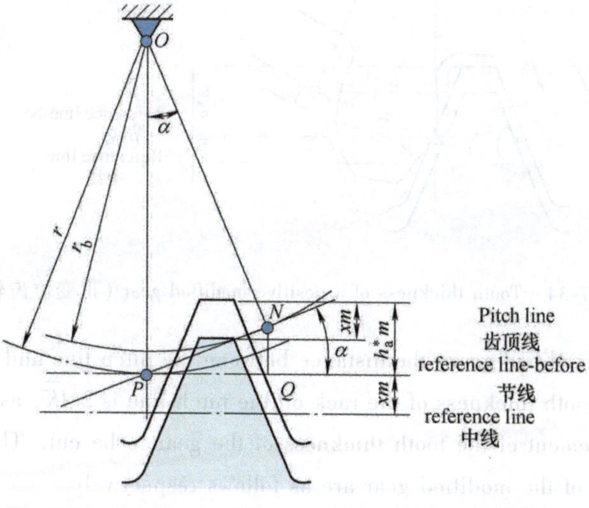


Fig. 7-33 Smallest coefficient of offset (最小变位系数)

Simultaneously solve it with equation (7-9), and we get

$$x \geqslant h_a^* - \frac{z}{2}\sin^2\alpha = h_a^*\left(1 - \frac{z}{z_{min}}\right)$$

The minimum coefficient of offset is then:

$$x_{min} = h_a^* - \frac{z}{2}\sin^2\alpha = h_a^*\left(1 - \frac{z}{z_{min}}\right) \tag{7-10}$$

When $z < 17$, then $x_{min} > 0$, this means that the rack is at least moved a distance $x_{min}m$ outwards from the gear center, and when $z > 17$, then $x_{min} < 0$, this means that the rack is at most moved a distance $x_{min}m$ towards to the gear center.

**3. Comparison of Standard Gear and Nonstandard Gear**

As mentioned previously, when a standard gear is generated by a rack cutter, the reference line of the rack is tangent to the reference circle of the gear. When the rack cutter is moved a distance $xm$ outwards from the gear center, the reference line of the rack is no longer tangent to the reference circle of the gear. A new pitch line on the rack will be tangent to the reference circle of the gear. The reference circle of the gear is always coincident with its pitch circle regardless of whether the gear being cut is standard or nonstandard.

The standard gear compares with corresponding nonstandard gear which has the same module, pressure angle and number of teeth, some of sizes vary, some of them are the same.

(1) No varied sizes The base circle, reference circle, circular pitch and base pitch are not varied. The involute curve is not changed; it is used on the different portion of the involute.

(2) Varied sizes The tooth thickness and width space are varied.

Fig. 7-34 shows a generating process of a standard gear and a positive modified gear.

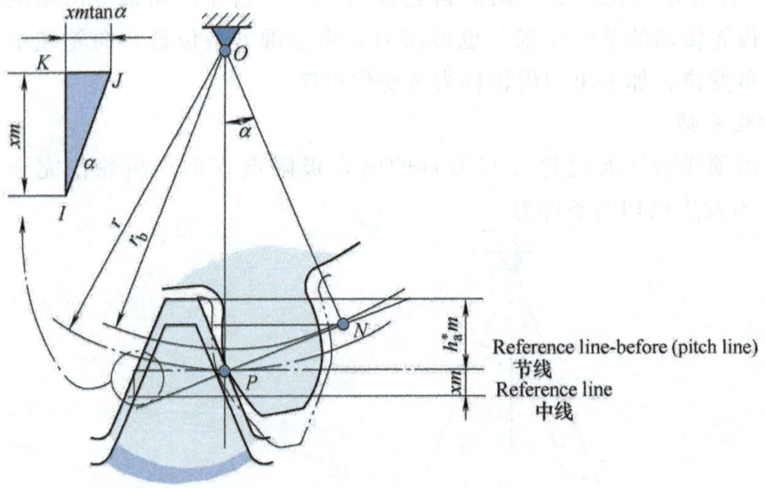


Fig. 7-34 Tooth thickness of a positive modified gear (正变位齿轮齿厚)

For the positive modified gear, the distance between the pitch line and the reference line is $xm$; the decrement of the tooth thickness of the rack on the pitch line is $2\,\overline{JK}$, as shown in Fig. 7-34, and it is equal to the increment of the tooth thickness of the gear to be cut. Therefore, the tooth thickness and width space of the modified gear are as follows respectively.

$$(h_a^* - x)m \leqslant \overline{NQ} = r\sin^2\alpha = \frac{mz}{2}\sin^2\alpha$$

$$h_a^* - x \leqslant \frac{z}{2}\sin^2\alpha$$

与式（7-9）联立求解得

$$x \geqslant h_a^* - \frac{z}{2}\sin^2\alpha = h_a^*\left(1 - \frac{z}{z_{min}}\right)$$

用标准齿条插刀切制小于最少齿数的齿轮不发生根切的最小变位系数为

$$x_{min} = h_a^* - \frac{z}{2}\sin^2\alpha = h_a^*\left(1 - \frac{z}{z_{min}}\right) \tag{7-10}$$

当 $z<17$ 时，$x_{min}>0$，说明为了避免根切，刀具应由标准位置向远离轮坯中心方向移动，移动的最小距离为 $x_{min}m$；当 $z>17$ 时，$x_{min}<0$，这说明刀具向轮坯中心方向移动一段距离也不会发生根切，移动的最大距离为 $x_{min}m$。

**3. 变位齿轮与标准齿轮的异同点**

变位齿轮与同参数的标准齿轮相比，它们的渐开线相同，只是使用同一条渐开线的不同部分。分度圆、基圆、齿距、基圆齿距不变，而齿顶圆、齿根圆，齿顶高、齿根高、分度圆齿厚和齿槽宽均发生了变化。

（1）齿厚与齿槽宽　如图 7-34 所示，对于正变位齿轮来说，刀具节线上的齿厚比中线上的齿厚减小了 $2\,\overline{JK}$，因此被切制齿轮分度圆上的齿槽宽将减小 $2\,\overline{JK}$。由几何关系可知 $\overline{JK} = xm\tan\alpha$。故正变位齿轮齿槽宽的计算公式为

$$e = \frac{\pi m}{2} - 2xm\tan\alpha$$

齿厚的计算公式为

$$s = \frac{\pi m}{2} + 2xm\tan\alpha$$

（2）齿顶高及齿根高　正变位齿轮的齿根高变小，齿顶高变大；负变位齿轮则相反。对于负变位齿轮，以上公式同样适用，只需注意变位系数为负值即可。

同参数的标准齿轮与变位齿轮的齿形比较如图 7-35 所示。

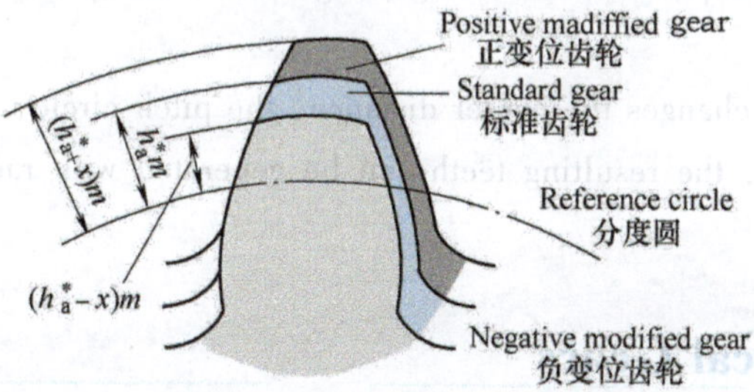


Fig. 7-35　Tooth profiles of the standard gear and modified gears（变位齿轮与标准齿轮的齿廓）

**4. 变位齿轮传动简介**

设一对互相啮合的变位齿轮的变位系数分别为 $x_1$、$x_2$，根据变位系数的不同，可以分为以下传动类型。

$$\overline{JK} = xm\tan\alpha$$

$$e = \frac{\pi m}{2} - 2xm\tan\alpha$$

$$s = \frac{\pi m}{2} + 2xm\tan\alpha$$

The addendum and dedendum are varied.

From the Fig. 7-35, we can observe that the addendum of a positive modified gear is larger than that of the standard gear, and its dedendum is smaller than that of the standard gear. In the same method, we can find the difference between the standard gear and negative modified gear in their addendum and dedendum.

**4. Brief Introduction of Nonstandard Gears**

Supposing that the coefficients of offset of the two gears are $x_1$ and $x_2$, the types of gear transmission can be classified as follows.

1) The coefficients of offset $x_1$ and $x_2$ are all zero, that is:

$x_1 + x_2 = 0$, and $x_1 = x_2 = 0$. It is a pair of standard gears.

2) The coefficients of offset $x_1$ and $x_2$ are equal in magnitude but opposite in direction; that is: $x_1 + x_2 = 0$, and $x_1 = -x_2 \neq 0$. The addendum of the pinion becomes longer, and the addendum of the other gear becomes shorter, so it is called the long-and-short addendum system or modified gearing of equal offset. It often happens in the design of machinery that the center distance between a pair of gears is fixed. To eliminate the undercutting of a pinion, the positive offset is adopted for the pinion and the negative offset is adopted for the large gear. In this long-and-short-addendum system, there is no change in the center distance, pitch circle and the pressure angle.

When designing the long-and-short-addendum system, the sum of the teeth of the two gears must be at least 34 for pressure angle 20°.

3) The sum of the coefficients of offset $x_1$ and $x_2$ is not equal zero, that is:

$$x_1 + x_2 \neq 0$$

This is called center-distance modification gears or angular modification gears.

When $x_1 + x_2 > 0$, it is positive modified gearing, and when $x_1 + x_2 < 0$, it is negative modified gearing.

Although this system changes the center distance, the pitch circle, tooth thickness, addendum and the angle of obliquity, the resulting teeth can be generated with rack cutters or with standard pinion shapers.

## 7.8 Parallel Helical Gears

The parallel-axis helical gears are simply called helical gears. The teeth are inclined to the axis of the gear, and they are used to transmit motion between the two parallel shafts.

**1. Shape of the Tooth of Helical Gear**

If a plane $S$ rolls on a base cylinder, a line $KK'$ parallel to the axis of the cylinder in the plane will generate an involute surface of a spur gear on the cylinder, shown in Fig. 7-36a. Thus, when a

1）$x_1+x_2=0$，且 $x_1=x_2=0$，此时为标准齿轮传动。

2）$x_1+x_2=0$，且 $x_1=-x_2\neq0$，此时为等移距变位齿轮传动，一般小齿轮采用正变位，大齿轮采用负变位。

3）$x_1+x_2\neq0$，此时为不等移距变位齿轮传动。其中：$x_1+x_2>0$ 时称为正传动；$x_1+x_2<0$ 时称为负传动。变位齿轮的设计请参考相关文献。

## 7.8 平行轴斜齿圆柱齿轮机构

平行轴斜齿圆柱齿轮机构简称斜齿轮机构。

**1. 斜齿轮齿廓曲面的形成**

直齿轮的轮齿与轴线平行，所以在垂直于齿轮轴线的任意平面上的齿廓形状及其啮合情况是完全一样的。

图 7-36a 所示，将直齿轮的齿廓形成扩展到空间，基圆成为基圆柱，发生线成为发生面，渐开线成为渐开面。当发生面 $S$ 在基圆柱上做纯滚动时，与基圆柱母线 $NN'$ 平行的直线 $KK'$ 的轨迹形成直齿轮的渐开线齿廓曲面。两齿廓啮合点成为啮合线 $KK'$，各啮合线均平行于基圆柱母线，直齿轮啮合传动中，轮齿是沿全齿宽同时进入啮合与同时退出啮合的。

斜齿轮齿廓曲面的形成与直齿轮类似，只不过直线 $KK'$ 不平行于 $NN'$，而与它成一偏斜角 $\beta_b$，如图 7-36b 所示。当发生面 $S$ 沿基圆柱做纯滚动时，直线 $KK'$ 上各点的轨迹仍为渐开线，各渐开线的起始点将在基圆柱上集合形成一条螺旋线 $AA'$，具有不同起始点的渐开线集合形成渐开线螺旋曲面。两斜齿轮的啮合过程是在从动轮齿顶的一点开始接触，然后啮合线由短变长，再由长变短，最后在靠近从动轮齿根的某一点分离。也就是说，斜齿轮的轮齿是在全齿宽方向逐渐进入啮合和退出啮合的，故传动平稳性好。

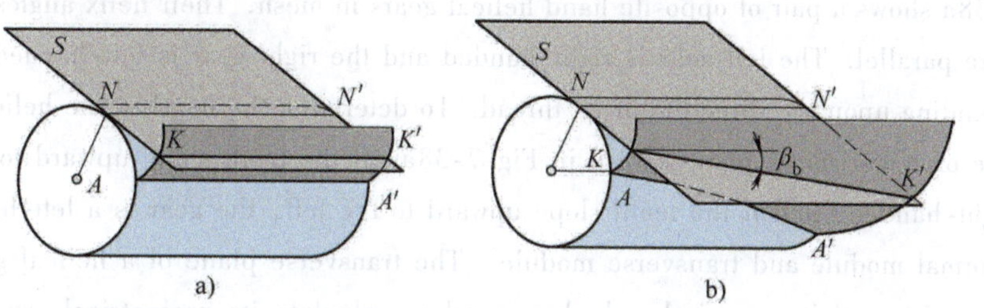


Fig. 7-36 Tooth surface of a spur gear and a helical gear（渐开线圆柱齿轮和斜齿轮轮齿的齿面）

**2. 斜齿轮的基本参数**

垂直于斜齿轮轴线的平面称为端面，端面是形成渐开线的基圆面，也是几何尺寸设计的基准面。垂直于斜齿轮螺旋线的平面称为法面，法面是斜齿轮传动的力作用面，也是齿廓加工的基准面。故斜齿轮有端面和法向两套参数，端面参数用下角标 t 表示，法向参数用下角标 n 表示。

（1）斜齿轮的螺旋角　将图 7-37a 所示的斜齿轮的分度圆柱延长到螺旋线的导程，其分度圆柱面展开成图 7-37b 所示的三角形，底边长 $\pi d$ 表示分度圆周长。阴影部分表示分度圆柱的展开图。$B$ 为斜齿轮的轴向宽度。螺旋线与轴线的夹角即螺旋角。将螺旋线延长到完整的导程，可有

pair of spur gears is in mesh, contact between the teeth is along the line parallel to the axis. However, if the line in the plane is inclined an angle $\beta_b$ to the axis of the cylinder, it will generates an involute helicoids, as shown in Fig. 7-36b.

The initial contact of spur gear teeth is a line extending all the way across the tooth face, and the initial contact of helical gear teeth is a point which changes into a line as the teeth come into more engagement. In helical gears, the line is diagonal across the tooth face, so it is gradual engagement of the teeth and the smooth transfer of load from one tooth to another. This gradual contact across the teeth results in less impact loading, and thus helical gears operate more quietly than spur gears.

**2. Basic Parameters of a Helical Gear**

(1) Helix angle  Fig. 7-37a shows a helical gear with right-hand thread. If we extend the width of the helical gear until it is equal to the lead of the helix, then develop the base cylinder and reference cylinder; the helix angles can be determined from Fig. 3-37b, so we have:

$$\tan\beta = \frac{\pi d}{P_z},\ \text{and}\ \tan\beta_b = \frac{\pi d_b}{P_z}$$

Where $d$ is the diameter of the reference circle, $d_b$ is the diameter of base circle, and $P_z$ is the lead of the helix. The intersection of the involute helicoids and base cylinder is a helix, and the helix angle is $\beta_b$. The intersection of the involute helicoids and reference cylinder is a helix too, but its helix angle is $\beta$.

The relationship between the $\beta_b$ and $\beta$ is as follows:

$$\tan\beta_b = \frac{d_b}{d}\tan\beta = \tan\beta\cos\alpha_t$$

Fig. 7-38a shows a pair of opposite hand helical gears in mesh. Their helix angles are equal and their axes are parallel. The left gear is right-handed and the right gear is left-handed. The hand of helix is depending upon the direction of its thread. To determine the hand of the helical gear, place its transverse on a horizontal plane shown in Fig. 7-38a. If the teeth slope upward to the right, the gear is a right-handed, and if the teeth slope upward to the left, the gear is a left-handed.

(2) Normal module and transverse module  The transverse plane of a helical gear is perpendicular to its axis, and it is a circle which is used to calculate its geometrical sizes. The normal plane is perpendicular to its tooth thread; it is an ellipse which is used to analyze force action. Therefore, the parameters of the helical gear can be divided into that of the transverse plane and normal plane.

The terminology used for the helical gear is very similar to that of spur gear. In fact, most of relations developed for spur gear are equally applicable to helical gear; the additional terms are the parameters on the normal and transverse plane.

For example, the circular pitch of a helical gear can be measured in two different ways: the transverse circular pitch $p_t$ is measured on the transverse plane; the normal circular pitch $p_n$ is measured on the normal plane, as shown in Fig. 7-38b.

The relationship between the normal pitch and transverse pitch is as follows:

$$p_n = p_t\cos\beta$$

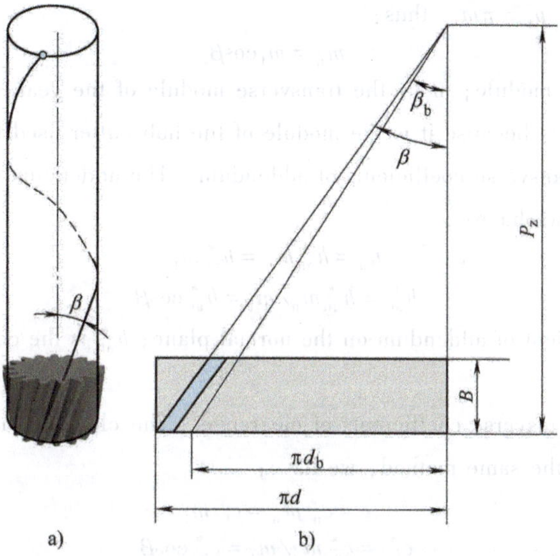


Fig. 7-37 Helix angles（螺旋角）

$$\tan\beta = \frac{\pi d}{P_z}$$

式中，$P_z$ 为螺旋线的导程。

由于斜齿轮各个圆柱面上的螺旋线的导程相同，所以基圆柱面上的螺旋角 $\beta_b$ 应为

$$\tan\beta_b = \frac{\pi d_b}{P_z}$$

由以上两式得

$$\tan\beta_b = \frac{d_b}{d}\tan\beta = \tan\beta\cos\alpha_t$$

根据螺旋线的走向，斜齿轮有左旋和右旋之分。以端面为基准，螺旋线从左向右升，为右旋齿轮；螺旋线从右向左升，为左旋齿轮。图 7-38a 所示为一对斜齿轮的旋向示意图。

（2）法向模数 $m_n$ 与端面模数 $m_t$　图 7-38b 所示的斜齿轮分度圆柱面展开图中，有剖面部分为轮齿，空白部分为齿槽。$\beta$ 为分度圆柱的螺旋角，$p_n$ 为法向齿距，$p_t$ 为端面齿距，根据几何关系可得

$$p_n = p_t\cos\beta$$

因为

$$p_n = \pi m_n, p_t = \pi m_t$$

所以，斜齿轮端面模数和法向模数的关系为

$$m_n = m_t\cos\beta$$

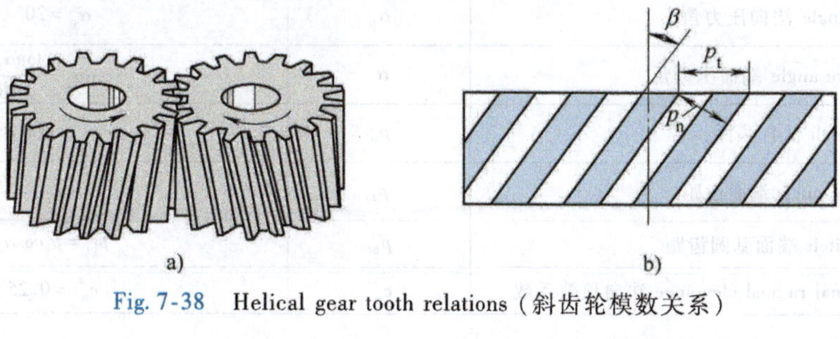


Fig. 7-38 Helical gear tooth relations（斜齿轮模数关系）

Because $p_n = \pi m_n$, $p_t = \pi m_t$, thus:

$$m_n = m_t \cos\beta$$

Where $m_n$ is the normal module; $m_t$ is the transverse module of the gear. The normal module is a very important parameter, because it is the module of the hob cutter used to manufacture the gear.

(3) Normal and transverse coefficients of addendum  The addendum is the same either normal or transverse plane, so we have:

$$h_a = h_{an}^* m_n = h_{at}^* m_t$$

$$h_{at}^* = h_{an}^* m_n / m_t = h_{an}^* \cos\beta$$

Where $h_{an}^*$ is the coefficient of addendum on the normal plane; $h_{at}^*$ is the coefficient of addendum on the transverse plane.

(4) Normal and transverse coefficients of clearance  The clearance is the same either normal or transverse plane. In the same method, we have:

$$c = c_n^* m_n = c_t^* m_t$$

$$c_t^* = c_n^* m_n / m_t = c_n^* \cos\beta$$

Where $c_n^*$ is the coefficient of clearance on the normal plane; $c_t^*$ is the coefficient of clearance on the transverse plane.

(5) Normal and transverse pressure angles  The spur gears are identified by means of one pressure angle; the geometry of helical gears requires the two pressure angles. Fig. 7-39 shows the tooth profile of a helical rack on the transverse plane and normal plane. A trigonometric relationship between the pressure angle and the helix angle of the rack can be obtained as follows:

$$\tan\alpha_n = \frac{\overline{a'c}}{\overline{a'b'}},\ \tan\alpha_t = \frac{\overline{ac}}{\overline{ab}} \text{ and } \cos\beta = \frac{\overline{a'c}}{\overline{ac}}$$

Lines $ab$ and $a'b'$ are all the height of the helical rack, so, $\overline{ab} = \overline{a'b'}$.

Therefore:

$\tan\alpha_n = \cos\beta \tan\alpha_t$

### 3. Sizes of Helical Gears

The dimensions of helical gears are shown in Tab. 7-3.

**Tab. 7-3 Formula for helical gear** (标准斜齿圆柱齿轮几何尺寸计算公式)

| Name 名称 | Symbol 符号 | Formula 计算公式 |
|---|---|---|
| Helix angle 螺旋角 | $\beta$ | $8° \sim 15°$ |
| Helix angle on a base cylinder 基圆柱螺角 | $\beta_b$ | $\tan\beta_b = \tan\beta\cos\alpha_t$ |
| Normal module 法向模数 | $m_n$ | table7-1 (按表 7-1 取标准值) |
| Transverse module 端面模数 | $m_t$ | $m_t = \frac{m_n}{\cos\beta}$ |
| Normal pressure angle 法向压力角 | $\alpha_n$ | $\alpha_n = 20°$ |
| Transverse pressure angle 端面压力角 | $\alpha_t$ | $\tan\alpha_t = \frac{\tan\alpha_n}{\cos\beta}$ |
| Normal circular pitch 法向齿距 | $p_n$ | $p_n = \pi m_n$ |
| Transverse circular pitch 端面齿距 | $p_t$ | $p_t = \pi m_t = \frac{p_n}{\cos\beta}$ |
| Transverse base pitch 端面基圆齿距 | $p_{bt}$ | $p_{bt} = p_t \cos\alpha_t$ |
| Coefficient of normal radical clearance 法向顶隙系数 | $c_n^*$ | $c_n^* = 0.25$ |

（续）

| Name 名称 | Symbol 符号 | Formula 计算公式 |
|---|---|---|
| Diameter of reference circle 分度圆直径 | $d$ | $d=m_t z=m_n z/\cos\beta$ |
| Diameter of base circle 基圆直径 | $d_b$ | $d_b=d\cos\alpha_t$ |
| Addendum 齿顶高 | $h_a$ | $h_a=h_{an}^* m_n=h_{at}^* m_t$ |
| Dedendum 齿根高 | $h_f$ | $h_f=(h_{an}^*+c_n^*)m_n=(h_{at}^*+c_t^*)m_t$ |
| Whole depth 全齿高 | $h$ | $h=h_a+h_f=(2h_{an}^*+c_n^*)m_n=(2h_{at}^*+c_t^*)m_t$ |
| Diameter of addendum circle 齿顶圆直径 | $d_a$ | $d_a=d+2h_a=(z+2h_{at}^*)m_t$ |
| Diameter of dedendum circle 齿根圆直径 | $d_f$ | $d_f=d-2h_f=(z-2h_{at}^*-2c_t^*)m_t$ |
| Clearance 顶隙 | $c$ | $c=c_n^* m_n=c_t^* m_t$ |
| Center distance 中心距 | $a$ | $a=(d_1+d_2)/2=(z_1+z_2)m_n/2\cos\beta$ |

（3）齿顶高系数　斜齿轮的齿顶高和齿根高，不论从法向或端面上看都是相同的，但齿顶高系数不同。法向齿顶高系数 $h_{an}^*$、端面齿顶高系数 $h_{at}^*$ 的换算关系如下：

$$h_a=h_{an}^* m_n=h_{at}^* m_t$$

$$h_{at}^*=h_{an}^* m_n/m_t=h_{an}^*\cos\beta$$

（4）顶隙系数　斜齿轮在法向和端面上的顶隙相同，但顶隙系数不同。法向顶隙系数 $c_n^*$、端面顶隙系数 $c_t^*$ 的关系如下：

$$c=c_n^* m_n=c_t^* m_t$$

$$c_t^*=c_n^* m_n/m_t=c_n^*\cos\beta$$

（5）压力角　斜齿轮压力角分为法向压力角 $\alpha_n$ 与端面压力角 $\alpha_t$。为简单起见，用斜齿条的端面压力角和法向压力角来定义斜齿轮的端面压力角和法向压力角。

图 7-39 所示齿条中，$bac$ 所在的面为端面，此面内的压力角为斜齿轮的端面压力角 $\alpha_t$，$b'a'c$ 所在的面为法面，此面内的压力角为斜齿轮的法向压力角 $\alpha_n$。

在直角三角形 $bac$、$b'a'c$ 及 $aa'c$ 中，有

$$\tan\alpha_n=\frac{\overline{a'c}}{\overline{a'b'}},\ \tan\alpha_t=\frac{\overline{ac}}{\overline{ab}}$$

因为$\overline{a'c}=\overline{ac}\cos\beta$，$\overline{ab}=\overline{a'b'}$，所以

$$\tan\alpha_n=\cos\beta\tan\alpha_t$$

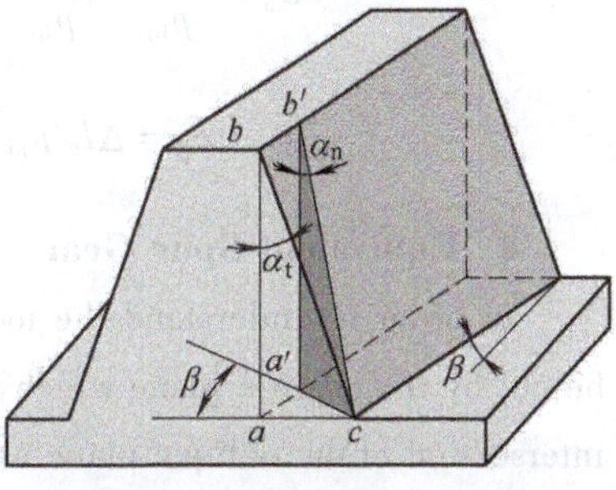

Fig. 7-39　Normal pressure angle and transverse pressure angle
（法向压力角和端面压力角）

**3. 斜齿轮的几何尺寸计算**

斜齿轮的几何尺寸与同样端面参数的直齿轮完全相同，相关计算公式与直齿轮完全一样，不过要换成斜齿轮的端面参数。当法向参数为标准值时，还需进一步用法向参数表达斜齿轮几何尺寸的计算公式。

计算公式见表 7-3。斜齿圆柱齿轮传动中心距的配凑可通过改变螺旋角 $\beta$ 来实现，因而变位斜齿轮很少使用。

**4. 斜齿轮传动的正确啮合条件**

一对斜齿圆柱齿轮正确啮合时，除满足两个齿轮的模数和压力角应分别相等外，它们的螺旋角还应匹配。因此，一对斜齿圆柱齿轮的正确啮合条件为

### 4. Conditions of Correctly Meshing for Helical Gears

The conditions of correctly engaging for a pair of helical gears are that the normal module and normal pressure angle of the two meshing gears must be the same respectively, or the transverse module and transverse pressure angle of the two meshing gears must be the same respectively; their helix angles must be equal and have opposite helical directions. They are:

$$m_{n1} = m_{n2} \text{ or } m_{t1} = m_{t2}$$

$$\alpha_{n1} = \alpha_{n2} \text{ or } \alpha_{t1} = \alpha_{t2}$$

$$\beta_1 = -\beta_2$$

### 5. Conditions of Continuous Transmission of Gears

The contact ratio of a pair of helical gears is greater than that of spur gears. This can be illustrated in Fig. 7-40. The upper figure is a developed base cylinder of a spur gear in which the transverse plane is coincident with the normal plane; the line $B_1B_2$ represents the contact length of the gears. The contact ratio is:

$$\varepsilon_a = \frac{\overline{B_1B_2}}{p_{bt}} = \frac{L}{p_{bt}}$$

The Fig. 7-40b is a developed base cylinder of a helical gear which has the same number of teeth and the transverse module with the spur gear shown in Fig. 7-40a.

The initial contact occurs at point $B_2$. The final contact point is at point $B_1'$, not at the point $B_1$. The increment of contact length is $\Delta L$.

$$\Delta L = B\tan\beta_b$$

The contact ratio is:

$$\varepsilon = \varepsilon_a + \varepsilon_\beta$$

$$\varepsilon_a = \frac{\overline{B_1B_2}}{p_{bt}} = \frac{L}{p_{bt}} = \frac{1}{2\pi}[z_1(\tan\alpha_{at1} - \tan\alpha_t') + z_2(\tan\alpha_{at2} - \tan\alpha_t')]$$

$$\varepsilon_\beta = \Delta L/p_{bt} = \frac{B\tan\beta_b}{p_{bt}} = \frac{B\tan\beta_b}{\pi m_t\cos\alpha_t} = \frac{B\tan\beta\cos\alpha_t}{\pi m_t\cos\alpha_t} = \frac{B\sin\beta}{\pi m_n}$$

### 6. Equivalent Spur Gear

In order to understand the tooth shape of the normal plane of a helical gear, a helical gear can be cut by the oblique plane which is perpendicular to the tooth direction, as shown in Fig. 7-41. The intersection of the oblique plane and the reference cylinder produces an ellipse whose semi-major axis is $a$, and semi-minor axis is $b$.

The semi-major axis and semi-minor axis are as follows:

$$a = \frac{r}{\cos\beta},\ b = r$$

The radius of curvature of the ellipse at point $P$ is given by the following formula.

$$\rho = \frac{a^2}{b} = \left(\frac{r}{\cos\beta}\right)^2\frac{1}{r} = \frac{r}{\cos^2\beta}$$

$$m_{n1}=m_{n2} \text{ 或 } m_{t1}=m_{t2}$$

$$\alpha_{n1}=\alpha_{n2} \text{ 或 } \alpha_{t1}=\alpha_{t2}$$

$$\beta_1=-\beta_2 (\text{“}-\text{”代表旋向相反})$$

**5. 斜齿轮连续传动的条件**

斜齿轮连续传动的条件仍由重合度的大小判断。斜齿轮的重合度由两部分组成，一是端面重合度，二是螺旋角引起的重合度增量。从端面看，斜齿轮啮合与直齿轮完全一样，因此用端面参数代入直齿轮重合度计算公式即可求得斜齿轮的端面重合度，即

$$\varepsilon_a=\frac{\overline{B_1B_2}}{p_{bt}}=\frac{L}{p_{bt}}=\frac{1}{2\pi}[z_1(\tan\alpha_{at1}-\tan\alpha_t')+z_2(\tan\alpha_{at2}-\tan\alpha_t')]$$

式中，$\alpha_{at1}$、$\alpha_{at2}$ 分别为齿轮 1、2 的端面齿顶圆压力角；$\alpha_t'$ 为端面啮合角。

图 7-40a 所示是齿轮宽度为 $B$ 的直齿轮基圆柱展开图，$\overline{B_1B_2}$ 为啮合线长度；图 7-40b 所示为斜齿轮基圆柱展开图。若啮合开始点在 $B_2$，当斜齿轮的轮齿上端在 $B_1$ 点脱离接触时，其下端仍在啮合，直到啮合到 $B_1'$ 点才脱离接触，所以其啮合线要增大 $\Delta L$。

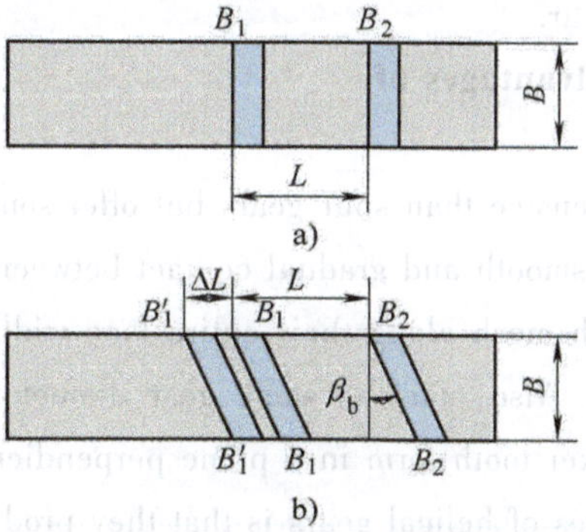


Fig. 7-40　Contact ratio for helical gears（斜齿轮传动重合度）

$$\Delta L=B\tan\beta_b$$

重合度增量为

$$\varepsilon_\beta=\Delta L/p_{bt}=\frac{B\tan\beta_b}{p_{bt}}=\frac{B\tan\beta_b}{\pi m_t\cos\alpha_t}=\frac{B\tan\beta\cos\alpha_t}{\pi m_t\cos\alpha_t}=\frac{B\sin\beta}{\pi m_n}$$

斜齿轮的总重合度为

$$\varepsilon=\varepsilon_a+\varepsilon_\beta$$

**6. 当量齿轮与当量齿数**

如图 7-41 所示，作斜齿轮分度圆柱螺旋线上点 $P$ 的法平面。该法平面与分度圆柱的交线为一椭圆，它的长轴长 $a=r/\cos\beta$，短轴长 $b=r$。椭圆上点 $P$ 的齿形为斜齿轮的法向齿形。以椭圆在点 $P$ 处的曲率半径 $\rho$ 画圆，该圆与点 $P$ 处椭圆弧段非常相近。该圆作为一个虚拟齿轮的分度圆，其模数、压力角均为斜齿轮的法向参数，称该虚拟齿轮为斜齿圆柱齿轮的当量齿轮。其齿数称为斜齿轮的当量齿数，用 $z_v$ 表示。椭圆在点 $P$ 处的曲率半径为

$$\rho=\frac{a^2}{b}=\left(\frac{r}{\cos\beta}\right)^2\frac{1}{r}=\frac{r}{\cos^2\beta}$$

当量齿数为

If a number of the normal teeth of the helical gear are arranged onto the circumference whose radius is $\rho$, we would obtain a equivalent spur gear. The normal tooth profile of the helical gear is the same as that of the equivalent spur gear. The number of teeth of the equivalent spur gear can be determined from the following relationship.

$$z_v = \frac{2\rho}{m_n} = \frac{d}{m_n \cos^2\beta} = \frac{m_t z}{m_n \cos^2\beta} = \frac{z}{\cos^3\beta}$$

Where $z_v$ is the number of teeth of the equivalent spur gear, $\beta$ is the helix angle of the helical gear, $m_n$ is the normal module of the helical gear, $m_t$ is the transverse module of the helical gear, and $d$ is the diameter of the reference circle of the helical gear.

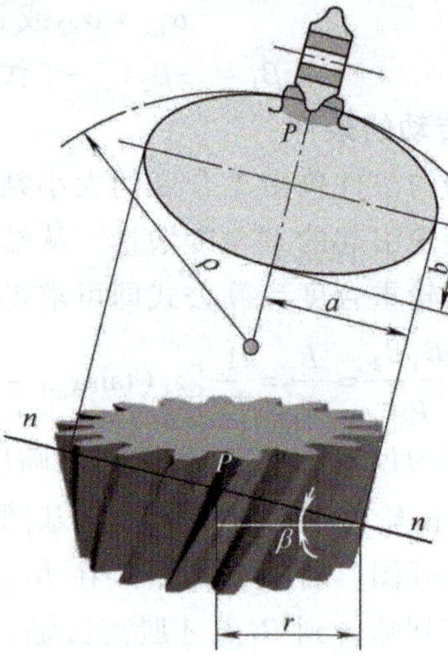

Fig. 7-41 Equivalent spur gear
(当量齿轮)

**7. Advantages and Disadvantages of Helical Gears**

Helical gears are more expensive than spur gears but offer some advantages. They operate quieter than spur gears because of smooth and gradual contact between their angle surfaces as the teeth come into mesh. Spur gear teeth mesh along their entire face width at once; the sudden impact of tooth on tooth causes vibration. Also, for the same gear diameter and module, a helical gear is stronger due to the slightly thicker tooth form in a plane perpendicular to the gear axis.

The one of the disadvantages of helical gears is that they produce an axial thrust force which is harmful to the bearings. Therefore, the helix angle is limited from 8°-15°.

When motion is to be transmitted between shafts which are not parallel, the spur gears can not be used. Two crossed-helical gears of the same hand can be meshed with their axes at an angle, the helix angles can be designed to accommodate any skew angle between the nonintersecting shafts.

## 7.9 Worm and Worm Gears

**1. Characteristics of Worm Gears**

The worm and worm gear are used to transmit motion between two shafts which are nonparallel, nonintersecting, usually at a shaft angle of 90°, as shown in Fig. 7-42.

The worm gears have many advantages, such as larger gear ratios, small package, carrying high loads. Perhaps the major advantage of the wormset is that it can be designed to be impossible to backdrive, that is to say, it can be designed as a self-locking wormset. Therefore, worm gears are most widely used in industry.

The major disadvantage is that the efficiency is lower, usually at 40% to 85%. The friction loss may result in overheating and serious wear.

**2. Types of Worm Gears**

$$z_v = \frac{2\rho}{m_n} = \frac{d}{m_n \cos^2\beta} = \frac{2}{m_n \cos^2\beta} \frac{m_t z}{2} = \frac{z}{\cos^3\beta}$$

式中，$z$ 为斜齿轮的齿数。

通过引入当量齿轮，可把研究斜齿轮法向齿形的问题转化为研究当量直齿轮的问题。一般情况下，当量齿数不是整数，供以后设计时选用。

**7. 斜齿轮传动的优缺点**

与直齿轮传动相比，斜齿轮传动的优点为：传动平稳，冲击、振动和噪声小，重合度大，承载能力强，结构紧凑，因而在大功率和高速齿轮传动中广泛应用。主要缺点为：因存在螺旋角 $\beta$，传动时齿面间会产生轴向推力。为了发挥斜齿轮的优点，又消除传动中轴向推力对轴承的不利影响，可采用齿向左、右完全对称的人字齿轮。所产生的轴向力可完全抵消，但人字齿轮制造比较困难。

斜齿轮的主要优缺点都与螺旋角 $\beta$ 有关，$\beta$ 越大优点越显著，缺点也越突出。通常斜齿轮的螺旋角 $\beta$ 在 8°~15°之间选取，人字齿轮的螺旋角 $\beta$ 可达 25°~40°。

## 7.9　蜗杆传动机构

**1. 蜗杆传动及其特点**

图 7-42 所示的蜗杆传动机构用于传递垂直交错的两轴间的运动和动力。蜗杆通常作为减速传动的主动件，蜗杆传动的传动比大，结构紧凑，工作平稳、噪声小，广泛应用于各类机械和仪器中。但蜗杆传动效率较低，故不适用于大功率长期连续工作。

Fig. 7-42　Worm and worm gear（蜗轮蜗杆）

**2. 蜗杆传动的类型**

蜗杆传动通常根据蜗杆形状和加工方法分类。根据蜗杆的形状可以分为图7-43a所示的圆柱面蜗杆传动、图 7-43b 所示的环面蜗杆传动和图 7-43c 所示的锥面蜗杆传动。环面蜗杆传动的承载能力大而且效率高，但其制造和安装精度要求高，成本也高。本章只讨论阿基米德圆柱蜗杆传动机构，其他各种类型的蜗杆传动可以参考相关资料和设计手册。

If the helix angle is increased sufficiently, the result will be a worm which has only one or more teeth wrapped continuously around its circumference a number of times, analogous to a screw thread. In fact, the tooth on a worm is often spoken of as thread. The meshing gear for a worm is designed as a worm wheel. However, the worm wheel is not a helical gear, because its face is made concave to fit the curvature of the worm in order to provide line contact instead of point contact.

According to the shapes of the worm, the worm gears can be divided into three types. Fig. 7-43a shows a cylinder worm, whose thread are wrapped on a cylinder, Fig. 7-43b shows an enveloping worms, and Fig. 7-43c shows a spiroid worm. Among them, the cylinder worms are used widely.

a) b) c)

Fig. 7-43 Worm Types (蜗杆传动类型)

### 3. Parameters and Sizes of Worm Gears

(1) Parameters of worm gears A meshing worm and a worm wheel with a shaft angle of 90° have the same helix hand, but the helix angles are different. The helix angle of the worm is quite large and very small on the worm wheel. The worm lead angle is equal to the helix angle of the worm wheel.

The common cylindrical worm is Archimedes worm which can be cut on a lathe or on a milling machine. In the main section which is perpendicular to the axis of the worm gear and contains the axis of the cylindrical worm, the conjugate action of worms is the same as a rack and a gear, as shown in Fig. 7-44.

The parameters for an Archimedes worm and worm gear with a 90° shaft angle are selected on the main plane.

1) Module. Transverse module of the worm gear $m_t$ is equal to the axial module of the worm $m_a$, and the module has been standardized (see Tab. 7-4).

2) Pressure angle. The pressure angle of the Archimedes worms is 20°.

3) Number of teeth and ratio. The worm teeth are often spoken of as threads, usually from 1 to 6 teeth. The more the number of teeth is, the more difficult the manufacture is. The gear ratio is as:

$$i = \frac{\omega_1}{\omega_2} = \frac{n_1}{n_2} = \frac{z_2}{z_1}$$

Where $z_1$ and $z_2$ are the number of worm teeth and the number of worm gear teeth respectively, $\omega_1$ and $\omega_2$ are the angular velocities of the worm and worm gear.

4) Quotient of worm diameter. If a complete revolution of thread on a worm, such as two

### 3. 蜗杆传动的主要参数和几何尺寸

（1）蜗杆传动的主要参数　图 7-44 所示为蜗轮与阿基米德圆柱蜗杆的啮合情况。通过蜗杆轴线的平面称为轴平面，垂直于蜗轮轴线的平面称为主截面。蜗杆的轴平面齿形是齿条。在主截面内蜗杆与蜗轮的啮合相当于齿条与渐开线齿轮的啮合，主截面内的参数为标准值，蜗杆传动的设计计算都以主截面的参数和几何关系为准。

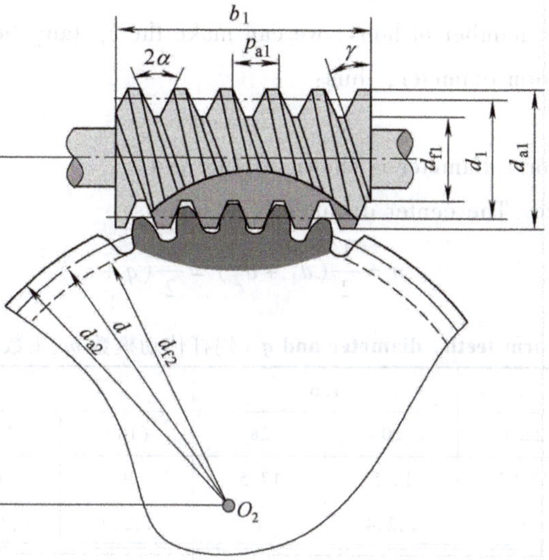


Fig. 7-44　Archimedes worm and worm gear
（阿基米德圆柱蜗杆传动）

1）模数。模数是蜗杆传动的主要参数，蜗杆的轴面模数和蜗轮的端面模数相等，且应该取标准数值。蜗杆模数系列与齿轮模数系列有所不同，国家标准规定的蜗杆模数系列见表 7-4。

2）压力角。阿基米德圆柱蜗杆传动的压力角的标准值为 20°。

3）齿数和传动比。蜗杆的轮齿绕蜗杆轴线形成螺旋形轮齿。因此，蜗杆的齿数通常很少。蜗杆齿数也称蜗杆的头数，推荐取蜗杆头数 $z_1=1$、2、4、6 。蜗杆传动的传动比随头数增加而减小，传动效率则增高。但蜗杆头数太多，会带来加工困难。

为了避免蜗轮轮齿发生根切，蜗轮齿数不应少于 26 。

设蜗杆头数为 $z_1$，蜗轮齿数为 $z_2$，则蜗杆主动时的传动比为

$$i=\frac{\omega_1}{\omega_2}=\frac{n_1}{n_2}=\frac{z_2}{z_1}$$

式中，$\omega_1$、$\omega_2$ 分别为蜗杆和蜗轮的角速度；$n_1$、$n_2$ 分别为蜗杆和蜗轮的转速。

4）蜗杆的直径系数 $q$。由蜗杆分度圆柱面展开图 7-45 可知

$$\tan\gamma=\frac{z_1P_z}{\pi d_1}=\frac{z_1\pi m}{\pi d_1}=\frac{z_1m}{d_1}$$

令 $q=z_1/\tan\gamma$，则 $d_1=qm$。称 $q$ 为蜗杆的直径系数。

蜗杆的模数 $m$ 与分度圆直径 $d_1$ 及直径系数 $q$ 的对应关系见表 7-4。

5）中心距 $a$。蜗杆传动的标准中心距为

$$a=\frac{1}{2}(d_1+d_2)=\frac{m}{2}(q+z_2)$$

threads, is unwrapped, the triangles result as shown in Fig. 7-45, and we can obtain the follows from this figure.

$$\tan\gamma = \frac{z_1 P_z}{\pi d_1} = \frac{z_1 \pi m}{\pi d_1} = \frac{z_1 m}{d_1}$$

However, this diameter of $d_1$ should be the same as the that of the hob used to cut the worm wheel, that is to say, each worm wheel needs a hob which has the same diameter. This is not economical. To reduce the number of hobs, we can make the $z_1/\tan\gamma$ be equal to a number of $q$, and call it as quotient of worm diameter, thus:

$$d_1 = qm$$

The quotient of worm diameter is shown in Tab. 7-4.

5) Center distance. The center distance is as follows:

$$a = \frac{1}{2}(d_1 + d_2) = \frac{m}{2}(q + z_2)$$

**Tab. 7-4 Module, worm teeth, diameter and $q$** (蜗杆传动模数 $m$、头数 $z_1$、分度圆直径 $d_1$ 及 $q$)(mm)

| | | | | | | | | |
|---|---|---|---|---|---|---|---|---|
| $m$ | 1.25 | | 1.6 | | 2 | | | |
| $d_1$ | 20 | 22.4 | 20 | 28 | (18) | 22.4 | (28) | 35.5 |
| $q$ | 16 | 17.92 | 12.5 | 17.5 | 9 | 11.2 | 14 | 17.75 |
| $z_1$ | 1 | 1 | 1,2,4 | 1 | 1,2,4 | 1,2,4,6 | 1,2,4 | 1 |
| $m$ | 2.5 | | | | 3.15 | | | |
| $d_1$ | (22.4) | 28 | (35.5) | 45 | (28) | 35.5 | (45) | 56 |
| $q$ | 8.96 | 11.2 | 14.2 | 18 | 8.889 | 11.27 | 14.286 | 17.778 |
| $z_1$ | 1,2,4 | 1,2,4,6 | 1,2,4 | 1 | 1,2,4 | 1,2,4,6 | 1,2,4 | 1 |
| $m$ | 4 | | | | 5 | | | |
| $d_1$ | (31.5) | 40 | (50) | 71 | (40) | 50 | (63) | 90 |
| $q$ | 7.875 | 10 | 12.5 | 17.75 | 8 | 10 | 12.6 | 18 |
| $z_1$ | 1,2,4 | 1,2,4,6 | 1,2,4 | 1 | 1,2,4 | 1,2,4,6 | 1,2,4 | 1 |
| $m$ | 6.3 | | | | 8 | | | |
| $d_1$ | (50) | 63 | (80) | 112 | (63) | 80 | (100) | 140 |
| $q$ | 7.036 | 10 | 12.698 | 17.778 | 87.875 | 10 | 12.5 | 17.5 |
| $z_1$ | 1,2,4 | 1,2,4,6 | 1,2,4 | 1 | 1,2,4 | 1,2,4,6 | 1,2,4 | 1 |
| $m$ | 10 | | | | 12.5 | | | |
| $d_1$ | (71) | 90 | (112) | 160 | (90) | 112 | (140) | 200 |
| $q$ | 7.1 | 9 | 11.2 | 16 | 7.2 | 8.96 | 11.2 | 16 |
| $z_1$ | 1,2,4 | 1,2,4,6 | 1,2,4 | 1 | 1,2,4 | 1,2,4 | 1,2,4 | 1 |
| $m$ | 16 | | | | 20 | | | |
| $d_1$ | (112) | 140 | (180) | 250 | (140) | 160 | (224) | 315 |
| $q$ | 7 | 8.75 | 11.25 | 15.625 | 7 | 8 | 11.2 | 15.75 |
| $z_1$ | 1,2,4 | 1,2,4 | 1,2,4 | 1 | 1,2,4 | 1,2,4 | 1,2,4 | 1 |

(2) Conditions of correctly meshing for worm gears The conditions of correctly engaging for

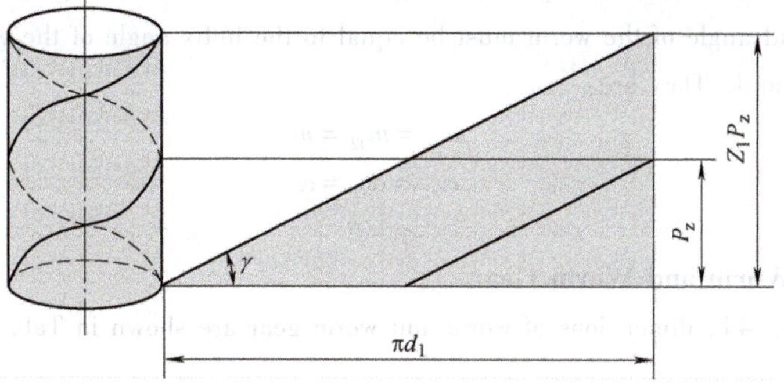


Fig. 7-45　Unwrapped threads on a worm（蜗杆分度圆柱展开图）

（2）正确啮合条件　蜗杆传动的正确啮合条件是：主截面内的模数和压力角分别相等且为标准值。蜗轮的螺旋角还应等于蜗杆的导程角。即

$$m_{a1} = m_{t2} = m$$

$$\alpha_{a1} = \alpha_{t2} = \alpha$$

$$\gamma = \beta \text{（旋向相同）}$$

式中，$m_{a1}$、$m_{t2}$分别为蜗杆的轴面模数和蜗轮的端面模数；$\alpha_{a1}$、$\alpha_{t2}$分别为蜗杆的轴面压力角和蜗轮的端面压力角；$\gamma$、$\beta$分别为蜗杆的导程角和蜗轮的螺旋角。

**4. 几何尺寸计算**

阿基米德圆柱蜗杆传动的各部分几何尺寸计算可参考表 7-5 和图 7-44。

**Tab. 7-5　Formula for Archimedes worm gear**（阿基米德圆柱蜗杆传动几何尺寸）

| Name 名称 | Symbol 符号 | Formula 计算公式 |
|---|---|---|
| Worm teeth 蜗杆头数 | $z_1$ | $z_1 = 1, 2, \cdots, 4$ |
| Worm gear teeth 蜗轮齿数 | $z_2$ | $z_2 = iz_1$ |
| Module 模数 | $m$ | Tab. 7-4 |
| Pressure angle 压力角 | $\alpha$ | $\alpha = 20°$ |
| Worm pitch diameter 蜗杆分度圆直径 | $d_1$ | $d_1 = mq$ |
| Quotient of worm diameter 蜗杆直径系数 | $q$ | $q = z_1/\tan\gamma$ |
| Lead angle of worm on the pitch circle 蜗杆分度圆导程角 | $\gamma$ | $\tan\gamma = z_1 m/d_1$ |
| Pitch diameter 蜗轮分度圆直径 | $d_2$ | $d_2 = mz_2$ |
| Worm diameter of addendum circle 蜗杆齿顶圆直径 | $d_{a1}$ | $d_{a1} = d_1 + 2h_a^* m$ |
| Worm diameter of dedendum circle 蜗杆齿根圆直径 | $d_{f1}$ | $d_{f1} = d_1 - 2(h_a^* + c^*)m$ |
| Worm gear diameter of addendum circle 蜗轮齿顶圆直径 | $d_{a2}$ | $d_{a2} = d_2 + 2h_a^* m$ |
| Worm gear diameter of dedendum circle 蜗轮齿根圆直径 | $d_{f2}$ | $d_{f2} = d_2 - 2(h_a^* + c^*)m$ |
| Center distance 标准中心距 | $a$ | $a = \frac{1}{2}(d_1 + d_2)$ |

worm gears are that the module and pressure angle of the worm and worm gear must be the same respectively; the lead angle of the worm must be equal to the helix angle of the worm gear, and have the same helical hand. They are:

$$m_{a1} = m_{t2} = m$$

$$\alpha_{a1} = \alpha_{t2} = \alpha$$

$$\gamma = \beta$$

**4. Sizes of Worm and Worm Gear**

Refer to Fig. 7-44, dimensions of worm and worm gear are shown in Tab. 7-5.

## 7.10 Bevel Gears

The bevel gears are used to transmit motion between two intersecting shafts, and the shaft angle usually is 90°. Straight bevel gears are easy to design, simple to manufacture and give very good result in service if they are mounted accurately.

Kinematically, bevel gears are equivalent to rolling pitch cones. These cones roll together without slipping. Actually, a bevel gear is a frustum of a cone; the dimensions on different transverse planes are different. For convenience, the parameters and dimensions at the large end of the cone are taken to be standard values.

**1. Pitch Angle and Gear Ratio**

Bevel gears must be mounted so that the apexes of both pitch cones are coincident because the pitch of the teeth depends upon the radial distance from the apex.

Fig. 7-46 shows a pair of straight bevel gears in mesh. The shaft angle is 90°. Some of the nomenclatures are as follows.

1) Pitch cones. The cones have the pitch surfaces of the bevel gears and roll together without slipping.

2) Apex. The intersection of the elements make up the pitch cones.

3) Cone distance. The length of a pitch cone element, denoted as $R$.

4) Pitch angle. The angle between an element of the pitch cone and the axis of the gear, denoted as $\delta_1$ for pinion, $\delta_2$ for another gear.

5) shaft angle. The angle between two axes of bevel gears, denoted as $\Sigma$. It is always equal to the sum of the pitch angles.

$$\Sigma = \delta_1 + \delta_2 = 90°$$

The gear ratio of bevel gears may be written as before.

$$i_{12} = \frac{\omega_1}{\omega_2} = \frac{z_2}{z_1} = \frac{r_2}{r_1} = \frac{R\sin\delta_2}{R\sin\delta_1} = \frac{\sin\delta_2}{\sin\delta_1}$$

$$\Sigma = 90°, \ i_{12} = \tan\delta_2 = \cot\delta_1$$

For the bevel gears, the pitch cones are always coincident with their reference cones, so the radii of $r_1$ and $r_2$ are on the reference circles at the large end of the bevel gears.

**2. Back Cone and Equivalent Gear**

When two pitch cones roll together without slipping, each point in a bevel gear remains a con-

## 7.10 锥齿轮机构

锥齿轮机构用于传递两相交轴之间的运动，一般取轴交角 $\Sigma = 90°$。锥齿轮的轮齿由大端至小端逐渐变小，为了计算和测量方便，取大端参数为标准值，几何尺寸也以大端为基础进行计算。

**1. 传动比与分度圆锥角**

一对锥齿轮的啮合传动相当于一对节圆锥做纯滚动，其分度圆锥与节圆锥重合。如图 7-46 所示，$\delta_1$、$\delta_2$ 分别为两锥齿轮的分度圆锥母线与各自轴线的夹角，称为分度圆锥角；$\Sigma = \delta_1 + \delta_2 = 90°$。$r_1$、$r_2$ 分别为两锥齿轮大端的分度圆半径；$\overline{OC}$ 为锥齿轮的锥距，用 $R$ 表示。

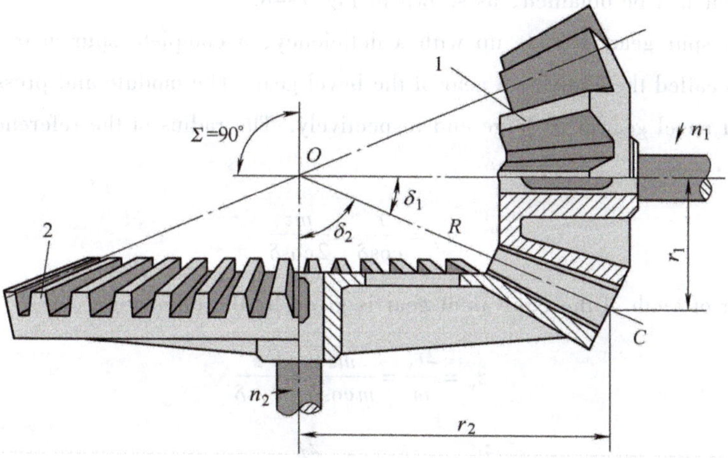


Fig. 7-46 Straight bevel gears（直齿锥齿轮传动）

锥齿轮传动的传动比为

$$i_{12} = \frac{\omega_1}{\omega_2} = \frac{z_2}{z_1} = \frac{r_2}{r_1} = \frac{R\sin\delta_2}{R\sin\delta_1} = \frac{\sin\delta_2}{\sin\delta_1}$$

$$\Sigma = 90°,\ i_{12} = \tan\delta_2 = \cot\delta_1$$

**2. 锥齿轮的背锥、当量齿轮和当量齿数**

锥齿轮齿廓曲面的形成与圆柱齿轮类似。如图 7-47 所示，圆发生面 $S$ 的圆心与锥顶 $O$ 重合，且与基圆锥相切于点 $P$。当该发生面绕基圆锥做纯滚动时，发生面上任意点 $K$ 的轨迹为球面渐开线，无数条半径不同的球面渐开线组成了球面渐开曲面。

锥齿轮的理论廓线为球面渐开线。由于无法将球面展开成平面，通常用近似方法来研究锥齿轮的理论廓线。

图 7-48 所示锥齿轮的轴剖面图中，$OAA'$ 为分度圆锥，过锥齿轮大端的 $A$ 点作 $OA$ 的垂线与锥齿轮的轴线交于点 $O_1$，以 $O_1$ 为锥顶、$O_1A$ 为母线、$OO_1$ 为轴线作一圆锥与锥齿轮大端球面在分度圆相切，该圆锥称为背锥。将锥齿轮大端的球面渐开线齿形投影到背锥上，背锥上的齿形与锥齿轮大端上的齿形十分接近，因此可近似地用背锥上的齿形来代替锥齿轮大端的齿形。

stant distance from the apex. Therefore, the pitch cones of a pair of bevel gears have spherical motion.

Imagine that a generating circular plane $S$ with radius $R$ whose center $O$ is coincident with the apex of the cone rolls on the cone without slipping, the point on the plane, such as the point $K$, will generates a spherical involute from the point $A$, as shown in Fig. 7-47.

The involute profile on a sphere is difficult to manufacture, thus an approximation in a plane is made. It is customary to make a cone which is called as the back cone, and the elements of the back cone are perpendicular to those of the pitch cone. The back cone is tangent to the sphere at the pitch circle, and then the teeth of the bevel gear on the sphere are projected on the back cone. They are almost the same, as long as the bevel gear has eight or more teeth; it is accurate enough for practical purposes. After the back cone is developed, a sector spur gear having a radius equal to the back cone element will be obtained, as shown in Fig. 7-48.

If the sector spur gear is made up with a deficiency, a complete spur gear will be obtained. This spur gear is called the equivalent gear of the bevel gear. The module and pressure angle are the same with that of bevel gear at its large end respectively. The radius of the reference circle of the equivalent gear is:

$$r_v = \frac{r}{\cos\delta} = \frac{mz}{2\cos\delta}$$

The number of teeth of the equivalent gear is given by the following:

$$z_v = \frac{2r_v}{m} = \frac{mz}{m\cos\delta} = \frac{z}{\cos\delta}$$

For bevel gears 1 and 2, $z_{v1} = \dfrac{z_1}{\cos\delta_1}$ and $z_{v2} = \dfrac{z_2}{\cos\delta_2}$.

**3. Parameters, Sizes and Properties**

(1) Parameters  Parameters are standardized at the large end of the bevel gears. The pressure angle is 20°, coefficient of addendum is $h_a^* = 1$, and coefficient of clearance is $c^* = 0.2$. The modules are shown in Tab. 7-6.

**Tab. 7-6  Modules of bevel gears (锥齿轮标准模数系列)** (mm)

| |
|---|
| … 1 1.125 1.25 1.375 1.5 1.75 2 2.5 2.75 3 3.25 3.5 3.75 4 4.5 5 5.5 6 6.5 7 8 9 10 … |

(2) Conditions of correctly meshing for bevel gears  The conditions of correctly engaging for bevel gears are that the module and pressure angle of bevel gear at the large end must be the same respectively. They are

$$m_1 = m_2$$

$$\alpha_1 = \alpha_2 = \alpha$$

(3) Condition of continuous transmission of bevel gears  The contact ratio of the bevel gears can be calculated as their equivalent spur gears. It is as follows:

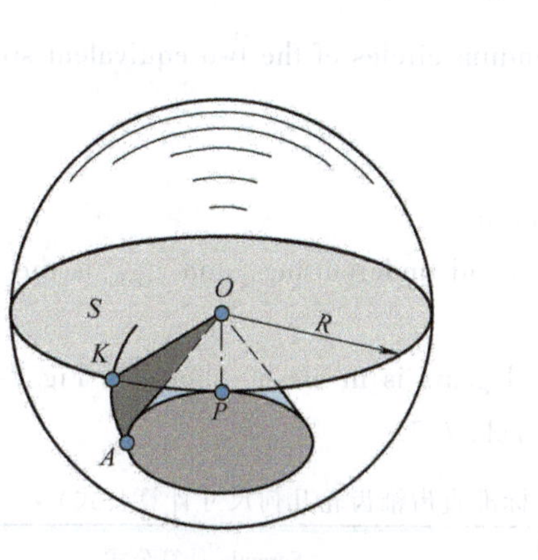


Fig. 7-47　Tooth surface of a bevel gear（锥齿轮的齿面）

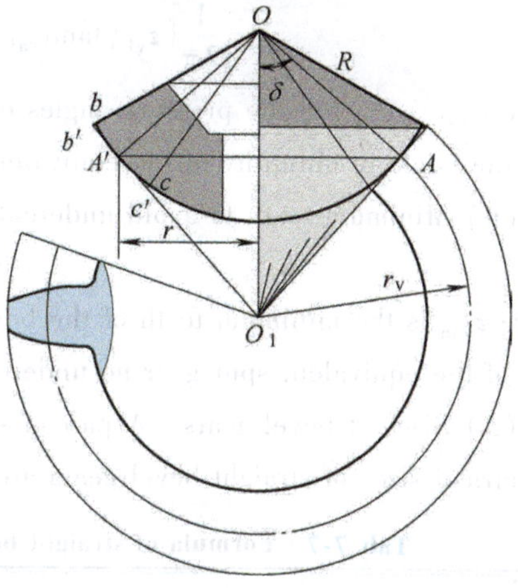


Fig. 7-48　Back cone and equivalent spur gear（背锥与当量齿轮）

背锥可展开成平面，展开后得到一扇形齿轮，其轮齿参数与锥齿轮大端轮齿参数基本相同，齿数为锥齿轮的齿数。将扇形缺口补齐成一圆形齿轮，该圆形齿轮称为锥齿轮的当量齿轮，其齿数 $z_v$ 为当量齿数。当量齿轮的半径为

$$r_v = \frac{r}{\cos\delta} = \frac{mz}{2\cos\delta}$$

故当量齿数 $z_v$ 与实际齿数 $z$ 的关系为

$$z_v = \frac{2r_v}{m} = \frac{mz}{m\cos\delta} = \frac{z}{\cos\delta}$$

**3. 锥齿轮的参数、几何尺寸计算及啮合特点**

（1）锥齿轮的基本参数　锥齿轮的大端轮齿参数为标准值。模数系列参见表 7-6，国家标准中规定锥齿轮大端轮齿的压力角 $\alpha = 20°$，齿顶高系数 $h_a^* = 1$，顶隙系数 $c^* = 0.2$。

（2）锥齿轮正确啮合条件　锥齿轮正确啮合的条件为：大端轮齿的模数与压力角分别相等，锥距分别相等。即

$$m_1 = m_2$$
$$\alpha_1 = \alpha_2 = \alpha$$

（3）连续传动的条件　仍用重合度的大小表示能否连续传动。直齿锥齿轮啮合传动的重合度可按其当量圆柱齿轮的重合度计算，即

$$\varepsilon = \frac{1}{2\pi}[z_{v1}(\tan\alpha_{va1} - \tan\alpha_v') + z_{v2}(\tan\alpha_{va2} - \tan\alpha_v')]$$

式中，$\alpha_{va1}$、$\alpha_{va2}$ 分别为当量圆柱齿轮 $z_{v1}$、$z_{v2}$ 的齿顶圆压力角。

（4）不发生根切的最小齿数　为了避免根切，锥齿轮不发生根切的最少齿数为

$$z_{min} = z_{vmin}\cos\delta$$

式中，$z_{min}$ 为锥齿轮不出现根切的最小齿数，当 $h_a^* = 1$，$\alpha = 20°$ 时，$z_{vmin} = 17$。故锥齿轮不出现根切的最小齿数为 17。

（5）几何尺寸计算　图 7-49 所示为一对直齿锥齿轮啮合简图，几何尺寸计算公式见表 7-7。

$$\varepsilon = \frac{1}{2\pi}[z_{v1}(\tan\alpha_{va1} - \tan\alpha_v') + z_{v2}(\tan\alpha_{va2} - \tan\alpha_v')]$$

Where $\alpha_{va1}$, $\alpha_{va2}$ are the pressure angles on the addendum circles of the two equivalent spur gears; $\alpha_v'$ is the angle of obliquity of the equivalent gear.

(4) Minimum teeth to avoid undercutting

$$z_{min} = z_{vmin}\cos\delta$$

Where $z_{min}$ is the minimum teeth of the bevel gear to avoid undercutting, and $z_{vmin}$ is the minimum teeth of the equivalent spur gear no undercutting.

(5) Sizes of bevel gears  A pair of straight bevel gears is in mesh, shown in Fig. 7-49. The geometrical sizes of straight bevel gears are shown in Tab. 7-7.

**Tab. 7-7 Formula of straight bevel gears (标准直齿锥齿轮几何尺寸计算公式)**

| Name 名称 | Symbol 符号 | Formula 计算公式 |
|---|---|---|
| Module 模数 | $m$ | Tab. 7-6 |
| Ratio 传动比 | $i$ | $i = \frac{z_2}{z_1} = \tan\delta_2 = \cot\delta_1$ |
| Addendum 齿顶高 | $h_a$ | $h_a = h_a^* m$ |
| Dedendum 齿根高 | $h_f$ | $h_f = (h_a^* + c^*)m$ |
| Whole depth 全齿高 | $h$ | $h = h_a + h_f = (2h_a^* + c^*)m$ |
| Diameter of pitch circle 分度圆直径 | $d_1$、$d_2$ | $d_1 = mz_1$, $d_2 = mz_2$ |
| Diameter of addendum circle 齿顶圆直径 | $d_{a1}$、$d_{a2}$ | $d_{a1} = d_1 + 2h_a\cos\delta_1$, $d_{a2} = d_2 + 2h_a\cos\delta_2$ |
| Diameter of dedendum circle 齿根圆直径 | $d_{f1}$ $d_{f2}$ | $d_{f1} = d_1 - 2h_f\cos\delta_1$, $d_{f2} = d_2 - 2h_f\cos\delta_2$ |
| Cone distance 锥距 | $R$ | $R = \sqrt{r_1^2 + r_2^2} = \frac{m}{2}\sqrt{z_1^2 + z_2^2} = \frac{d_1}{2\sin\delta_1} = \frac{d_2}{2\sin\delta_2}$ |
| Face width 齿宽 | $b$ | $b \leqslant \frac{R}{3}$ |
| Dedendum angle 齿根角 | $\theta_f$ | $\theta_f = \arctan\frac{h_f}{R}$ |
| Addendum angle 齿顶角 | $\theta_a$ | $\theta_a = \arctan\frac{h_a}{R}$ |
| Root angle 根锥角 | $\delta_{f1}$、$\delta_{f2}$ | $\delta_{f1} = \delta_1 - \theta_f$, $\delta_{f2} = \delta_2 - \theta_f$ |
| Face angle 顶锥角 | $\delta_{a1}$、$\delta_{a2}$ | $\delta_{a1} = \delta_1 + \theta_a$, $\delta_{a2} = \delta_2 + \theta_a$ |

Some of the terms used in the bevel gears are as follows.

Addendum angle: It is an angle between the elements of the pitch cone and face cone, denoted as $\theta_a$.

Dedendum angle: It is an angle between the elements of the pitch cone and root cone, denoted as $\theta_f$.

Face angle: It is an angle between the element of the face cone and the axis of the gear, denoted as $\delta_a$.

Root angle: It is an angle between the element of the root cone and the axis of the gear, denoted as $\delta_f$.

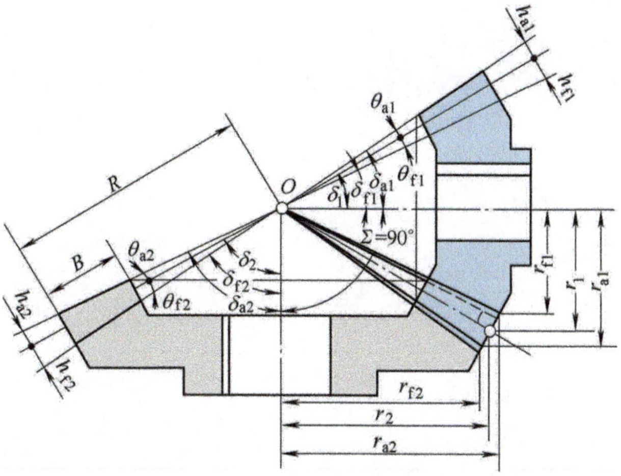


Fig. 7-49 Parameters and dimensions of straight bevel gears（锥齿轮的参数和几何尺寸）

# Chapter 8

# Design of Gear Trains

# 第8章

# 轮系及其设计

## 8.1 Classification of Gear Trains

A gear train is a combination of gears used to transmit motion from one shaft to another. Even a single pair of gears is, strictly speaking, a gear train, though the term usually suggests that there are three or more moving gears. The gear train becomes necessary when it is required to obtain large speed ratio with a small space.

Fig. 8-1a shows a gear-box in an automobile; there are many pairs of gears to transmit different motion. Fig. 8-1b shows a quartz watch in which the hour hand, minute hand and second hand rotate as their definite gear ratios to indicate the times.

Fig. 8-1 Application of gear trains (轮系的应用)
1、2—clutch (离合器) 3、4—driving shaft (输入轴) 5—gear train (轮系)

A gear train is any collection of two or more meshing gears, such as spur gears, bevel gears, worm gears and their combination of different kinds of gears.

If the gear axes are all parallel to each other, it is a planar gear train, otherwise, it is a spatial gear train.

If all the gear axes remain fixed relative to the frame, the train is an ordinary gear train. If one of the gear axes rotates relative to the frame, this gear train is an epicyclic gear train.

**1. Ordinary Gear Trains**

There are three types of ordinary gear trains.

1) Simple ordinary gear train. It is a gear train in which each shaft carries only one gear, as shown in Fig. 8-2a.

2) Compound gear train. It is a gear train in which at least one shaft carries more than one gear, as shown in Fig. 8-2b. This will be a parallel or series-parallel arrangement, rather than the pure series connections of the simple gear trains.

3) Reverted gear train. If the axes of input gear and output gear in a compound gear train are coincident, this is called a reverted gear train, as shown in Fig. 8-2c. In some cases, such as automobile transmissions, it is desirable to have the output shaft concentric with the input shaft. But the design of a reverted gear train is more complicated because of the additional constraint that center

## 8.1　轮系及其分类

一系列互相啮合的齿轮所构成的系统称为轮系。图 8-1a 所示为汽车中多级齿轮组成的变速齿轮机构；图 8-1b 所示为钟表的齿轮机构，各组齿轮按传动比设计，使时针、分针和秒针的运动具有一定的比例关系。

一个轮系中可以同时包含圆柱齿轮、锥齿轮和蜗杆传动等各种类型的齿轮机构。若轮系中齿轮的轴线全部平行，则该轮系称为平面轮系；包含空间齿轮机构的轮系称为空间轮系。根据轮系运转时各齿轮几何轴线在空间的相对位置关系是否变动，将轮系分为定轴轮系和周转轮系两种基本类型。

**1. 定轴轮系**

轮系在运转过程中，每个齿轮的几何轴线位置相对于机架的位置均固定不动，则称该轮系为定轴轮系，图 8-1 所示轮系均为定轴轮系。

根据组成情况，定轴轮系可分为：

1）单式轮系。每根轴上只安装一个齿轮所构成的简单轮系，称为单式轮系。图 8-2a 所示为单式轮系。

2）复式轮系。有的轴上安装有两个以上齿轮的轮系，称为复式轮系。图 8-2b 所示的二级齿轮减速器即为复式轮系。复式轮系应用最为广泛。

3）回归轮系。输出齿轮和输入齿轮共轴线的轮系，称为回归轮系。图 8-2c 所示为典型的回归轮系。

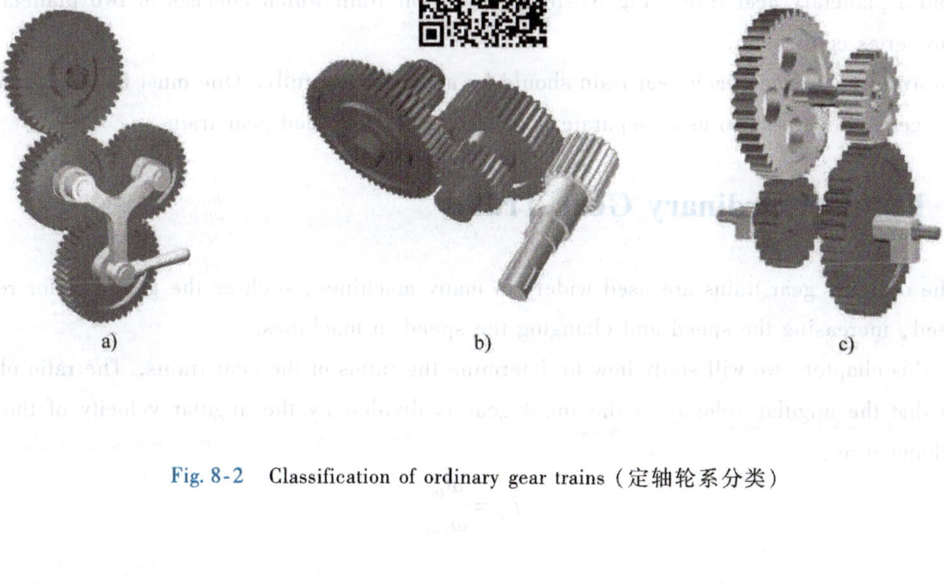

a)　　b)　　c)

Fig. 8-2　Classification of ordinary gear trains（定轴轮系分类）

distance of the stages must be equal.

**2. Epicyclic Gear Trains**

A gear train having a relative motion of axes is called an epicyclic gear train or planetary gear train, as shown in Fig. 8-3a. If the axes of the gears in an epicyclic gear train are fixed, these gears are called sun gears or central gears, such as gears 1 and 3 in Fig. 8-3a, denoted as K. The gear whose axis rotates about the axes of the sun gears is the planet gear, such as the gear 2. The link which is pivoted to the frame at the point $O$ and carries the planet gear 2 to maintain gears 1, 2 and 3 in mesh is the arm or planet carrier, denoted as H.

If a sun gear is fixed, either gear 1 or gear 3, this is a planet gear train which has one degree of freedom, as shown in Fig. 8-3c. If no sun gear is fixed, this is a differential gear train which has two degrees of freedom, as shown in Fig. 8-3b.

To produce a better force balance, more planet gears are added, as shown in Fig. 8-3a, but the additional planet gears contribute nothing to the kinematic performance.

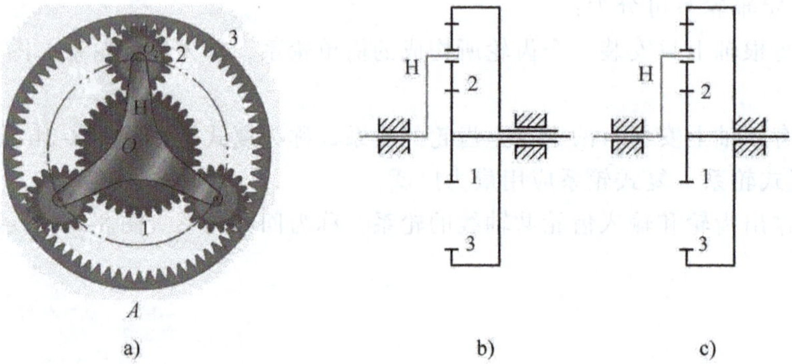


Fig. 8-3 Epicyclic gear train (周转轮系)

**3. Combined Gear Trains**

Quite often a gear train may contain the combination of ordinary gear trains and epicyclic gear trains or some planetary gear trains. Fig. 8-4a shows a gear train which consists of an ordinary gear train and a planetary gear train. Fig. 8-4b shows a gear train which consists of two planetary gear trains in series connection.

To avoid confusion, each gear train should be analysed carefully. One must be able to visualize how the gear train works so as to separate it out from the combined gear train.

## 8.2 Ratio of Ordinary Gear Trains

The ordinary gear trains are used widely in many machines, such as the gear-box for reducing the speed, increasing the speed and changing the speed in machines.

In this chapter, we will study how to determine the ratios of the gear trains. The ratio of a gear train is that the angular velocity of the input gear is divided by the angular velocity of the output gear, denoted as:

$$i_{io} = \frac{\omega_{in}}{\omega_{out}}$$

**2. 周转轮系**

轮系运转时，如果某齿轮的轴线位置相对于机架的位置是转动的，则该轮系称为周转轮系。在图 8-3a 所示的轮系中，齿轮 2 一方面绕其自身轴线 $O_2$ 自转，另一方面又随着构件 H 一起绕固定轴线 $O$ 公转，这种既自转又公转的齿轮称为行星轮；支承并带动行星轮 2 做公转的构件 H 称为系杆或转臂；齿轮 1 与 3 的轴线相对机架的位置固定不动，称为太阳轮。轴线相对不动的太阳轮和系杆称为周转轮系的基本构件。

根据周转轮系自由度的不同，周转轮系可进一步分为图 8-3b 所示的差动轮系和图 8-3c 所示的行星轮系。行星轮系的自由度为 1，差动轮系的自由度为 2。在周转轮系中，太阳轮常用 K 表示，系杆用 H 表示。

**3. 混合轮系**

工程中的轮系有时既包含定轴轮系，又包含周转轮系，或直接由几个周转轮系组合而成。由定轴轮系和周转轮系或由两个以上的周转轮系构成的复杂轮系称为混合轮系或复合轮系，图 8-4a 所示为由定轴轮系和行星轮系组成的混合轮系，图 8-4b 所示为由周转轮系组成的混合轮系。

a)

b)

Fig. 8-4　Combined gear trains（混合轮系）

## 8.2　定轴轮系传动比的计算

定轴轮系在机械中有广泛的应用，包括减速、增速、变速，实现运动和动力的传递与变换。

轮系中首、末两轮的转动速度之比，称为轮系的传动比，表示为

$$i_{io} = \frac{\omega_{in}}{\omega_{out}}$$

式中，$\omega_{in}$ 为轮系中首轮角速度；$\omega_{out}$ 为末轮角速度。由于角速度有方向性，因此轮系的传动比包括大小和方向两个参数。

（1）传动比大小的计算　以图 8-5 所示的平面定轴轮系为例，讨论其传动比的计算方法。已知各轮齿数，主动齿轮 1 为首轮，从动齿轮 5 为末轮，则该轮系的总传动比为

Where $\omega_{in}$ is the angular velocity of the input gear, and $\omega_{out}$ is the angular velocity of the output gear.

Note that the ratio of a gear train has two factors. They are magnitude and rotating direction of the output gear.

In an ordinary gear train we can observe the following:

A pair of external gears always rotates in opposite direction; an external pinion and an internal gear in mesh have the same direction of rotation. For a pair of bevel gears, the directions of rotation can be indicated by arrows on a sketch. For worm gears, we can use the method of velocity analysis to determine the rotating direction of the worm gear.

(1) Parallel-axis ordinary gear trains The following example will provide us with a general rule to determine the ratio of ordinary gear trains.

Fig. 8-5 shows a parallel-axis ordinary gear train in which the numbers of teeth of gears are $z_1$, $z_2$, $z_3$, $z_{3'}$, $z_4$, $z_{4'}$, $z_5$ respectively. Determine the ratio $i_{15}$.

$$i_{15}=\frac{\omega_1}{\omega_5}$$

First, we can determine the ratios of angular velocities from the first pair of gears to the last pair of gears. They are as follows:

$$i_{12}=\frac{\omega_1}{\omega_2}=-\frac{z_2}{z_1}$$

$$i_{23}=\frac{\omega_2}{\omega_3}=-\frac{z_3}{z_2}$$

$$i_{3'4}=\frac{\omega_{3'}}{\omega_4}=\frac{z_4}{z_{3'}}$$

$$i_{4'5}=\frac{\omega_{4'}}{\omega_5}=-\frac{z_5}{z_{4'}}$$

Making the product of the left of these equations be equal to the product of the right, we have:

$$i_{12}i_{23}i_{3'4}i_{4'5}=\frac{\omega_1}{\omega_2}\frac{\omega_2}{\omega_3}\frac{\omega_{3'}}{\omega_4}\frac{\omega_{4'}}{\omega_5}=\frac{\omega_1}{\omega_5}$$

$\omega_3=\omega_{3'}$, $\omega_4=\omega_{4'}$

Rearranging the above equations, we have:

$$i_{15}=\frac{\omega_1}{\omega_5}=i_{12}i_{23}i_{3'4}i_{4'5}=-\frac{z_2z_3z_4z_5}{z_1z_2z_{3'}z_{4'}}=-\frac{z_3z_4z_5}{z_1z_{3'}z_{4'}}$$

The general expression can be written as:

$$i_{1k}=\frac{\omega_1}{\omega_k}=\frac{n_1}{n_k}=(-1)^m\frac{z_2\cdots z_k}{z_1\cdots z_{k-1}}=(-1)^m\frac{\text{product of driven tooth number}}{\text{product of driving tooth number}} \tag{8-1}$$

The negative sign " − " expresses that the output gear and the input gear have the different rotation.

Equation (8-1) can be written as:

$$i_{1k}=\frac{\omega_1}{\omega_k}=(-1)^m\frac{z_2\cdots z_k}{z_1\cdots z_{k-1}} \tag{8-2}$$

This formula can be employed to any parallel ordinary gear trains.

(2) Spatial ordinary gear trains The formula of calculating the ratio is the same with the parallel gear trains, but the $(-1)^m$ can not be used to determine the rotating direction of the output

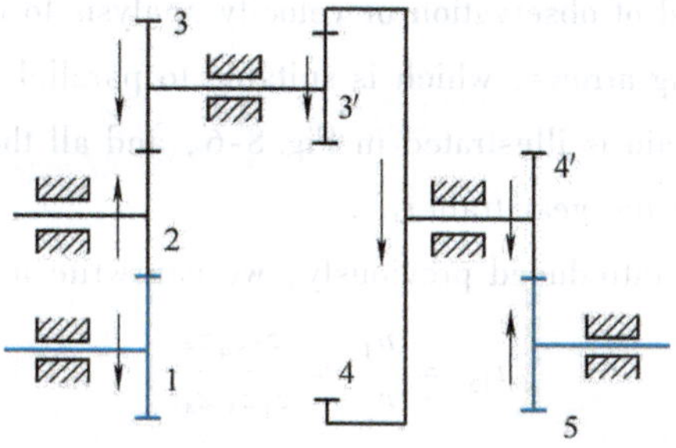


Fig. 8-5 Planar ordinary gear train（平面定轴轮系）

$$i_{15}=\frac{\omega_1}{\omega_5}$$

从首轮到末轮之间的传动，是通过一系列相互啮合的齿轮组合实现的，各对互相啮合齿轮传动比的大小如下

$$i_{12}=\frac{\omega_1}{\omega_2}=\frac{z_2}{z_1}$$

$$i_{23}=\frac{\omega_2}{\omega_3}=\frac{z_3}{z_2}$$

$$i_{3'4}=\frac{\omega_{3'}}{\omega_4}=\frac{z_4}{z_{3'}}$$

$$i_{4'5}=\frac{\omega_{4'}}{\omega_5}=\frac{z_5}{z_{4'}}$$

由于齿轮 3 与 3′同轴，4 与 4′同轴，所以 $\omega_3=\omega_{3'}$、$\omega_4=\omega_{4'}$。将上述各式两边分别连乘，得

$$i_{12}i_{23}i_{3'4}i_{4'5}=\frac{\omega_1}{\omega_2}\frac{\omega_2}{\omega_3}\frac{\omega_{3'}}{\omega_4}\frac{\omega_{4'}}{\omega_5}=\frac{\omega_1}{\omega_5}$$

即

$$i_{15}=\frac{\omega_1}{\omega_5}=i_{12}i_{23}i_{3'4}i_{4'5}=\frac{z_2z_3z_4z_5}{z_1z_2z_{3'}z_{4'}}=\frac{z_3z_4z_5}{z_1z_{3'}z_{4'}}$$

该轮系中，齿轮 2 既为前一对齿轮机构中的从动轮，同时又为后一对齿轮机构中的主动轮。齿轮 2 的作用仅为改变齿轮 3 的转向，并不影响传动比的大小，我们称该齿轮为介轮。

由此可知：定轴轮系的传动比等于组成该轮系的各对啮合齿轮传动比的连乘积，其大小等于轮系中所有从动轮齿数的连乘积与所有主动轮齿数的连乘积之比，其通式为

$$i_{1k}=\frac{\omega_1}{\omega_k}=\frac{n_1}{n_k}=\frac{z_2\cdots z_k}{z_1\cdots z_{k-1}}=\frac{\text{所有从动轮齿数的连乘积}}{\text{所有主动轮齿数的连乘积}} \tag{8-1}$$

式（8-1）为计算定轴轮系传动比的公式。

（2）首、末轮转向关系的确定　平面轮系与空间轮系中，传动比的大小计算方法相同，但首、末轮的转向判别不同。

1）平面轮系。当定轴轮系各轮几何轴线互相平行时，首、末两轮的转向不是相同就是

gear. We must use the method of observation or velocity analysis to determine the rotating direction of the output gear by indicating arrows, which is suitable to parallel gear trains.

A spatial ordinary gear train is illustrated in Fig. 8-6, and all the numbers of teeth of gears are known. Determine the ratio of the gear train $i_{15}$.

According to the formula introduced previously, we can write it directly.

$$i_{15} = \frac{n_1}{n_5} = \frac{z_3 z_4 z_5}{z_1 z_{3'} z_{4'}}$$

(3) Summary

1) The angular velocity ratio of an ordinary gear train is equal to the product of driven tooth number divided by the product of driving tooth number.

2) The angular velocity ratio of an ordinary gear train is also equal to the product of the angular velocity ratio of each pair of meshing gears.

3) The idle gear, whose number of teeth is canceled in equation calculating the ratio, hence is used to change the direction of rotation. In Fig. 8-5 and Fig. 8-6, the gear 2 is an idle gear. For the pair of gears 1 and 2, the gear 2 is a driven gear, and for gears 2 and 3, the gear 2 is a driving gear. It does not affect the speed value, but affects the rotating direction of the output gear.

Idle gears can be used for two purposes: the first is to connect gears where a large center distance is required; the second is to control the direction of the rotation between the input gear and output gear.

4) If all the shafts are parallel in the gear train, the sign of the train value depends on the number of the external gears. However, placing arrows on gear train can be used in any ordinary gear train to determine the rotating direction of the output gear.

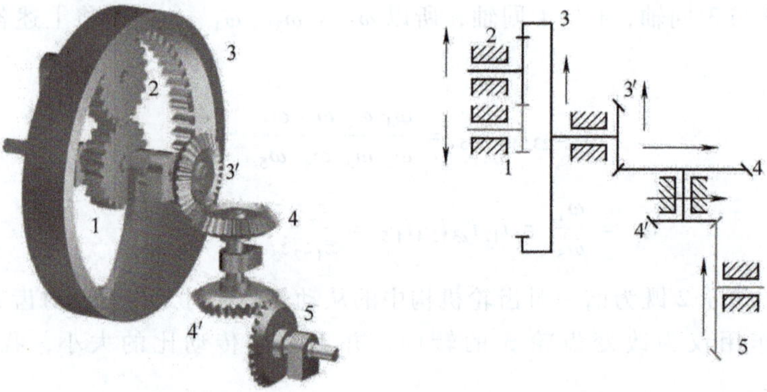


Fig. 8-6 Spatial ordinary gear train (空间定轴轮系)

**Example 8-1** Fig. 8-7a illustrates a spatial gear train, in which all the numbers of teeth of gears are known. The worm is the driving gear and has a right hand. Determine the angular velocity ratio $i_{15}$.

**Solution**

Referring equation (8-2), we have:

$$i_{15} = \frac{n_1}{n_5} = \frac{z_2 z_3 z_5}{z_1 z_{2'} z_{3'}}$$

The rotating direction of the worm can be determined in Fig. 8-7b by using the method of velocity

相反，因此在传动比数值前加上“+”“-”号来表示两轮转向关系。由于一对内啮合圆柱齿轮的转向相同，而一对外啮合圆柱齿轮的转向相反，因此每经过一次外啮合就改变一次方向，若用 $m$ 表示轮系中外啮合齿轮的对数，则可用 $(-1)^m$ 来确定轮系传动比的“+”“-”号。即

$$i_{1k}=\frac{\omega_1}{\omega_k}=(-1)^m\frac{z_2\cdots z_k}{z_1\cdots z_{k-1}} \tag{8-2}$$

式（8-2）为计算平面定轴轮系传动比的公式。

若计算结果为“+”，表明首、末两轮的转向相同；反之，则转向相反。

2）空间轮系。空间轮系中，不能用外啮合的次数 $(-1)^m$ 判别首、末轮的转向，只能用标注箭头法确定。图 8-6 所示的空间轮系中，在图上按传动顺序用箭头逐一标出各轮转向，最后判断末轮转向。

利用画箭头的方法判别首、末轮的转向，不仅适合空间轮系，也适合平面轮系。

图 8-6 所示的空间定轴轮系中，齿轮 1 为主动轮，该轮系传动比的大小为

$$i_{15}=\frac{n_1}{n_5}=\frac{z_3z_4z_5}{z_1z_{3'}z_{4'}}$$

通过画箭头可知：首、末轮转动方向相反。

（3）结论

1）定轴轮系的传动比等于组成该轮系的所有从动轮齿数连乘积除以所有主动轮齿数的连乘积。

2）定轴轮系的传动比还等于组成该轮系的各对齿轮传动比的连乘积。

3）轮系中的介轮不影响传动比的大小，但影响末轮转向。

4）平面轮系可按外啮合的次数 $(-1)^m$ 判别末轮转向，也可用画箭头的方法判别末轮转向。空间轮系只能用画箭头的方法判别末轮转向。

**例 8-1** 图 8-7 所示的空间定轴轮系中，已知各轮齿数，蜗杆 1 为主动轮，右旋，求传动比 $i_{15}$。

**解** 蜗杆传动中，蜗轮转向可以用运动分析方法判别。

该轮系传动比的大小为 $$i_{15}=\frac{n_1}{n_5}=\frac{z_2z_3z_5}{z_1z_{2'}z_{3'}}$$

从主动轮 1 起，依次在图 8-7 中用箭头标出各轮转向，由此可知齿轮 5 逆时针方向转动。

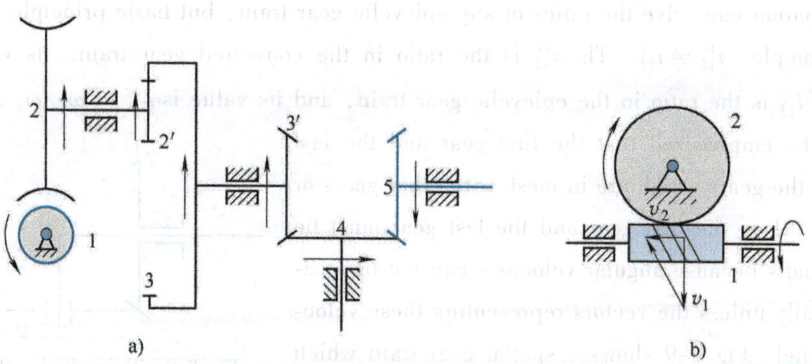


Fig. 8-7 Ratio of the spatial ordinary gear train（空间定轴轮系的传动比）

analysis. The direction of the rotation of the gear 5 can be determined by using the indicating arrows.

## 8.3 Ratio of Epicyclic Gear Trains

The planet gear does not rotate about a fixed gear center. The rule developed for ordinary gear trains can not be applied to the epicyclic gear trains, but we will use the rule to help solve epicyclic train problem.

**1. Converted Gear Train**

Consider the epicyclic gear train shown in Fig. 8-8a, in which the absolute velocities of gears 1, 2, 3 and the arm H are $\omega_1$, $\omega_2$, $\omega_3$ and $\omega_H$ respectively. Now, suppose that the train is inverted at an angular velocity which is equal to that of the arm but the direction is opposite, then the arm is considered to be in stationary, see Fig. 8-8b. We call this inverted train a converted gear train, which is an imaginary ordinary gear train. Therefore, the rules of ordinary gear trains can be used to solve the ratios of epicyclic gear trains.

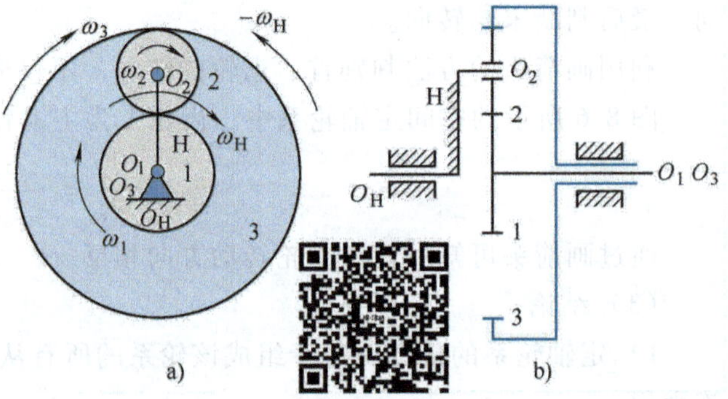


Fig. 8-8 Ratio of the epicyclic gear train (周转轮系传动比)

The angular velocities of gears 1, 2, 3 and the arm H in the converted train are $\omega_1^H$, $\omega_3^H$, $\omega_2^H$ and $\omega_H^H$ respectively. The absolute angular velocities of all the members and the angular velocities relative to the arm are shown in Tab. 8-1. The superscript H means that the arm is considered to be in stationary.

In the converted gear train, if the gears 1 and 3 are input and output gears, the ratio $i_{13}^H$ is as follows:

$$i_{13}^H = \frac{\omega_1^H}{\omega_3^H} = \frac{\omega_1 - \omega_H}{\omega_3 - \omega_H} = -\frac{z_2 z_3}{z_1 z_2} = -\frac{z_3}{z_1}$$

The general equation may be written as:

$$i_{1k}^H = \frac{\omega_1^H}{\omega_k^H} = \frac{\omega_1 - \omega_H}{\omega_k - \omega_H} = \pm \frac{\text{product of driven tooth number}}{\text{product of driving tooth number}} \tag{8-3}$$

This equation can solve the ratios of any epicyclic gear train, but basic principle must be made clear, for example, $i_{1k}^H \neq i_{1k}$. The $i_{1k}^H$ is the ratio in the converted gear train. Its value is $i_{1k}^H = \omega_1^H / \omega_k^H$, and $i_{1k}$ is the ratio in the epicyclic gear train, and its value is $i_{1k} = \omega_1 / \omega_k$.

It must be emphasized that the first gear and the last gear must be the gears which are in mesh with other gears or planet gears. Also, the first gear and the last gear must be on parallel shafts because angular velocities can not be treated algebraically unless the vectors representing these velocities are parallel. Fig. 8-9 shows a spatial gear train which consists of bevel gears. We can write the following formula,

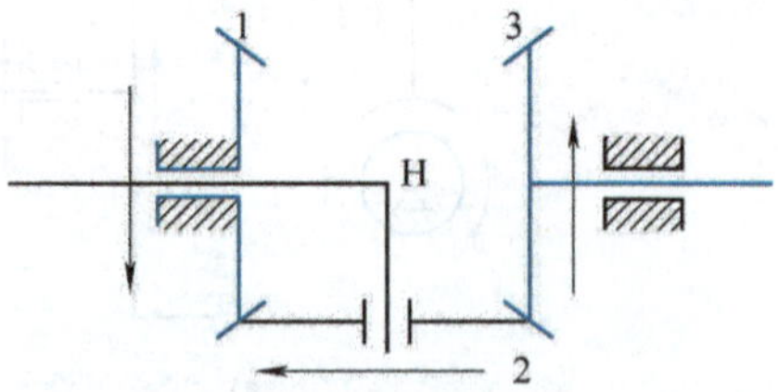


Fig. 8-9 Epicyclic bevel gear train (锥齿轮周转轮系)

## 8.3　周转轮系传动比的计算

### 1. 周转轮系的转化轮系

在周转轮系中，由于支承行星齿轮的系杆绕太阳轮轴线转动，所以不能用计算定轴轮系传动比的方法来计算周转轮系的传动比。若将周转轮系中支承行星轮的系杆 H 固定，周转轮系便转化为定轴轮系，传动比的计算问题也就迎刃而解。如果给周转轮系施加一个反向转动，反向转动角速度等于系杆的角速度，则系杆静止不动，原周转轮系转化为一个假想的定轴轮系，称为周转轮系的转化轮系。转化轮系是定轴轮系，可按定轴轮系列出传动比公式。

图 8-8a 所示轮系中，设 $\omega_1$、$\omega_2$、$\omega_3$、$\omega_H$ 分别为太阳轮 1、行星轮 2、太阳轮 3 和系杆 H 的绝对角速度。给整个周转轮系施加 $-\omega_H$ 的反转角速度后，系杆 H 相对固定不动，原周转轮系转化为假想的定轴轮系，如图 8-8b 所示。这时转化轮系中各构件的角速度分别变为 $\omega_1^H$、$\omega_3^H$、$\omega_2^H$、$\omega_H^H$，它们与原周转轮系中各齿轮的角速度的关系见表 8-1。

**Tab. 8-1　Angular velocity of the epicyclic gear train and the converted gear train**

（周转轮系与转化轮系中各构件的角速度）

| Symbol of members<br>构件代号 | Angular velocity of the epicyclic gear train<br>原周转轮系各构件的角速度 | Angular velocity of the converted gear train<br>转化轮系中各构件的角速度 |
|---|---|---|
| 1 | $\omega_1$ | $\omega_1^H = \omega_1 - \omega_H$ |
| 2 | $\omega_2$ | $\omega_2^H = \omega_2 - \omega_H$ |
| 3 | $\omega_3$ | $\omega_3^H = \omega_3 - \omega_H$ |
| H | $\omega_H$ | $\omega_H^H = \omega_H - \omega_H = 0$ |

根据定轴轮系传动比的公式，可写出转化轮系传动比 $i_{13}^H$ 为

$$i_{13}^H = \frac{\omega_1^H}{\omega_3^H} = \frac{\omega_1 - \omega_H}{\omega_3 - \omega_H} = -\frac{z_2 z_3}{z_1 z_2} = -\frac{z_3}{z_1}$$

式中，“-”号表示在转化机构中 $\omega_1^H$ 和 $\omega_3^H$ 转向相反。

在转化轮系的传动比表达式中，包含着原周转轮系中的各轮角速度，所以可利用转化轮系求解周转轮系的传动比。

对于周转轮系中任意两轴线平行的齿轮 1 和齿轮 $k$，它们在转化轮系中的传动比为

$$i_{1k}^H = \frac{\omega_1^H}{\omega_k^H} = \frac{\omega_1 - \omega_H}{\omega_k - \omega_H} = \pm \frac{\text{从动轮齿数连乘积}}{\text{主动轮齿数连乘积}} \tag{8-3}$$

式（8-3）为计算周转轮系传动比的公式。

计算周转轮系传动比时应注意以下问题：

1）转化轮系的传动比表达式中，含有原周转轮系的各轮绝对角速度，可从中找出待求值。

2）齿数比前的“+”“-”号按转化轮系的判别方法确定。

3）$i_{1k}^H \neq i_{1k}$，因 $i_{1k}^H = \omega_1^H / \omega_k^H$，其大小和转向按定轴轮系传动比的方法确定；而 $i_{1k} = \omega_1 / \omega_k$，其大小和转向由计算结果确定。

4）上述公式仅适用于主、从动轴平行的情况。对于图 8-9 所示的空间周转轮系，其转

because the axes of gears 1 and 3 are parallel to each other.

$$i_{13}^{H} = \frac{\omega_1 - \omega_H}{\omega_3 - \omega_H} = -\frac{z_3}{z_1}$$

The axes of the gear 1 and gear 2 are not parallel, they are inclined at an angle; their angular velocity can not be added algebraically.

Therefore, $i_{12}^{H} \neq \frac{\omega_1 - \omega_H}{\omega_2 - \omega_H}$.

**2. Some Examples**

**Example 8-2** Fig. 8-10 shows a planetary gear train, in which the numbers of teeth of gears are $z_1 = 100$, $z_2 = 101$, $z_{2'} = 100$, $z_3 = 99$. Determine the gear ratio $i_{H1}$.

**Solution**

The train has been inverted at an angular velocity $-\omega_H$. The arm is considered to be in stationary, and it is an imaginary ordinary gear train, so we can write:

$$i_{13}^{H} = \frac{\omega_1^H}{\omega_3^H} = \frac{\omega_1 - \omega_H}{\omega_3 - \omega_H} = +\frac{z_2 z_3}{z_1 z_{2'}} = +\frac{101 \times 99}{100 \times 100}$$

The absolute angular velocities of gears, such as $\omega_1$, $\omega_3$ and $\omega_H$, in the planetary gear train are included in the converted ordinary gear train. The absolute angular velocity of gear 3 is zero, because it is a fixed gear.

Substituting $\omega_3 = 0$ into the above equation, we have:

$$i_{13}^{H} = \frac{\omega_1 - \omega_H}{0 - \omega_H} = 1 - \frac{\omega_1}{\omega_H} = 1 - i_{1H} = +\frac{101 \times 99}{100 \times 100}$$

$$i_{1H} = 1 - \frac{101 \times 99}{100 \times 100} = +\frac{1}{10000}$$

$$i_{H1} = \frac{\omega_H}{\omega_1} = \frac{1}{i_{1H}} = +10000$$

In this simple example of an epicyclic gear train, only a few gears can obtain a large gear ratio. This is an advantage of epicyclic gear trains.

**Example 8-3** Fig. 8-11 shows a spatial planetary gear train, in which the numbers of teeth of gears are as follows:

$z_1 = 35$, $z_2 = 48$, $z_{2'} = 55$, $z_3 = 70$

The rotating speeds of input gears are $n_1 = 250$r/min, $n_3 = 100$r/min; determine the rotating speed of the arm.

**Solution**

This is a differential gear train which has two degree of freedom; the speed ratio of gear 1 to gear 3 in the converted gear train is given by:

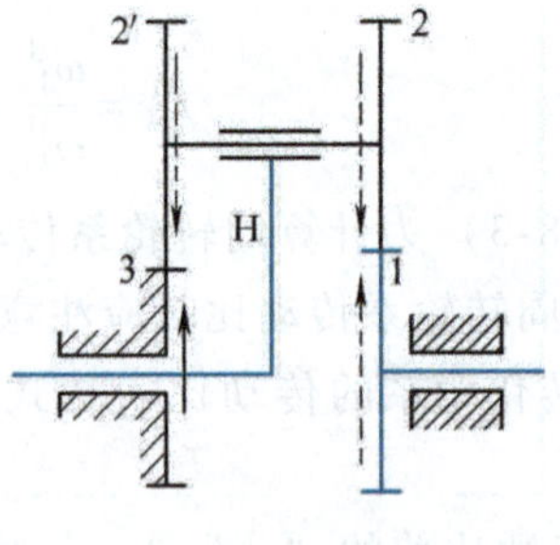

Fig. 8-10 Train with two pairs of external gears (双排外啮合周转轮系)

化轮系传动比可写为

$$i_{13}^{H}=\frac{\omega_1-\omega_H}{\omega_3-\omega_H}=-\frac{z_3}{z_1}$$

由于齿轮 1、2 的轴线不平行，故

$$i_{12}^{H}\neq\frac{\omega_1-\omega_H}{\omega_2-\omega_H}$$

**2. 例题**

**例 8-2** 图 8-10 所示的双排外啮合周转轮系中，已知：$z_1=100$，$z_2=101$，$z_{2'}=100$，$z_3=99$。求传动比 $i_{H1}$。

**解** 施加 $-\omega_H$ 反转后，假想系杆 H 静止不动，则 $z_1$、$z_2$、$z_{2'}$、$z_3$ 成为假想的定轴轮系。该转化轮系的传动比为

$$i_{13}^{H}=\frac{\omega_1^H}{\omega_3^H}=\frac{\omega_1-\omega_H}{\omega_3-\omega_H}=+\frac{z_2z_3}{z_1z_{2'}}=+\frac{101\times 99}{100\times 100}$$

将 $\omega_3=0$ 代入上式得

$$i_{13}^{H}=\frac{\omega_1-\omega_H}{0-\omega_H}=1-\frac{\omega_1}{\omega_H}=1-i_{1H}=+\frac{101\times 99}{100\times 100}$$

$$i_{1H}=1-\frac{101\times 99}{100\times 100}=+\frac{1}{10000}$$

$$i_{H1}=\frac{\omega_H}{\omega_1}=\frac{1}{i_{1H}}=+10000$$

$i_{H1}$ 为“+”，说明齿轮 1 与系杆 H 转向相同。

例 8-2 表明：周转轮系中，仅用少数齿轮就能获得相当大的传动比。若将齿轮 2′减去一个齿，则 $i_{H1}=-100$。这说明同一结构类型的周转轮系，齿数仅发生微小变动，对传动比的影响很大，输出构件的转向也随之改变。这是周转轮系与定轴轮系的显著区别。

**例 8-3** 图 8-11 所示空间轮系中，已知：$z_1=35$，$z_2=48$，$z_{2'}=55$，$z_3=70$，$n_1=250\text{r/min}$，$n_3=100\text{r/min}$，转向如图所示。试求系杆 H 的转速 $n_H$ 的大小和转向。

**解** 这是周转轮系中的差动轮系，首先要计算其转化轮系的传动比。

由
$$i_{13}^{H}=\frac{n_1^H}{n_3^H}=\frac{n_1-n_H}{n_3-n_H}=-\frac{z_2z_3}{z_1z_{2'}}=-\frac{48\times 70}{35\times 55}=-1.75$$

导出

$$n_H=\frac{n_3i_{13}^{H}-n_1}{i_{13}^{H}-1}=\frac{-1.75n_3-n_1}{-1.75-1}=\frac{1.75n_3+n_1}{2.75}$$

由于 $n_1$、$n_3$ 转向相反，若令 $n_1$ 为正值，则 $n_3$ 应以负值代入，于是有

$$i_{13}^{H} = \frac{n_1^H}{n_3^H} = \frac{n_1 - n_H}{n_3 - n_H} = -\frac{z_2 z_3}{z_1 z_{2'}} = -\frac{48 \times 70}{35 \times 55} = -1.75$$

Where $n_H$ is the rotating speed of the arm.

The minus sign is determined by placing arrows in the sketch.

The numerator and denominator of the above equation are divided by $n_H$ , and rearranging them, we have:

$$n_H = \frac{n_3 i_{13}^H - n_1}{i_{13}^H - 1} = \frac{-1.75 n_3 - n_1}{-1.75 - 1} = \frac{1.75 n_3 + n_1}{2.75}$$

Substituting $n_1$ and $n_3$ , then we have:

$$n_H = \frac{1.75 \times (-100) + 250}{2.75} \text{r/min} = 27.27 \text{ r/min}$$

Summary

In order to determine the speed ratio of an epicyclic gear train, we begin the problem by rotating the entire train at an angular velocity which is equal to that of the arm in opposite direction. Therefore, the arm is stationary, that is to say, we obtain a converted gear train.

Write the speed ratio of the converted gear train according to the rule of ordinary gear train. In the formula, the input and output gears usually are two sun gears, from which we can solve the desired value.

## 8.4 Ratio of Combined Gear Trains

We must be able to visualize how the gear train works so as to separate it out from the combined gear train.

Before calculating the speed ratio of a combined gear train, some rules must be cleared.

1) How many elementary gear trains are there in the combined gear train?

2) Analyzing the connection of the elementary gear trains.

3) Write the speed ratio of the elementary gear trains independently, then simultaneously solve these equations.

First, the ordinary gear trains must be found out by inspection if there are some ordinary gear trains, then the epicyclic gear train. Finally, it is necessary to find out how the several gear trains are connected together.

When several gears are in mesh, and their axes are fixed, it is an ordinary gear train.

When several gears are in mesh, in which there are one or more planet gears and they are in mesh with sun gears, it is an epicyclic gear train.

Usually, the tandem connection, closed and superimpose connection are often used in combined gear trains.

**1. Tandem Combined Gear Trains**

In this gear train, the output shaft of the first train is the input shaft of the latter train. Therefore the total ratio is the product of the ratios of all the gear trains.

$$n_H = \frac{1.75 \times (-100) + 250}{2.75}\text{r/min} = 27.27\text{ r/min}$$

计算结果为“+”，说明 $n_H$ 与 $n_1$ 转向相同。

图 8-11 中虚线所标出的箭头方向只表示转化轮系的齿轮转向，并不是周转轮系各齿轮的真实转向。

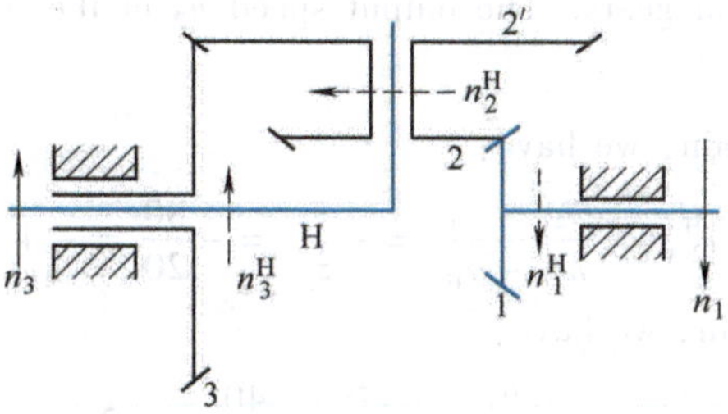


Fig. 8-11　Spatial planetary gear train（空间轮系）

## 8.4　混合轮系传动比的计算

混合轮系可以是定轴轮系与周转轮系的组合，也可以是周转轮系的组合。在计算混合轮系的传动比时，不能将其作为一个整体用反转法求解。应按以下原则求解：

1）分析混合轮系的组成，分别找出其中的基本轮系，如定轴轮系、周转轮系。

2）弄清楚各基本轮系之间的连接关系。

3）分别列出各基本轮系的传动比表达式，然后联立求解。

在分析轮系组成时，先找出定轴轮系。凡是轴线位置固定不动的一系列互相啮合的齿轮，组成定轴轮系。

判别周转轮系时，要先找出轴线转动的行星轮及支承行星轮的系杆。与行星轮相啮合且轴线固定的齿轮为太阳轮。这些行星轮、太阳轮、系杆就构成周转轮系。最后再判别各轮系的连接方法，一般情况下，各轮系之间可用串联、封闭连接和叠加连接组成混合轮系。

**1. 串联组合的混合轮系**

前一个单自由度轮系的输出构件与后一个单自由度轮系的输入构件连接，组成串联型混合轮系。其结构特点是前面轮系的输出转速等于后一个轮系的输入转速。因此，整个混合轮系传动比等于所串联的基本轮系传动比的连乘积，分别列出各基本轮系的传动比关系式，可求出混合轮系的传动比。

**例 8-4**　图 8-12 所示轮系中，已知各轮齿数为：$z_1 = 20$，$z_2 = 40$，$z_{2'} = 20$，$z_3 = 30$，$z_4 = 80$，$n_1 = 300\text{r/min}$，试求系杆 H 的转速 $n_H$。

**解**　该轮系中齿轮 1、2、2′、4 的几何轴线相对机架固定，但齿轮 1、2 相啮合，故齿轮 1、2 构成定轴轮系。齿轮 2′、3、4 啮合，但齿轮 3 既有自转，又有绕齿轮 2′轴线的公转，所以是行星轮。齿轮 2′、4 与行星轮 3 啮合，支承齿轮 3 的构件 H 为系杆。因此，齿轮 2′、3、4 和系杆 H 组成周转轮系中的行星轮系。在该轮系中，其传动比为

**Example 8-4** Fig. 8-12 shows a combined gear train in which the numbers of teeth of gears are $z_1 = 20$, $z_2 = 40$, $z_{2'} = 20$, $z_3 = 30$, $z_4 = 80$, and the driving speed of the gear 1 is $n_1 = 300\text{r/min}$. Determine the speed of the arm.

**Solution**

The ordinary gear train consists of gears 1 and 2 in which their shafts are all fixed, and the planetary gear train consists of gears 2′, 3, 4 and the arm H, in which gear 3 is the planet gear, and gears 2′ and 4 are the sun gears. The output speed $n_2$ of the first train is equal to the input speed $n_{2'}$ of the latter train.

For the planetary gear train, we have:

$$i_{2'4}^{H} = \frac{n_{2'} - n_H}{n_4 - n_H} = -\frac{z_4}{z_{2'}} = -\frac{80}{20} = -4$$

For the ordinary gear train, we have:

$$i_{12} = \frac{n_1}{n_2} = -\frac{z_2}{z_1} = \frac{40}{20} = -2$$

Simultaneously solving the two equations and substituting $n_2 = n_{2'}$ and $n_4 = 0$, the result is as follows:

$$n_H = -30\text{r/min}$$

The minus sign indicates that the direction of rotation of the arm is opposite to that of the driving gear 1.

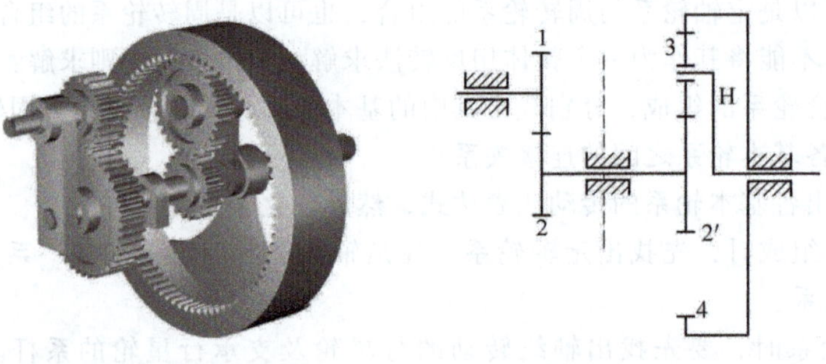

Fig. 8-12 Tandem combined gear train (串联式混合轮系)

### 2. Closed Combined Gear Trains

If two output members, or two input members, or an input member and an output member of a differential gear train are connected by an ordinary gear train, such a combined train is said to be a closed combined gear train.

We can write the speed ratio formula of the differential gear train and the ordinary gear train respectively; then solve them simultaneously.

**Example 8-5** Fig. 8-13 shows a reducer of an electric hoister, in which the numbers of teeth of gears are $z_1 = 24$, $z_2 = 52$, $z_{2'} = 21$, $z_3 = 78$, $z_{3'} = 18$, $z_4 = 30$, $z_5 = 78$. Determine the gear ratio $i_{15}$.

**Solution**

Gears 1, 2, 2′, 3 and 5 (arm H) constitute a differential gear train, in which the gear 3 and the gear 5 (arm H) are the two output gears. Gears 3′, 4 and 5 constitute an ordinary gear train. The gear 3 and the gear 5, two output gears, of the differential gear train are connected by the ordi-

$$i_{2'4}^{H}=\frac{n_{2'}-n_H}{n_4-n_H}=-\frac{z_4}{z_{2'}}=-\frac{80}{20}=-4$$

在定轴轮系中 $$i_{12}=\frac{n_1}{n_2}=-\frac{z_2}{z_1}=\frac{40}{20}=-2$$

$n_2=n_{2'}$，$n_4=0$，联立以上两式求解得

$$n_H=-30\text{r/min}$$

式中，“ - ”号说明 $n_H$ 与 $n_1$ 转向相反。

**2. 封闭组合式混合轮系**

差动轮系的两个构件被自由度为 1 的轮系封闭连接，形成一个自由度为 1 的混合轮系，该轮系称为封闭型混合轮系。

**例 8-5** 图 8-13 为一电动卷扬机减速器的运动简图。已知各轮齿数为：$z_1=24$，$z_2=52$，$z_{2'}=21$，$z_3=78$，$z_{3'}=18$，$z_4=30$，$z_5=78$。试求传动比 $i_{15}$。

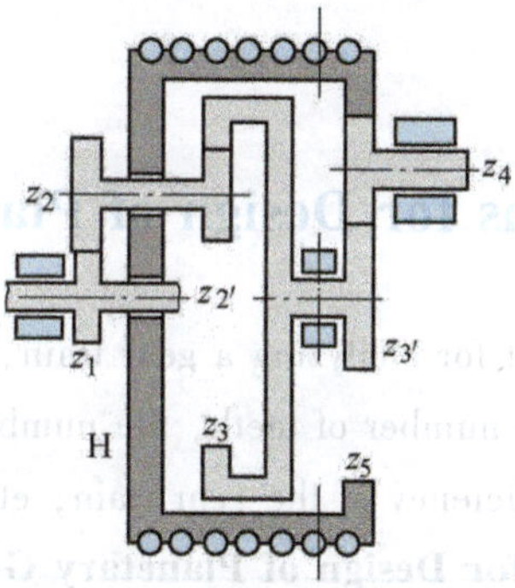


Fig. 8-13 Reducer of the electric hoister（电动卷扬机减速器）

**解** 该轮系中，齿轮 2 - 2′的几何轴线绕内齿轮 5（卷筒 H）的轴线转动，是行星轮；卷筒 H 与内齿轮 5 连成一体，就是系杆 H。与行星轮相啮合的齿轮 1 和 3 是太阳轮。因此，齿轮 1、2 - 2′、3 和系杆 H 组成一个差动轮系。其余齿轮 3′、4、5 构成定轴轮系。差动轮系的两个输出构件，齿轮 3 与系杆 H 被齿轮 3′、4、5 组成的定轴轮系封闭连接，组成封闭型混合轮系。分别写出差动轮系和定轴轮系的传动比表达式。

在差动轮系中，传动比表达式为

$$i_{13}^{H}=i_{13}^{5}=\frac{n_1-n_5}{n_3-n_5}=\frac{\dfrac{n_1}{n_5}-1}{\dfrac{n_3}{n_5}-1}=-\frac{z_2z_3}{z_1z_{2'}}=-\frac{52\times 78}{24\times 21}=-8.05$$

在定轴轮系中，传动比表达式为

$$i_{3'5}=\frac{n_{3'}}{n_5}=-\frac{z_5}{z_{3'}}=-\frac{78}{18}$$

$n_3=n_{3'}$，联立以上两式求解得

nary gear train. It is a closed combined gear train.

From the differential gear train, we have:

$$i_{13}^{H} = i_{13}^{5} = \frac{n_1 - n_5}{n_3 - n_5} = \frac{\dfrac{n_1}{n_5} - 1}{\dfrac{n_3}{n_5} - 1} = -\frac{z_2 z_3}{z_1 z_{2'}} = -\frac{52 \times 78}{24 \times 21} = -8.05$$

From the ordinary gear train, we have:

$$i_{3'5} = \frac{n_{3'}}{n_5} = -\frac{z_5}{z_{3'}} = -\frac{78}{18}$$

Simultaneously solving the two equations and substituting $n_3 = n_{3'}$, the result is as follows:

$$i_{13}^{H} = \frac{\dfrac{n_1}{n_5} - 1}{\dfrac{n_3}{n_5} - 1} = \frac{\dfrac{n_1}{n_5} - 1}{-\dfrac{78}{18} - 1} = -8.05$$

Thus: $i_{15} = \dfrac{n_1}{n_5} = +43.9$

## 8.5 Some Considerations for Design of Planetary Gear Trains

The gear ratio is very important for analyzing a gear train, but there are many matters needing attention, such as how to select the number of teeth, the number of planet gears, how to mount the planet gear and how to calculate efficiency of the gear train, etc.

**1. Fundamental Conditions for Design of Planetary Gear Trains**

In this chapter, we only deal with the planetary gear train of 2K-H which consists of two sun gears, three planetary gears and an arm, as shown in Fig. 8-14.

(1) Conditions of ratio for design of planetary gear trains The gear train must be designed to satisfy the desired ratio $i_{1H}$, therefore, we must select the numbers of teeth of the gears correctly. From the previous formula, we have:

$$i_{13}^{H} = \frac{\omega_1 - \omega_H}{\omega_3 - \omega_H} = -\frac{z_3}{z_1} = \frac{\omega_1 - \omega_H}{0 - \omega_H} = 1 - i_{1H}$$

$$i_{1H} = 1 - i_{13}^{H} = 1 + \frac{z_3}{z_1}$$

$$\frac{z_3}{z_1} = i_{1H} - 1$$

The relationship between the number of teeth on the sun gear 1 and the number of teeth on the sun gear 3 is given by:

$$z_3 = (i_{1H} - 1) z_1$$

(2) Concentric condition of the sun gears The sun gears and the arm must rotate about the same axis, so we have:

$$r'_3 = r'_1 + 2r'_2$$

If the gears are standard spur gears or long-and-short-addendum gears, their pitch circles and

$$i_{13}^{H}=\frac{\frac{n_1}{n_5}-1}{\frac{n_3}{n_5}-1}=\frac{\frac{n_1}{n_5}-1}{-\frac{78}{18}-1}=-8.05$$

$$i_{15}=\frac{n_1}{n_5}=+43.9$$

式中，“+”号说明 $n_5$ 与 $n_1$ 转向相同。

## 8.5　周转轮系设计中的若干问题

传动比计算是周转轮系设计中的重要内容。此外，轮系中的齿数选择、行星齿轮个数的选择以及安装条件、周转轮系的效率与传动比关系等，都是行星轮系设计中的基本问题。

**1. 周转轮系设计的基本问题**

设计周转轮系时，为了平衡行星轮的惯性力，一般在系杆上均匀分布多个行星轮。因此，行星轮的个数以及各轮齿数的选配必须满足下述条件（本书仅讨论图 8-14 所示的单排 2K-H 型周转轮系）。

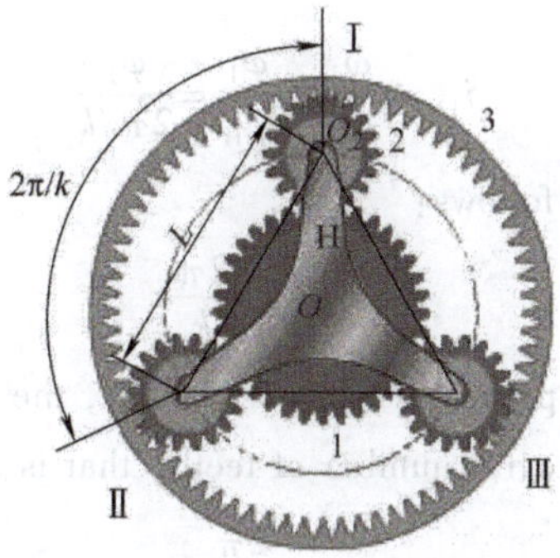


Fig. 8-14　Planetary gear train of 2K-H（2K-H 行星轮系）

（1）传动比条件　周转轮系必须能实现所要求的传动比 $i_{1H}$，或者在其公差范围内。

由
$$i_{13}^{H}=\frac{\omega_1-\omega_H}{\omega_3-\omega_H}=-\frac{z_3}{z_1}=\frac{\omega_1-\omega_H}{0-\omega_H}=1-i_{1H}$$

得
$$i_{1H}=1-i_{13}^{H}=1+\frac{z_3}{z_1}$$

即
$$\frac{z_3}{z_1}=i_{1H}-1$$

太阳轮的齿数关系应满足

$$z_3=(i_{1H}-1)z_1$$

（2）同心条件　周转轮系中各基本构件的回转轴线重合，这就是同心条件。因此，各轮的节圆半径之间必须符合一定的关系，即 $r'_3=r'_1+2r'_2$。如行星轮和太阳轮均为标准齿轮或为高度变位传动，则 $r_3=r_1+2r_2$，$z_3=z_1+2z_2$，考虑到传动比条件，可有

reference circles are coincident respectively, therefore, we have:

$$r_3 = r_1 + 2r_2$$

Rearranging the above equations, we obtain:

$$z_3 = z_1 + 2z_2$$

Substituting $z_3 = (i_{1H} - 1)z_1$ into the above formula, we can get the following:

$$z_2 = \frac{z_3 - z_1}{2} = \frac{z_1(i_{1H} - 2)}{2}$$

(3) Assembly condition To balance the centrifugal force of the planet gear, more planet gears are often mounted between the two sun gears.

If there are $k$ planet gears to be assembled, and they are separated from each other by an angle $\varphi_H$ shown in Fig. 8-14, then we have:

$$\varphi_H = \frac{2\pi}{k}$$

Suppose that the first planet gear has been mounted at the pivot $O_2$, the position Ⅰ, as shown in Fig. 8-14, and then we rotate the arm H to pass through an angle $\varphi_H$ and the arm will arrive at the position Ⅱ. The sun gear 1 rotates an angle $\varphi_1$ with a ratio $i_{1H}$. The relationship between $\varphi_1$ and $\varphi_H$ is given by:

$$i_{1H} = \frac{\omega_1}{\omega_H} = \frac{\varphi_1}{\varphi_H} = \frac{\varphi_1}{2\pi/k}$$

The angle of the gear 1 is as follows:

$$\varphi_1 = i_{1H}\frac{2\pi}{k}$$

In order to mount the second planet gear at position Ⅰ, the angle $\varphi_1$ which the gear 1 has been rotated must contain an exactly integral number of teeth, that is:

$$\varphi_1 = \frac{2\pi}{z_1}n$$

Where $n$ is an integer. When we compare the two equations about angle $\varphi_1$, we get:

$$\varphi_1 = i_{1H}\frac{2\pi}{k} = \left(1 + \frac{z_3}{z_1}\right)\frac{2\pi}{k} = \frac{2\pi}{z_1}n$$

The integer $n$ is as follows:

$$n = \frac{z_1 + z_3}{k}$$

This expresses that if the planet gears of $k$ are mounted between the sun gears, the sum of the teeth of the sun gears must be divided exactly by the number of the planet gears of $k$.

Rearranging the previous equations, and using $z_1$ to express $z_2$ and $z_3$, we have:

$$z_1 : z_2 : z_3 : n = z_1 : \frac{(i_{1H} - 2)}{2}z_1 : (i_{1H} - 1)z_1 : \frac{i_{1H}}{k}z_1$$

(4) Adjacent condition There are not too many planet gears in a planetary gear train, otherwise their tooth top will be overlap each other. It is necessary to determine the maximum number of

$$z_2=\frac{z_3-z_1}{2}=\frac{z_1(i_{1H}-2)}{2}$$

即两个太阳轮的齿数应同时为奇数或偶数。

(3) 装配条件　当轮系中有两个以上的行星轮时，每一个行星轮应均匀地装入两太阳轮之间。因此，应使行星轮的数目和各轮的齿数之间满足一定的条件，即满足装配条件。

如图 8-14 所示，设 $k$ 为行星轮个数，相邻两行星轮所夹的中心角为 $\frac{2\pi}{k}$。先将第一个行星轮在位置Ⅰ装入，这时太阳轮 1 和 3 的相对位置便已确定。为了能在位置Ⅱ和Ⅲ顺利地装入行星轮，使系杆 H 逆时针方向转动 $\varphi_H=\frac{2\pi}{k}$，到达位置Ⅱ。这时的太阳轮 1 将按传动比 $i_{1H}$ 转过角 $\varphi_1$。

由于
$$i_{1H}=\frac{\omega_1}{\omega_H}=\frac{\varphi_1}{\varphi_H}=\frac{\varphi_1}{2\pi/k}$$

则
$$\varphi_1=i_{1H}\frac{2\pi}{k}$$

如果在位置Ⅰ再安装第二个行星轮，角 $\varphi_1$ 必须是太阳轮 1 的 $n$ 个轮齿所对的中心角，刚好包含 $n$ 个齿距，故 $\varphi_1=\frac{2\pi}{z_1}n$。

联立求解得
$$\varphi_1=i_{1H}\frac{2\pi}{k}=\left(1+\frac{z_3}{z_1}\right)\frac{2\pi}{k}=\frac{2\pi}{z_1}n$$

整理后得
$$n=\frac{z_1+z_3}{k}$$

即欲将 $k$ 个行星轮均布安装的装配条件是：周转轮系中两太阳轮的齿数之和应为行星轮数的整数倍。

将前面公式中的 $z_2$、$z_3$ 用 $z_1$ 来表示，得到 2K-H 型周转轮系设计的配齿公式为

$$z_1:z_2:z_3:n=z_1:\frac{(i_{1H}-2)}{2}z_1:(i_{1H}-1)z_1:\frac{i_{1H}}{k}z_1$$

(4) 邻接条件　相邻两行星轮顶部不发生碰撞的条件即为邻接条件。行星轮的个数不能过多，否则会造成相邻两轮的齿顶发生碰撞，为避免产生这种现象，需使两行星轮中心距 $L$ 大于其齿顶圆半径之和。对于标准齿轮传动，根据图 8-14，可有

$$L=2(r_1+r_2)\sin\frac{\pi}{k}$$

$$2(r_1+r_2)\sin\frac{\pi}{k}>2(r_2+h_a^*m)$$

即
$$(z_1+z_2)\sin\frac{\pi}{k}>(z_2+h_a^*m)$$

设计时先用配齿公式初步确定各轮齿数，再验算是否满足邻接条件，如不满足，则应通过增减齿轮齿数等方法重新进行设计。

planet gears that can be used without overlapping. To avoid the overlap between two adjacent planet gears, the center distance $L$ of them must be greater than the sum of outside radii of the planet gears, so this can be written as:

$$L=2(r_1+r_2)\sin\frac{\pi}{k}$$

$$2(r_1+r_2)\sin\frac{\pi}{k}>2(r_2+h_a^* m)$$

$$(z_1+z_2)\sin\frac{\pi}{k}>(z_2+h_a^* m)$$

When we design a planetary gear train, the type of trains, the numbers of teeth of gears, train ratio and the number of planet gears are very important.

**2. Efficiency of Planetary Gear Trains**

The general definition of efficiency is that of the output power of a machine divided by the input power, and it is expressed as a decimal %. Suppose the input power is $P_d$, the output power is $P_r$ and the friction power is $P_f$, then the efficiency $\eta$ can be written as

$$\eta=\frac{P_d-P_f}{P_d}=1-\frac{P_f}{P_d} \tag{8-4}$$

The friction power is then:

$$P_f=(1-\eta)P_d \tag{8-5}$$

(1) Efficiency of ordinary gear trains

Fig. 8-15 shows a flow chart of a tandem ordinary gear train, in which the efficiencies of pairs of gears are denoted as $\eta_1$, $\eta_2$, $\eta_3$, …, $\eta_k$. Their efficiency values are as follows:

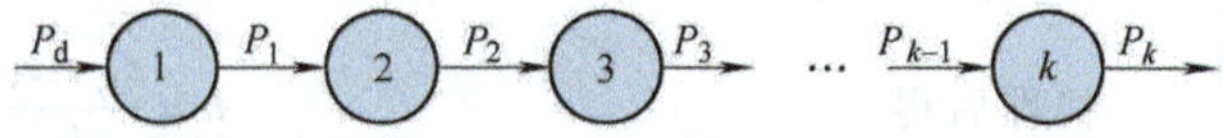

Fig. 8-15 Efficiency of tandem ordinary gear train (串联定轴轮系效率)

$$\eta_1=\frac{P_1}{P_d},\ \eta_2=\frac{P_2}{P_1},\ \eta_3=\frac{P_3}{P_2},\ \dots,\ \eta_k=\frac{P_k}{P_{k-1}}$$

Making the product of the left of these equations be equal to the product of the right, we have

$$\eta=\frac{P_k}{P_d}=\eta_1\eta_2\eta_3\cdots\eta_k \tag{8-6}$$

The power loss per gearset is only 1% to 2% depending on such factors as tooth finish and lubrication. An external gearset will have an efficiency of about 0.98 or better and an external-internal gearset about 0.99 or better.

The overall efficiency of an ordinary gear train will be the product of efficiencies of all its stages. The more the meshing gears, the lower the total efficiency.

(2) Efficiency of planetary gear train  Planetary gear train can have higher overall efficiency than conventional trains, but, if the planetary gear train is poorly designed, its efficiency can be so low that it will generate excessive heat and even may be unable to operate at all.

The computation of the overall efficiency of a planetary gear train is much more complicated

**2. 轮系的机械效率**

轮系的机械效率与轮齿啮合效率、轴承效率、搅油损失效率有关。本节仅讨论轮齿间的啮合效率。

（1）定轴轮系的传动效率　对于任何机械来说，输入功率 $P_d$ 等于输出功率 $P_r$ 和摩擦损失功率 $P_f$ 之和。则机械效率 $\eta$ 可按式（8-4）计算

$$\eta = \frac{P_d - P_f}{P_d} = 1 - \frac{P_f}{P_d} \tag{8-4}$$

损失功率为

$$P_f = (1-\eta)P_d \tag{8-5}$$

当已知机械中的输入功率 $P_d$ 或输出功率 $P_r$ 时，只要能求出摩擦损失功率 $P_f$，就可根据以上公式计算出机械效率 $\eta$。

轮系的效率与轮系中齿轮机构的组合形式有关，其中串联组合的定轴轮系应用最为广泛。

图 8-15 所示为 $k$ 对齿轮传动组成的串联型定轴轮系，设轮系输入功率为 $P_d$，输出功率为 $P_k$，则其总效率为

$$\eta = \frac{P_k}{P_d}$$

设各对齿轮的啮合效率分别为 $\eta_1$，$\eta_2$，…，$\eta_k$，则

$$\eta_1 = \frac{P_1}{P_d},\ \eta_2 = \frac{P_2}{P_1},\ \eta_3 = \frac{P_3}{P_2},\ \dots,\ \eta_k = \frac{P_k}{P_{k-1}}$$

各式两边分别相乘后得

$$\eta = \frac{P_k}{P_d} = \eta_1\eta_2\eta_3\dots\eta_k \tag{8-6}$$

式（8-6）为计算串联定轴轮系传动效率的公式。

串联轮系的效率等于各级齿轮传动效率的乘积。由于 $\eta_1$，$\eta_2$，…，$\eta_k$ 均小于 1，故啮合对数越多，传动的总效率越低。

（2）周转轮系的效率　周转轮系中具有既自转又公转的行星轮，不能直接用定轴轮系效率公式进行计算。但可利用转化轮系法来找出周转轮系和定轴轮系效率之间的内在联系，从而得到周转轮系传动效率的计算公式。

利用转化轮系法将周转轮系转化为定轴轮系以后，各构件之间的相对运动关系没有发生变化，轮系中各运动副间的作用力以及摩擦因数也没有发生变化。所以在不考虑各构件离心惯性力的影响时，可以假设周转轮系中的摩擦损失功率 $P_f$ 与其转化轮系中的摩擦损失功率 $P_f^H$ 近似相等，即 $P_f = P_f^H$。这样就把周转轮系的效率与其转化轮系的效率联系起来，为计算周转轮系的传动效率提供了理论依据。

图 8-16 所示的 2K-H 型周转轮系中，齿轮 1 的角速度为 $\omega_1$，转矩为 $M_1$，则齿轮 1 所传递的功率为

than that of the conventional trains.

In this chapter, we will calculate the efficiency of a planetary gear train by using the converted gear train.

In a converted gear train which is an imaginary ordinary gear train, the relative motion, reaction force acting on the pairs and the coefficients of friction are the same as the planetary gear train.

Suppose that the friction power $P_f$ of the planetary gear train is equal to the friction power $P_f^H$ of the converted gear train, that is $P_f = P_f^H$. This is the base to calculate the efficiency of planetary gear trains.

Fig. 8-16 shows a planetary gear train, denoted as 2K-H. The gear 1 rotates about its axis at constant angular velocity $\omega_1$, and torque acting on it is $M_1$, so the power through the gear 1 is:

$$P_1 = M_1\omega_1$$

Where $P_1$ is the power through the gear 1.

In the converted gear train, the angular velocity of the gear 1 is $\omega_1^H = \omega_1 - \omega_H$; the power through the gear 1 will become as follows:

$$P_1^H = M_1(\omega_1 - \omega_H)$$

Where $P_1^H$ is the power though the gear 1 in the converted gear train.

The relationship between the two powers is as follows:

$$\frac{P_1^H}{P_1} = \frac{M_1(\omega_1 - \omega_H)}{M_1\omega_1} = 1 - \frac{1}{i_{1H}}$$

There are two possible cases:

If $1 - \frac{1}{i_{1H}} > 0$, the power flows of $P_1^H$ and $P_1$ are the same, that is to say, the gear 1 has the same function either in the planetary gear train or in the converted gear train.

The frictional power in the converted gear train is as follows:

$$P_f^H = P_1^H(1 - \eta^H) = M_1(\omega_1 - \omega_H)(1 - \eta^H) \tag{8-7}$$

Where $\eta^H$ is the efficiency of the converted gear train.

If $1 - \frac{1}{i_{1H}} < 0$, the power flows of $P_1^H$ and $P_1$ are opposite, that is to say, the gear 1 is a driving gear in the planetary gear train and a driven gear in the converted gear train.

The frictional power in the converted gear train is as follows:

$$P_f^H = \frac{P_1^H(1 - \eta^H)}{\eta^H} = \frac{M_1(\omega_1 - \omega_H)(1 - \eta^H)}{\eta^H} \tag{8-8}$$

Comparing the equations (8-7) and (8-8), we find that the difference between their calculating results is very small, so gear 1 can be taken as a driving gear no matter it is a driving gear or a driven gear. Therefore, the input power of the planetary gear train is given by:

$$P_1^H = M_1(\omega_1 - \omega_H)$$

The friction power of the converted gear train is as before.

$$P_f^H = P_1^H(1 - \eta^H) = M_1(\omega_1 - \omega_H)(1 - \eta^H)$$

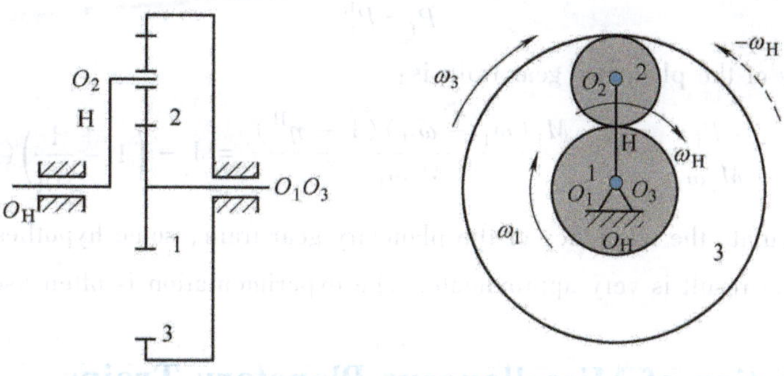


Fig. 8-16　Efficiency of 2K-H planetary gear train（2K-H 型周转轮系）

$$P_1 = M_1\omega_1$$

在转化轮系中，齿轮 1 的角速度 $\omega_1^H = \omega_1 - \omega_H$，故在 $M_1$ 保持不变的情况下，转化轮系中齿轮 1 传递的功率为

$$P_1^H = M_1(\omega_1 - \omega_H)$$

两者关系为

$$\frac{P_1^H}{P_1} = \frac{M_1(\omega_1 - \omega_H)}{M_1\omega_1} = 1 - \frac{1}{i_{1H}}$$

当 $1 - \frac{1}{i_{1H}} > 0$ 时，$P_1^H$ 与 $P_1$ 符号相同，说明齿轮 1 在行星轮系和其转化轮系中的作用不变，同为主动轮或同为从动轮。此时，转化轮系的摩擦损失功率为

$$P_f^H = P_1^H(1 - \eta^H) = M_1(\omega_1 - \omega_H)(1 - \eta^H) \tag{8-7}$$

式中，$\eta^H$ 为转化轮系的效率。

当 $1 - \frac{1}{i_{1H}} < 0$ 时，$P_1^H$ 与 $P_1$ 符号相反，说明齿轮 1 由原机构中的主动轮改变成转化轮系中的从动轮。

此时，转化轮系的摩擦损失功率为

$$P_f^H = \frac{P_1^H(1 - \eta^H)}{\eta^H} = \frac{M_1(\omega_1 - \omega_H)(1 - \eta^H)}{\eta^H} \tag{8-8}$$

式（8-7）与式（8-8）的计算结果相差较小，所以在计算 2K-H 型周转轮系的效率时，不再考虑齿轮 1 在转化轮系中是主动轮还是从动轮，均按主动轮状态进行计算，并取其功率的绝对值。此时，齿轮 1 输入，系杆 H 输出的机械效率 $\eta_{1H}$ 计算过程如下：

转化轮系的输入功率为

$$P_1^H = M_1(\omega_1 - \omega_H)$$

转化轮系的摩擦损失功率为

$$P_f^H = P_1^H(1 - \eta^H) = M_1(\omega_1 - \omega_H)(1 - \eta^H)$$

由于转化轮系是一个定轴轮系，效率 $\eta^H$ 按定轴轮系效率公式进行计算。在外力矩 $M_1$ 相同的情况下，$P_f = P_f^H$，则推出周转轮系的效率 $\eta_{1H}$ 计算公式为

$$\eta_{1H} = 1 - \frac{P_f}{M_1\omega_1} = 1 - \frac{M_1(\omega_1 - \omega_H)(1 - \eta^H)}{M_1\omega_1} = 1 - \left(1 - \frac{1}{i_{1H}}\right)(1 - \eta^H) \tag{8-9}$$

由于用转化轮系法计算效率时作了一些假设，其计算结果不能真实反映实际效率状况，

$$P_f = P_f^H$$

The efficiency of the planetary gear train is:

$$\eta_{1H} = 1 - \frac{P_f}{M_1\omega_1} = 1 - \frac{M_1(\omega_1 - \omega_H)(1 - \eta^H)}{M_1\omega_1} = 1 - \left(1 - \frac{1}{i_{1H}}\right)(1 - \eta^H) \qquad (8\text{-}9)$$

When we calculate the efficiency of the planetary gear train, some hypotheses have been made by designers, so the result is very approximate. The experimentation is often used in engineering.

## 8.6 Introduction of Miscellaneous Planetary Trains

In addition to the ordinary gear trains and the conventional epicyclic gear trains, some special planetary gear trains are often used in machines.

**1. Planetary Gear Train with Small Teeth Difference**

Fig. 8-17 shows a planetary gear train with small teeth difference, in which the internal sun gear 1 is fixed, the arm H is an input and the planet gear 2 is an output gear. The train gear ratio is as follows:

$$i_{21}^H = \frac{n_2 - n_H}{n_1 - n_H} = \frac{z_1}{z_2}$$

Since, $n_1 = 0$, so $1 - \frac{n_2}{n_H} = \frac{z_1}{z_2}$.

Rearranging the above equations, we have:

$$i_{H2} = -\frac{z_2}{z_1 - z_2}$$

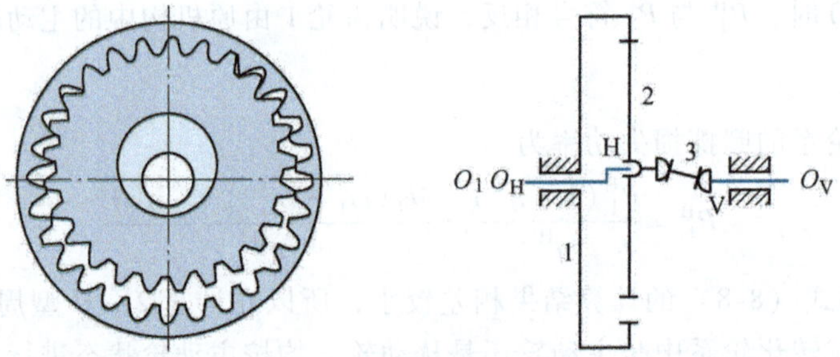


Fig. 8-17 Planetary gear train with small teeth difference (少齿差行星轮系)

To transform the planetary motion of the planet gear, a constant angular velocity mechanism can be used, as shown in Fig. 8-18.

There are several holes with radius of $r_W$ and they are equispaced in the planetary gear plate 2 at a radius of $\rho$. The same number of pins with radius of $r_P$ are equispaced in another plate 3 at the same radius $\rho$.

The difference between the radii of the hole and pin is $e$, and the distance is equal to the offset between the centers of sun gear and the planet gear, that is:

$$e = r_W - r_P, \quad O_2O_V = e = r_1 - r_2$$

The linkage $O_VO_2O_2'O_V'$ is a parallelogram form; the crank $O_2O_2'$ representing the planetary gear

所以常用于轮系效率的定性分析。工程设计时，还需要用实验的方法进行效率测定。

## 8.6 其他类型的周转轮系简介

除上述轮系之外，工程中还经常应用一些特殊的周转轮系。

**1. 渐开线少齿差行星轮系**

渐开线少齿差行星轮系如图 8-17 所示，通常太阳轮 1 固定，系杆 H 为输入轴，V 为输出轴。输出轴 V 与行星轮 2 通过等角速比机构 3 相连接，所以输出轴 V 的转速始终与行星轮 2 的绝对转速相同。由于太阳轮 1 和行星轮 2 都是渐开线齿轮，齿数差很少，故称为渐开线少齿差行星轮系。其传动比为

$$i_{21}^{H} = \frac{n_2 - n_H}{n_1 - n_H} = \frac{z_1}{z_2}$$

$n_1 = 0$，则有

$$1 - \frac{n_2}{n_H} = \frac{z_1}{z_2}$$

$$i_{2H} = \frac{n_2}{n_H} = 1 - \frac{z_1}{z_2} = -\frac{z_1 - z_2}{z_2},\quad i_{H2} = -\frac{z_2}{z_1 - z_2}$$

故 $i_{HV} = i_{H2} = -\dfrac{z_2}{z_1 - z_2}$

齿数差越小，传动比越大。齿数差 $z_1 - z_2 = 1$ 时，称为一齿差行星轮系，此时传动比达到最大值。

少齿差行星轮系通常采用销孔输出机构作为等角速比机构。图 8-18 所示行星轮 2 上，沿半径为 $\rho$ 的圆周均布开孔，孔半径为 $r_W$。在输出圆盘上，沿半径为 $\rho$ 的圆周上安装相同数目的圆柱销，并将其分别插入行星轮 2 的圆孔中，使行星轮和输出轴连接起来。销孔与销轴间距等于行星轮轴线与输出轴轴线间的距离，即 $e = r_W - r_P$。

$O_2$ 至圆孔中心 $O_2'$ 和 $O_V$ 至销轴中心 $O_V'$ 等距且平行，$O_V O_2 O_2' O_V'$ 总是保持平行四边形机构。$O_2 O_2'$ 代表行星轮 2 的运动，$O_V O_V'$ 代表输出构件的运动，两者永远相等，保证输出轴 V 的转速始终与行星轮的绝对转速相同。

渐开线少齿差行星传动的特点是传动比大、结构简单紧凑、齿轮易加工、装配方便，在机械工程领域有广泛应用。当齿数差过小时，渐开线少齿差行星传动容易发生齿廓重叠干

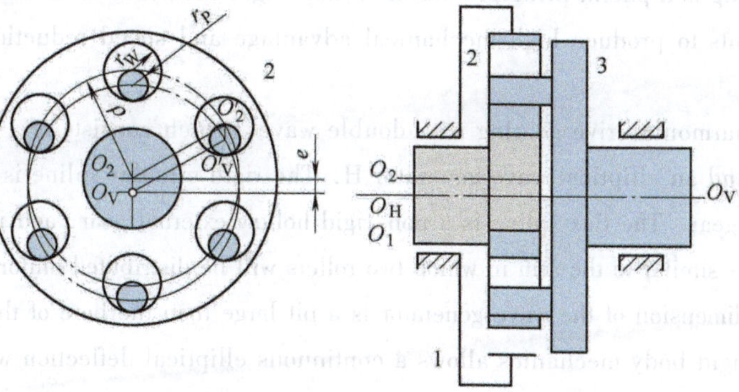


Fig. 8-18 Output mechanism with the same velocity（等角速比输出机构）

and crank $O_V O'_V$ representing the output plate have the same angular velocity, therefore, the motion of the planet gear can be transformed by the output shaft, denoted as V, then the train ratio can be written as:

$$i_{HV} = i_{H2} = -\frac{z_2}{z_1 - z_2}$$

The less the teeth difference is, the greater the train ratio is.

**2. Cycloidal-pin Wheel Planetary Gears**

The principle of cycloidal-pin wheel planetary gears is similar to the planetary gear train with small teeth difference, but the teeth difference is one. In the cycloidal-pin wheel planetary gears, the fixed internal gear is a sleeve pin wheel and the planet gear is one of having epicycloids, as shown in Fig. 8-19.

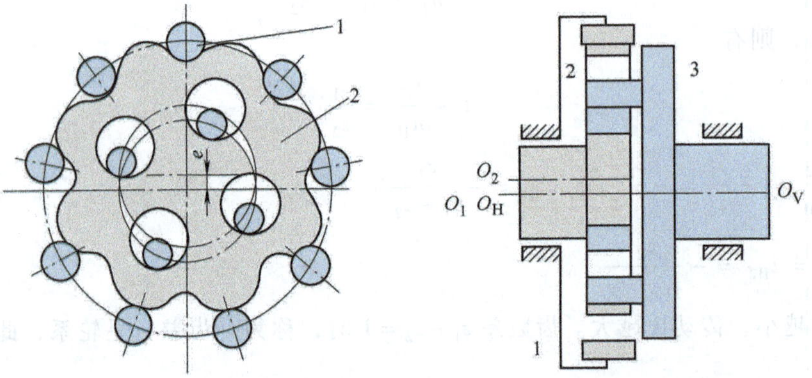


Fig. 8-19 Cycloidal-pin wheel planetary train (摆线针轮行星轮系)
1—sleeve pin (针轮) 2—epicycloids wheel (摆线轮) 3—V-shaft (V 轴)

The output mechanism is the same with that of the planetary gear train with small teeth difference, and it has more advantages.

1) The minimum teeth to occur undercutting is less than that of involute tooth profiles; it is about number seven.

2) There are more pairs of teeth to contact, so it has a larger contact ratio.

3) The rolling friction will occur between their teeth, so the transmitting capacity is high.

**3. Harmonic Drive Gearing**

Harmonic drive gearing is a patent principle based on non-rigid body mechanics. It employs the three concentric components to produce high mechanical advantage and speed reduction, as shown in Fig. 8-20a.

Fig. 8-20a shows a harmonic drive gearing with double wave, which consists of a rigid circular spline 1, a flex spline 2 and an elliptical wave generator H. The rigid circular spline is a rigid internal gear and it is a fixed gear. The flex spline is a non-rigid hollow external gear, and it is an output gear. The wave generator is similar to the arm in which two rollers will be distributed uniformly, and it is an input arm. The radial dimension of the wave generator is a bit large than the bore of the flex spline.

The use of the non-rigid body mechanics allows a continuous elliptical deflection wave to be induced in a non-rigid external gear, thereby providing a continuous rolling mesh with a rigid internal

涉，必须按变位齿轮进行设计。

**2. 摆线针轮行星轮系**

摆线针轮行星轮系的原理和结构与渐开线少齿差行星轮系基本相同，但行星轮采用外摆线齿廓，太阳轮轮齿为圆柱形，故称为摆线针轮行星传动，如图 8-19 所示。

这种传动与渐开线少齿差行星传动的差别之处在于齿廓曲线不同。摆线针轮行星轮系的优点是齿廓间为滚动摩擦，因此承载力大、传动平稳、轮齿磨损小、使用寿命长；但加工较复杂，精度要求高，必须用专用机床和刀具来加工摆线针轮。摆线针轮行星轮系广泛应用于各种机械设备上。

**3. 谐波齿轮行星轮系**

图 8-20a 所示谐波齿轮行星轮系是由波发生器 H、刚轮 1 和柔轮 2 组成的。其中柔轮为一薄壁构件，外壁有齿，内壁孔径略小于波发生器的尺寸。在相当于系杆的波发生器 H 作用下，相当于行星轮的柔轮产生弹性变形而呈椭圆形。其椭圆长轴两端的轮齿插进刚轮的齿槽中，而短轴两端的轮齿则与钢轮脱开。

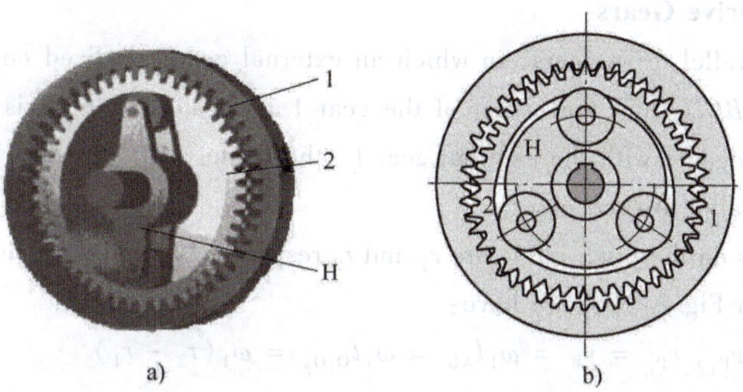


Fig. 8-20　Harmonic driving planetary gears（谐波齿轮行星轮系）
1—rigid circular spline（刚轮）　2—non-rigid flexspline（柔轮）　H—wave generator（波发生器）

一般刚轮固定不动，当波发生器 H 回转时，柔轮与刚轮的啮合区跟着发生转动。由于在传动过程中柔轮产生的弹性波形近似于谐波，故称为谐波齿轮行星轮系，也称谐波传动。

由于柔轮比刚轮少（$z_1-z_2$）个齿，所以 H 转一周，柔轮相对刚轮沿相反方向转过（$z_1-z_2$）个齿的角度，其传动比为

$$i_{H2}=\frac{n_H}{n_2}=-\frac{z_2}{z_1-z_2}$$

根据波发生器 H 上安装的滚轮数不同，谐波齿轮行星轮系传动可分为图 8-20a 所示的双波传动和图 8-20b 所示的三波传动等，最常用的是双波传动。谐波齿轮行星轮系传动的齿数差应等于波数或波数的整数倍。为了加工方便，谐波齿轮的齿形多采用渐开线齿廓。

谐波传动装置不需要等角速比机构，因此结构简单；传动比大、体积小、重量轻、效率高；啮合齿数多，承载力大，传动平稳；它的齿侧间隙小，适用于反向传动。但柔轮周期性发生变形，容易发热，需用疲劳强度很高的材料，且对加工、热处理要求都很高。

**4. 平动齿轮传动**

图 8-21a 所示平行四边形 *ABCD* 的连杆 *BC* 与齿轮 $z_1$ 固接在一起，齿轮中心 $O_1$ 位于连

gear. The teeth of the nonrigid flexspline and the rigid circular spline are in continuous engagement.

If the flexspline has two fewer teeth than the circular spline, one revolution of the input causes relative motion between the flexspline and the circular spline equal to two teeth.

Suppose that the rigid circular spline has teeth of $z_1$, and the nonrigid flexspline has teeth of $z_2$, so the teeth difference between the two gears is $(z_1 - z_2)$.

When the wave generator rotates, the rollers which is similar to the arm of a planetary gear train will make the teeth of the flexspline engage with the rigid circular spline, and the flexspline is similar to the planet gear, so the train ratio is given by the following:

$$i_{H2} = \frac{n_H}{n_2} = -\frac{z_2}{z_1 - z_2}$$

Where $n_H$ is the speed of the wave generator, and $n_2$ is the speed of the flexspline.

If there are two rollers in the wave generator, it is a two-wave generator shown in Fig. 8-20a, and if there are three rollers, it is a three-wave generator shown in Fig. 8-20b.

**4. Parallel Motion Drive Gears**

Fig. 8-21a shows a parallel drive gears, in which an external gear 1 is fixed on the coupler of the parallelogram linkage $ABCD$, and the center of the gear 1 is located at the axis of the coupler $BC$. If an internal gear 2 engages with the external gear 1, the center distance must be equal to the length of the crank and parallel to each other.

If the radii of the gears on the pitch circle are $r_1$ and $r_2$ respectively, and the point $P$ is their instantaneous center shown in Fig. 8-21b, we have:

$$v_{P_1} = v_{P_2},\ v_{P_1} = v_B = \omega_1 l_{AB} = \omega_1 l_{O_1O_2} = \omega_1 (r_2 - r_1)$$

$$v_{P_2} = \omega_2 r_2$$

$$\omega_1 (r_2 - r_1) = \omega_2 r_2$$

$$i_{12} = \frac{\omega_1}{\omega_2} = \frac{r_2}{r_2 - r_1} = \frac{z_2}{z_2 - z_1}$$

When the teeth difference is small, we can obtain a large gear ratio.

杆轴线上，随同连杆 BC 做无自转的平动。两齿轮的中心距$\overline{O_1O_2}=\overline{AB}=\overline{CD}$，且 $O_1O_2$ 与 $AB$、$CD$ 平行。

如图 8-21b 所示，设齿轮 1、2 的分度圆半径分别为 $r_1$、$r_2$，利用两齿轮的瞬心 $P$ 为速度重合点的概念，可推导出其传动比。即

$$v_{P_1}=v_{P_2},\ v_{P_1}=v_B=\omega_1 l_{AB}=\omega_1 l_{O_1O_2}=\omega_1(r_2-r_1)$$

$$v_{P_2}=\omega_2 r_2$$

$$\omega_1(r_2-r_1)=\omega_2 r_2$$

$$i_{12}=\frac{\omega_1}{\omega_2}=\frac{r_2}{r_2-r_1}=\frac{z_2}{z_2-z_1}$$

当齿数差很小时，可获得较大的传动比。

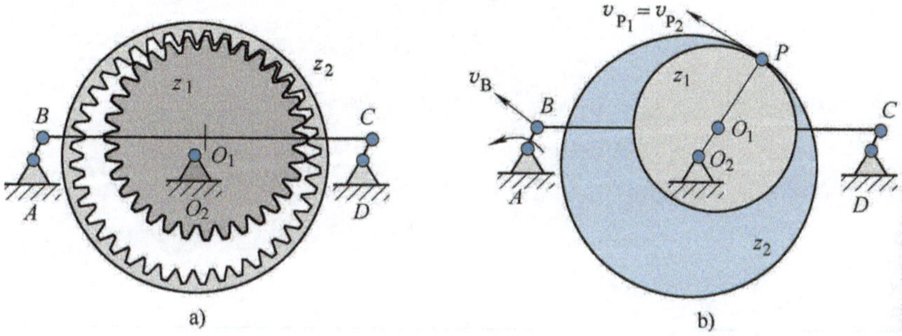


Fig. 8-21　Parallel moving gear train（平动齿轮传动）

# Chapter 9

# Introduction of Screws, Hook's Couplings and Intermittent Mechanisms

# 第9章

# 螺旋机构、万向联轴器和间歇运动机构简介

In addition to linkages, gears and cams, there are many special mechanisms such as screw mechanisms, Hooke's coupling mechanisms, indexing mechanisms and so on.

## 9.1 Screw Mechanisms

Screw mechanisms can convert the rotational motion into the rectilinear motion; they are frequently used as machine tool drives. If a screw with a single thread engages a nut which is not permitted to rotate, the nut will move relative to the thread a distance equal to the pitch for each screw rotation. Fig. 9-1a shows a screw mechanism which consists of screw 1, nut 2 and frame 3, in which the pairs $A$, $B$ and $C$ are screw pair, turning pair and prismatic pair respectively.

Suppose the lead of screw is $l$, and when the screw rotates an angle $\varphi$, the distance moved by the nut is as follows:

$$s = l\frac{\varphi}{2\pi} \tag{9-1}$$

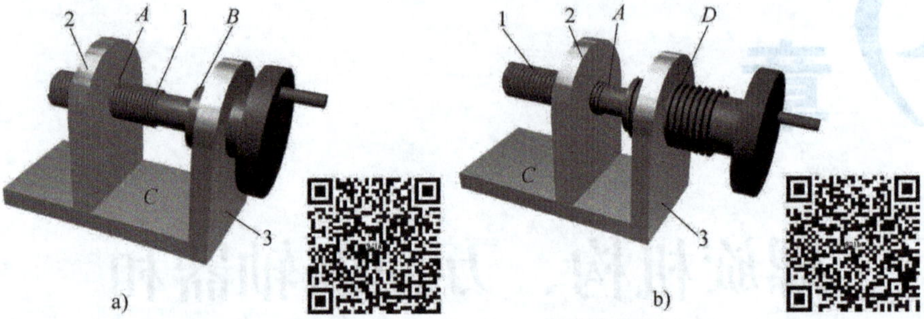


Fig. 9-1 Screw mechanisms (螺旋机构)

1—screw (螺杆) 2—nut (螺母) 3—frame (机架) $A$—screw pair (螺旋副)
$B$—turning pair (转动副) $C$—sliding pair (移动副) $D$—screw pair (螺旋副)

If the turning pair $B$ is changed to a screw pair $D$, we obtain a differential screw mechanism in which the leads of the screws are $l_A$ and $l_D$ respectively, as shown in Fig. 9-1b. When the screw bar rotates an angle $\varphi$, the distance moved by the nut is as follows:

$$s = (l_A \pm l_D)\frac{\varphi}{2\pi} \tag{9-2}$$

The plus sign expresses that the two threads have the opposite hands of helix; we can use this screw mechanism to obtain a quick motion.

The minus sign expresses that the two threads have the same hands of helix. If the difference between the leads $l_A$ and $l_D$ is very small, we can obtain a lower motion of the nut. Therefore, the differential screws are often used as fine measuring instruments and fine adjustment devices.

According to the friction property between the thread and nut, the screws can be classified as sliding screws and rolling screws. Square thread and acme thread are used to screws by which power is to be transmitted.

Fig. 9-2 shows a roller screw in which rollers are located between the threads of the screw and nut. It is called as a roller screw mechanism which has high efficiency and long life and it is used widely as a power screw. Its equivalent friction angle $\varphi_v$ is very small, usually $\varphi_v$ is less than the

齿轮机构、凸轮机构、连杆机构是应用最为广泛的机构。此外，在机械中还有一些机构因其特殊功能也有很多应用。本章将对这些机构作简单介绍。

## 9.1　螺旋机构

螺旋机构是把转动变换为移动的常用机构，有广泛应用。螺旋机构由螺杆 1、螺母 2 和机架 3 组成，它是利用螺旋副来传递运动和动力的机构。

图 9-1a 中，$A$ 为螺旋副，其导程为 $l$；$B$ 为转动副，$C$ 为移动副。当螺杆 1 转动 $\varphi$ 角时，螺母 2 的位移 $s$ 为

$$s = l\frac{\varphi}{2\pi} \tag{9-1}$$

如果将图 9-1a 中的转动副 $B$ 也换成螺旋副，便得到图 9-1b 所示的螺旋机构。现假设 $A$、$D$ 两段螺旋的导程分别为 $l_A$、$l_D$，当螺杆 1 转动 $\varphi$ 角时，螺母 2 的位移 $s$ 为

$$s = (l_A \pm l_D)\frac{\varphi}{2\pi} \tag{9-2}$$

式中，“－”号用于两段螺旋旋向相同时，而“＋”号用于两段螺旋旋向相反时。

由式（9-2）可知，当两段螺旋旋向相同时，若 $l_A$ 与 $l_D$ 相差很小，则螺母 2 的位移可以很小，这种螺旋机构称为差动螺旋机构或微动螺旋机构；当两段螺旋旋向相反时，螺母 2 可以产生快速移动，这种螺旋机构称为复式螺旋机构。

根据螺杆与螺母之间摩擦状态的不同，螺旋传动可分为滑动螺旋传动和滚动螺旋传动。滑动螺旋传动中，螺杆与螺母的螺旋面直接接触，其摩擦状态为滑动摩擦。滚动螺旋传动中，螺杆与螺母的螺纹滚道间有滚动体，称之为滚珠螺旋传动机构。由于当量摩擦角 $\varphi_v$ 很小，远小于其螺旋升角，故滚珠螺旋传动机构一般不能自锁。图 9-2 所示滚珠丝杠传动机构中，滚珠沿螺旋槽滚动，并借助于导向装置将滚珠导入返回滚道，然后进入工作滚道中，如此循环往复，使滚珠形成一个闭合的循环回路。滚珠螺旋传动机构具有传动效率高、起动力矩小、传动灵敏平稳、工作寿命长等优点。

螺旋机构的优点是结构简单，制造方便，工作平稳，无噪声，可以传递很大的轴向力，能将回转运动变换为直线运动，运动准确性高，并且有很大的降速比。如果选择合适的螺纹升角，还可以满足自锁要求。

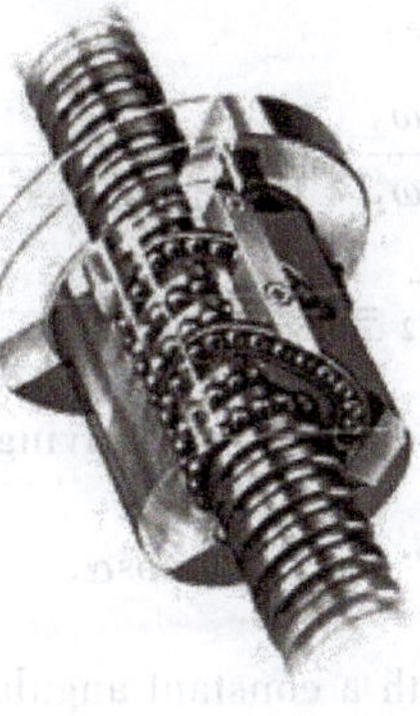

Fig. 9-2　Roller screw mechanism（滚珠丝杠传动机构）

inclined angel of the roller screw, so the self-locking can not occur.

## 9.2 Universal Joints

Universal joints are used to connect two intersecting shafts. They are also known as Hook's couplings, and can be classified as single universal joints and double universal joints.

**1. Single Universal Joint**

Fig. 9-3a shows a single universal joint which consists of the driving yoke, driven yoke and cross piece which connects the two yokes. The axes of the driving shaft and driven shaft intersect at the point $O$ which is the rotating center of the cross piece, and the acute angle between the two shafts is denoted as $\alpha$. Although both shafts must complete a revolution in the same length of time, the angular velocity ratio of the shafts is not a constant during the revolution, but varies as a function of the angles $\alpha$ and $\varphi_1$.

As shafts I and II rotate, the points $A$ and $B$ describe two circles shown in Fig. 9-3b.

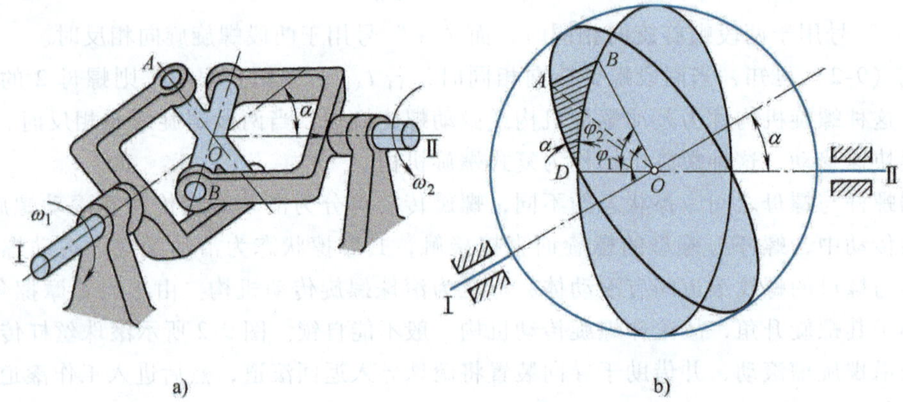


Fig. 9-3 Single universal joint (单万向联轴器)

The two circles intersect at the point $D$ and are separated by the shaft angle $\alpha$. Let the point $A$ travel a distance $\varphi_1$ from the intersection $D$, and the point $B$ travel a distance $\varphi_2$ from the point $D$. According to a right-triangle formula from spherical trigonometry, we have:

$$\tan\varphi_1 = \tan\varphi_2 \cos\alpha$$

Differentiating both sides, we have:

$$\frac{\omega_2}{\omega_1} = \frac{\cos\alpha}{1 - \sin^2\alpha\cos^2\varphi_1}$$

$$\omega_2 = \frac{\cos\alpha}{1 - \sin^2\alpha\cos^2\varphi_1}\omega_1$$

Where $\omega_1$ and $\omega_2$ are angular velocities of the driving shaft I and driven shaft II.

If, $\varphi_1 = 0°$, $\omega_2 = \dfrac{\omega_1}{\cos\alpha}$, $\varphi_1 = 90°$, $\omega_2 = \omega_1\cos\alpha$.

When the driving shaft I rotates with a constant angular velocity $\omega_1$, the angular velocity $\omega_2$ will fluctuate periodically, and the range is as follows:

## 9.2　万向联轴器

万向联轴器是用于传递两相交轴或平行轴之间运动和动力的机构，在传动过程中，两轴之间的夹角可以变动。万向联轴器包括单万向联轴器和双万向联轴器两种形式。

**1. 单万向联轴器**

单万向联轴器由两个固接于轴Ⅰ和Ⅱ端部的叉形接头、一个十字形构件和机架组成，如图 9-3a 所示。十字形构件的中心 $O$ 与两轴轴线的交点重合，轴Ⅰ和Ⅱ所夹的锐角 $\alpha$ 称为万向联轴器的轴角。主动轴Ⅰ每转一周，从动轴Ⅱ也随之转动一周，但是两轴的瞬时传动比却因位置的不同而随时变动。

图 9-3b 是两个叉面球面运动示意图。轴Ⅰ的叉面由 $OD$ 转动 $\varphi_1$，到达 $OA$；轴Ⅱ的叉面转过 $\varphi_2$ 到达 $OB$。轴Ⅰ和Ⅱ的角速度分别为 $\omega_1$ 和 $\omega_2$。

在球面直角三角形 $ABD$ 中，$\angle ADB=\alpha$，轴Ⅰ转过微小角度 $\varphi_1$，则轴Ⅱ转过微小角度 $\varphi_2$。有

$$\tan\varphi_1=\tan\varphi_2\cos\alpha$$

两边求导数后

$$\frac{\omega_2}{\omega_1}=\frac{\cos\alpha}{1-\sin^2\alpha\cos^2\varphi_1}$$

$$\omega_2=\frac{\cos\alpha}{1-\sin^2\alpha\cos^2\varphi_1}\omega_1$$

$$\varphi_1=0°,\ \omega_2=\frac{\omega_1}{\cos\alpha},\ \varphi_1=90°,\ \omega_2=\omega_1\cos\alpha$$

从动轴的角速度变化范围为

$$\omega_1\cos\alpha\leqslant\omega_2\leqslant\frac{\omega_1}{\cos\alpha}$$

从动轴Ⅱ的角速度 $\omega_2$ 变化的幅度与两轴间夹角 $\alpha$ 的大小有关。因此，应用单万向联轴器时，一般取 $\alpha\leqslant45°$。

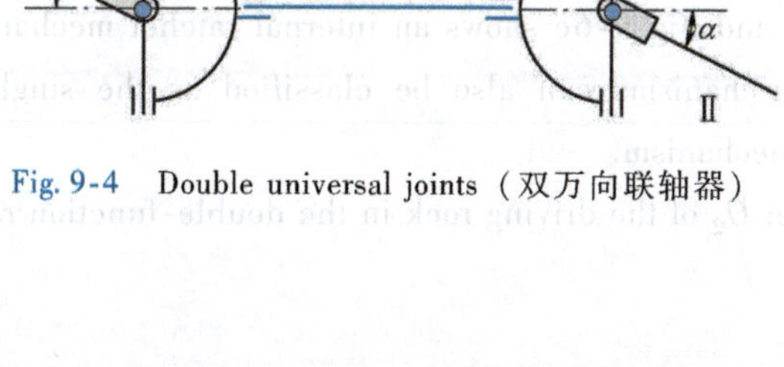


Fig. 9-4　Double universal joints（双万向联轴器）

$$\omega_1 \cos\alpha \leqslant \omega_2 \leqslant \frac{\omega_1}{\cos\alpha}$$

To reduce the speed fluctuation of the driven shaft, the shaft angle is usually less than 45°. A solution to this problem can be obtained by using two universal joints.

**2. Double Universal Joints**

It is possible to connect two shafts by two universal joints in which the second joint compensates for the variations in speed produced by the first. Fig 9-4 shows a double universal joints.

To have the same angular velocities of the input shaft Ⅰ and output shaft Ⅱ, the following conditions must be satisfied.

1) The driving shaft Ⅰ, driven shaft Ⅱ and the intermediate shaft Ⅲ must lie in the same plane.

2) The two yokes of the intermediate shaft must lie in the same plane.

3) The angle between the driving shaft and intermediate shaft must be equal to the angle between the driven shaft and the intermediate shaft.

The double universal joints are used widely on most rear-wheel-drive cars.

## 9.3 Ratchet Mechanisms

Ratchet mechanisms are used to transmit motion of rotation or translation into intermittent rotation or translation.

Fig. 9-5 shows a ratchet mechanism which consists of a driving rock 1, a ratchet wheel 3, pawls 2、4 and a spring 5. As the driving rock 1 is made to oscillate counterclockwise about the center of the ratchet, pawl levers, the ratchet 3 will rotate counterclockwise with an intermediate motion. When the driving rock oscillates clockwise, the driving pawl will slide over the tooth back of the ratchet, the holding pawl will move the tooth space to prevent the ratchet from reversing.

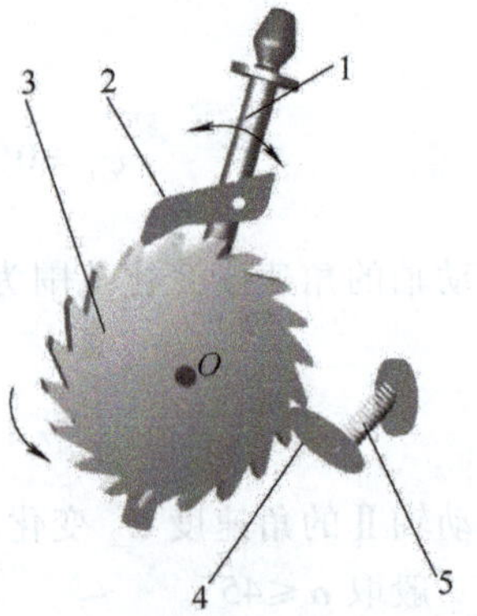


Fig. 9-5 Ratchet mechanism (棘轮机构)
1—driving rocker (主动摇杆)
2—driving pawl (棘爪) 3—ratchet (棘轮)
4—holding pawl (止回棘爪) 5—spring (弹簧)

**1. Types and Applications of Ratchet Mechanisms**

According to structures, ratchet mechanisms can be classified as two types: one is the tooth ratchet mechanism, and the other is the silent ratchet mechanism.

(1) Tooth ratchet mechanism  The teeth of ratchet are often made into the shapes of oblique triangle, trapezoid and rectangle. Fig. 9-6a shows an external ratchet mechanism, Fig. 9-6b shows a ratchet rack mechanism, and Fig. 9-6c shows an internal ratchet mechanism.

The tooth ratchet mechanisms can also be classified as the single-function mechanism and double-function ratchet mechanism.

The oscillating center $O_2$ of the driving rock in the double-function ratchet mechanism is not co-

**2. 双万向联轴器**

单万向联轴器传动时，从动轴角速度有波动，将产生周期性的附加动载荷。为避免这一缺点，工程中常将单万向联轴器成对使用，构成双万向联轴器，即用中间轴Ⅲ把两个单万向联轴器连接起来，如图 9-4 所示。

在双万向联轴器中，为使主、从动轴的角速度始终保持相等，应满足如下条件：

1）主、从动轴Ⅰ、Ⅱ与中间轴Ⅲ应位于同一平面内。

2）中间轴两端的叉面应位于同一平面内。

3）主动轴与中间轴的夹角应等于从动轴与中间轴的夹角。

双万向联轴器能连接两轴交角较大的相交轴或偏距较大的平行轴，且在运转时轴交角或偏距可以不断改变，因此在机械中得到广泛应用。

## 9.3　棘轮机构

棘轮机构是使从动轴实现间歇运动的机构。

图 9-5 所示的棘轮机构由主动摇杆 1、棘爪 2、棘轮 3、止回棘爪 4 和机架等部分组成。弹簧 5 用来使止回棘爪 4 和棘轮 3 保持接触。主动摇杆 1 空套在与棘轮 3 固连的从动轴 $O$ 上，并与棘爪 2 用转动副相连。当主动摇杆做逆时针方向摆动时，棘爪 2 便插入棘轮 3 的齿槽内，推动棘轮转动一定的角度，此时止回棘爪 4 在棘轮的齿背上滑过。当主动摇杆顺时针方向摆动时，止回棘爪阻止棘轮顺时针方向转动，棘爪 2 在棘轮的齿背上滑过，棘轮 3 保持静止不动。这样，当主动件做连续的往复摆动时，棘轮做单向的间歇转动。

**1. 棘轮机构的类型及其应用**

根据结构特点，将棘轮机构分为齿式棘轮机构和摩擦式棘轮机构两类。

（1）齿式棘轮机构　齿式棘轮的轮齿一般采用三角形齿、梯形齿或矩形齿，分为外齿棘轮和内齿棘轮。当外齿棘轮的直径为无穷大时，棘轮变为棘条。图 9-6a 所示为外齿棘轮机构，图 9-6b 所示为棘条机构，图 9-6c 所示为内齿棘轮机构。根据驱动爪的数目，棘轮机构还可分为单动式棘轮机构和双动式棘轮机构。

Fig. 9-6　Tooth ratchet mechanisms（齿式棘轮机构）

图 9-6 所示机构为单动式棘轮机构。当主动摇杆向一个方向摆动时，棘轮沿同一方向转过一定角度；而当主动摇杆反向摆动时，棘轮静止不动。

incident with the rotating center $O_1$ of the ratchet wheel. They are shown in Fig. 9-7. The driving rock can drive the ratchet to rotate intermediately in the same way in both directions of oscillation.

In Fig. 9-7a, the driving pawls will push the ratchet to rotate, and in Fig. 9-7b, the pawls will pull the ratchet to rotate.

Fig. 9-7 Double-function ratchet mechanisms (双动式棘轮机构)

Fig. 9-8 shows some ratchet mechanisms in which the ratchets can rotate in double directions. These ratchet wheels usually have rectangular teeth, while the pawl should be reversible. In Fig. 9-8a and Fig. 9-8b, the pawl can inverse itself about its center $O$, and in Fig. 9-8c and Fig. 9-8d, the pawl must be lift first, then rotate 180° about its axis and drop down.

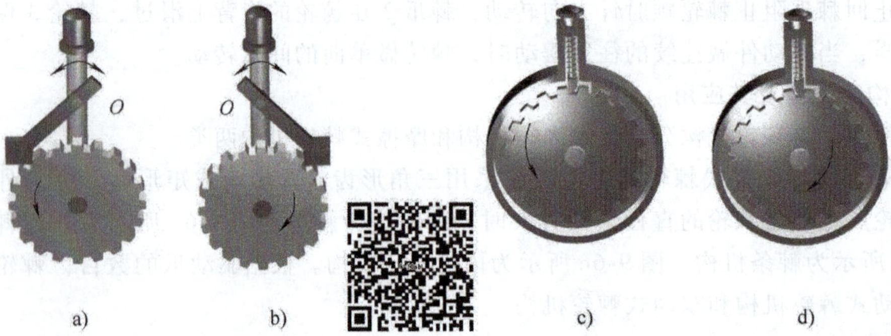


Fig. 9-8 Ratchet mechanisms in which the rotating direction can be changed (双向棘轮机构)

Fig. 9-9 shows a table-feed system of a shaper. This system consists of a ratchet mechanism, a gear mechanism and a linkage.

(2) Silent ratchet mechanism  Silent ratchet mechanisms are shown in Fig. 9-10. There are no teeth on the ratchet, and the devices depend upon the wedging of smooth surface between the pawl and ratchet.

In Fig. 9-10a, the intermittent rotation of the ratchet depends upon the frictional force between the pawl and the ratchet surface, so it has no noise to transmit motion.

Fig. 9-10b and Fig. 9-10c are internal ratchet mechanisms. In Fig. 9-10c, the pawls are rollers located between the star wheel 1 and the toroid 3. If the star wheel rotates counterclockwise, and the frictional force causes the rollers to move towards the narrow end of the notch, the toroid will rotate at the same angular velocity. When the star wheel rotates clockwise, and the frictional force causes the rollers to move towards the large end of the notch, the toroid will be stationary. When the star

图 9-7a 所示的机构为双动式棘轮机构。主动摇杆 1 不是绕棘轮转动中心 $O_1$ 摆动，而是绕 $O_2$ 轴摆动，摇杆上分别装有两个棘爪。当主动摇杆往复摆动一次时，两个棘爪分别推动棘轮沿同一方向间歇转动一次。当载荷较大，齿数较少，摇杆摆角小于齿距角时，需采用双动式棘轮机构。图 9-7b 所示双动式棘轮机构中，棘爪拉动棘轮实现间歇运动。

双向棘轮机构的棘轮能做正反两个方向的间歇运动。图 9-8a、b 所示的机构为摆动棘爪双向棘轮机构。棘爪安放在图 9-8a 所示位置时，将推动棘轮沿逆时针方向做单向间歇转动；棘爪安放在图 9-8b 所示位置时，推动棘轮沿顺时针方向做单向间歇转动。图 9-8c、d 所示的机构为直动棘爪双向棘轮机构。棘爪安放在图 9-8c 所示位置时，推动棘轮沿逆时针方向做单向间歇转动。棘爪向上提起，转过 180°，如图 9-8d 所示位置，推动棘轮沿顺时针方向做单向间歇转动。

双向棘轮机构一般采用矩形齿或对称梯形齿。

图 9-9 所示牛头刨床工作台的横向进给机构中，即采用了图 9-8c、d 所示的双向棘轮机构以及曲柄摇杆机构和齿轮机构的组合，从而实现了工作台的双向进给运动。

Fig. 9-9　Table-feed mechanism of a shaper（牛头刨床工作台横向进给机构）

(2) 摩擦式棘轮机构　摩擦式棘轮机构是依靠棘爪和棘轮之间的摩擦力实现间歇运动的。

图 9-10a 所示的摩擦式棘轮机构中，依靠棘爪 2 和棘轮 3 之间的摩擦力，将主动摇杆 1 的往复摆动转换成棘轮 3 的单向间歇转动。它克服了齿式棘轮机构中棘爪在棘轮齿面滑行时引起的噪声大、传动平稳性差以及棘轮每次转过角度的大小不能无级调节等缺点，但其运动准确性较差。

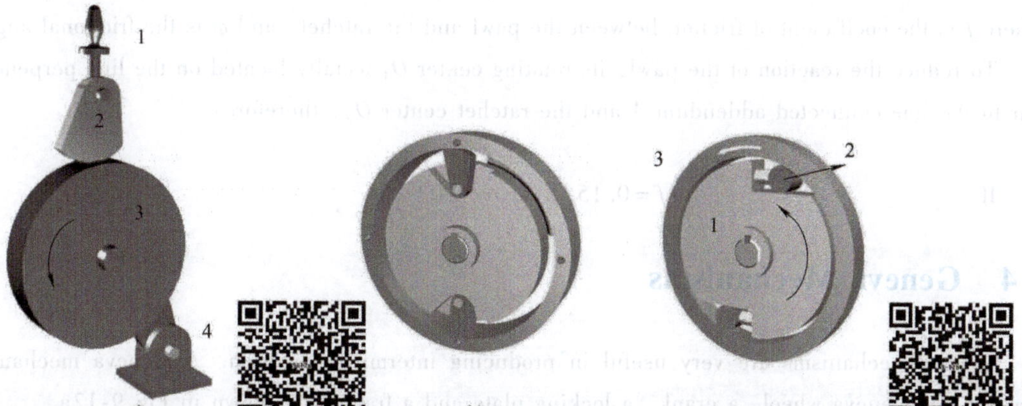


Fig. 9-10　Silent ratchet mechanisms（摩擦式棘轮机构）

图 9-10b、c 所示为内摩擦式棘轮机构，其中图 9-10c 所示机构中的棘爪 2 为滚柱形。当主动星轮 1 逆时针方向转动时，由于摩擦力的作用，滚柱 2 楔紧在主动星轮 1 和套筒 3 之间的空隙小端，从而带动套筒 3 随主动星轮 1 以相同的转速回转。当主动星轮 1 顺时针方向转动时，同样由于摩擦力的作用，滚柱 2 滚向主动星轮 1 和套筒 3 之间的空隙大端，此时套筒 3 静止不动，故该机构常用作单向离合器。此外，当主动星轮 1 逆时针方向转动时，如果

wheel and the toroid wheel are all rotating counterclockwise, and the angular velocity of toroid is larger than that of the star wheel, they may rotate respectively. The internal ratchet mechanism becomes an overrunning clutch.

**2. Design Outlines of Ratchet Mechanisms**

Take the tooth ratchet mechanism for example, the design outlines are as follows.

(1) Module and number of teeth  The dimensions of ratchet wheel are depended upon the module like gears, and the numbers of teeth of the ratchet are depended upon the minimum rotating angle $\theta_{min}$.

$$z \geqslant \frac{2\pi}{\theta_{min}}$$

(2) Oblique angle of the tooth surface and action angle of the pawl  Fig. 9-11 shows a tooth ratchet mechanism, in which the line of action $n$—$n$ of the driving pawl and tooth must pass through the line $O_1O_2$, so that the pawl remains in contact with the tooth. In this figure, the angle $\alpha$ between the tooth surface $AB$ and the radial line $O_2A$ is called the oblique angle of the tooth; the angle $\delta$ between the line connected the addendum $A$ and the pawl center $O_1$ and the line of $n$—$n$ passing through the point $A$ is called the action angle of the pawl.

Suppose $O_1A = l$, normal force acting on the pawl is $F_n$, and tangent force acting on the pawl is $F_t$. To maintain the pawl in contact with the tooth, the moment of the normal force about the pawl center must be greater than that of tangent force. Thus we have:

$$F_n l\sin\delta > F_t l\cos\delta$$

$$\tan\delta > F_t / F_n$$

$$F_t / F_n = f = \tan\varphi$$

$$\delta > \varphi$$

Where $f$ is the coefficient of friction between the pawl and the ratchet, and $\varphi$ is the frictional angle.

To reduce the reaction of the pawl, its rotating center $O_1$ usually located on the line perpendicular to the line connected addendum $A$ and the ratchet center $O_2$, therefore:

$$\alpha = \delta$$

If  $f = 0.15\text{-}0.2$, $\alpha = 10°\text{-}15°$.

## 9.4 Geneva Mechanisms

Geneva mechanisms are very useful in producing intermittent motion. A Geneva mechanism consists of a Geneva wheel, a crank, a locking plate and a frame, as shown in Fig. 9-12a.

The crank usually is a driving member and rotates continuously at a constant angular velocity. It contain a driving roller $G$ that engages in a slot of the Geneva wheel. The slots are positioned so that the roller enters and leaves them tangentially in order to reduce shock.

When the roller $G$ just enters the slot to drive the Geneva wheel, the locking plate $\overset{\frown}{nn}$ will leave the concave surface of the wheel $\overset{\frown}{mm}$, and the wheel will rotate a fractional part of revolution. When the roller leaves the slot, the convex surface of the locking plate will mate with the concave surface of the wheel, and the wheel is in stationary.

套筒 3 的逆时针方向转速超过主动星轮 1 的转速，星轮 1 和套筒 3 脱开，并以各自的速度转动，此时该机构用作超越离合器。

**2. 棘轮机构的设计要点**

以齿式棘轮机构为例，介绍棘轮机构的设计要点。

（1）模数、齿数的选择　同齿轮一样，棘轮有关尺寸也是以模数为基本参数来确定的，模数的标准值可在手册中查取。棘轮的齿数一般根据所要求的棘轮最小转角 $\theta_{min}$ 来确定，即

$$z \geqslant \frac{2\pi}{\theta_{min}}$$

（2）棘轮齿面偏斜角和棘爪轴心位置角的确定　如图 9-11 所示，棘轮工作齿面 $AB$ 与径向线 $O_2A$ 的夹角 $\alpha$ 称为齿面偏斜角，棘爪轴心 $O_1$ 与棘齿顶点 $A$ 的连线 $O_1A$ 与过点 $A$ 的齿面法线 $n$—$n$ 的夹角 $\delta$ 称为棘爪轴心位置角。

设 $\overline{O_1A}=l$，棘齿对棘爪的法向反作用力为 $F_n$，棘齿作用于棘爪的摩擦力为 $F_t$，它与棘爪滑入棘轮根部的趋势方向相反。为保证棘轮机构正常地工作，应使棘爪能顺利地滑入棘轮的齿根部并自动啮紧。法向反作用力 $F_n$ 对 $O_1$ 所产生的力矩必须大于摩擦力 $F_t$ 对 $O_1$ 所产生的力矩，即

$$F_n l\sin\delta > F_t l\cos\delta$$

$$\tan\delta > F_t/F_n$$

$$F_t/F_n = f = \tan\varphi$$

故

$$\delta > \varphi$$

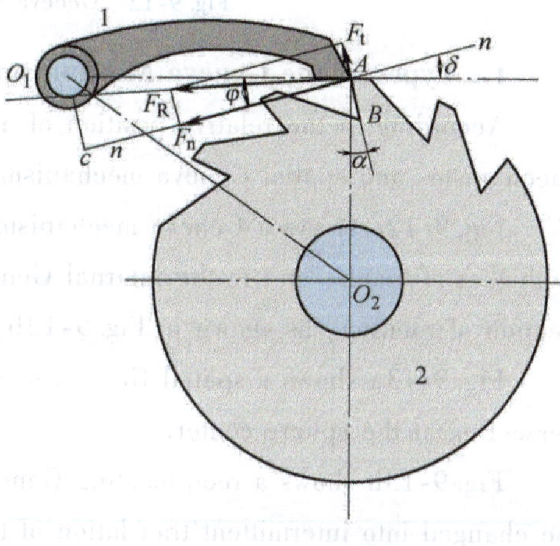


Fig. 9-11　Design of ratchet mechanism
（棘轮机构的设计）

式中，$f$ 和 $\varphi$ 分别为棘爪与棘轮齿面间的摩擦因数和摩擦角。

由以上分析可知，棘爪能顺利地滑入棘轮齿根部并自动啮紧的条件是：棘爪轴心位置角 $\delta$ 大于棘爪与棘轮齿面间的摩擦角 $\varphi$，即在接触点处棘齿对棘爪的总反力 $F_R$ 的作用线与轴心连线 $O_1O_2$ 的交点必须在 $O_1$、$O_2$ 之间。为了使棘爪受力尽可能小，通常将棘爪轴心 $O_1$ 选取在棘轮齿顶 $A$ 的径向线 $O_2A$ 的垂线上，则有

$$\alpha = \delta$$

当摩擦因数 $f=0.15\sim0.2$ 时，齿面偏斜角 $\alpha$ 通常取 $10°\sim15°$，所以棘轮常采用锐角齿形。

## 9.4　槽轮机构

槽轮机构是具有一种分度性质的间歇运动机构，由具有圆销的主动销轮 1、具有若干径向槽的从动槽轮 2 及机架组成，如图 9-12a 所示。主动销轮 1 以等角速度 $\omega_1$ 连续转动时，从动槽轮 2 便做单向间歇转动。当主动销轮 1 上的圆销 $G$ 进入槽轮 2 的径向槽时，销轮外凸的锁止 $\widehat{nn}$ 弧和槽轮内凹的锁止弧 $\widehat{mm}$ 脱开，圆销 $G$ 拨动槽轮 2 做顺时针方向的转动；当圆销 $G$ 与槽轮脱开时，槽轮因其内凹的锁止弧被销轮外凸的锁止弧锁住而静止。这样，就把销轮的连续回转运动转换为槽轮的单向间歇转动。

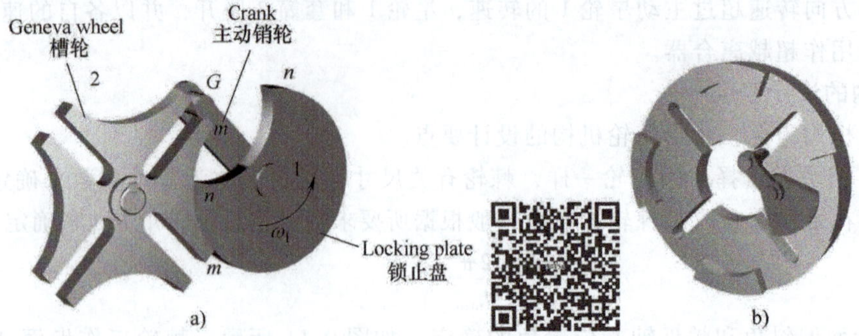


Fig. 9-12 Geneva wheel mechanisms 1 (槽轮机构 1)

## 1. Types of the Geneva Mechanisms

According to the relative position of axes, this mechanism can be classified as planar Geneva mechanisms and spatial Geneva mechanisms.

Fig. 9-12a shows a Geneva mechanism in which the rotating direction of the wheel is opposite with that of crank; and in the internal Geneva mechanism, the wheel and crank have the same direction of rotation, as shown in Fig. 9-12b.

Fig. 9-13a shows a spatial Geneva mechanism in which the axes of the roller and crank are intersecting at the sphere center.

Fig. 9-13b shows a reciprocating Geneva mechanism. The continuous rotation of the crank can be changed into intermittent translation of the Geneva rack.

The Geneva mechanism has many advantages, such as simple structures, small dimensions, easy to manufacture, steady motion and so on, but it can not be used at a high speed of crank.

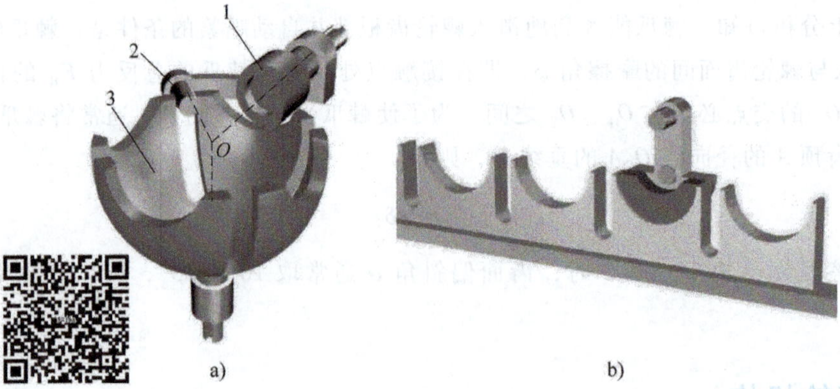


Fig. 9-13 Geneva wheel mechanisms 2 (槽轮机构 2)

1—Crank (主动销轮) 2—Roller (圆销) 3—Geneva wheel (槽轮)

Sometimes, when the crank rotates one revolution, and the intermittent period of the wheel is not the same, we can use the mechanisms shown in Fig. 9-14.

If the rollers are distributed unevenly, the times for each dwell of the Geneva will be different, as shown in Fig. 9-14a. If we want to increase the smoothness of the wheel, we can use the Geneva wheel in which the slot is made to be a curve shape, as shown in Fig. 9-14b.

**1. 槽轮机构的类型**

按照主、从动槽轮轴线的相对位置不同，可将槽轮分为平面槽轮机构和空间槽轮机构两大类。平面槽轮机构用来传递平行轴运动，它又分为两种形式：一种是外槽轮机构（图 9-12a），其主动销轮与从动槽轮转向相反；另一种是内槽轮机构（图 9-12b），其主动销轮与从动槽轮转向相同。与外槽轮机构相比，内槽轮机构传动较平稳，机构也较为紧凑。

空间槽轮机构用来传递相交轴的间歇运动。图 9-13a 所示为垂直相交轴间的球面槽轮机构，槽轮呈半球形，主动销轮 1、球面槽轮 3 以及圆销 2 的轴线都通过球心，当主动销轮 1 连续转动时，球面槽轮 3 做单向间歇转动。图 9-13b 所示为移动型槽轮机构，可实现圆弧齿条的间歇移动。

槽轮机构的特点是结构简单、工作可靠，在圆销进入啮合和退出啮合时，传动较平稳，能准确控制转动角度。但由于槽轮在起动和停止时加速度变化大，有冲击，且随槽数减少、转速增高而加剧，因此不适用于高速场合。

当要求销轮在转动一周的时间内槽轮多次停歇时间互不相等时，可以将圆销不均匀地分布在主动销轮的圆周上，如图 9-14a 所示。为提高分度过程中的平稳性，也可采用图 9-14b 所示的曲线槽轮机构。

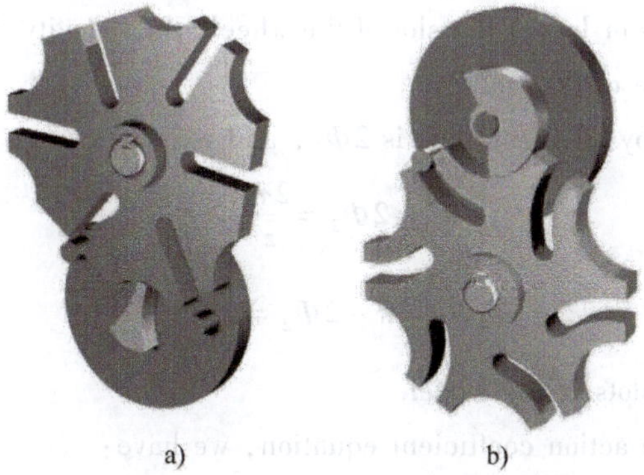

a) b)

Fig. 9-14 Geneva wheel mechanisms having special functions
（有特殊功能的槽轮机构）

**2. 槽轮机构的设计要点**

槽轮机构设计的内容主要包括确定槽轮的槽数和柱销的数目以及计算槽轮机构的基本尺寸。

（1）槽轮机构的运动系数 如图 9-15 所示，从动槽轮 2 做周期性的间歇运动。设从动槽轮 2 每运动一次所需要的时间为 $t_2$，主动销轮 1 转动一周的时间为 $t_1$。两者之比称为槽轮机构的运动系数，用$\tau$表示，即

$$\tau = \frac{t_2}{t_1}$$

时间 $t_2$、$t_1$ 对应的主动销轮 1 的转角为 $2\Phi_1$ 与 $2\pi$，故运动系数可表示为

$$\tau = \frac{t_2}{t_1} = \frac{2\Phi_1}{2\pi}$$

**2. Design Outlines of Geneva Mechanisms**

The design of a Geneva mechanism is initiated by specifying the crank radius, the roller diameter, the number of slots and other dimensions.

(1) Action coefficient of the wheel  In Fig. 9-15, when the crank rotates one revolution with a time $t_1$, the time with which the wheel have a intermittent motion is $t_2$, the ratio between the time $t_2$ and $t_1$ is defined as the action ratio, denoted as $\tau$.

$$\tau = \frac{t_2}{t_1}$$

Suppose the times $t_2$ and $t_1$ correspond to angles $2\Phi_1$ and $2\pi$ of the driving crank, so we have:

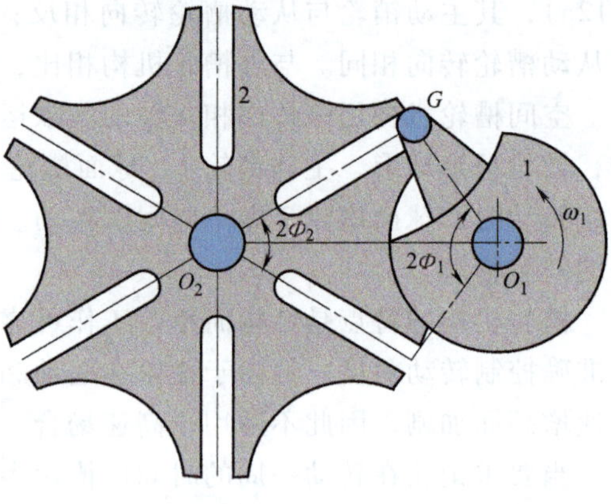


Fig. 9-15 Design of Geneva wheel (槽轮机构设计)

$$\tau = \frac{t_2}{t_1} = \frac{2\Phi_1}{2\pi}$$

When the roller enters or leaves the slot of the wheel, the velocity direction of the point $G$ must pass through the center line of the slot.

The angle subtended by adjacent slots is $2\Phi_2$, and we have.

$$2\Phi_2 = \frac{2\pi}{z}$$

$$2\Phi_1 = \pi - 2\Phi_2 = \pi - \frac{2\pi}{z}$$

Where $z$ is the number of slots of the wheel.

Substituting it into the action coefficient equation, we have:

$$\tau = \frac{t_2}{t_1} = \frac{\pi - \dfrac{2\pi}{z}}{2\pi} = \frac{z-2}{2z}$$

(2) The number of teeth of the wheel  Since the ratio $\tau$ is always greater than zero, the number of teeth must be greater than three.

(3) Dimensions  The dimension of the Geneva mechanism can be consulted from some handbook.

Geneva mechanisms are widely used in indexing mechanisms. Fig. 9-16 shows a Geneva mechanism applied to a beehive-coal machine.

## 9.5 Indexing Cam Mechanisms

An indexing cam mechanism consists of a driving cam, a rotating disk and a frame; it is a spatial indexing mechanism.

为了避免槽轮在开始转动和停止转动时产生碰撞和冲击，圆销 $G$ 在开始进入或退出径向槽时，圆销中心的线速度方向应沿着径向槽的中心线方向。因此，当槽轮 2 每转过 $2\Phi_2 = 2\pi/z$ 角度时，销轮 1 所对应的转角为

$$2\Phi_1 = \pi - 2\Phi_2 = \pi - \frac{2\pi}{z}$$

式中，$z$ 为槽轮的槽数。

将以上关系式代入运动系数$\tau$的定义表达式，可以得到

$$\tau = \frac{t_2}{t_1} = \frac{\pi - \frac{2\pi}{z}}{2\pi} = \frac{z-2}{2z}$$

（2）槽轮机构的槽数　为保证销轮能驱动槽轮，运动系数$\tau$应大于零，槽轮径向槽的数目应大于或等于 3，并且$\tau$总小于 0.5。说明在此种槽轮机构中，槽轮的运动时间总小于其停歇时间。

（3）尺寸设计　槽轮机构的尺寸设计可参阅相关设计手册。图 9-16 所示为槽轮机构在蜂窝煤机中的应用实例。

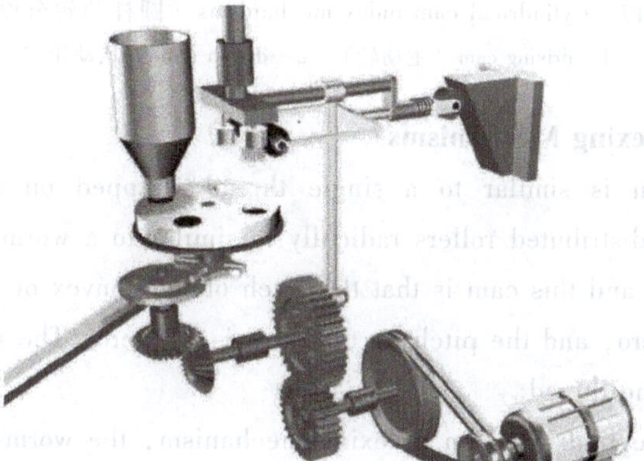

Fig. 9-16　Application of Geneva mechanism（槽轮机构的应用）

## 9.5　凸轮式间歇运动机构

凸轮式间歇运动机构一般由主动凸轮、从动转盘和机架组成。凸轮式间歇运动机构一般有以下两种形式。

**1. 圆柱凸轮间歇运动机构**

如图 9-17 所示，圆柱凸轮间歇运动机构的主动轮 1 是带有曲线沟槽或凸脊的圆柱凸轮，从动轮 2 为带有柱销的圆盘。图 9-17a 所示是带有曲线沟槽的圆柱凸轮，图 9-17b 所示是带有凸脊的圆柱凸轮。当圆柱凸轮转动时，通过其曲线沟槽或凸脊拨动柱销，使从动圆盘做间歇运动。这种机构多用于两交错轴间的分度运动。

**1. Cylindrical Cam Indexing Mechanisms**

Fig. 9-17a shows a cylindrical cam indexing mechanism in which the cam is a cylinder with a groove curve, and the rollers on the turning plate mate with the groove. Fig. 9-17b shows a cylindrical cam indexing mechanism in which the cam is a cylinder with a ridge curve, and the rollers on the turning plate mate with the spaces of the ridge.

When the cylindrical cam rotates about its axis, the plate will perform an intermittent motion.

In order to ensure accurate indexing, the width of the groove or ridge of the cam must be designed such that both faces are always in contact with the rollers on the driven plate.

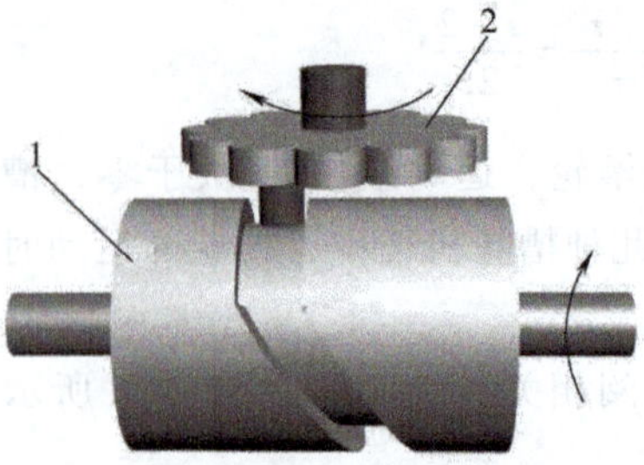


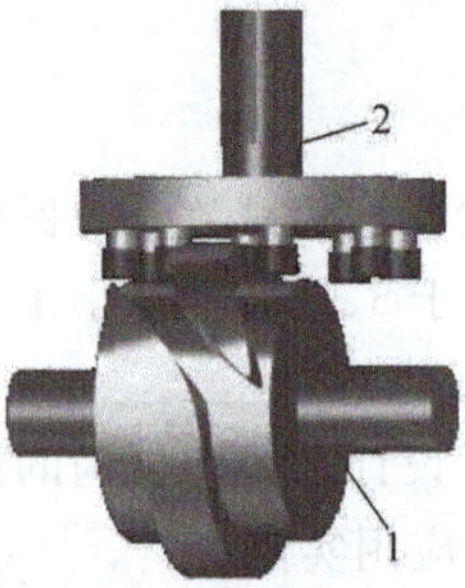


Fig. 9-17 Cylindrical cam index mechanisms (圆柱凸轮分度机构)

1—driving cam (主动轮) 2—driven plate (从动轮)

**2. Worm-cam Indexing Mechanisms**

This cam mechanism is similar to a single thread wrapped on a cylinder, as shown in Fig. 9-18, and the plate distributed rollers radically is similar to a worm gear. The difference between the cylindrical cam and this cam is that the pitch of the convex or concave curve surfaces on the cylindrical cam are zero, and the pitch on this cam is not zero. The rotation of the driven plate depends on the pitch of the thread.

Comparing with the cylindrical cam indexing mechanism, the worm-cam indexing mechanism has higher indexing accuracy and better dynamical characteristics.

## 9.6 Intermittent Gear Mechanisms

This mechanism is an intermittent device in which the driving gear has several teeth, and the driven gear will have a number of tooth spaces to produce the required indexing angle. Fig. 9-19a shows an intermittent gear mechanisms in which the driving gear carries three teeth and the driven gear carries six locking arcs and six parts of teeth, and each part has three tooth spaces.

A locking device must be employed to prevent the driven gear from rotating when not indexing.

The convex surface of the driving wheel mates with the concave surface between the tooth spaces on the driven gear.

When the driving gear rotates one revolution, the driven gear has six dwells, and has one-sixth of revolution.

In Fig. 9-19b, the driven gear has four locking arcs. When the driving gear rotates one revolu-

**2. 蜗杆形凸轮间歇运动机构**

如图 9-18 所示，蜗杆形凸轮间歇运动机构的主动轮 1 为蜗杆形的凸轮，其上有一条凸脊，就像一个变螺旋角的蜗杆，从动轮 2 为带有径向均布柱销的圆盘，相当于蜗轮。当蜗杆凸轮转动时，通过其上的凸脊推动转盘上的柱销，从而使从动圆盘做间歇运动。同圆柱凸轮间歇运动机构类似，蜗杆形凸轮间歇运动机构也多用于两交错轴间的分度运动。这种机构具有良好的动力学性能，适用于高速精密传动，但加工制造较为困难。

圆柱形凸轮间歇运动机构与蜗杆形凸轮间歇运动机构的差别是：圆柱形凸轮曲线的节距为零，蜗杆形凸轮曲线的节距不为零。

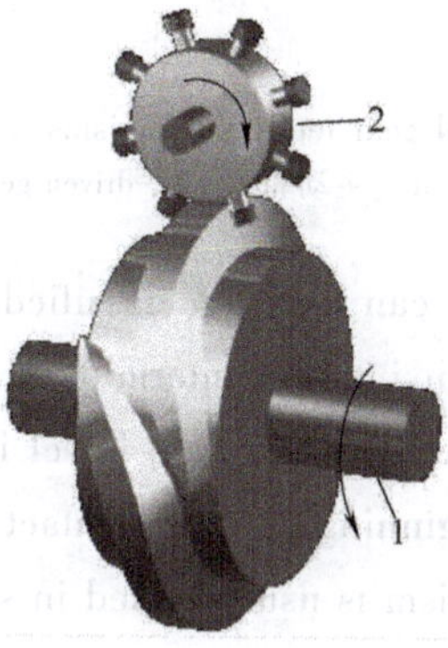


Fig. 9-18　Worm-cam index mechanism（蜗杆式凸轮分度机构）
1—driving cam（主动轮）　2—driven plate（从动轮）

## 9.6　不完全齿轮机构

不完全齿轮机构也是一种间歇运动机构。图 9-19 为外啮合不完全齿轮机构示意图。其与一般齿轮机构的主要区别是：不完全齿轮机构主动轮上的轮齿不是布满在整个圆周上，主动轮 1 上只有一个或几个齿，其余部分为锁止弧。根据运动时间与停歇时间的要求，在从动轮 2 上加工有与主动轮相啮合的轮齿。

在不完全齿轮机构中，主动轮 1 连续转动，当轮齿进入啮合时，从动轮 2 开始转动；当轮齿退出啮合时，由于主动轮 1 和从动轮 2 上锁止弧的密合定位作用，使得从动轮 2 处于停歇位置，从而实现了从动轮 2 的间歇转动。

在图 9-19a 所示的不完全齿轮机构中，从动轮上分布有 6 段锁止弧和 6 段轮齿，每段轮齿有 3 个齿槽与主动轮上 3 个轮齿相啮合。主动轮每转一周，从动轮只转 1/6 周。

在图 9-19b 所示的不完全齿轮机构中，从动轮上分布有 4 段锁止弧和 4 段轮齿，每段轮齿有 4 个齿槽与主动轮上 4 个轮齿相啮合。主动轮每转一周，从动轮只转 1/4 周。

不完全齿轮机构有外啮合、内啮合及齿轮齿条三种形式，图 9-20a 所示为内啮合不完全齿轮机构，图 9-20b 所示为齿条型不完全齿轮机构，图 9-20c 所示为锥齿轮组成的不完全齿轮机构。

与其他间歇运动机构相比，不完全齿轮机构具有结构简单、设计灵活等特点。主动轮每转一周，从动轮运动角的幅度、停歇的次数及每次停歇的时间，都比棘轮机构和槽轮机构有更宽的选择范围。但不完全齿轮机构中的从动轮在转动开始和终止时，角速度有突变，引

tion, the driven gear has four dwells, and has one-fourth of revolution.

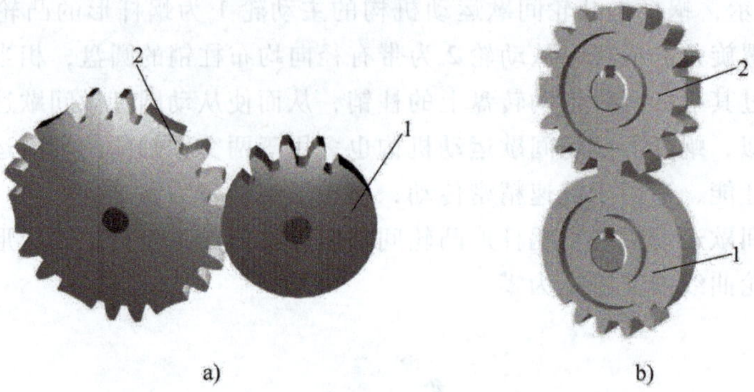


Fig. 9-19 Partial external gear teeth mechanisms (外啮合不完全齿轮机构)
1—driving gear (主动轮) 2—driven gear (从动轮)

The intermittent gear mechanism can also be classified as external gears, internal gears and rack mechanisms. Fig. 9-20a shows an internal intermittent gear mechanism, Fig. 9-20b shows an intermittent rack mechanism, and Fig. 9-20c shows a bevel intermittent gear mechanism.

The gear ratio is varied at the beginning point of contact and the end point of contact, so shock and vibration will occur. This mechanism is usually used in some indexing mechanisms which run at a lower speed.

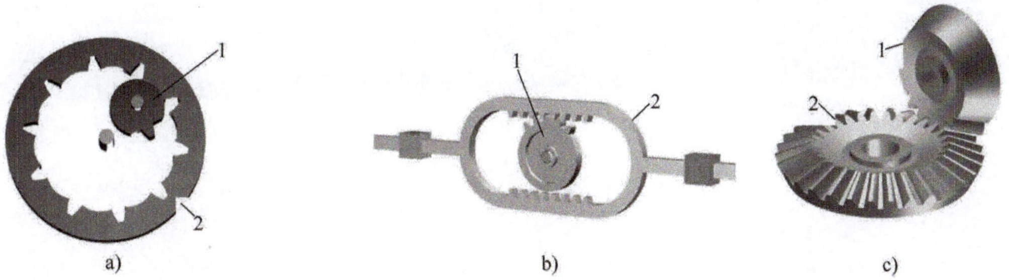


Fig. 9-20　The other intermittent gear mechanisms（其他不完全齿轮机构）
1—driving gear（主动轮）　2—driven gear（从动轮）

起刚性冲击，因此一般只适用于低速轻载的工作场合。

不完全齿轮机构与一般的齿轮机构的区别不仅在于轮齿的分布不同，而且在啮合过程中，当首齿进入啮合及末齿退出啮合时，其轮齿不在基圆的内公切线上接触传动，所以在此期间不能保持定传动比传动。

# Chapter 10

# Spatial Mechanisms and Robotic Mechanisms

# 第10章

# 空间连杆机构及机器人机构概述

## 10.1 Introduction of Spatial Mechanisms

Spatial mechanisms have simple and compact structures, which are used widely in special applications. Besides, spatial mechanisms are also the fundamentals of robotic mechanisms.

**1. Spatial Kinematical Pairs**

In three-space, each unconnected link has six degrees of freedom. A kinematic pair can offer a maximum constraint of 5 and a minimum constraint of 1. Therefore, kinematic pairs can be classified by the number of constraints offered by the pairs. For examples, a class Ⅰ kinematic pair offers one constraint, a class Ⅱ kinematic pair offers two constraints, a class Ⅲ kinematic pair offers three constraints, a class Ⅳ kinematic pair offers four constraints, and a class Ⅴ kinematic pair offers five constraints.

(1) Class Ⅰ kinematic pairs  A sphere is placed on an even plane shown in Fig. 10-1a. They form a sphere-plane pair, which has only one constraint along the normal $n$—$n$ and five degrees of freedom. This is a class Ⅰ pair, denoted as SE, and is rarely used in engineering.

(2) Class Ⅱ kinematic pairs  Each class Ⅱ kinematic pair offers two constraints and has four degrees of freedom. Fig. 10-1b shows a cylinder-plane pair which constrains the translational motion along the $z$-axis and the rotational motion about the $x$-axis. The cylinder-plane pair is denoted as CE. Fig. 10-1c shows a sphere-groove pair which constrains the translational motions along the $x$- and $z$-axes. The sphere-groove pair is dented as SG. Class Ⅱ kinematic pairs are rarely used in engineering, too.

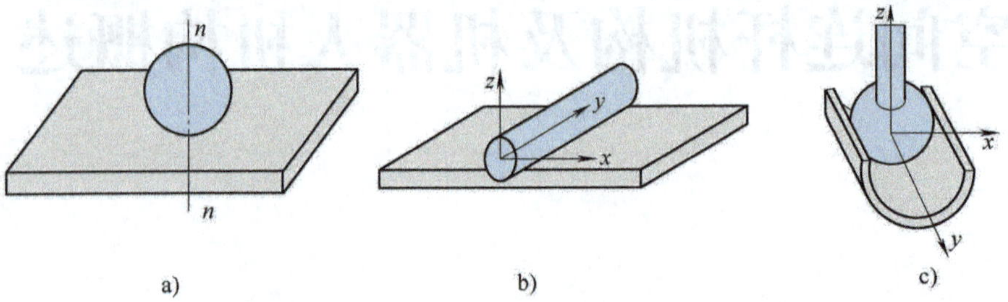


Fig. 10-1  Spatial kinematic pairs 1(空间运动副 1)

(3) Class Ⅲ kinematic pairs  Each class Ⅲ kinematic pair offers three constraints and three degrees of freedom. Fig. 10-2a shows a spherical pair which constrains translational motions along the $x$-, $y$-and $z$-axes while offers three rotational motions about these axes, denoted as S. Fig. 10-2b shows the sketch diagram of the spherical pair. This pair is widely used in spatial mechanisms.

The pair shown in Fig. 10-2c is formed by two contacting planes and it offers three constraints which are one motion along $z$-axis and rotational motions about the $x$-and $y$-axes. This pair is rarely used in engineering, too.

(4) Class Ⅳ kinematic pairs  Every class Ⅳ kinematic pair offers four constraints and two degrees of freedom. Fig. 10-3a shows a sphere-pin pair which has only two rotational degrees of freedom because of the constraint of the pin. The sphere-pin pair is denoted as S′. Fig. 10-3b is its sketch diagram. Fig. 10-3c shows a cylinder pair which permits the translational motion along its axes and the rotational motion about the same axis. Fig. 10-4d shows its sketch diagram and its symbol is denoted as C. Class Ⅳ kinematic pairs are widely used in spatial mechanisms.

## 10.1 空间连杆机构概述

空间连杆机构结构简单、紧凑，不但有独特的应用，而且是机器人机构的设计基础。

**1. 空间连杆机构中的运动副**

在三维空间中，每个构件有 6 个自由度。连接两构件的运动副最多提供 5 个约束，最少提供 1 个约束。因此可按运动副提供的约束对运动副进行分类。提供 1 个约束的运动副称为Ⅰ类副，提供 2 个约束的运动副称为Ⅱ类副，提供 3 个约束的运动副称为Ⅲ类副，提供 4 个约束的运动副称为Ⅳ类副，提供 5 个约束的运动副称为Ⅴ类副。

（1）Ⅰ类副　图 10-1a 所示的球放在平面上，形成点接触的高副，仅提供沿两者公法线 $n—n$ 方向的一个约束。Ⅰ类副用 SE 表示，工程中很少应用。

（2）Ⅱ类副　具有 2 个约束、4 个自由度。图 10-1b 所示的圆柱平面副中，提供沿 $z$ 轴移动和绕 $x$ 轴转动的 2 个约束，用 CE 表示圆柱平面副。图 10-1c 所示的球槽副中，提供沿 $z$ 轴移动和沿 $x$ 轴移动 2 个约束，用 SG 表示球槽副，它们是典型的Ⅱ类副。Ⅱ类副也很少应用。

（3）Ⅲ类副　具有 3 个约束和 3 个自由度的运动副，图 10-2a 中的球置于球面槽中，形成典型的球面副，用 S 表示球面副。球面副限制了沿 $x$ 轴、$y$ 轴、$z$ 轴的移动，保留绕 3 个轴转动的自由度。球面副在空间机构中应用广泛。图 10-2b 所示为球面副的代表符号。图 10-2c 所示为两平面接触形成的平面副，用 E 表示，提供了沿 $z$ 轴移动及绕 $x$ 轴、$y$ 轴转动的 3 个约束，应用很少。

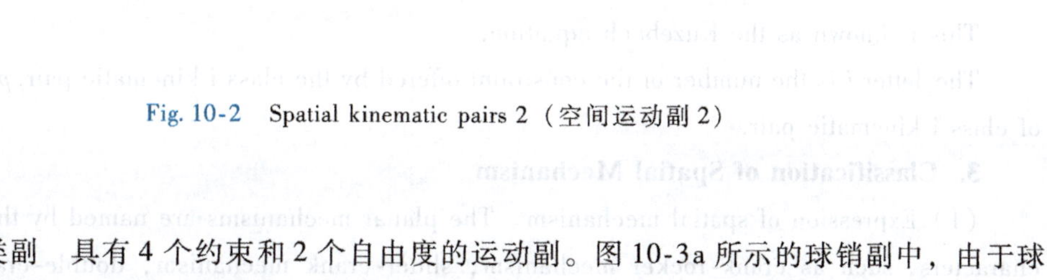


Fig. 10-2　Spatial kinematic pairs 2（空间运动副 2）

（4）Ⅳ类副　具有 4 个约束和 2 个自由度的运动副。图 10-3a 所示的球销副中，由于球销的约束，仅保留 2 个转动自由度。运动副符号如图 10-3b 所示，名称用 S′表示。图 10-3c 所示的圆柱副中，仅保留沿轴线的移动和绕轴线的转动自由度，运动副符号如图 10-3d 所示，名称用 C 表示。Ⅳ类运动副在空间机构中应用较广泛。

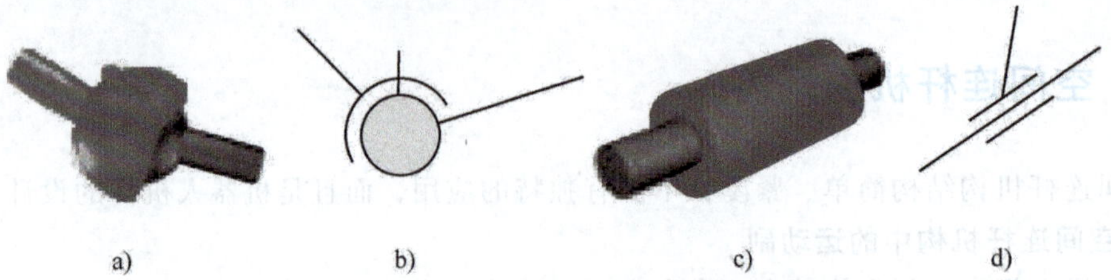


Fig. 10-3 Spatial kinematic pairs 3(空间运动副 3)

(5) Class V kinematic pairs A class V kinematic pair offers five constraints and one degree of freedom. Fig. 10-4a shows a revolute pair which has only one rotational degree of freedom. Fig. 10-4b is its sketch, denoted as R.

Fig. 10-4c shows a prismatic pair which has only one degree of freedom that permits it to execute translational motion along the axes, denoted as P. Fig. 10-4d shows its sketch.

Fig. 10-4e shows a helical pair which permits the translational motion along its axis and the rotational motion about the same axis. However, these two kinds of motions are correlated, so a helical pair has only one degree of freedom. Fig. 10-4d is its sketch diagram, denoted as H.

Class V kinematic pairs are very important to design planar mechanisms and spatial mechanisms.

**2. Degrees of Freedom of Spatial Mechanisms**

In the three-dimensional space, each free link will have six degrees of freedom. Therefore, a cluster of such links of $n$ would have degrees of freedom of $6n$. But after these links are connected by kinematic pairs, the motions of these links will be constrained by these pairs. The difference between the total number of the degree of freedom of the free links and the sum of the total number of the constraints offered by all pairs is the degree of freedom of the mechanism.

We assume that there are class I kinematic pairs of $p_1$, so they can offer constraints of $1p_1$. Similarly, class II kinematic pairs of $p_2$ can offer constraints of $2p_2$, class III kinematic pairs of $p_3$ can offer constraints of $3p_3$, class IV kinematic pairs of $p_4$ can offer constraints of $4p_4$, and class V kinematic pairs of $p_5$ can offer constraints of $5p_5$. Therefore, the degrees of freedom of spatial mechanisms can be written as follows:

$$F = 6n - (p_1 + 2p_2 + 3p_3 + 4p_4 + 5p_5) = 6n - \sum ip_i$$

This is known as the Kuzebach equation.

The letter $i$ is the number of the constraint offered by the class i kinematic pair. $p_i$ is the number of class i kinematic pair.

**3. Classification of Spatial Mechanism**

(1) Expression of spatial mechanism The planar mechanisms are named by their kinematical characters, such as crank-rocker mechanism, slider-crank mechanism, double-crank mechanism and so on. Generally, we name the spatial mechanisms by the names of their kinematic pairs. The first alphabet is the symbol of the kinematic pair which connects the driving link and frame, then we can arrange the symbols of other kinematic pairs orderly. Fig. 10-5 shows an aircraft undercarriage which can be named as the spatial mechanism of SPSR.

(2) Classification of spatial mechanisms The spatial mechanisms are classified as closed

（5）Ⅴ类副　具有 5 个约束、1 个自由度的运动副。图 10-4a 所示的转动副中，仅有一个绕轴线的转动自由度，运动副代表符号如图 10-4b 所示，名称用 R 表示。图 10-4c 所示移动副中，仅有 1 个沿导路方向的移动自由度，运动副代表符号如图 10-4d 所示，名称用 P 表示。图 10-4e 所示的螺旋副中，沿轴线的移动和绕轴线的转动线性相关，所以只有 1 个移动自由度，代表符号如图 10-4f 所示，名称用 H 表示。Ⅴ类运动副在平面机构结构分析中已经有详细介绍，也是应用非常广泛的运动副。

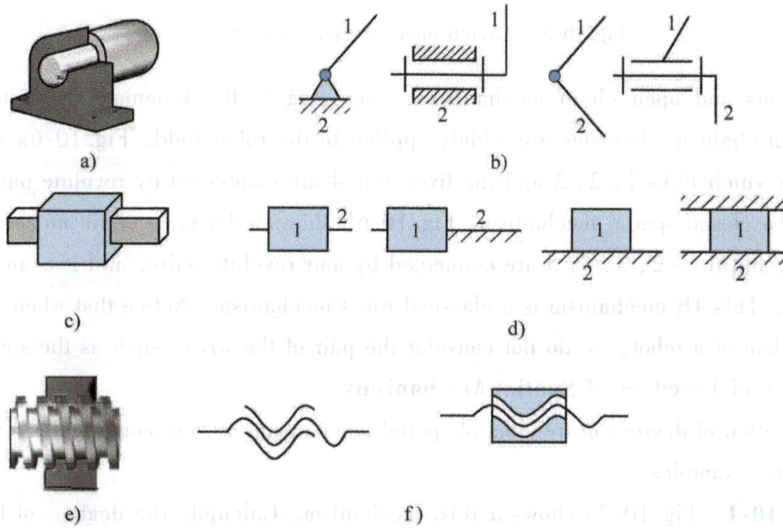


Fig. 10-4　Spacial kinematic pairs 4（空间运动副 4）

**2. 空间连杆机构的自由度**

三维空间中的每个自由构件有 6 个自由度，$n$ 个构件则有 $6n$ 个自由度。这些构件用运动副连接组成机构后，构件的运动就会受到运动副的约束。$n$ 个构件的自由度总数 $6n$ 减去各运动副的约束总数，就是机构的自由度数。

设机构中的Ⅰ类副数目为 $p_1$，则其提供的约束为 $1p_1$；Ⅱ类副数目为 $p_2$ 个，则其提供的约束为 $2p_2$ 个；Ⅲ类副的数目为 $p_3$，则其提供的约束为 $3p_3$；Ⅳ类副的数目为 $p_4$，则其提供的约束为 $4p_4$；Ⅴ类副的数目为 $p_5$，则其提供的约束为 $5p_5$。则机构自由度为

$$F = 6n - (p_1 + 2p_2 + 3p_3 + 4p_4 + 5p_5) = 6n - \sum ip_i$$

式中，$i$ 表示第 $i$ 类运动副的约束数目；$p_i$ 为 i 类运动副的个数。

上式表明，空间机构的自由度等于各活动构件自由度之和减去各运动副约束之和。

**3. 空间连杆机构分类**

（1）空间连杆机构表示方法　平面连杆机构名称是按其运动特性确定的。如曲柄摇杆机构、曲柄滑块机构、双曲柄机构等。空间机构的名称则用运动副名称表示。第一个字母一般是原动件与机架连接的运动副的名称，然后按顺序依次排列。图 10-5 所示飞机起落架机构可称为 SPSR 空间连杆机构。

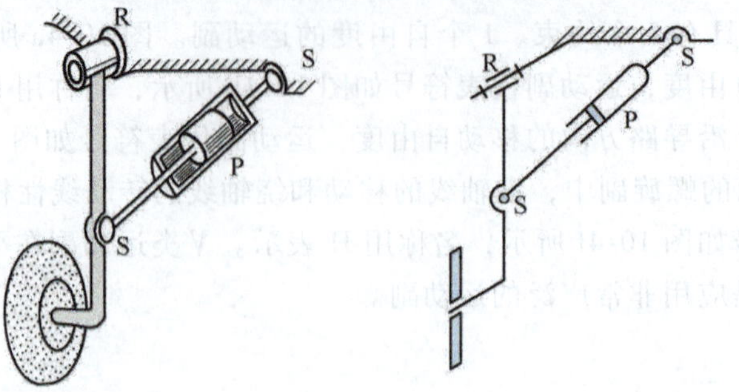


Fig. 10-5 Aircraft undercarriage(飞机起落架)

chain mechanisms and open chain mechanisms, according to the kinematical chain is closed or open. The open chain mechanisms are widely applied to the robot field. Fig. 10-6a shows a RSSR mechanism, in which links 1, 2, 3 and the fixed link 4 are connected by revolute pairs and spherical pairs. It is a closed spatial mechanism. Fig. 10-6b shows a 4R open chain mechanism in which the fixed link 1 and links 2, 3, 4, 5 are connected by four revolute pairs, and it is an unclosed spatial mechanism. This 4R mechanism is a classical robot mechanism. Notice that when calculating the degrees of freedom of a robot, we do not consider the pair of the wrist, such as the spherical pair S.

**4. Degrees of Freedom of Spatial Mechanisms**

The calculation of degrees of freedom of spatial mechanisms is very complicated. Here, we only give some simple examples.

**Example 10-1** Fig. 10-7a shows a R3C mechanism. Calculate the degrees of freedom of the mechanism.

**Solution**

$$F = 6n - \sum ip_i$$

There are three moving links, one revolute pair and three cylindrical pairs in this mechanism, so we have:

$$n = 3, p_5 = 1, p_4 = 3$$

No other type of pairs, therefore:

$$F = 6 \times 3 - (5 \times 1 + 4 \times 3) = 1$$

This mechanism has only one degree of freedom; it is a rotation about the axis of the revolute pair.

**Example 10-2** Fig.10-7b shows a RSSR mechanism. Calculate the degrees of freedom of this mechanism.

**Solution**

$$F = 6n - \sum ip_i$$

The spatial linkage consists of three moving links, two turning pairs and two spherical pairs, so $n = 3$, $p_5 = 2$, $p_3 = 2$. No other type of pairs, we have:

$$F = 6 \times 3 - (5 \times 2 + 3 \times 2) = 2$$

The result is correct from the viewpoint of the formula but misleading. One degree of freedom is the idle rotation of the link 2 about its axis. It is called the redundant degree of freedom and has no effect to the motion of the linkage. So it should be subtracted from the result of the equation, then

（2）空间连杆机构分类　按组成空间连杆机构的运动链是否封闭，空间连杆机构分为闭链空间连杆机构和开链空间连杆机构。开链机构在机器人领域应用广泛。图 10-6a 所示 RSSR 机构中，构件 1 ~4 通过转动副和球面副连接，形成一个封闭运动链，构件 4 为机架。图 10-6b 所示机构中，构件 1 ~5 通过转动副连接，形成一个不封闭的运动链，构件 1 为机架，则组成 4R 型空间开链机构。该机构是典型的机器人机构。在研究开链机器人机构时，往往不计末端执行器处铰链和抓取手指部分。

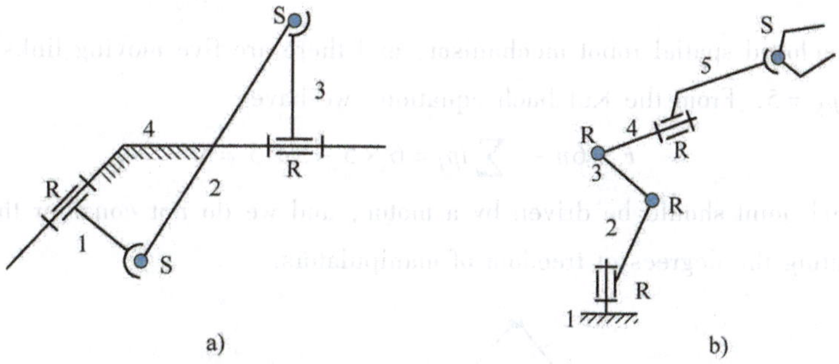


Fig. 10-6　Classification of spatial mechanisms（空间机构分类）

#### 4. 空间连杆机构自由度的计算

空间连杆机构自由度的计算比较复杂，这里仅对一些简单的空间连杆机构进行自由度计算。

**例 10-1**　计算图 10-7a 所示 R3C 机构的自由度。

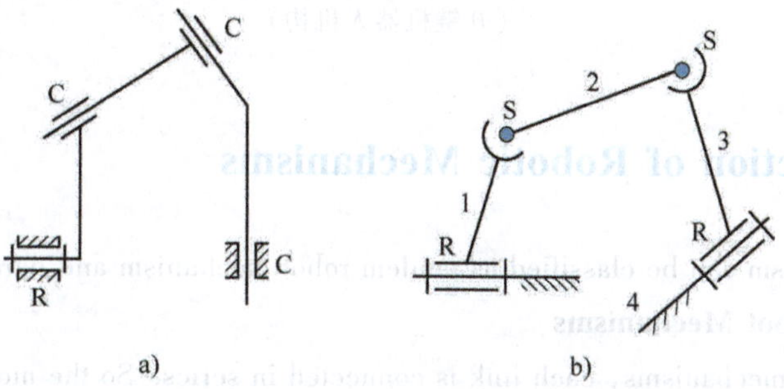


Fig. 10-7　D. o. F of spatial mechanisms（空间机构的自由度）

**解**

$$F=6n-\sum ip_i$$

$n=3$，转动副（Ⅴ类副）1 个，圆柱副（Ⅳ类副）3 个，故

$$F=6\times3-(5\times1+4\times3)=1$$

该机构仅有一个自由度，即绕 R 副轴线的转动自由度。

**例 10-2**　计算图 10-7b 所示 RSSR 机构的自由度。

**解**　空间连杆机构自由度计算公式为

$$F=6n-\sum ip_i$$

$n=3$，R 副 2 个，即Ⅴ类副 2 个。球面副 2 个，即Ⅲ类副 2 个。则

$$F=6\times3-(5\times2+3\times2)=2$$

实际上该机构只有 1 个自由度，出现自由度为 2 的情况是因为构件 2 绕自身轴线转动的

$$F = 6 \times 3 - (5 \times 2 + 3 \times 2) - 1 = 1$$

There are a great many examples of redundant degree of freedom in spatial linkage, so it is advisable that we must consider this condition carefully when calculating the degrees of freedom. In Fig. 10-5, there is a redundant degree of freedom in the undercarriage mechanism.

**Example 10-3** Fig. 10-8 shows an open link robot mechanism. Calculate the degrees of freedom of this mechanism.

**Solution**

This is an unclosed spatial robot mechanism, and there are five moving links and five revolute pairs, so $n = 5$, $p_5 = 5$. From the Kutzbach equation, we have:

$$F = 6n - \sum ip_i = 6 \times 5 - 5 \times 5 = 5$$

Actually, each joint should be driven by a motor, and we do not consider the wrist joint as a pair when calculating the degrees of freedom of manipulators.

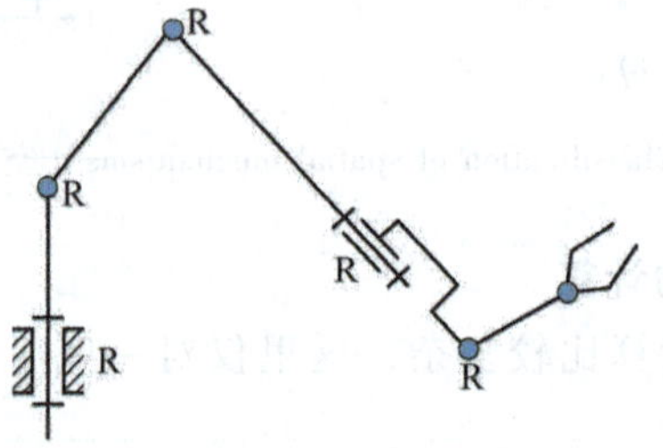


Fig. 10-8 Open link robot mechanism
(开链机器人机构)

## 10.2 Introduction of Robotic Mechanisms

Robotic mechanism can be classified as tandem robot mechanism and parallel robot mechanism.

**1. Tandem Robot Mechanisms**

In tandem robot mechanisms, each link is connected in series. So the motion is transferred from one link to the next link. Tandem robots usually are unclosed mechanisms, and they can operate in plane motion, or spatial motion.

Fig. 10-9a shows a 3R robot, which consists of three revolute pairs and three links. A tandem robot usually has five parts; they are base, waist, shoulder, elbow and wrist. Accordingly, there are four kinematic pairs which contain waist, shoulder, elbow and wrist joint. Fig. 10-9b is the sketch diagram.

The tandem robots have some advantages such as large workspace and simple structure, while they also have some disadvantages such as low rigidity and accumulated error.

Generally, the wrist of tandem robot is considered as the end effector.

The analysis which contains the position, velocity, acceleration of the end effector is the task of the kinematics of robots, while the dynamics of robot is mainly about the force, rigidity, flexibility and balance.

自由度对机构运动没有影响，应视为局部自由度除去。此时

$$F=6\times3-(5\times2+3\times2)-1=1$$

空间连杆机构出现局部自由度的情况很多，在计算其自由度时应加以注意。图 10-5 所示的飞机起落架机构中，就存在一个局部自由度。

**例 10-3**　计算图 10-8 所示开链机器人机构的自由度。

**解**　该机构中，活动构件数目 $n=5$，转动副共 5 个，即Ⅴ类副 5 个，则机构自由度为

$$F=6n-\sum ip_i=6\times5-5\times5=5$$

说明该机器人的 5 个关节处都需要安装驱动电动机。

值得注意的是，在计算机械手或开链机器人机构的自由度时，腕部铰链（末端执行器处铰链）及手指部分不计入构件数和运动副数。

计算空间连杆机构自由度时，还有许多注意事项，可参阅相关著作。

## 10.2　机器人机构概述

常用机器人可分为串联机器人和并联机器人两大类。

### 1. 串联机器人机构

串联机器人中，各构件都是串联的，即后一个构件的运动是由前一个构件传递而来的。串联机器人大都是开链机构，图 10-9a 所示机器人是由 3 个转动副、3 个构件组成的串联机器人，也简称 3R 串联机器人。串联机器人机构可以是平面开链机构，也可以是空间开链机构。串联机器人一般由底座、腰部、大臂、小臂和腕部组成，分别对应腰关节、肩关节、肘关节和腕关节。图 10-9b 为其机构简图。从串联机器人的结构特点可以看出，其刚度较小，累积运动误差较大，但运动空间很大。

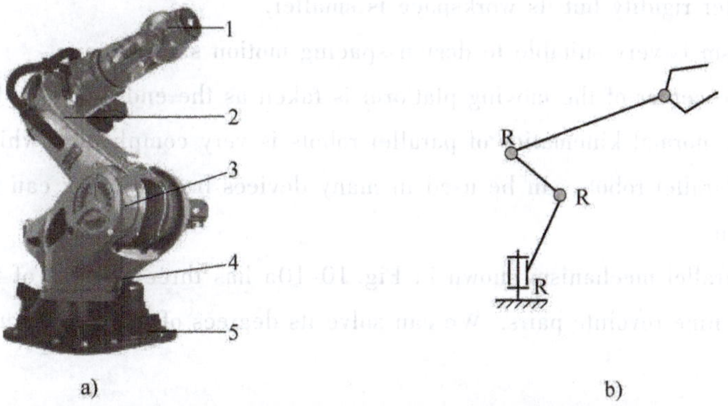


Fig. 10-9 Tandem robot mechanism（串联机器人机构）

1—Wrist joint（腕关节）　2—Elbow joint（肘关节）　3—Shoulder joint（肩关节）　4—Waist joint（腰关节）　5—Base（底座）

一般情况下常把串联机器人腕部作为机器人的末端操作器，研究末端操作器的位姿变化、速度与加速度变化是机器人运动学的主要任务，而研究构件的受力、弹性变形、刚度等则是动力学的主要任务。

在研究机器人末端操作器的位姿时，常遇到正解与逆解的问题。机器人的运动学正解是指已知某一时刻机器人各主动件的运动位置，求解机构末端操作器位姿的过程。机器人的逆解是指已知某一时刻机器人末端操作器的位姿，反求机器人各输入构件位置的过程。虽然机

We often deal with the normal solution and inverse solution of a robot when designing a robot. The normal kinematics is that parameters of the joints of a robot are given, and then we should solve the positions, velocities and accelerations of the end effector.

The inverse kinematics is that parameters of kinematics of end effector are given, and then we should solve the input angles of joints.

Although the normal kinematics and inverse kinematics of a robot are both complicated transformations of matrixes, normal kinematics is easier than inverse kinematics.

We can use the Kutzbach mobility equation to calculate the degree of freedom of tandem robots, for example, calculate the degrees of freedom of the tandem robot shown in Fig. 10-9b.

Kutzbach mobility equation is as follows:

$$F = 6n - \sum ip_i = 6 \times 3 - 5 \times 3 = 3$$

This means that the robot needs three driving joints to provide three rotational degrees of freedom.

**2. Parallel Robot Mechanisms**

In parallel robot mechanisms, two or more driving links drive an output link simultaneously, and usually the output link is the end effector.

Parallel robot mechanism can be classified as planar parallel robot mechanism and spatial parallel robot mechanism. Fig. 10-10a shows a planar parallel mechanism in which three input links drive the moving platform 1 to perform desired motion. This is called 3RRR planar parallel mechanism, and it has a wide use in micro-mechanisms.

Fig. 10-10b shows a spatial parallel robot mechanism. It is called 3RPS parallel robot mechanism. There is one degree of freedom in every branch. Compared with the tandem robot, this parallel robot has a better rigidity but its workspace is smaller.

This mechanism is very suitable to design spacing motion simulators.

Generally, the center of the moving platform is taken as the end effector. It is contrary to tandem robots that the normal kinematics of parallel robots is very complicated while the inverse kinematics is easier. Parallel robots can be used in many devices because they can realize very complicated spatial motion.

The planar parallel mechanism shown in Fig. 10-10a has three degrees of freedom. It consists of seven links and nine revolute pairs. We can solve its degrees of freedom by using Gruebler equation.

$$F = 3n - 2p_l - p_h = 3 \times 7 - 2 \times 9 = 3$$

The spatial parallel mechanism shown in Fig. 10-10b has three degrees of freedom. It consists of three branches with pairs of RPS and a moving platform. There are three turning pairs, three prismatic pairs and three spherical pairs. We can solve its degrees of freedom by using Kuzebach equation.

$$F = 6n - \sum ip_i = 6 \times 7 - (3 \times 5 + 3 \times 5 + 3 \times 3) = 3$$

Generally, the three prismatic pairs are driven by three hydraulic cylinders in order to realize

器人的正解与逆解过程都是复杂的矩阵变换过程，但是，对于串联机器人，正解比较容易，逆解十分困难。

串联机器人机构自由度计算方法与空间机构相同。图 10-9b 所示的串联机器人机构的自由度为

$$F = 6n - \sum ip_i = 6 \times 3 - 5 \times 3 = 3$$

说明该机器人需要 3 个驱动件。

**2. 并联机器人机构**

并联机器人中，各构件形成多个封闭的构件系统，由多个输入构件共同驱动一个输出构件运动。并联机器人分为平面并联机器人和空间并联机器人。图 10-10a 所示为 3 自由度平面并联机器人，3 个连架杆为驱动件，共同驱动平台 1 运动。该机器人简称为平面 3RRR 并联机器人，在微动机构中有广泛应用。图 10-10b 所示为 3 自由度空间并联机器人。按照每个支链的运动副结构，又称为 3RPS 空间并联机器人。这种机器人相对串联机器人而言，具有刚度大的优点，因而承载能力大；但其运动空间较小，在空间运动模拟器中有广泛应用。

一般情况下，常把并联机器人的动平台形心处作末端操作器。与串联机器人相反，并联机器人的正解十分困难，但逆解比较容易。并联机器人中，由于动平台可以实现复杂的空间运动，已经广泛应用在各种高科技的机电设备中。

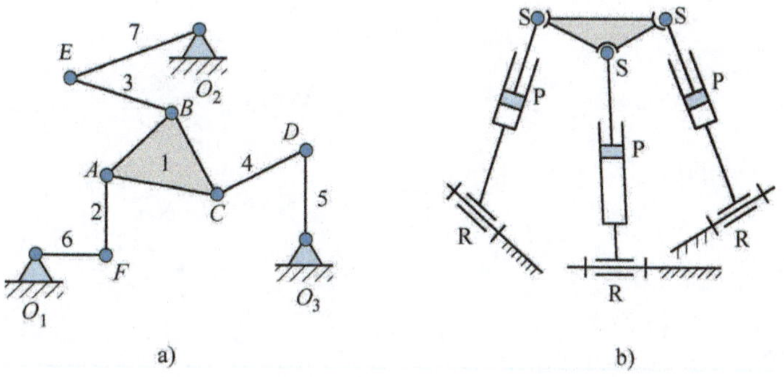


Fig. 10-10　Parallel robot mechanisms（并联机器人机构）

图 10-10a 所示 3RRR 平面并联机器人机构中，$n=7$，有 9 个转动副，没有高副。可按平面机构自由度计算公式计算，即

$$F = 3n - 2p_l - p_h = 3 \times 7 - 2 \times 9 = 3$$

图 10-10b 所示 3RPS 空间并联机器人机构中，$n=7$，有 3 个Ⅴ类转动副、3 个Ⅴ类移动副、3 个Ⅲ类副（球面副）。则

$$F = 6n - \sum ip_i = 6 \times 7 - (3 \times 5 + 3 \times 5 + 3 \times 3) = 3$$

the complicated spatial motions of the moving platform.

**3. Applications of Robotic Mechanisms**

Robots are typical electromechanical products which have a very wide application. Robots can be classified by their working purposes, such as industrial robots, medical robots, service robots, military robots, toy robots and so on.

Industrial robots have been used in automatic production field. Medical robots have been widely used in surgery. Service robots have been used in normal families. Military robots have been widely applied to battle field, such as fighting, mine clearance and logistics guarantee. And toy robots have been used by children.

Robot technology will prompt the progress of human being.

一般情况下，3 个移动副采用 3 个液压缸驱动，以实现动平台的空间复杂运动。

**3. 机器人应用**

机器人是典型的机电一体化产品，应用非常广泛。按用途可分为工业机器人、医用机器人、服务机器人、军用机器人和玩具机器人等。工业机器人广泛应用在自动化生产领域；医用机器人也应用在人体的手术中；服务机器人正在走向家庭；军用机器人在作战、排雷和后勤保证中也得到广泛应用；玩具机器人已经普及。未来的机器人将会促进人类社会的快速发展。

# Chapter 11

## Design of Mechanism Systems

# 第11章

## 机构系统设计

## 11.1 Introduction of Mechanism Systems

In the previous chapters, we have discussed linkage mechanisms, gear mechanisms, cam mechanisms, screws, Hooke's coupling and intermittent mechanisms. In practical engineering, however, a single mechanism is rarely used. A mechanism system which consists of several mechanisms is used widely in many machines. In this chapter, we will discuss the composition of mechanism systems.

The mechanism systems can be classified as three types shown in the following.

**1. Simple Mechanism System Consisting of a Single Mechanism**

A simple machine consists of only one fundamental mechanism. For instance, the actuating mechanism in an air compressor only contains one slider-crank mechanism. The ore crusher and radar steering machine all contain a kind of rocker-crank mechanism which works as their actuating mechanisms. Hoister only consists of gear mechanism. We can apply the knowledge acquired in previous chapters to solve this kind of problems for designing simple mechanical systems.

For performing the same work, the simpler the mechanism system is, the better the machine is. So we suggest using simple machine to perform a desired motion.

When designing a simple mechanism system, methods of evolution and mutation could provide an approach for creating new machines. For example, Fig. 11-1b shows the sketch diagram of a shearing machine. The pin of the revolute pair $B$ and the dimension of the prismatic pair $C$ are enlarged enough to obtain a shearing machine showed in Fig. 11-1a. Therefore, the strength and stiffness of the crank will increase; the mechanical property of the shear will be improved a lot.

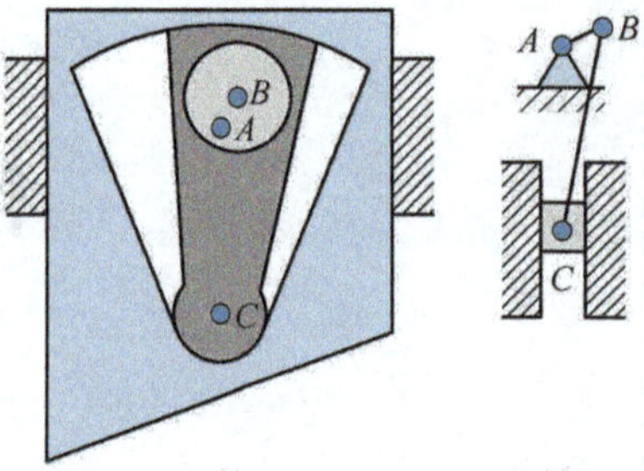

a) b)

Fig. 11-1 Shearing machine
(剪床机构)

**2. Complex Mechanism System Consisting of Independent Mechanisms**

Several simple mechanisms work independently, in other words there is no structural connection between every two mechanisms. However, these mechanisms must work orderly and coordinately in order to achieve the expected motion.

Selecting the mechanism types and harmonization design of the mechanism motions are necessary to design these kinds of machines.

Harmonization of motions can be realized by using mechanical or auto-control method.

Fig. 11-2 shows a hydraulic mechanism system in which the hydrocylinders 1 and 2 will operate independently. The hydrocylinder 1 will transfer a workpiece from position 1 to position 2, then the hydrocylinder 2 will transfer the workpiece to position 3. At this time hydrocylinder 1 must go back to its initial position. So the motions of the two hydrocylinders must be harmonized orderly and the work can be done well. The motion harmonization may be realized by using travel switches which control the reciprocating motion of the hydrocylinders.

## 11.1　机构系统设计概述

前面各章介绍了连杆机构、齿轮机构、凸轮机构以及其他各类常用机构的设计。但实际机械中很少应用单一的机构，绝大部分机械都是由各类机构通过不同的方式组合成一个系统，从而实现各种功能的。本章介绍机构系统的设计方法。

机构系统可分为三大类，下面分别作简单介绍。

**1. 简单机构组成的机构系统**

仅包含一个基本机构的机构系统可组成简单机械。如空气压缩机中仅有一个曲柄滑块机构作为主体机构，矿石破碎机、雷达转向机等机械仅含有曲柄摇杆机构作为主体运动机构，卷扬机仅由齿轮机构组成，使用前面各章知识就可以解决这类机构系统的设计问题。

在完成预期工作任务的前提下，建议优先使用简单机构，使机械系统更加简单。

使用简单机构进行机械系统运动方案设计时，利用机构的演化与变异原理，可提高机械的力学性能和使用寿命。图 11-1b 所示的剪床机构中，扩大转动副 $B$ 和移动副 $C$，得到图 11-1a 所示的剪床机构，可提高曲柄强度并提高剪刀的力学性能。

**2. 多个独立工作的简单机构组成的机构系统**

多个简单机构，各自独立工作，各机构之间没有任何结构上的连接。但各简单机构的运动必须互相协调，才能满足预期的工作要求。简单机构的选型和运动协调设计是这类机构系统的设计重点。运动协调手段可以通过机械方式和控制方式实现。

图 11-2 所示液压机构系统中，液压缸 1 和 2 是两个独立的简单机构。液压缸 1 把工件送到位置 2，触动液压缸 2 的开关后，即刻返回原位。液压缸 2 再把工件送到位置 3。两个液压缸的协调运动才能满足即定的工作要求，其运动协调依靠行程开关控制液压换向油路即可实现。

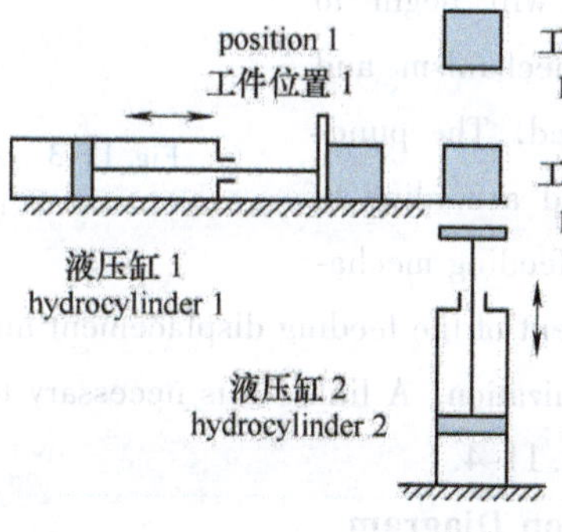


Fig. 11-2　Motion harmonization of mechanism

（机构运动的协调）

**3. Complex Mechanism System Consisting of Mechanisms Connected by Each Other**

There are many methods to design this kind of mechanisms system, such as adding links group, tandem connection, parallel connection, superpositional connection and closed connection. In practical engineering, these complex mechanical systems have been most widely used.

Fig. 11-3 shows a complex mechanism system, in which belt mechanism, worm mechanism, cam mechanism, linkage mechanism are connected in series. It could clip and feed automatically.

In engineering, most of mechanism systems consist of several simple mechanisms which are connected by each other to realize various functions.

## 11.2 Harmonization Design of Mechanism Motions

**1. Motion Harmonization of a Mechanism System**

There are two different ways for motion harmonization of a mechanism system.

One is to control the actuator or other controllable elements to realize the motion harmonization of a mechanical system. This is easy and practical, but its reliability is not the best. The second is to use mechanical methods to realize the motion harmonization of a mechanical system. This way is also easy, practical and more reliable. Therefore, we only deal with the second method to realize the motion harmonization of a mechanical system.

Fig. 11-4 shows a punching machine in which the action of the slider-crank mechanism *ABC* is used to punch and the slider-crank mechanism *FGH* is used to feed. When the punching head goes up after finishing a punching process, the feeding mechanism starts to feed. Besides, the feeding mechanism must complete feeding and return to its initial position, and then the punching head will begin to punch. The motions of punching mechanism and feeding mechanism must be harmonized. The punching mechanism *ABC* can be designed according to the punching requirement, while the feeding mechanism *FGH* should satisfy the requirement of the feeding displacement and its sizes and positions should also fit the condition of motion harmonization. A link *DE* is necessary to connect the two cranks of the two mechanisms; this is shown in Fig. 11-4.

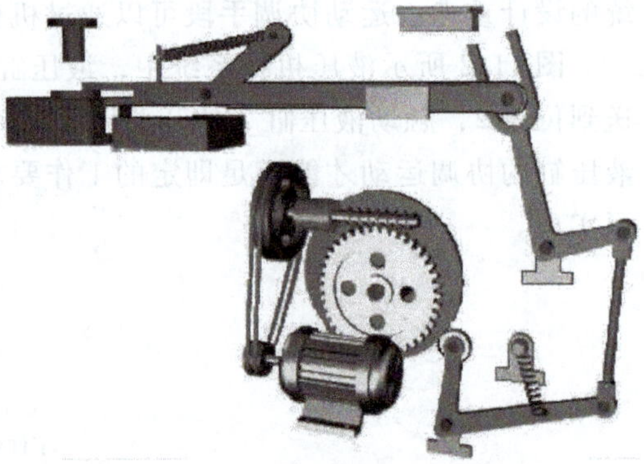

Fig. 11-3 Automatic feeding machine (自动送料机)

**2. Design of Motion Circulation Diagram**

Motion circulation diagram is a graph which can indicate the motion relationships of all actuators in an operating period. In order to make all actuators work orderly, it is necessary to design motion circulation diagram.

We can use a rectangular coordinate system to describe the motion circulation diagram. Fig. 11-5 shows the motion circulation diagram of a punching mechanism. In the figure, the abscissa axis indicates the motion cycle of the actuators and the ordinate axis indicates the motion status of the

**3. 简单机构连接组成的机构系统**

各种简单机构通过连接杆组、串联组合、并联组合、叠加组合和封闭组合，可形成一系列的复杂机构系统。实际机械中，这类机构系统应用最为广泛。图 11-3 所示的机构是由带传动机构、蜗杆机构、凸轮机构、铰链四杆机构和正切机构串联组成的，可实现零件的自动夹取与送料运动。

工程中的各类机械运动系统大部分都是由多个简单机构连接在一起，组成机构系统，实现各种功能目标的。

## 11.2　机构系统的运动协调设计

**1. 机构系统的运动协调**

运动协调设计有两种途径：其一是通过对电动机或其他可控元件的时序控制，实现机械的运动协调设计，这类方法简单、实用，但可靠性差些；其二是通过机械手段实现机械的运动协调设计，这类方法同样简单、实用，但可靠性好些。本节主要介绍通过机械手段实现机械运动协调设计方法。

图 11-4 所示压力机机构简图中，机构 *ABC* 为冲压机构，机构 *FGH* 为送料机构。要求在冲压结束后，冲压头回升过程中开始送料，到冲压头下降过程的某一时刻完成送料并返回原位。冲压机构与送料机构的动作必须协调。冲压机构 *ABC* 的设计可按冲压要求设计，送料机构 *FGH* 不但要满足送料位移要求，其尺寸与位置必须满足运动协调的条件。设计时可通过连杆 *DE* 连接两个机构。

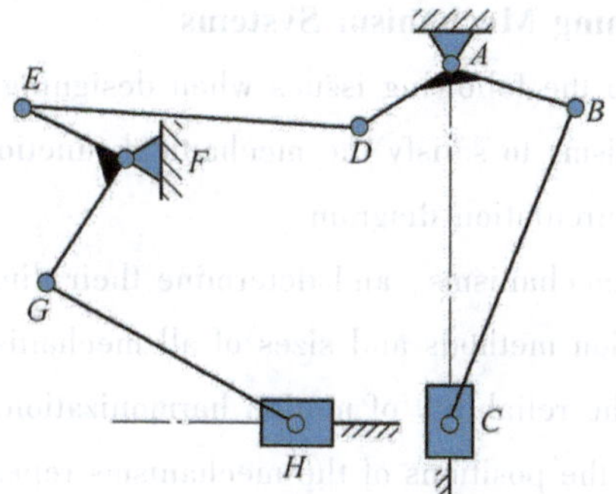


Fig. 11-4　Punching mechanism
（压力机机构）

**2. 运动循环图的设计**

表明机械一个工作循环中各执行机构的运动配合关系的图形，称为机械运动循环图。为了使各执行机构能按照工艺动作有序地互相配合，必须进行运动循环图的设计。

执行机构的运动循环图可用直角坐标表示，图 11-5 所示为压力机机构的运动循环图，横坐标表示执行机构的运动周期，纵坐标表示执行机构的运动状态。每一个机构的运动状态均可在循环图上表示，通过合理设计可以实现它们之间的工作协调。

图 11-5 所示的压力机机构的运动循环图中，*AB* 为工作行程，*BC* 为回程，其中 *GF* 为冲压过程。送料机构的运动循环图中，*EC* 为开始送料阶段，*AD* 为退出送料阶段。在冲压阶

actuators. Every motion status of mechanisms can be expressed in this motion circulation diagram. We can harmonize these motions of mechanisms by designing the motion circulation diagram.

In the punching process, the line $AB$ is the stage of working travel, the line $BC$ is the stage of return travel and the segment $GF$ is the actual punching progress. In the feeding diagram, $EC$ is the stage of feeding travel and $AD$ is the stage of return travel. In the segment $DE$, the feeding mechanism stays a short time to avoid motion interference. So it must satisfy the condition $T_5 > T_2$. According to this motion circulation diagram, we can design the feeding mechanism.

Motion circulation diagram can be designed differently; therefore, it is available to find an optimum solution to make the mechanical system work perfectly.

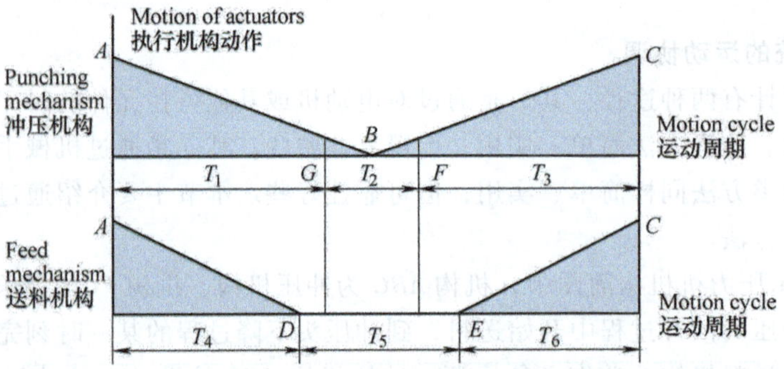


Fig. 11-5 Motion circulation diagram of a punching machine

(压力机机构的运动循环图)

### 3. Key Points for Designing Mechanism Systems

We should pay attention to the following issues when designing a mechanical system.

1) Choose simple mechanisms to satisfy the mechanical functions.

2) Sketch up the motion circulation diagram.

3) Synthesize the simple mechanisms, and determine their dimensions.

4) Determine the connection methods and sizes of all mechanisms.

5) Simulate and inspect the reliability of motion harmonization.

6) Optimize the sizes and the positions of the mechanisms repeatedly, until they can satisfy the requirements.

## 11.3 Combined Methods of Mechanism Systems

There are many methods to design a mechanism system. Here we introduce some methods widely used.

1) Connection of basic link groups for designing a mechanism system.

2) Tandem connection of simple mechanisms for designing a mechanism system.

3) Parallel connection of simple mechanisms for designing a mechanism system.

4) Superposition of simple mechanisms for designing a mechanism system.

5) Closed connection of simple mechanisms for designing a mechanism system.

段，送料机构在 *DE* 阶段不动，使其运动不发生干涉，其条件必须满足 $T_5 > T_2$。根据拟订的运动循环图，设计送料机构的尺寸和位置。

运动循环图的设计结果不是唯一的，设计过程中，要使机构之间的运动协调实现最佳配合。

**3. 机构系统设计要点**

机构系统的设计应注意以下问题：

1）按机械功能目标选择各简单机构。

2）拟订运动循环图。

3）进行各简单机构的尺度综合，确定各机构的尺寸。

4）确定各机构的连接方法与连接件尺寸。

5）进行计算机仿真，检验运动协调的可靠性。

6）反复进行机构尺寸与位置的修订，直到满意为止。最后进行结构设计。

## 11.3 机构系统的组合方法

简单机构组合成机构系统的方法很多，这里介绍几种常用方法：连接基本杆组、简单机构串联、简单机构并联、简单机构叠加、简单机构封闭连接。

**1. 连接基本杆组**

把基本杆组的外接副分别连接到原动件和机架上，可组成串联机构；把基本杆组的外接副全部连接到原动件上，可组成并联机构。在新机构的基础上，再连接基本杆组，可组成更复杂的机构系统。基本杆组类型很多，连接方法也多样化，连接杆组法是创新设计机构系统的重要方法之一。

如图 11-6a 所示，Ⅱ级杆组 *BCD* 的外接副 *B*、*D* 连接到原动件 *AB* 和机架上，组成四杆机构 *ABCD*。再把一个Ⅱ级杆组中的外接副 *E*、*F* 连接到四杆机构 *ABCD* 的杆 *DC* 和机架上，组成了一个机构系统。以此类推，可设计出更复杂的机构系统。各连接点的位置可通过机构综合方法求取。如果把Ⅱ级杆组中的外接副 *B*、*D* 连接到两个原动件上，组成了图 11-6b 所示的 2 自由度并联机构。

Fig. 11-6 Connecting basic link groups（基本杆组）

**1. Connection Basic Link Groups**

As we know previously, in a basic link group, the pairs which connect inside links of the link group are called inside pairs, and the pairs which are used to connect the other links are called outside pairs.

A tandem mechanism may be obtained by connecting the driving links and frame to the outside pairs of a basic link group. If we connect another basic link group to a moving link of the mechanism and frame, a more complicated mechanism system will be obtained too.

If we connect all the driving links to outside pairs of the basic link group, a parallel mechanism can be obtained.

There are many kinds of basic link groups and connection methods, so this method is an important way to innovate mechanism systems.

In Fig. 11-6a, the outside pairs *B* and *D* of the class Ⅱ link group *BCD* are connected to the driving link *AB* and the frame, and we obtain a four-bar linkage *ABCD*. If the outside pairs *E* and *F* of another class Ⅱ link group are connected to the link *DC* and the frame, a six-link mechanism would be obtained showed in Fig. 11-6a. Similarly, more complicated mechanism systems can be obtained in this way. The location of the connected point, such as the point *E*, can be solved through the synthesis method. If the outside pairs *B* and *D* are connected to two driving links, we can get a parallel mechanism with two degrees of freedom shown in Fig. 11-6b.

**2. Tandem Connection of Simple Mechanisms**

The output link of a mechanism with one degree of freedom is connected rigidly to the input link of another mechanism with single degree of freedom; this is called tandem connection. The mechanisms which are used to connect the others could be the same type or different.

Fig. 11-7a shows a gear-linkage system, in which the driven gear is connected rigidly to the driving crank of the crank-rocker mechanism. This mechanism system can reduce the speed of the linkage mechanism. Fig. 11-7b shows another gear-linkage system, in which the internal gear of the gears is connected rigidly to the coupler of the parallelogram mechanism *ABCD*. Because of the parallel translation of the internal gear $z_1$, the external gear $z_2$ would rotate at a very low speed, when it satisfies the conditions, $O_1O_2 = AB = CD$ and they are parallel to each other.

Fig. 11-8a shows a tandem mechanism system in which the guider of the crank-guider bar mechanism is connected to the crank of the slider-crank mechanism. The slider would have a special motion characteristic. Fig. 11-8b shows a tandem system consisting of a gear mechanism and a cam mechanism. Tandem connection is an important way for designing new mechanical systems. In engineering, most of machines contain tandem mechanical systems.

We can use various mechanisms with single degree of freedom in tandem connection.

**3. Parallel Connection of Simple Mechanisms**

There are two kinds of parallel connections. The first is that one kind of motion is resolved into many kinds of motions or many kinds of motions are resulted in one kind of motion. The second is that one kind of motion is resolved into many kinds of motions and then they are resulted in another kind of motion.

Parallel connections are widely used in engineering. In Fig. 11-9a, two parallel hydraulic cylin-

**2. 简单机构串联**

前一个单自由度机构的输出构件与后一个单自由度机构的输入构件刚性连接在一起，称为串联组合。串联组合中的各机构可以是同类型的机构，也可以是不同类型的机构。

图 11-7a 所示齿轮机构的从动轮与连杆机构的主动曲柄刚性连接，形成了齿轮连杆机构组合系统。该系统可降低连杆机构的运转速度；图 11-7b 中，平行四边形机构 $ABCD$ 的连杆与齿轮机构的内齿轮刚性连接，形成了连杆机构和齿轮机构的组合系统。由于内齿轮 $z_1$ 做平动。当满足$\overline{O_1 O_2} \underline{\parallel} \overline{AB} \underline{\parallel} \overline{CD}$时，该系统的外齿轮 $z_2$ 做大速比减速输出。

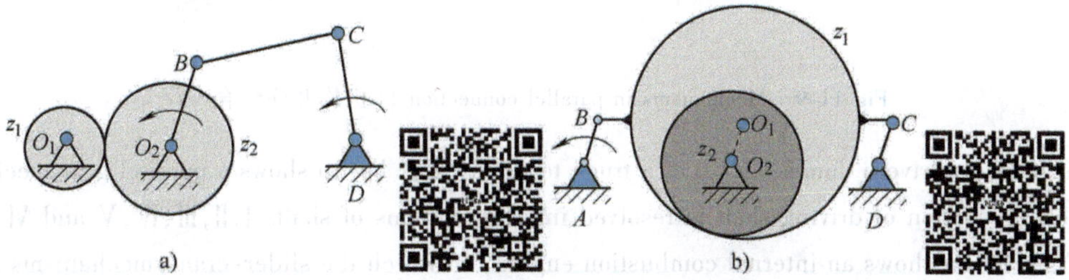


Fig. 11-7　Mechanisms in tandem connection 1（机构的串联组合 1）

图 11-8a 中，摆动导杆机构的输出摆杆与曲柄滑块机构的曲柄连接，可得到滑块的特殊运动规律。图 11-8b 所示为齿轮机构与凸轮机构串联而成的机构系统。机构串联是设计新机构系统的重要途径，工程中的大部分机械都含有串联的机构系统。

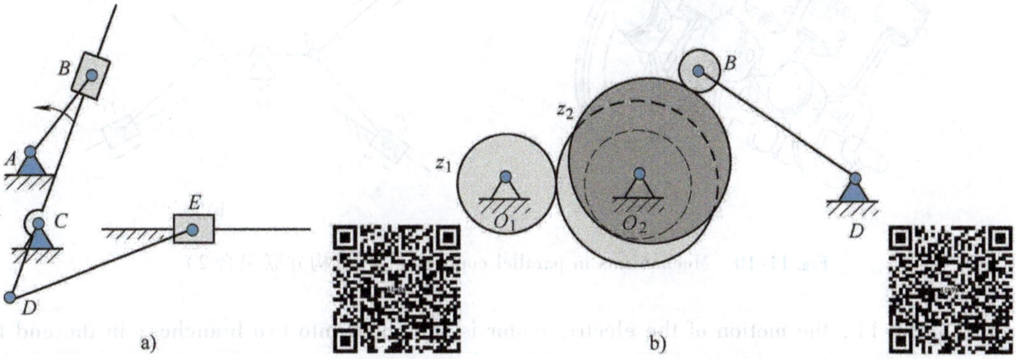


Fig. 11-8　Mechanisms in tandem connection 2（机构的串联组合 2）

在串联组合中，由于可以使用各种不同机构进行连接，因而可实现多种功能目标，是重要的创新方法。

**3. 简单机构并联**

并联组合方法有两种：其一是将一种运动分解为若干种运动，或将若干种运动合成为一种运动；其二是将一种运动分解为若干种运动后再合成为一种运动。

并联组合也是最为常见的机构组合方法。图 11-9a 为两个并联的导杆机构共同驱动汽车翻斗的示意图；图 11-9b 为将主轴运动分流到轴Ⅰ、Ⅱ、Ⅲ、Ⅳ、Ⅴ、Ⅵ的并联齿轮机构示意图。

图 11-10a 所示机构系统为轴向并联布置的内燃机。连杆机构的活塞为主动件，8 套呈Ⅴ形布置的曲柄滑块机构共同驱动曲轴转动，实现了动力的合成。图 11-10b 所示为径向并

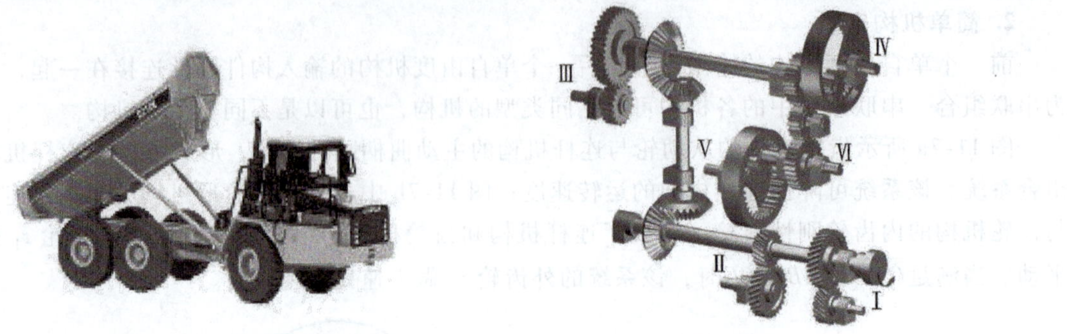


a) b)

Fig. 11-9 Mechanisms in parallel connection 1(机构并联组合 1)

der mechanisms drive a dump-bucket of a truck together. Fig. 11-9b shows a parallel gear mechanism in which the motion of driving shaft is resolved into the motions of shafts Ⅰ,Ⅱ,Ⅲ,Ⅳ,Ⅴ and Ⅵ.

Fig. 11-10a shows an internal combustion engine, in which the slider-crank mechanisms are in-line arrangement to V. The pistons are driving links and drive the crank shaft to rotate. Fig. 11-10b shows another combustion engine, in which the slider-crank mechanisms are radial arrangement. Four slider-crank mechanisms drive the crank shaft to rotate.

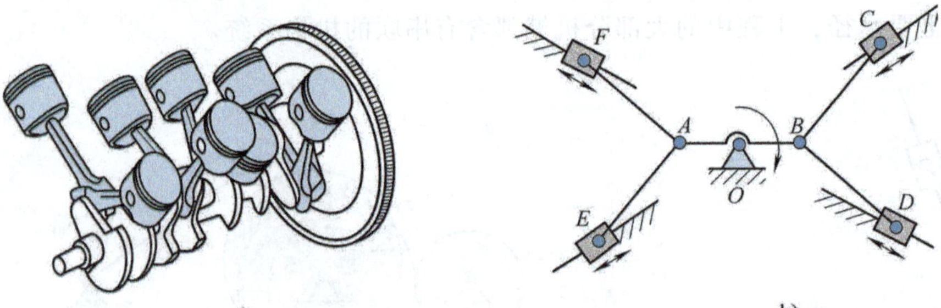


a) b)

Fig. 11-10 Mechanisms in parallel connection 2(机构并联组合 2)

In Fig. 11-11, the motion of the electric motor is separated into two branches; in the end they will be resulted in one rotation of the worm gear again. This parallel mechanism can enhance the output capacity of the worm gear.

In engineering, we usually take tandem mechanisms as the branches of parallel mechanism systems. Fig. 11-11 shows an example that both gear mechanisms are tandem connections. Combinations of tandem connections and parallel mechanisms are very common in mechanism systems.

**4. Superposition of Simple Mechanisms**

Locating a mechanism on a moving link of the other mechanism is also a common way for designing a mechanism system.

Fig. 11-12a shows a hydraulic excavator, in which a guider-bar mechanism is located on the turntable 1. The arm 4 is driven by hydraulic cylinder 2 and rod 3. Another hydraulic mechanism is superposed on the arm 4, and it drives the upper arm 7 by using hydraulic cylinder 5 and rod 6. In the same way, there is the third hydraulic mechanism which is superposed on the upper arm 7 to

联布置的内燃机简图。4 套曲柄滑块机构共同驱动曲轴转动，实现了动力的合成。

图 11-11 所示的机构系统把电动机的运动分解为两路传动，然后再合成一个蜗杆运动，形成并联机构组合系统。这种并联组合系统可提高蜗轮的输出动力。

并联组合的各个支路中，经常应用串联机构系统。图 11-11中，电动机左、右两个齿轮机构系统均是串联机构系统。机构的串联组合和并联组合经常混用，在机构系统中很常见。

Fig. 11-11 Mechanisms in parallel connection 3（机构并联组合 3）

**4. 简单机构叠加**

一个机构安装在另一个机构的运动构件之上，是叠加组合的基本途径。叠加组合也是设计机构系统的常用方法。图 11-12a所示的液压挖掘机中，转塔 1 上安装摆杆机构。通过液压缸 2、驱动杆 3，再驱动大臂 4 摆动。大臂 4 上又安装一套同类液压机构，通过液压缸 5、杆 6 驱动小臂 7 摆动；小臂 7 上还安装一套液压缸机构，通过铰链机构驱动铲斗工作。图 11-12b 所示的升降机构中，在一个平行四边形机构上叠加另外一个平行四边形机构，工作平台在升降过程中保持一个稳定姿态。

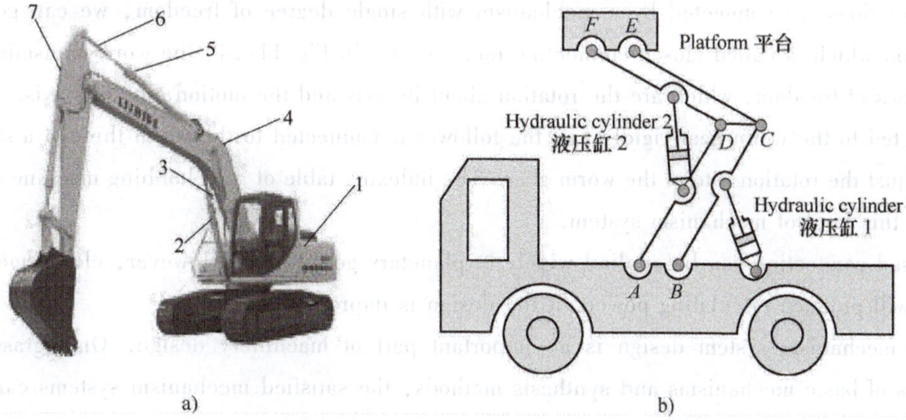


Fig. 11-12 Mechanisms in superposed connection 1（机构叠加组合 1）

1—turntable（转塔） 2、5—hydraulic cylinder（液压缸） 3、6—rod（杆） 4—arm（大臂） 7—upper arm（小臂）

叠加组合机构系统在工程机械及军事装备中有广泛应用。

图 11-13 所示机构是在周转轮系系杆上安装一个单头蜗杆机构，由蜗轮给行星轮提供输入运动，带动系杆缓慢转动。蜗杆驱动扇叶转动，又可驱动系杆做 360°的慢速转动，实现风扇的全方位运动。系杆转动速度可按轮系传动比计算，即

$$n_{\mathrm{H}}=\frac{z_3}{z_2 z_4}n_1$$

式中，$n_1$ 为电动机转速；$n_{\mathrm{H}}$ 为系杆转速。调整齿轮的齿数可调整系杆的转速。

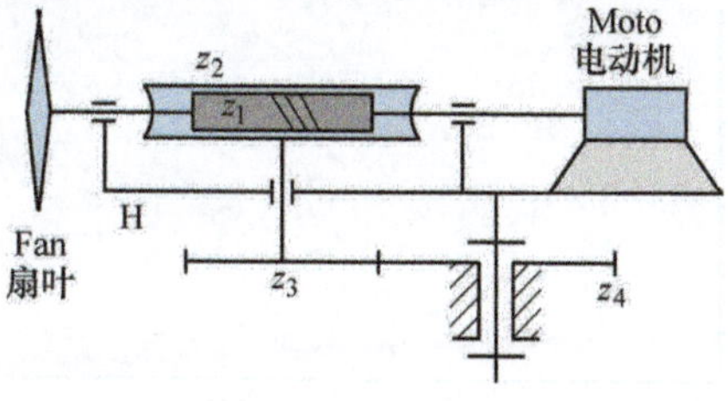


Fig. 11-13 Mechanisms in superposed connection 2（机构叠加组合 2）

由机构叠加组合而成的机构系统具有很多优点，可满足复杂的运动要求，机构的传力功能较好，但设

drive the bucket.

Fig. 11-12b shows a hydraulic lifter in which a parallelogram linkage *DCEF* is superposed on the coupler of the parallelogram linkage *ABCD*, so that the platform can move up and down and keep balance when working.

In Fig. 11-13, a worm mechanism is superposed on the arm of the planetary gear train, and the output shaft of the worm gear is connected to the planet gear. In this system, the worm gear supplies power to the planetary gear and then drives the planet carrier to rotate slowly. The worm can also drive the fan and the planet carrier to rotate. The angular velocity of the planet carrier can be calculated as follows:

$$n_H = \frac{z_3}{z_2 z_4} n_1$$

Where $n_1$ = speed of the motor; $n_H$ = speed of the planet carrier.

Superposed mechanism has a great many advantages. For examples, it can realize complicated motion and has better mechanical property. However, it is hard to design.

**5. Closed connection of Simple Mechanisms**

There are two independent motions in a mechanism with two degrees of freedom. If two independent motions are connected by a mechanism with single degree of freedom, we can get a new mechanism which is called closed connection mechanism. In Fig. 11-14, the worm transmission has two degrees of freedom, which are the rotation about its axis and the motion along its axis. The cam is connected to the worm gear rigidly and the follower is connected to the worm through a slip ring. It can adjust the rotation rate of the worm gear. The indexing table of gear hobbing machine is an example of this kind of mechanism system.

Closed connection can be applied widely to planetary gear trains. However, closed mechanism systems will produce circulating power, if the design is improper.

The mechanism system design is an important part of machinery design. Only grasping the properties of basic mechanisms and synthesis methods, the satisfied mechanism systems can be designed out.

计构思难度较大。

**5. 简单机构封闭连接**

2 自由度的机构，共有两个独立运动。如果用一个单自由度的机构连接其中两个独立运动，就形成新的自由度为 1 的机构系统，称为封闭连接机构系统。图 11-14 所示机构中，蜗杆传动为一个 2 自由度的机构，即蜗杆绕轴线的转动和沿轴线的移动。单自由度的凸轮机构中，凸轮与蜗轮连接，推杆与蜗杆通过滑环连接，并可推动蜗杆沿轴线移动，起到调整蜗轮转速的作用。齿轮加工机床分度台的差动运动就是通过这种机构组合系统实现的。

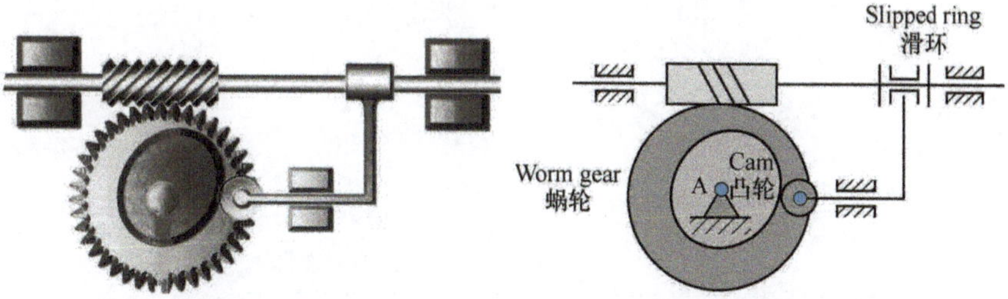


Fig. 11-14 Mechanisms in closed connection（机构封闭组合）

封闭机构组合机构系统具有优良的运动特性，在行星传动中有广泛应用。如果设计不当，有时会产生机构系统内部的封闭功率流，降低机械效率。

机构系统设计是机械系统运动方案设计的主体内容，是机械创新设计的重要途径。只有在充分了解机构性能的基础上，运用机构系统设计的基本方法，才能设计出满足功能要求的机构系统。

# Chapter 12

## Fluctuation and Regulation in Speed of Machines

# 第12章

## 机械系统的运转及速度波动的调节

## 12.1 Operating Analysis of Machinery

The operating process of machinery is relevant to the forces acting on the machinery, the link masses and inertia moments of links.

**1. Forces Acting on a Machine**

The forces acting on a machine can be divided into two types. They are working resistance and driving force.

(1) Working resistance  Working resistance refers to the types of machines.

1) Working resistance is constant, such as in cranes and rolling mills.

2) Working resistance varies with displacement, such as in compressors.

3) Working resistance varies with speed, such as in air blowers and centrifugal pumps.

4) Working resistance varies with times, such as in ball mills, flour-rubbing machines.

(2) Driving force  The different prime motors have different operating peculiarities. The relationship between the driving force and its speed is often represented by the peculiar curve of a machine. Fig. 12-1a shows a peculiar curve of an internal-combustion engine, and Fig. 12-1b shows a peculiar curve of an *AC* electromotor.

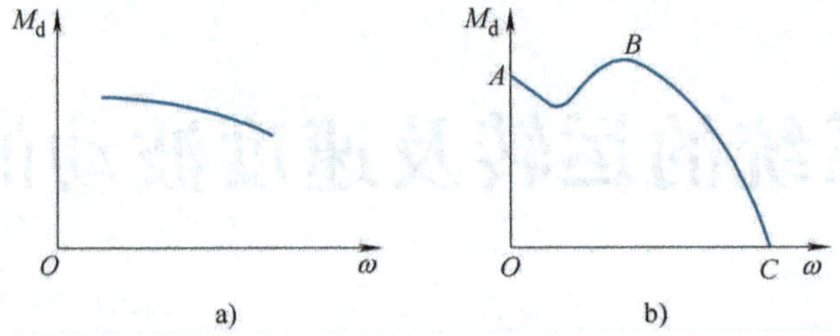


Fig. 12-1  Mechanical behavior curve of prime machines (原动机的机械特性曲线图)

Most machines run at higher speed and low moment of torque or run at lower speed and large moment of torque. The internal-combustion engine shown in Fig. 12-1a has not self-adjust ability. The *AC* electromotor shown in Fig. 12-1b has self-adjust ability along the curve *BC*.

**2. Operating Process of Machines**

The operation of a machine from the moment it is set in motion up to the moment of stopping can be divided into three periods: start-up period, steady running period and stopping period.

(1) Start-up period  The running speed of a machine is from zero to work running speed in this period, and the work done by the driving force is greater than the work done by the resistance.

The relationship of the works is as follows:

$$W_d = W_r + \Delta E$$

Where, $W_d$ is the work done by the driving force. $W_r$ is the work done by the resistance. $\Delta E$ is the increment of the kinetic energy.

To reduce the time in the start-up period, the machine runs usually at the moment of no working resistance.

(2) Steady running period  In the steady running period, the work done by the driving force is

## 12.1 机械运转过程分析

机械的运转过程与作用在机械上的外力、构件的质量和转动惯量有关。

**1. 作用在机械上的力**

(1) 工作阻力 工作阻力是机械正常工作时必须克服的外载荷。不同的机械，其工作阻力的性质不相同。

1) 工作阻力是常量，即 $F_r = C$，如起重机、轧钢机等机械的工作阻力均为常量。

2) 工作阻力随位移而变化，即 $F_r = f(s)$，如空气压缩机的工作阻力随位移而变化。

3) 工作阻力随速度而变化，即 $F_r = f(\omega)$，如鼓风机、离心泵等机械的工作阻力均随叶片的转速而变化。

4) 工作阻力随时间而变化，即 $F_r = f(t)$，如球磨机、揉面机等机械的工作阻力均随时间的增加而变化。

工作阻力的特性要根据具体的机械来确定。

(2) 驱动力 原动机不同，驱动力的特性不相同，常用机械特性曲线表示驱动力与其运转速度之间的关系。工程中常用内燃机、电动机作原动机，它们的驱动力特性不同。

图 12-1a 所示为内燃机的机械特性曲线，图 12-1b 所示为三相交流异步电动机的机械特性曲线。

许多机械在工作过程中要求满足高转速、小转矩，低转速、大转矩的工作要求。图 12-1a所示内燃机的机械特性曲线中，当工作载荷增加而导致机械转速降低时，其驱动力矩不能相应大幅增加，不能自动平衡外载荷的变化，故内燃机无自调性。图 12-1b 所示的三相交流异步电动机的机械特性曲线中，$BC$ 段曲线具有自调性，所以电动机工作阶段必须在 $BC$ 段工作。

**2. 机械的运转过程**

机械的运转过程一般都要经历起动、稳定运转和停车三个阶段，如图 12-3 所示。其中稳定运转阶段是机械的工作阶段，是机械工作性能优劣的具体表现阶段。

(1) 机械的起动阶段 机械由零转速逐渐上升到正常工作转速的过程称为机械的起动阶段，该阶段中驱动力所做的功 $W_d$ 大于阻抗力所做的功 $W_r$，两者之差为机械起动阶段的动能增量 $\Delta E$。即

$$W_d = W_r + \Delta E$$

为减少机械起动的时间，一般在空载下起动，即 $W_r = 0$。则

$$W_d = \Delta E$$

(2) 机械的稳定运转阶段 当驱动力所做的功 $W_d$ 和阻抗力所做的功 $W_r$ 相平衡时，动能增量 $\Delta E$ 为零，机械主轴的平均运转角速度保持不变。该阶段称为等速稳定运转阶段，也是机械的工作阶段。图 12-2a 所示的曲柄压力机在冲压过程中，阻抗力急剧增加，导致机械主轴的角速度迅速减小。在冲压完毕的返回行程中，阻抗力减小，机械主轴的角速度又恢复到原来的数值，周而复始。其瞬时角速度做周期性的波动，但其平均值 $\omega_m$ 保持不变。这种类型的机械稳定运转称为周期性变速稳定运转。内燃机、曲柄压力机、刨床等许多机械，在稳定运转过程中，其角速度做周期性的波动，但角速度的平均值为常量。

在周期性变速稳定运转过程中，某一时刻驱动力所做的功不等于阻抗力所做的功。例如：

equal to the work done by the resistance in a period. The increment of the kinetic energy is zero. We have:

$$W_{dp} = W_{rp}$$

Where, $W_{dp}$ is the work done by the driving force in a whole period. $W_{rp}$ is the work done by the resistance in a whole period.

Fig. 12-2 shows a crank punch which is cutting holes on the metal sheet. When the machine presses the hole, the resistance increases and the angular velocity of the machine decreases. After punching, the resistance decreases, and the angular velocity of the machine increases (see Fig. 12-2). The angular velocity varies in the running period.

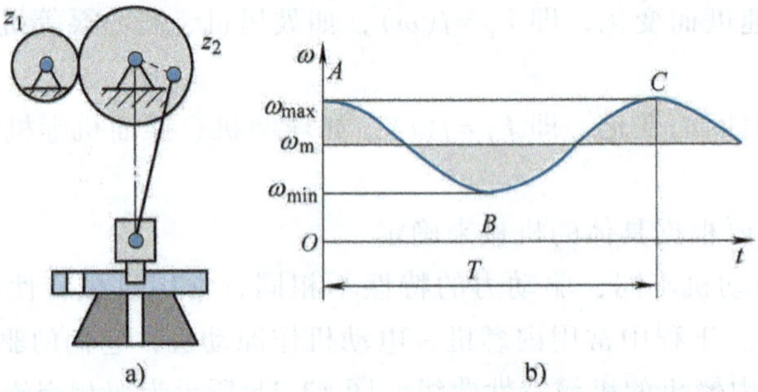


Fig. 12-2 Crank punch (曲柄压力机)

In the steady running period, if the operating speed is a constant, it is said to be a constant steady running. If the operating speed is variable and the average speed is a constant, it is said to be a no-uniform steady running.

A flywheel is used to control the variations in speed during each cycle of the machine. When the supply of energy is more than required, the flywheel stores energy and the speed will be reduced. During the requirement is more than the supply, it releases energy, and the speed will be increased.

To restrict variations of speed of the machine to narrow limits, a flywheel is fitted to act as a source of kinetic energy. The angular velocity $\omega_{max}$ and $\omega_{min}$ can be restricted to some limits.

The average speed $\omega_m$ is as follows:

$$\omega_m = \frac{1}{2}(\omega_{max} + \omega_{min}) \tag{12-1}$$

The permissible variation in speed can be described as the coefficient of speed fluctuation and is defined as:

$$\delta = \frac{\omega_{max} - \omega_{min}}{\omega_m} \tag{12-2}$$

The values of the coefficient of fluctuation can be found in Tab. 12-1.

From the equation (12-1) and the equation (12-2), we have:

$$\omega_{max} = \omega_m\left(1 + \frac{\delta}{2}\right) \quad \omega_{min} = \omega_m\left(1 - \frac{\delta}{2}\right)$$

$$\omega_{max}^2 - \omega_{min}^2 = 2\delta\omega_m^2 \tag{12-3}$$

(3) Stopping period In the stopping period, the driving force will cease to work, thus $W_d =$

图 12-2b 中的 $AB$ 工作段，角速度呈下降趋势，驱动力所做的功小于阻抗力所做的功，即 $W_d < W_r$。在 $BC$ 段，驱动力所做的功大于阻抗力所做的功，即 $W_d > W_r$，角速度增大。由于在一个运转周期的始末两点的角速度相等，即 $\omega_A = \omega_C$，说明在一个运转周期的始末两点的机械动能相等，或者说在一个运转周期内驱动力所做的功 $W_{dp}$ 等于阻抗力所做的功 $W_{rp}$。即

$$W_{dp} = W_{rp}$$

尽管周期性变速稳定运转过程中的平均角速度 $\omega_m$ 为常量，但过大的速度波动会影响机械的工作性能。因此，必须把周期性变速稳定运转过程中的速度波动调节到许用范围之内。

角速度 $\omega_m$ 的平均值可近似为

$$\omega_m = \frac{1}{2}(\omega_{max} + \omega_{min}) \tag{12-1}$$

角速度的差值（$\omega_{max} - \omega_{min}$）可反映机械运转过程中速度波动的绝对量，但不能反映机械运转的不均匀程度。因此，工程上用速度波动的绝对量与平均角速度的比值来表示机械运转的不均匀程度，用 $\delta$ 表示，并称之为机械运转的不均匀系数。即

$$\delta = \frac{\omega_{max} - \omega_{min}}{\omega_m} \tag{12-2}$$

由式（12-1）、式（12-2）可得

$$\omega_{max} = \omega_m\left(1 + \frac{\delta}{2}\right),\quad \omega_{min} = \omega_m\left(1 - \frac{\delta}{2}\right)$$
$$\omega_{max}^2 - \omega_{min}^2 = 2\delta\omega_m^2 \tag{12-3}$$

当 $\omega_m$ 一定时，机械的运转不均匀系数 $\delta$ 越小，$\omega_{max}$ 与 $\omega_{min}$ 的差值就越小，表明机械运转越平稳。机械的运转不均匀系数的大小反映了机械运转过程中的速度波动的大小，部分机械的许用运转不均匀系数参见表 12-1。

**Tab. 12-1　Allowable coefficient of speed fluctuation of machines**（部分机械的许用运转不均匀系数）

| Machines 机械名称 | Allowable coefficient of speed fluctuation 许用运转不均匀系数 [δ] | Machines 机械名称 | Allowable coefficient of speed fluctuation 许用运转不均匀系数 [δ] |
|---|---|---|---|
| Rock crusher 石料破碎机 | $\frac{1}{20} \sim \frac{1}{5}$ | Paper machine, loom 造纸机、织布机 | $\frac{1}{50} \sim \frac{1}{40}$ |
| Agricultural machine 农业机械 | $\frac{1}{50} \sim \frac{1}{5}$ | Compressor 压缩机 | $\frac{1}{100} \sim \frac{1}{50}$ |
| Punching, shearing, forging Machine 压力机、剪床、锻床 | $\frac{1}{10} \sim \frac{1}{7}$ | Spinning machine 纺纱机 | $\frac{1}{100} \sim \frac{1}{60}$ |
| Rolling mill 轧钢机 | $\frac{1}{25} \sim \frac{1}{10}$ | Internal-combustion engine 内燃机 | $\frac{1}{150} \sim \frac{1}{80}$ |
| Machine tools 金属切削机床 | $\frac{1}{50} \sim \frac{1}{20}$ | DC generator 直流发电机 | $\frac{1}{200} \sim \frac{1}{100}$ |
| Automobile, tractor 汽车、拖拉机 | $\frac{1}{60} \sim \frac{1}{20}$ | AC generator 交流发电机 | $\frac{1}{300} \sim \frac{1}{200}$ |
| Pump, blower 水泵、鼓风机 | $\frac{1}{50} \sim \frac{1}{30}$ | Turbogenerator 汽轮发电机 | $\leq \frac{1}{200}$ |

（3）机械的停车阶段　停车阶段是指机械由稳定运转的工作转速下降到零转速的过程。要停止机械运转必须首先撤去机械的驱动力，即使 $W_d = 0$。这时阻抗力所做的功用于克服机械在稳定运转过程中积累的惯性动能 $\Delta E$，即

0. The motion continues for a certain time owing to the inertia kinetic energy of the machine.

The relationship of the works becomes:

$$W_r = \Delta E$$

In order to reduce the stopping time, the parasitic resistance must be increased. A brake is often used to reduce the dwell period. The total operating process of machines is shown as Fig. 12-3. In the stopping period, the dotted line $B$ is the stopping curve fitted a brake.

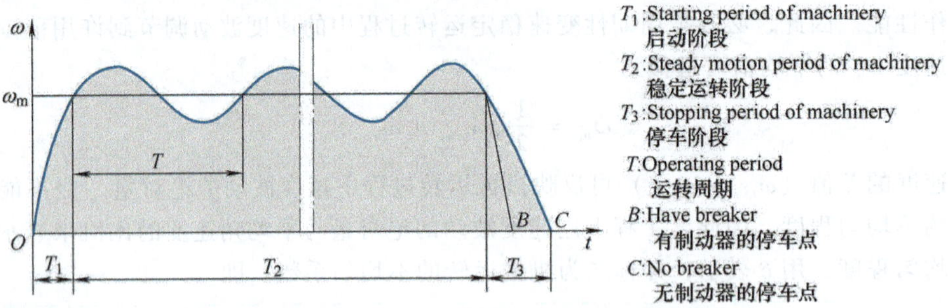


Fig. 12-3 Operating process of machinery (机械的运转过程)

## 12.2 Equivalent Kinetic Model of Mechanism Systems

### 1. The Method of Researching Running Process of Machines

In order to find the actual motion of a mechanism shown in Fig. 12-4, we must solve the driving torque acting on the link 1 and the angular velocity. The dynamic force analysis of the slider-crank mechanism has a large number of unknowns to solve. This mechanism has three moving links. Links 1 and 2 provide three equations respectively; the link 3 provides two equations, altogether eight equations. The unknowns are eight. We should expect to have eight equations in eight unknowns for this problem. This is very tedious. In order to simplify the problem, we can use an imaginary link in place of the mechanism, and then all forces and masses of links on the mechanism are reduced to the link. This link is called equivalent link.

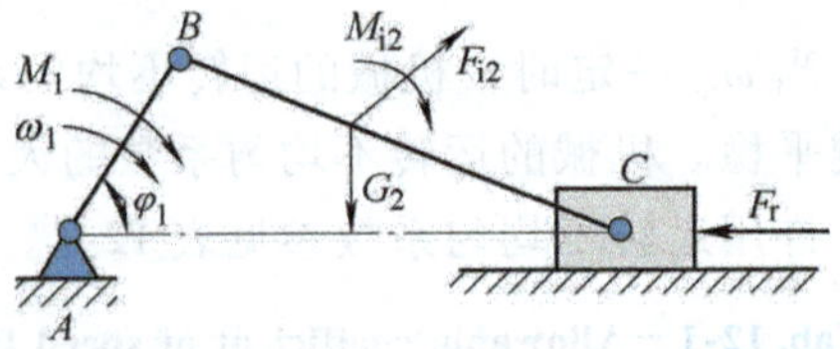


Fig. 12-4 Force analysis of a crank punch (曲柄压力机的受力分析)

We often select the rotating link or reciprocating link as an equivalent link. Fig. 12-5 shows those equivalent links.

The mass of the equivalent link is called equivalent mass, denoted as $m_e$.

The moment acting on the equivalent link is called equivalent moment, denoted as $M_e$.

The moment of inertia of the equivalent link is called equivalent moment of inertia, denoted as $J_e$.

The force acting on the equivalent link is called equivalent force, denoted as $F_e$.

If the motion of the equivalent link will be the same with that of machine, the following conditions must be satisfied.

1) The work done by the equivalent force or moment must be equal to the sum of the work done by all the forces acting on the links of the machine.

$$W_r = \Delta E$$

由于停车阶段一般要撤去阻抗力，仅靠摩擦力做功去克服惯性动能导致停车时间过长。为了缩短停车时间，一般要在机械中安装制动器，加速消耗机械的惯性动能，减少停车时间。图 12-3 中，无制动器的停车时间在 $C$ 点，加制动器的停车时间在 $B$ 点，很明显，加装制动器缩短了停车时间。

## 12.2　机械系统的等效动力学模型

**1. 研究机械运转的方法**

机械的运转与作用在机械上的力及各力做功情况密切相关。例如：研究图 12-4 所示的曲柄压力机的运转情况时，以滑块为分离体，可以建立两个平衡方程，以连杆为分离体，可以建立三个平衡方程，再以曲柄为分离体，又可以建立三个平衡方程，共计八个平衡方程。而未知数有七个约束反力 $F_{ij}$（$A$、$B$、$C$ 铰链处的约束反力和机架给滑块的约束反力）和作用在曲柄上的平衡力矩 $M_1$，共计八个未知数。求解出作用在曲柄上的平衡力矩 $M_1$ 以后，再根据功率 $P = M_1\omega_1$，可求解曲柄的角速度 $\omega_1$。每求解一个位置的角速度都要求解八个方程，十分繁锁。因此，需要研究解决机械运转的有效方法。

对于单自由度的机械系统，给定一个构件的运动后，其余各构件的运动也随之确定。因此可以用机械中的一个构件的运动代替整个机械系统的运动。把这个能代替整个机械系统运动的构件称为等效构件。为使等效构件的运动和机械系统的真实运动一致，等效构件具有的动能应和整个机械系统的动能相等。也就是说，作用在等效构件上的外力所做的功应和整个机械系统中各外力所做的功相等。另外，等效构件上的外力在单位时间内所做的功也应等于整个机械系统中各外力在单位时间内所做的功，即等效构件上的瞬时功率等于整个机械系统的瞬时功率。这样就把研究复杂的机械系统的运动问题简化为研究一个简单的等效构件的运动问题。

为使问题简化，常取机械系统中做简单运动的构件为等效构件，即取做定轴转动的构件或做往复移动的构件为等效构件。

对等效构件进行分析时，常用到下面几个名词术语：

1）等效转动惯量。等效构件绕其质心轴的转动惯量，用 $J_e$ 表示。

2）等效质量。等效构件的质量，用 $m_e$ 表示。

3）等效力矩。作用在等效构件上的力矩，用 $M_e$ 表示。

4）等效力。作用在等效构件上的力，用 $F_e$ 表示。

当选择定轴转动的构件为等效构件时，常用到等效转动惯量和等效力矩。当选择往复移动的构件为等效构件时，常用到等效质量和等效力。

等效构件的示意图如图 12-5 所示。

为建立等效构件的动力学方程，必需求解出等效构件的转动惯量或等效构件的质量、作

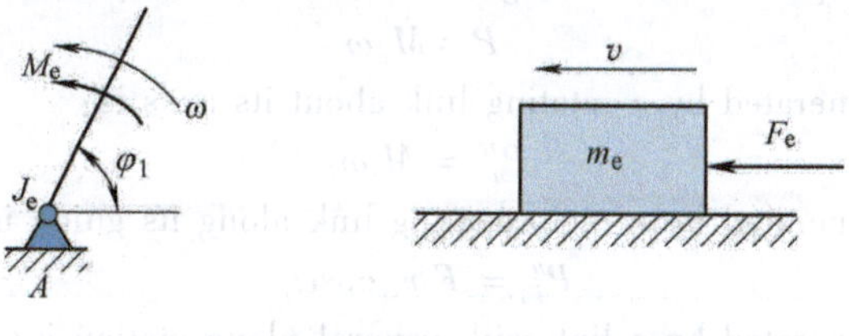


Fig. 12-5　Equivalent links（等效构件）

2) The instant power generated by the equivalent link must be equal to the sum of the power generated by all the links of the machine.

**2. The Parameter of The Equivalent Link**

Now, we discuss how to calculate the equivalent force, the equivalent moment, the equivalent mass and the equivalent moment of inertia.

(1) Rotating equivalent link When the equivalent link is rotating about an axis passing through the mass center with a constant angular velocity, the kinetic energy is as follows:

$$E = \frac{1}{2} J_e \omega^2$$

There are several moving links in a machine; some of links rotate about their axes respectively, some of them translate along their guide respectively and some of them have general plane motions. Their kinetic energies are as follows respectively.

Rotating links:

$$E_i = \frac{1}{2} J_{si} \omega_i^2$$

Translating links:

$$E_i = \frac{1}{2} m_i v_{si}^2$$

General plane motion:

$$E_i = \frac{1}{2} J_{si} \omega_i^2 + \frac{1}{2} m_i v_{si}^2$$

The total kinetic energy of the machine is given by:

$$E = \sum_{i=1}^{n} \frac{1}{2} J_{si} \omega_i^2 + \sum_{i=1}^{n} \frac{1}{2} m_i v_{si}^2$$

Where $E$ is the kinetic energy of the machine; $E_i$ is the kinetic energy of a rotating link; $J_e$ is the moment of inertia of the equivalent link; $J_{si}$ is the moment of inertia about its mass center of a rotating link; $\omega_i$ is the angular velocity of a rotating link; $m_i$ is the mass of a moving link; $v_{si}$ is the linear velocity of a moving link.

The kinetic energy of the equivalent link is equal to the kinetic energy of the whole machine, so we have:

$$\frac{1}{2} J_e \omega^2 = \sum_{i=1}^{n} \frac{1}{2} J_{si} \omega_i^2 + \sum_{i=1}^{n} \frac{1}{2} m_i v_{si}^2$$

Dividing both sides by $\frac{1}{2}\omega^2$, we obtain:

$$J_e = \sum_{i=1}^{n} J_{si} \left( \frac{\omega_i}{\omega} \right)^2 + \sum_{i=1}^{n} m_i \left( \frac{v_{si}}{\omega} \right)^2 \tag{12-4}$$

The instant power generated by the equivalent link must be equal to the sum of the power generated by all the links of the machine.

If the equivalent link performs a rotating motion, the instant power is as follows:

$$P = M_e \omega$$

The instant power generated by a rotating link about its axes is:

$$P_i' = M_i \omega_i$$

The instant power generated by a reciprocating link along its guide is:

$$P_i'' = F_i v_{si} \cos\alpha_i$$

The instant power generated by a link with general plane motion is:

用在等效构件上的外力或外力矩。

**2. 等效参量的计算**

等效构件的转动惯量或等效质量可根据等效构件的动能与机械系统动能相等的条件来求解。

（1）定轴转动的等效构件　如等效构件以角速度 $\omega$ 做定轴转动，则其动能为

$$E = \frac{1}{2}J_e\omega^2$$

组成机械系统的各构件或做定轴转动，或做往复直线移动，或做平面运动，各类不同运动形式的构件动能分别为

$$E_i = \frac{1}{2}J_{si}\omega_i^2 \text{（转动构件）}$$

$$E_i = \frac{1}{2}m_i v_{si}^2 \text{（移动构件）}$$

$$E_i = \frac{1}{2}J_{si}\omega_i^2 + \frac{1}{2}m_i v_{si}^2 \text{（平面运动构件）}$$

整个机械系统的动能为

$$E = \sum_{i=1}^{n}\frac{1}{2}J_{si}\omega_i^2 + \sum_{i=1}^{n}\frac{1}{2}m_i v_{si}^2$$

式中，$E_i$ 为第 $i$ 个构件的动能；$J_e$ 为等效转动惯量；$\omega_i$ 为第 $i$ 个构件的角速度；$m_i$ 为第 $i$ 个构件的质量；$J_{si}$ 为第 $i$ 个构件绕其质心轴的转动惯量；$v_{si}$ 为第 $i$ 个构件质心处的速度。

由于等效构件的动能与机械系统的动能相等，则有

$$\frac{1}{2}J_e\omega^2 = \sum_{i=1}^{n}\frac{1}{2}J_{si}\omega_i^2 + \sum_{i=1}^{n}\frac{1}{2}m_i v_{si}^2$$

方程两边同除以 $\frac{1}{2}\omega^2$，可求解出等效转动惯量为

$$J_e = \sum_{i=1}^{n}J_{si}\left(\frac{\omega_i}{\omega}\right)^2 + \sum_{i=1}^{n}m_i\left(\frac{v_{si}}{\omega}\right)^2 \tag{12-4}$$

由等效构件的瞬时功率与机械系统的瞬时功率相等，可求解等效力矩。

做定轴转动的等效构件的瞬时功率为

$$P = M_e\omega$$

机械系统中各类不同运动形式的构件的瞬时功率分别为

$$P_i' = M_i\omega_i \text{（转动构件）}$$

$$P_i'' = F_i v_{si}\cos\alpha_i \text{（移动构件）}$$

$$P_i''' = P_i' + P_i'' = M_i\omega_i + F_i v_{si}\cos\alpha_i \text{（平面运动构件）}$$

整个机械系统的瞬时功率为

$$P = \sum_{i=1}^{n}M_i\omega_i + \sum_{i=1}^{n}F_i v_{si}\cos\alpha_i$$

等效构件的瞬时功率与机械系统的瞬时功率相等，即

$$M_e\omega = \sum_{i=1}^{n}M_i\omega_i + \sum_{i=1}^{n}F_i v_{si}\cos\alpha_i$$

$$P'''_i = M_i\omega_i + F_i v_{si}\cos\alpha_i$$

The instant power generated by the equivalent link is equal to the ones generated by the whole machine, so we have:

$$M_e\omega = \sum_{i=1}^{n} M_i\omega_i + \sum_{i=1}^{n} F_i v_{si}\cos\alpha_i$$

Dividing both sides by $\omega$, we obtain:

$$M_e = \sum_{i=1}^{n} M_i\left(\frac{\omega_i}{\omega}\right) + \sum_{i=1}^{n} F_i\left(\frac{v_{si}}{\omega}\right)\cos\alpha_i \tag{12-5}$$

(2) Reciprocating equivalent link

If the equivalent link performs a translating motion, the kinetic energy is as follows:

$$E = \frac{1}{2}m_e v^2$$

In the same way, we have:

$$\frac{1}{2}m_e v^2 = \sum_{i=1}^{n}\frac{1}{2}J_{si}\omega_i^2 + \sum_{i=1}^{n}\frac{1}{2}m_i v_{si}^2$$

Dividing both sides by $\frac{1}{2}v^2$, we obtain:

$$m_e = \sum_{i=1}^{n} J_{si}\left(\frac{\omega_i}{v}\right)^2 + \sum_{i=1}^{n} m_i\left(\frac{v_{si}}{v}\right)^2 \tag{12-6}$$

Where $m_e$ is the mass of the equivalent link; $v$ is the translating velocity of the equivalent link.

If the equivalent link performs a translating motion, the instant power is as follows:

$$P = F_e v$$

The instant power generated by the machine is:

$$P = \sum_{i=1}^{n} M_i\omega_i + \sum_{i=1}^{n} F_i v_{si}\cos\alpha_i$$

The equivalent force is as follows:

$$F_e = \sum_{i=1}^{n} M_i\left(\frac{\omega_i}{v}\right) + \sum_{i=1}^{n} F_i\left(\frac{v_{si}}{v}\right)\cos\alpha_i \tag{12-7}$$

Where $P$ is the instantaneous power generated by the machine; $F_e$ is the force acting on the equivalent link; $F_i$ is the force acting on a link; $\alpha_i$ is the angle between the direction of force acting on the link mass-center and the velocity of the mass-center.

All the driving forces or moments acting on the machine are replaced by one force or moment applied to the equivalent link, this is called the equivalent driving force or driving moment. All the resistances acting on the machine are replaced by one force or moment applied to the equivalent link, this is called the equivalent resistant force or resistant moment.

Obviously, we can write them as follows:

$$M_e = M_{ed} - M_{er}$$

$$F_e = F_{ed} - F_{er}$$

Where $M_{ed}$ and $F_{ed}$ are the equivalent driving moment and equivalent driving force; $M_{er}$ and $F_{er}$ are the equivalent resistant moment and equivalent resistance.

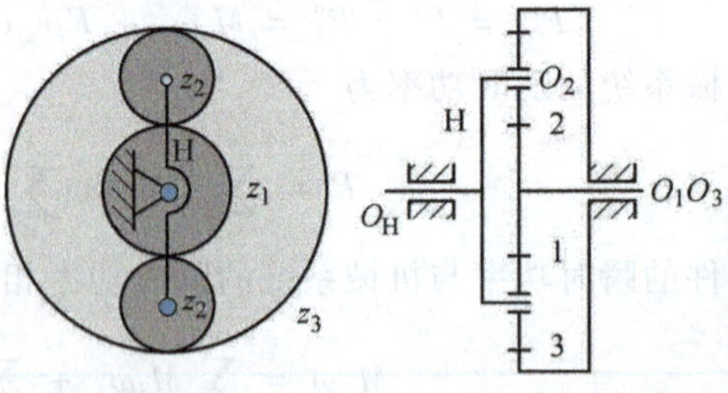


Fig. 12-6 Planetary gear train(行星轮系)

**Example 12-1** Fig. 12-6 shows a

方程两边同除以 $\omega$，可求解出等效力矩为

$$M_e = \sum_{i=1}^{n} M_i\left(\frac{\omega_i}{\omega}\right) + \sum_{i=1}^{n} F_i\left(\frac{v_{si}}{\omega}\right)\cos\alpha_i \tag{12-5}$$

式中，$M_i$ 为第 $i$ 个构件上的力矩；$F_i$ 为第 $i$ 个构件上的力；$\alpha_i$ 为第 $i$ 个构件质心处的速度 $v_{si}$与作用力 $F_i$ 之间的夹角。

（2）直线移动的等效构件　如等效构件为移动件，则其动能为

$$E = \frac{1}{2}m_e v^2$$

由于等效构件的动能与机械系统的动能相等，则有

$$\frac{1}{2}m_e v^2 = \sum_{i=1}^{n}\frac{1}{2}J_{si}\omega_i^2 + \sum_{i=1}^{n}\frac{1}{2}m_i v_{si}^2$$

方程两边同除以$\frac{1}{2}v^2$，可求解出等效质量为

$$m_e = \sum_{i=1}^{n} J_{si}\left(\frac{\omega_i}{v}\right)^2 + \sum_{i=1}^{n} m_i\left(\frac{v_{si}}{v}\right)^2 \tag{12-6}$$

同理，由等效构件的瞬时功率与机械系统的瞬时功率相等，可求解等效力。

等效构件作往复移动，其瞬时功率为

$$P = F_e v$$

等效构件的瞬时功率与机械系统的瞬时功率相等，即

$$F_e v = \sum_{i=1}^{n} M_i\omega_i + \sum_{i=1}^{n} F_i v_{si}\cos\alpha_i$$

方程两边同除以 $v$，可求解出等效力为

$$F_e = \sum_{i=1}^{n} M_i\left(\frac{\omega_i}{v}\right) + \sum_{i=1}^{n} F_i\left(\frac{v_{si}}{v}\right)\cos\alpha_i \tag{12-7}$$

由以上计算可知，等效转动惯量、等效质量、等效力矩、等效力的数值均与构件的速度比值有关，而构件的速度又与机构位置有关，故等效转动惯量、等效质量、等效力矩、等效力均为机构位置的函数。

这里的等效力矩是指作用在等效构件上的等效驱动力矩 $M_{ed}$ 和等效阻抗力矩 $M_{er}$ 之和；等效力是指作用在等效构件上的等效驱动力 $F_{ed}$ 与等效阻抗力 $F_{er}$ 的和。即

$$M_e = M_{ed} - M_{er}$$

$$F_e = F_{ed} - F_{er}$$

工程上有时需要求解某一个力的等效力或等效力矩。

驱动力的等效驱动力可按驱动力的瞬时功率等于等效驱动力的瞬时功率来求解，驱动力矩的等效驱动力矩可按驱动力矩的瞬时功率等于等效驱动力矩的瞬时功率来求解。

阻抗力的等效阻抗力可按阻抗力的瞬时功率等于等效阻抗力的瞬时功率来求解，阻抗力矩的等效阻抗力矩可按阻抗力矩的瞬时功率等于等效阻抗力矩的瞬时功率来求解。

**例 12-1**　图 12-6 所示轮系中，已知各齿轮的齿数分别为 $z_1$、$z_2$、$z_3$，各齿轮与系杆 H 的质心与其回转中心重合，绕质心的转动惯量分别为 $J_1$、$J_2$、$J_H$。有两个行星轮，每个行星轮的质量均为 $m_2$。若等效构件设置在齿轮 1 处，求其等效转动惯量 $J_e$。

planetary gear train. The numbers of teeth of gears are $z_1$, $z_2$, $z_3$ respectively. The mass-centers of these gears are coefficient with their rotating centers, and the moments of inertia of gears 1, 2 and the arm about their centers are $J_1$, $J_2$, $J_H$. The mass of the planetary gear is $m_2$. We assume the gear 1 to be the equivalent link. Determine the equivalent moment of inertia.

**Solution**

The kinetic energy of the equivalent link is:

$$E = \frac{1}{2}J_e\omega_1^2$$

The kinetic energy of the train is:

$$E = \frac{1}{2}J_1\omega_1^2 + 2\left(\frac{1}{2}J_2\omega_2^2 + \frac{1}{2}m_2 v_{s2}^2\right) + \frac{1}{2}J_H\omega_H^2$$

Therefore, we have:

$$J_e = J_1 + 2\left[J_2\left(\frac{\omega_2}{\omega_1}\right)^2 + m_2\left(\frac{v_{s2}}{\omega_1}\right)^2\right] + J_H\left(\frac{\omega_H}{\omega_1}\right)^2$$

$$J_e = J_1 + 2\left[J_2\left(\frac{\omega_2}{\omega_1}\right)^2 + m_2\left(\frac{\omega_H r_H}{\omega_1}\right)^2\right] + J_H\left(\frac{\omega_H}{\omega_1}\right)^2$$

Calculating the ratio of the train, we have:

$$\frac{\omega_2}{\omega_1} = \frac{z_2 - z_3}{z_1 + z_3}\frac{z_1}{z_2}, \quad \frac{\omega_H}{\omega_1} = \frac{z_1}{z_1 + z_3}$$

Rearranging the above equations, we obtain:

$$J_e = J_1 + 2J_2\left[\frac{z_1(z_2 - z_3)}{z_2(z_1 + z_3)}\right]^2 + (2m_2 r_H^2 + J_H)\left(\frac{z_1}{z_1 + z_3}\right)^2$$

**Example 12-2** Fig. 12-7 shows a Scotch-yoke mechanism. The crank has a length of $l_1$, and the moment of inertia about its rotating center $A$ is $J_1$. The link 2 and 3 have masses of $m_2$, $m_3$. The resistance acting on the link 3 is $F_3$. Determine the equivalent moment of inertia, when the crank is the equivalent link, and equivalent resistance moment acting on the crank.

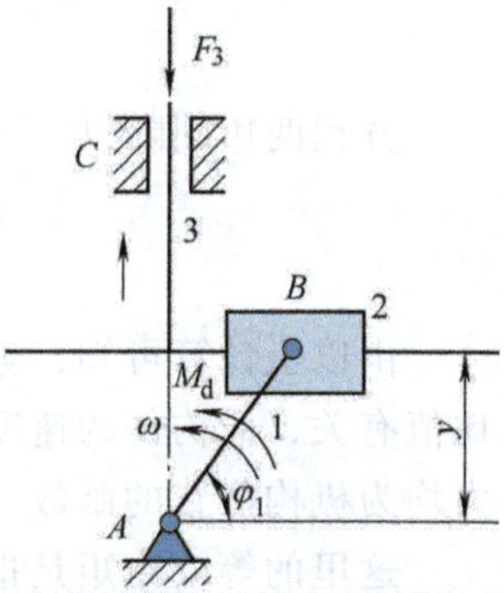

Fig. 12-7 Scotch-yoke mechanism (正弦机构)

**Solution**

The kinetic energy of the equivalent link 1 is equal to the sum of the kinetic energy of all links of the machine.

$$\frac{1}{2}J_e\omega_1^2 = \frac{1}{2}J_1\omega_1^2 + \frac{1}{2}m_2 v_B^2 + \frac{1}{2}m_3 v_C^2$$

Each side of the equation is divided by $\frac{\omega_1^2}{2}$, and we obtain:

$$J_e = J_1 + m_2\left(\frac{v_B}{\omega_1}\right)^2 + m_3\left(\frac{v_C}{\omega_1}\right)^2$$

Where $v_B = \omega_1 l_1$, $v_C = y' = (l_1\sin\varphi_1)' = l_1\omega_1\cos\varphi_1$.

Thus, we have:

$$J_e = J_1 + m_2 l_1^2 + m_3 l_1^2\cos^2\varphi_1 = J_C + J_v$$

Where $J_C = J_1 + m_2 l_1^2$, $J_C$ is a constant. $J_v = m_3 l_1^2\cos^2\varphi_1$, $J_v$ is the function of the link position, and a variation.

The instantaneous power supplied by the equivalent resistant moment acting on the equivalent link

**解**　等效构件在齿轮 1 处，其动能为

$$E = \frac{1}{2}J_e\omega_1^2$$

机构系统的动能为

$$E = \frac{1}{2}J_1\omega_1^2 + 2\left(\frac{1}{2}J_2\omega_2^2 + \frac{1}{2}m_2v_{s2}^2\right) + \frac{1}{2}J_H\omega_H^2$$

由于两者动能相等，两边同除以$\frac{1}{2}\omega_1^2$并整理得

$$J_e = J_1 + 2\left[J_2\left(\frac{\omega_2}{\omega_1}\right)^2 + m_2\left(\frac{v_{s2}}{\omega_1}\right)^2\right] + J_H\left(\frac{\omega_H}{\omega_1}\right)^2$$

$$J_e = J_1 + 2\left[J_2\left(\frac{\omega_2}{\omega_1}\right)^2 + m_2\left(\frac{\omega_H r_H}{\omega_1}\right)^2\right] + J_H\left(\frac{\omega_H}{\omega_1}\right)^2$$

由轮系传动比可有

$$\frac{\omega_2}{\omega_1} = \frac{z_2 - z_3}{z_1 + z_3}\frac{z_1}{z_2}, \quad \frac{\omega_H}{\omega_1} = \frac{z_1}{z_1 + z_3}$$

再行整理得

$$J_e = J_1 + 2J_2\left[\frac{z_1(z_2 - z_3)}{z_2(z_1 + z_3)}\right]^2 + (2m_2r_H^2 + J_H)\left(\frac{z_1}{z_1 + z_3}\right)^2$$

由例 12-1 可知：传动比为常量的机械系统，其等效转动惯量也为常量。

**例 12-2**　图 12-7 所示正弦机构中，已知曲柄长为 $l_1$，绕 $A$ 轴的转动惯量为 $J_1$，构件 2 和 3 的质量分别为 $m_2$、$m_3$，作用在构件 3 上的阻抗力为 $F_3$。若等效构件设置在构件 1 处，求其等效转动惯量 $J_e$，并求出阻抗力 $F_3$ 的等效阻抗力矩 $M_{er}$。

**解**　根据动能相等的条件，有

$$\frac{1}{2}J_e\omega_1^2 = \frac{1}{2}J_1\omega_1^2 + \frac{1}{2}m_2v_B^2 + \frac{1}{2}m_3v_C^2$$

$$J_e = J_1 + m_2\left(\frac{v_B}{\omega_1}\right)^2 + m_3\left(\frac{v_C}{\omega_1}\right)^2$$

由运动分析可知

$$v_B = \omega_1 l_1, \quad v_C = y' = (l_1\sin\varphi_1)' = l_1\omega_1\cos\varphi_1$$

将其代入上述方程中可解出 $J_e$

$$J_e = J_1 + m_2l_1^2 + m_3l_1^2\cos^2\varphi_1 = J_C + J_v$$

式中，$J_C = J_1 + m_2l_1^2$，$J_v = m_3l_1^2\cos^2\varphi_1$。

例 12-2 说明机械系统含有连杆机构时，其等效转动惯量由常量和变量两部分组成。由于工程中的连杆机构常安装在低速级，等效转动惯量中的变量部分有时可以忽略不计。

由阻抗力的瞬时功率等于等效阻抗力的瞬时功率，可得

$$M_{er}\omega_1 = F_3v_C\cos180°$$

is equal to the power supplied by resistant force acting on the machine.

We obtain:

$$M_{er}\omega_1 = F_3 v_C \cos 180^\circ$$

$$M_{er} = -\frac{F_3 l_1 \omega_1}{\omega_1}\cos\varphi_1 = -F_3 l_1 \cos\varphi_1$$

The negative sign means that the direction of $M_{er}$ is opposite to that of $\omega_1$.

## 12.3 Kinetic Equations of Mechanism Systems

After an actual machine has been reduced to an equivalent link, it is convenient to build the kinetic equations of the machine.

**1. Kinetic Equations of the Equivalent Link**

We suppose that the kinetic energy of the equivalent link increases by d$E$ in an interval time d$t$, and the elementary work done by all the external forces acting on the link at the same time is d$W$. We have:

$$\mathrm{d}E = \mathrm{d}W$$

Suppose the equivalent link performs a rotating motion, the change in kinetic energy in differential form is as follows:

$$\mathrm{d}\left(\frac{1}{2}J\omega^2\right) = M\mathrm{d}\varphi \tag{12-8}$$

If the equivalent link performs a translating motion, the change in kinetic energy in differential form is as follows:

$$\mathrm{d}\left(\frac{1}{2}mv^2\right) = F\mathrm{d}s \tag{12-9}$$

Dividing the equation (12-8) by d$\varphi$, we have:

$$\frac{\mathrm{d}\left(\frac{1}{2}J\omega^2\right)}{\mathrm{d}\varphi} = M \tag{12-10}$$

The left side of the equation (12-10) is derived with respect to angle $\varphi$, then:

$$J\frac{\omega\mathrm{d}\omega}{\mathrm{d}\varphi} + \frac{\omega^2}{2}\frac{\mathrm{d}J}{\mathrm{d}\varphi} = M = M_d - M_r \tag{12-11}$$

$$\frac{\mathrm{d}\omega}{\mathrm{d}\varphi} = \frac{\mathrm{d}\omega}{\mathrm{d}t}\frac{\mathrm{d}t}{\mathrm{d}\varphi} = \frac{\mathrm{d}\omega}{\mathrm{d}t}\frac{1}{\omega}$$

Rearrange the above equations, we have:

$$J\frac{\mathrm{d}\omega}{\mathrm{d}t} + \frac{\omega^2}{2}\frac{\mathrm{d}J}{\mathrm{d}\varphi} = M = M_d - M_r \tag{12-12}$$

The equation (12-12) is called the kinetic equation in differential form when the equivalent link is rotating about its axis.

Dividing the equation (12-9) by d$s$, and deriving with respect to $s$, then:

$$m\frac{v\mathrm{d}v}{\mathrm{d}s} + \frac{v^2}{2}\frac{\mathrm{d}m}{\mathrm{d}s} = F = F_d - F_r \tag{12-13}$$

$$\frac{\mathrm{d}v}{\mathrm{d}s} = \frac{\mathrm{d}v}{\mathrm{d}t}\frac{\mathrm{d}t}{\mathrm{d}s} = \frac{\mathrm{d}v}{\mathrm{d}t}\frac{1}{v}$$

$$M_{er} = -\frac{F_3 l_1 \omega_1}{\omega_1}\cos\varphi_1 = -F_3 l_1 \cos\varphi_1$$

## 12.3 机械系统的运动方程及其求解

引入等效构件的概念后，就可以把研究机械系统的运动规律问题简化为研究等效构件的运动规律问题。只要建立等效构件的运动方程，就可以确定机械系统的运动规律。

**1. 等效构件的运动方程**

在研究等效构件的运动方程时，为简化书写格式，在不引起混淆的情况下，略去表示等效概念的下角标 e。根据动能定理，在 d$t$ 时间内，等效构件上的动能增量 d$E$ 应等于该瞬时等效力或等效力矩所做的功 d$W$，即

$$\mathrm{d}E = \mathrm{d}W$$

如等效构件做定轴转动，则有

$$\mathrm{d}\left(\frac{1}{2}J\omega^2\right) = M\mathrm{d}\varphi \tag{12-8}$$

如等效构件做往复移动，则有

$$\mathrm{d}\left(\frac{1}{2}mv^2\right) = F\mathrm{d}s \tag{12-9}$$

由式（12-8）可得

$$\frac{\mathrm{d}\left(\frac{1}{2}J\omega^2\right)}{\mathrm{d}\varphi} = M \tag{12-10}$$

由于等效转动惯量、等效力、等效力矩及角速度均是机构位置的函数，实际上 $J = J(\varphi)$，$F = F(\varphi)$，$M = M(\varphi)$，$\omega = \omega(\varphi)$。

整理式（12-10），得

$$J\frac{\omega \mathrm{d}\omega}{\mathrm{d}\varphi} + \frac{\omega^2}{2}\frac{\mathrm{d}J}{\mathrm{d}\varphi} = M = M_d - M_r \tag{12-11}$$

由于 $$\frac{\mathrm{d}\omega}{\mathrm{d}\varphi} = \frac{\mathrm{d}\omega}{\mathrm{d}t}\frac{\mathrm{d}t}{\mathrm{d}\varphi} = \frac{\mathrm{d}\omega}{\mathrm{d}t}\frac{1}{\omega}$$

将其代入式（12-11），得

$$J\frac{\mathrm{d}\omega}{\mathrm{d}t} + \frac{\omega^2}{2}\frac{\mathrm{d}J}{\mathrm{d}\varphi} = M = M_d - M_r \tag{12-12}$$

式（12-12）称为做定轴转动的等效构件的微分方程。

等效构件做往复移动时的微分方程推导如下。

整理式（12-9），得

$$m\frac{v\mathrm{d}v}{\mathrm{d}s} + \frac{v^2}{2}\frac{\mathrm{d}m}{\mathrm{d}s} = F = F_d - F_r \tag{12-13}$$

将 $\frac{\mathrm{d}v}{\mathrm{d}s} = \frac{\mathrm{d}v}{\mathrm{d}t}\frac{\mathrm{d}t}{\mathrm{d}s} = \frac{\mathrm{d}v}{\mathrm{d}t}\frac{1}{v}$ 代入式（12-13），得

$$m\frac{\mathrm{d}v}{\mathrm{d}t} + \frac{v^2}{2}\frac{\mathrm{d}m}{\mathrm{d}s} = F = F_d - F_r \tag{12-14}$$

式（12-14）称为等效构件做往复移动的微分方程。

Rearranging the above equations, we have:

$$m\frac{dv}{dt} + \frac{v^2}{2}\frac{dm}{ds} = F = F_d - F_r \tag{12-14}$$

The equation (12-14) is called the kinetic equation in differential form when the equivalent link is translating along its path.

If integrating the equation (12-8), and considering the border conditions, we obtain:

$$\frac{1}{2}J\omega^2 - \frac{1}{2}J_0\omega_0^2 = \int_{\varphi_0}^{\varphi} M d\varphi = \int_{\varphi_0}^{\varphi} (M_d - M_r) d\varphi \tag{12-15}$$

Where $\omega_0$ is the initial angular velocity; $\omega$ is the angular velocity at the end of the interval; $\varphi_0$ is the initial angular position; $\varphi$ is the angular position at the end of the interval; $J_0$ is the equivalent moment of inertia, when $\varphi = \varphi_0$; $J$ is the equivalent moment of inertia at the end of the interval.

The equation (12-15) is called the kinetic equation in integrating form when the equivalent link is rotating about its axis.

If integrating the equation (12-9), and considering the border conditions, we obtain:

$$\frac{1}{2}mv^2 - \frac{1}{2}m_0 v_0^2 = \int_{s_0}^{s} F ds = \int_{s_0}^{s} (F_d - F_r) ds \tag{12-16}$$

Where $v_0$ is the initial linear velocity; $v$ is the linear velocity at the end of the interval; $s_0$ is the initial displacement; $s$ is the displacement at the end of the interval; $m_0$ is the equivalent mass, when $s = s_0$; $m$ is the equivalent mass at the end of the interval.

The equation (12-16) is called the kinetic equation in integrating form when the equivalent link is translating along its path.

**2. Solution of the Kinetic Equations**

Generally, the equivalent mass or the equivalent moment of inertia is a function of position of the equivalent link, and the equivalent force or equivalent moment is also a function of position. We will discuss a few cases below.

(1) The equivalent moment of inertia and equivalent moment are constants This is suitable for machines with constant ratio and constant force or moment, such as a gear hoist.

Since the equivalent moment of inertia and equivalent moment are constants, the equation (12-12) can be written as:

$$J\frac{d\omega}{dt} = M$$

Furthermore, we have:

$$\frac{d\omega}{dt} = \frac{M}{J} = \alpha$$

Where $\alpha$ is the angular acceleration of the equivalent link.

Integrating both side, we have:

$$\int_{\omega_0}^{\omega} d\omega = \int_{t_0}^{t} \alpha dt$$

Therefore:

$$\omega = \omega_0 + \alpha(t - t_0)$$

$$\varphi = \varphi_0 + \omega_0(t - t_0) + \frac{\alpha}{2}(t - t_0)^2$$

如果对式（12-8）两边积分，并取边界条件为：$t=t_0$，$\varphi=\varphi_0$，$\omega=\omega_0$，$J=J_0$，则有

$$\frac{1}{2}J\omega^2-\frac{1}{2}J_0\omega_0{}^2=\int_{\varphi_0}^{\varphi}M\mathrm{d}\varphi=\int_{\varphi_0}^{\varphi}(M_\mathrm{d}-M_\mathrm{r})\mathrm{d}\varphi \tag{12-15}$$

式中，$\omega_0$、$\omega$ 分别为等效构件在初始位置和任意位置的角速度；$\varphi_0$、$\varphi$ 分别为等效构件在初始位置和任意位置的角位移；$J_0$、$J$ 分别为等效构件在初始位置和任意位置的等效转动惯量。

式（12-15）称为等效构件做定轴转动的积分方程。

如果对式（12-9）两边积分，并取边界条件为：$t=t_0$，$s=s_0$，$v=v_0$，$m=m_0$，则有

$$\frac{1}{2}mv^2-\frac{1}{2}m_0v_0^2=\int_{s_0}^{s}F\mathrm{d}s=\int_{s_0}^{s}(F_\mathrm{d}-F_\mathrm{r})\mathrm{d}s \tag{12-16}$$

式中，$v_0$、$v$ 分别为等效构件在初始位置和任意位置的线速度；$s_0$、$s$ 分别为等效构件在初始位置和任意位置的位移；$m_0$、$m$ 分别为等效构件在初始位置和任意位置的等效质量。

式（12-16）称为等效构件做往复移动的积分方程。

在描述等效构件的运动时，有微分方程和积分方程两种形式。具体应用时要看使用哪个方程更为方便。

**2. 运动方程的求解**

不同机械的驱动力和工作阻力特性不同，它们可能是时间的函数，也可能是机构位置或速度的函数，等效转动惯量可能是常数也可能是机构位置的函数，等效力或等效力矩可能是机构位置的函数，也可能是速度的函数。因此，运动方程的求解方法也不尽相同。

工程上常选做定轴转动的构件为等效构件，故下面仅讨论等效构件做定轴转动的简单情况。

（1）等效转动惯量与等效力矩均为常数的运动方程求解　等效转动惯量与等效力矩均为常数是定传动比机械系统中的常见情况。在这种情况下运转的机械大都属于等速稳定运转，使用力矩方程求解该类问题要方便些。

由于 $J=$ 常数，$M=$ 常数，故式（12-12）可改写为

$$J\frac{\mathrm{d}\omega}{\mathrm{d}t}=M$$

$$\frac{\mathrm{d}\omega}{\mathrm{d}t}=\frac{M}{J}=\alpha$$

$\mathrm{d}\omega=\alpha\mathrm{d}t$，两边积分后得

$$\int_{\omega_0}^{\omega}\mathrm{d}\omega=\int_{t_0}^{t}\alpha\mathrm{d}t$$

$$\omega=\omega_0+\alpha(t-t_0)$$

$$\varphi=\varphi_0+\omega_0(t-t_0)+\frac{\alpha}{2}(t-t_0)^2$$

**例 12-3**　图 12-8 所示的机械系统中，已知电动机 $A$ 的转速为 1440r/min，减速箱的传动比 $i=2.5$，选 $B$ 轴为等效构件，等效转动惯量 $J_\mathrm{e}=0.5\mathrm{kg}\cdot\mathrm{m}^2$。要求 $B$ 轴制动后 3s 停车，求解等效制动力矩。

**Example 12-3** Fig. 12-8 shows a mechanism system in which the electric-motor $A$ rotates at 1440r/min, and the ratio of the gear reducer is 2.5. The shaft $B$ is selected as an equivalent link, and its equivalent moment of inertia $J_e$ is 0.5kg · m$^2$. If the shaft $B$ has been braked, the hoist must be stopped within three seconds. Determine the braking moment supplied to the shaft $B$.

**Solution**

$$\omega_B = \frac{1440}{2.5} \times \frac{2\pi}{60}\text{rad/s} = 60.32\text{rad/s}$$

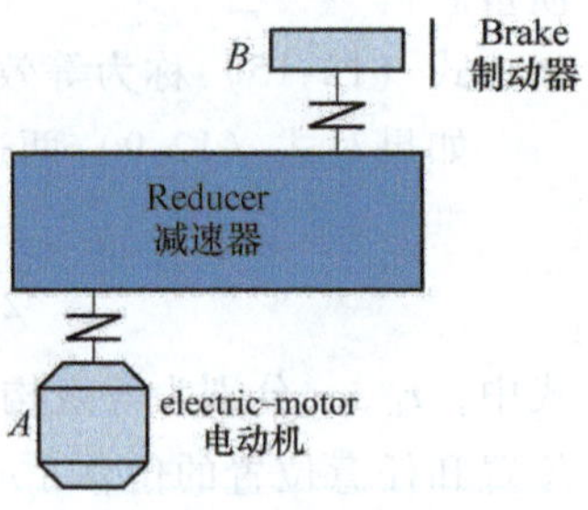


Fig. 12-8 Simple mechanical system(简单的机械系统)

Because of $\omega = \omega_0 + \alpha(t - t_0)$, and $\omega_0 = \omega_B$, $\omega = 0$, $t = 3$, $t_0 = 0$.

Thus, the angular acceleration $\alpha$ is:

$$\alpha = \frac{\omega - \omega_0}{t - t_0} = \frac{0 - 60.32}{3} = -20.1\text{rad/s}^2$$

When braking the shaft $B$, the driving force must be abolished, therefore:

$$M = M_d - M_r = -M_r$$

$M_r$ is the braking moment.

Since $\dfrac{d\omega}{dt} = \dfrac{M}{J} = \alpha$, we have:

$$M_r = -\alpha J = -20.1 \times 0.5 = -10.05\text{N} \cdot \text{m}$$

(2) The equivalent moment of inertia and equivalent moment are functions of position When the equivalent moment of inertia and equivalent moment are analytical forms, we can use the integrating equation to solve this problem.

From the equation:

$$\frac{1}{2}J\omega^2 - \frac{1}{2}J_0\omega_0^2 = \int_{\varphi_0}^{\varphi} M d\varphi$$

We have:

$$\omega = \sqrt{\frac{J_0}{J}\omega_0^2 + \frac{2}{J}\int_{\varphi_0}^{\varphi} M d\varphi}$$

## 12.4 Periodic Speed Fluctuation and Regulation in a Machine

### 1. Work and Energy in a Steady Running Period

The characteristic of the operating period is that the work done by the equivalent driving moment or force must be equal to the work done by the equivalent resistant moment or force in an operating cycle. But at some instant the work done by the equivalent driving moment or force is not equal to the work done by the equivalent resistant moment or force. Fig. 12-9 shows a diagram of equivalent driving moment $M_d$ and equivalent resistant moment $M_r$.

When $M_d > M_r$, the kinetic energy of the machine increases, and the angular velocity increases too. When $M_d < M_r$, the kinetic energy of the machine decreases, and the angular velocity decreases too.

**解**
$$\omega_B = \frac{1440}{2.5} \times \frac{2\pi}{60} \text{rad/s} = 60.32 \text{rad/s}$$

由 $\omega = \omega_0 + \alpha(t - t_0)$，$\omega_0 = \omega_B$，$\omega = 0$，$t = 3$，$t_0 = 0$，得

$$\alpha = \frac{\omega - \omega_0}{t - t_0} = \frac{0 - 60.32}{3} \text{rad/s}^2 = -20.1 \text{rad/s}^2$$

制动时要取消驱动力矩和工作阻力，$M = M_d - M_r = -M_r$，此处 $M_r$ 为制动力矩。

由$\dfrac{d\omega}{dt} = \dfrac{M}{J} = \alpha$，可知

$$M_r = -\alpha J = -20.1 \times 0.5 = -10.05 \text{N} \cdot \text{m}$$

（2）等效转动惯量与等效力矩均为等效构件位置函数的运动方程的求解　当 $J = J(\varphi)$ 和 $M = M(\varphi)$ 可用解析式表示时，用积分方程求解方便些。

由

$$\frac{1}{2}J\omega^2 - \frac{1}{2}J_0\omega_0^2 = \int_{\varphi_0}^{\varphi} M d\varphi$$

可解出

$$\omega = \sqrt{\frac{J_0}{J}\omega_0^2 + \frac{2}{J}\int_{\varphi_0}^{\varphi} M d\varphi}$$

当等效转动惯量与等效力矩不能写成函数式时，可用数值解法求解。

## 12.4　周期性速度波动及飞轮设计

周期性变速稳定运转过程中，在一个运转周期内，等效驱动力矩做的功等于等效阻抗力矩做的功。但在运转周期内的任一时刻，等效驱动力矩做的功不等于等效阻抗力矩做的功，从而导致了机械运转过程中的速度波动。

### 1. 周期性变速稳定运转过程中的功能关系

图 12-9 为等效力矩的变化线图。等效驱动力矩和等效阻抗力矩均为机构位置的函数。$M_d = M_d(\varphi)$，$M_r = M_r(\varphi)$。$\varphi_a$、$\varphi_e$ 分别为运转周期的开始位置和终止位置，运转周期为 $2\pi$。

在 $\varphi_a \sim \varphi_b$ 区间，$M_d > M_r$，动能增量 $\Delta E_1 > 0$。机械动能增加，角速度上升。

在 $\varphi_b \sim \varphi_c$ 区间，$M_d < M_r$，动能增量 $\Delta E_2 < 0$。机械动能减小，角速度下降。

在 $\varphi_c \sim \varphi_d$ 区间，$M_d > M_r$，动能增量 $\Delta E_3 > 0$。机械动能增加，角速度上升。

在 $\varphi_d \sim \varphi_e$ 区间，$M_d < M_r$，动能增量 $\Delta E_4 < 0$。机械动能减小，角速度下降。

在一个运转周期内，等效驱动力矩所做的功 $W_{dp}$ 等于等效阻抗力矩做的功 $W_{rp}$。即

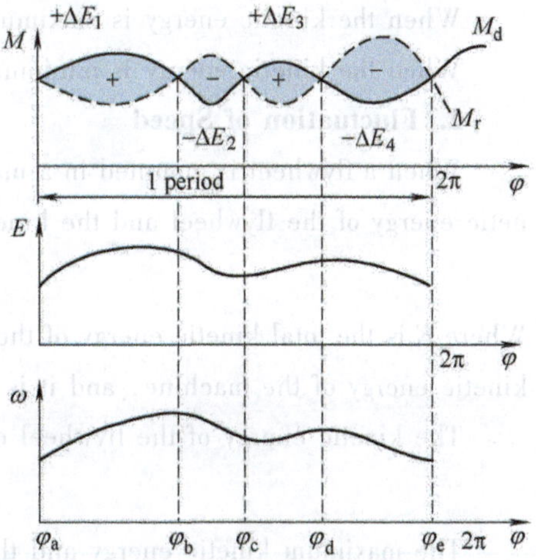


Fig. 12-9　Equivalent moment diagram
（等效力矩线图）

$$W_{dp} = \int_{\varphi_a}^{\varphi_e} M_d(\varphi) d\varphi$$

$$W_{rp} = \int_{\varphi_a}^{\varphi_e} M_r(\varphi) d\varphi$$

From Fig. 12-9, the work done by the equivalent driving moment is as follows:

$$W_{dp} = \int_{\varphi_a}^{\varphi_e} M_d(\varphi)\,d\varphi$$

The work done by the equivalent resistant moment is determined in the same way.

$$W_{rp} = \int_{\varphi_a}^{\varphi_e} M_r(\varphi)\,d\varphi$$

The above equations can be written as:

$$\int_{\varphi_a}^{\varphi_e} M_d(\varphi)\,d\varphi - \int_{\varphi_a}^{\varphi_e} M_r(\varphi)\,d\varphi = \int_{\varphi_a}^{\varphi_e} [M_d(\varphi) - M_r(\varphi)]\,d\varphi = 0 \tag{12-17}$$

We can obtain:

$$\sum_{i=1}^{n} \Delta E_i = 0$$

Where areas $\Delta E$ are called variation of kinetic energy. The areas $\Delta E$ lying above the $M_r$ curve are called increase works, such as the areas $\Delta E_1$, $\Delta E_3$. The areas $\Delta E$ lying above the $M_d$ curve are called decrease works, such as the areas $\Delta E_2$, $\Delta E_4$. Therefore, we can also say that all the sum of the increase works and decrease works are zero.

Suppose the kinetic energy at the start in a steady running motion is $E_a$, the energies of the equivalent link corresponding to positions $a$, $b$, $c$, $d$, $e$ are as follows:

$$E_b = E_a + \Delta E_1$$

$$E_c = E_b - \Delta E_2 = E_a + \Delta E_1 - \Delta E_2$$

$$E_d = E_c + \Delta E_3 = E_a + \Delta E_1 - \Delta E_2 + \Delta E_3$$

$$E_e = E_d - \Delta E_4 = E_a + \Delta E_1 - \Delta E_2 + \Delta E_3 - \Delta E_4$$

We can determine the maximum kinetic energy and the minimum kinetic energy from the above formulas.

When the kinetic energy is maximum, the speed is maximum.

When the kinetic energy is minimum, the speed is minimum.

**2. Fluctuation of Speed**

When a flywheel is mounted in a machine spindle, the total kinetic energy is the sum of the kinetic energy of the flywheel and the kinetic energy of the machine. So we have:

$$E = E_f + E_e$$

Where $E$ is the total kinetic energy of the machine; $E_f$ is the kinetic energy of the flywheel; $E_e$ is the kinetic energy of the machine, and it is the kinetic energy of the equivalent link.

The kinetic energy of the flywheel can be written as:

$$E_f = E - E_e$$

The maximum kinetic energy and the minimum kinetic energy are as follows:

$$E_{fmax} = (E - E_e)_{max} \tag{12-18}$$

$$E_{fmin} = (E - E_e)_{min} \tag{12-19}$$

Suppose the moment of inertia of the flywheel is $J_f$, the maximum kinetic energy $E_{fmax}$ and the minimum kinetic energy $E_{fmin}$ of the flywheel are as follows:

$$\int_{\varphi_a}^{\varphi_e} M_d(\varphi)\,d\varphi - \int_{\varphi_a}^{\varphi_e} M_r(\varphi)\,d\varphi = \int_{\varphi_a}^{\varphi_e} [M_d(\varphi) - M_r(\varphi)]\,d\varphi = 0 \tag{12-17}$$

$W_{dp}$为曲线 $M_d(\varphi)$所包围的面积，$W_{rp}$为曲线 $M_r$（$\varphi$）所包围的面积。由式（12-17）可知

$$\sum_{i=1}^{n} \Delta E_i = 0$$

设机械系统在稳定运转周期开始位置的动能为 $E_a$，则图 12-9 所示各位置的动能为

$$E_b = E_a + \Delta E_1$$

$$E_c = E_b - \Delta E_2 = E_a + \Delta E_1 - \Delta E_2$$

$$E_d = E_c + \Delta E_3 = E_a + \Delta E_1 - \Delta E_2 + \Delta E_3$$

$$E_e = E_d - \Delta E_4 = E_a + \Delta E_1 - \Delta E_2 + \Delta E_3 - \Delta E_4$$

计算出一系列的动能后，可从中选出动能的最大值与最小值。

当等效转动惯量为常量，机械动能处于最大值 $E_{max}$时，其角速度 $\omega$ 也达到最大值 $\omega_{max}$。机械动能处于最小值 $E_{min}$时，其角速度 $\omega$ 也下降到最小值 $\omega_{min}$。所以，可通过控制机械的最大动能 $E_{max}$与最小动能 $E_{min}$来限制角速度 $\omega$ 的波动。

**2. 周期性变速稳定运转过程中速度波动的调节**

周期性变速稳定运转过程中的速度波动可通过在机械中安装具有较大转动惯量的飞轮来进行调节。当速度升高时，飞轮的惯性阻止其速度增加，飞轮储存能量，限制了 $\omega_{max}$的升高。当速度降低时，飞轮的惯性又阻止其速度减小，飞轮释放能量，限制了 $\omega_{min}$的降低，从而达到了调节速度波动的目的。

在机械系统的等效构件上安装飞轮后，机械系统的总动能 $E$ 为飞轮动能 $E_f$ 和机械系统中各构件的动能 $E_e$ 之和。即

$$E = E_f + E_e$$

飞轮动能为

$$E_f = E - E_e$$

飞轮动能的最大值和最小值分别为

$$E_{fmax} = (E - E_e)_{max} \tag{12-18}$$

$$E_{fmin} = (E - E_e)_{min} \tag{12-19}$$

若飞轮的转动惯量为 $J_f$，则其动能的最大值与最小值为

$$E_{fmax} = \frac{1}{2} J_f \omega_{max}^2$$

$$E_{fmin} = \frac{1}{2} J_f \omega_{min}^2$$

$$E_{fmax} = \frac{1}{2}J_f\omega_{max}^2$$

$$E_{fmin} = \frac{1}{2}J_f\omega_{min}^2$$

Rearranging the above equations, we have:

$$E_{fmax} - E_{fmin} = (E - E_e)_{max} - (E - E_e)_{min}$$

The difference between the maximum kinetic energy and minimum kinetic energy is known as the maximum fluctuation of energy, and the difference between the greatest speed and least speed is known as the maximum fluctuation of speed.

$$E_{fmax} - E_{fmin} = \frac{1}{2}J_f\omega_{max}^2 - \frac{1}{2}J_f\omega_{min}^2 = \frac{1}{2}J_f(\omega_{max}^2 - \omega_{min}^2) = J_f\,\delta\omega_m^2$$

$$J_f = \frac{(E - E_e)_{max} - (E - E_e)_{min}}{\delta\omega_m^2} \tag{12-20}$$

Because the kinetic energy supplied by the flywheel is far greater than the kinetic energy supplied by the machine, so the kinetic energy $E_e$ can be neglected, thus:

$$J_f = \frac{E_{max} - E_{min}}{\delta\omega_m^2} \tag{12-21}$$

This is called a simplified formula.

The formula is very approximate, but it is convenient to use in practical engineering.

As we know, the moment of inertia consists of two portions: one portion is constant and the other is various. The kinetic energy of the equivalent link can be written as:

$$E_e = \frac{1}{2}J_e\omega^2 = \frac{1}{2}(J_C + J_v)\omega^2$$

Usually, the variable portion of the equivalent moment of inertia is smaller than its constant, so it can be neglected, thus:

$$E_e = \frac{1}{2}J_C\omega^2$$

Rearranging the equation (12-20), we have:

$$J_f = \frac{\left(E - \frac{1}{2}J_C\omega^2\right)_{max} - \left(E - \frac{1}{2}J_C\omega^2\right)_{min}}{\delta\omega_m^2} \tag{12-22}$$

Because when the kinetic energy is maximum the speed is maximum, and when the kinetic energy is minimum the speed is minimum. So we have:

$$\left(E - \frac{1}{2}J_C\omega^2\right)_{max} = E_{max} - \frac{1}{2}J_C\omega_{max}^2$$

$$\left(E - \frac{1}{2}J_C\omega^2\right)_{min} = E_{min} - \frac{1}{2}J_C\omega_{min}^2$$

Rearranging these equations, we have:

$$J_f = \frac{E_{max} - \frac{1}{2}J_C\omega_{max}^2 - E_{min} + \frac{1}{2}J_C\omega_{min}^2}{\delta\omega_m^2} = \frac{E_{max} - E_{min}}{\delta\omega_m^2} - \frac{\frac{1}{2}J_C(\omega_{max}^2 - \omega_{min}^2)}{\delta\omega_m^2}$$

$$E_{\text{fmax}} - E_{\text{fmin}} = (E - E_e)_{\max} - (E - E_e)_{\min}$$

$$E_{\text{fmax}} - E_{\text{fmin}} = \frac{1}{2}J_f\omega^2_{\max} - \frac{1}{2}J_f\omega^2_{\min} = \frac{1}{2}J_f(\omega^2_{\max} - \omega^2_{\min}) = J_f\,\delta\omega^2_m$$

$$J_f = \frac{(E - E_e)_{\max} - (E - E_e)_{\min}}{\delta\omega^2_m} \tag{12-20}$$

因等效构件的动能与飞轮的动能相差较大，简单计算时可以忽略不计，即 $E_e=0$，则式(12-20)简化为

$$J_f = \frac{E_{\max} - E_{\min}}{\delta\omega^2_m} \tag{12-21}$$

该式为计算飞轮转动惯量的简便公式。

机械中各构件的动能或者说等效构件的动能为

$$E_e = \frac{1}{2}J_e\omega^2 = \frac{1}{2}(J_C + J_v)\omega^2$$

当忽略等效转动惯量中的变量部分时，$J_v=0$，则

$$E_e = \frac{1}{2}J_C\omega^2$$

代入式（12-20），得

$$J_f = \frac{\left(E - \frac{1}{2}J_C\omega^2\right)_{\max} - \left(E - \frac{1}{2}J_C\omega^2\right)_{\min}}{\delta\omega^2_m} \tag{12-22}$$

由于认为 $\omega_{\max}$ 近似地发生在 $E_{\max}$ 处，$\omega_{\min}$ 近似地发生在 $E_{\min}$ 处，而机械总动能又远大于等效构件的动能，则有

$$\left(E - \frac{1}{2}J_C\omega^2\right)_{\max} = E_{\max} - \frac{1}{2}J_C\omega^2_{\max}$$

$$\left(E - \frac{1}{2}J_C\omega^2\right)_{\min} = E_{\min} - \frac{1}{2}J_C\omega^2_{\min}$$

将其代入式（12-22）中并整理，得

$$J_f = \frac{E_{\max} - \frac{1}{2}J_C\omega^2_{\max} - E_{\min} + \frac{1}{2}J_C\omega^2_{\min}}{\delta\omega^2_m} = \frac{E_{\max} - E_{\min}}{\delta\omega^2_m} - \frac{\frac{1}{2}J_C(\omega^2_{\max} - \omega^2_{\min})}{\delta\omega^2_m}$$

$$J_f = \frac{E_{\max} - E_{\min}}{\delta\omega^2_m} - J_C \tag{12-23}$$

式（12-23）为计算飞轮转动惯量的近似公式。

上述方法是按飞轮安装在等效构件上计算的。如果飞轮没有安装在等效构件上，仍按安装在等效构件上计算，然后再把计算结果转换到安装飞轮的构件上。

$$J_{\mathrm{f}} = \frac{E_{\max} - E_{\min}}{\delta \omega_{\mathrm{m}}^2} - J_{\mathrm{C}} \tag{12-23}$$

This is called approximate formula.

The moment of inertia of the flywheel is based on that the flywheel is mounted to the equivalent link. If the flywheel is fixed to the other link, such as $x$ link, rather than the equivalent link, its moment of inertia $J_x$ can be determined according to the same energy supplied by the flywheel. This can be written as:

$$\frac{1}{2} J_x \omega_x^2 = \frac{1}{2} J_{\mathrm{f}} \omega^2$$

$$J_x = J_{\mathrm{f}} \left( \frac{\omega}{\omega_x} \right)^2$$

Where $x$ is the shaft fixed the flywheel.

**3. Dimensions of Flywheels**

There are two shapes of flywheels: the one is disk, and the other is webbed disk.

The shape of flywheels is often made of disk.

Fig. 12-10 shows these flywheels. The inertia of a flywheel is provided by the hub, web and the rim. However, the inertia due to the hub and the web is very small, usually it is ignored.

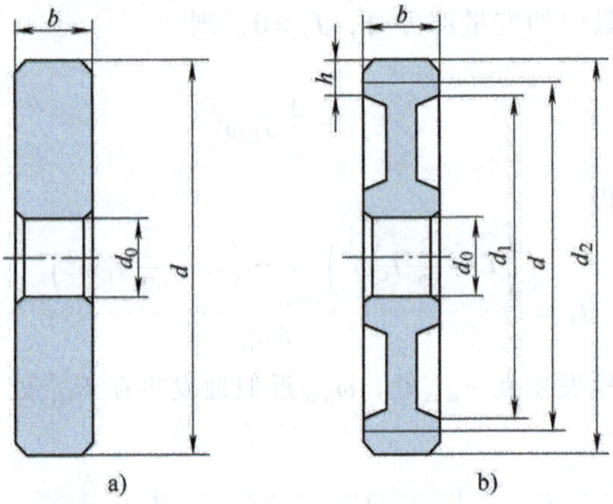


Fig. 12-10 Dimensions of flywheels(飞轮尺寸)

Consider a disk of flywheel shown in Fig. 12-10a, and let: $m$ = mass of the flywheel, $d$ = mean diameter.

We have:

$$J_{\mathrm{f}} = \frac{1}{2} m \left( \frac{d}{2} \right)^2 = \frac{1}{8} m d^2$$

$$m d^2 = 8 J_{\mathrm{f}}$$

Where, $md^2$ is called the moment of flywheel.

According to the Fig. 12-10b, we have:

$$J_{\mathrm{f}} = \frac{1}{4} m d^2$$

$$m d^2 = 4 J_{\mathrm{f}}$$

无论飞轮安装在哪个构件上，所提供的调速动能是一样的。设飞轮安装在 $x$ 轴上，转动惯量为 $J_x$，则飞轮提供的动能为

$$J_f = \frac{1}{2}J_x\omega_x^2$$

式中，$\omega_x$ 为 $x$ 轴的角速度。

由于安装在 $x$ 轴上的飞轮的动能与安装在等效构件上的动能相等，则有

$$\frac{1}{2}J_x\omega_x^2 = \frac{1}{2}J_f\omega^2$$

$$J_x = J_f\left(\frac{\omega}{\omega_x}\right)^2$$

由于飞轮的转动惯量是常量，$\omega/\omega_x$ 的值也必须是常量，这就是说，安装飞轮的轴与等效构件的轴之间的传动链必须是定传动比机构。从减小飞轮的尺寸角度出发，将飞轮安装在高速轴上是有利的。

**3. 飞轮尺寸的设计**

求出飞轮的转动惯量后，进而可设计飞轮的尺寸。工程中常把飞轮做成圆盘状或腹板状。图 12-10a 所示为直径为 $d$、宽度为 $b$、质量为 $m$ 的飞轮，图 12-10b 所示为腹板状飞轮。图 12-10a 所示为圆盘状飞轮，由理论力学可知，圆盘状飞轮对其转轴的转动惯量 $J_f$ 为

$$J_f = \frac{1}{2}m\left(\frac{d}{2}\right)^2 = \frac{1}{8}md^2$$

$$md^2 = 8J_f$$

称 $md^2$ 为飞轮矩，设定飞轮直径 $d$ 后，可求出飞轮的质量 $m$。直径越大，其质量越小。但过大的直径会导致飞轮的尺寸过大，使其圆周速度和离心力增大。为防止发生飞轮破裂事故，所选择的飞轮直径与对应的圆周速度要小于工程上规定的许用值。

对于图 12-10b 所示的腹板状飞轮，有

$$J_f = \frac{1}{4}md^2$$

则

$$md^2 = 4J_f$$

**例 12-4**　某刨床的主轴为等效构件，平均转速 $n = 60\text{r/min}$。在一个运转周期内的等效阻抗力矩 $M_r$ 如图 12-11a 所示，$M_r = 600\text{N}\cdot\text{m}$。等效驱动力矩 $M_d$ 为常数。运转不均匀系数 $\delta = 0.1$。若不计飞轮以外构件的转动惯量，试计算安装在主轴上的飞轮转动惯量。

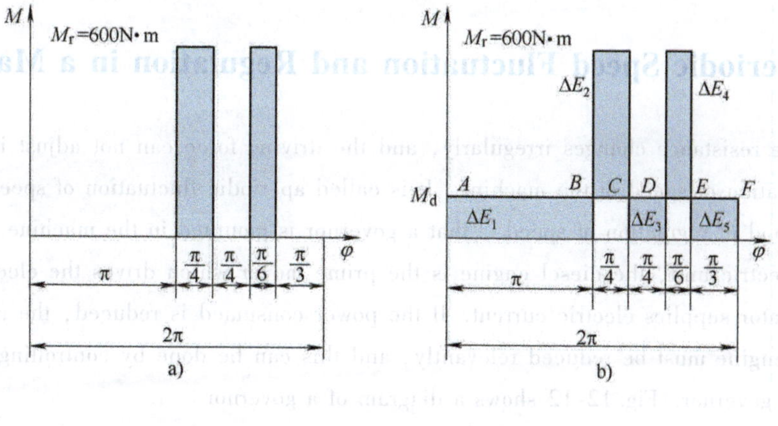


Fig. 12-11　The diagram of Equivalent moment of the shaper（刨床等效力矩图）

**Example 12-4** The equivalent link is selected as the crank shaft of a shaper, and its average speed is 60r/min. The operating period of steady motion is $2\pi$. The allowable coefficient of speed fluctuation $\delta$ is 0.1. The diagram of the equivalent resistant moment $M_r$ versus angle $\varphi$ is shown as Fig. 12-11, and the equivalent driving moment $M_d$ is a constant. If the moment of inertia of the machine has been ignored, determine the moment of inertia of the flywheel.

**Solution**

The work done by the $M_d$ is equal to the work done by the $M_r$.

$$M_d \times 2\pi = 600 \times \frac{\pi}{4} + 600 \times \frac{\pi}{6}$$

$$M_d = 125\text{N} \cdot \text{m}$$

We can find the equivalent driving moment $M_d$, and then determine the maximum kinetic energy and the minimum kinetic energy.

Suppose the starting kinetic energy at the operating period is $E_A$, we have:

$$E_B = E_A + \Delta E_1 = E_A + 125\pi$$

$$E_C = E_B - \Delta E_2 = E_A + 125\pi - (600 - 125) \times \frac{\pi}{4} = E_A + 6.25\pi$$

$$E_D = E_C + \Delta E_3 = E_A + 6.25\pi + 125 \times \frac{\pi}{4} = E_A + 37.5\pi$$

$$E_E = E_D - \Delta E_4 = E_A + 37.5\pi - (600 - 125) \times \frac{\pi}{6} = E_A - 41.67\pi$$

$$E_F = E_E + \Delta E_5 = E_A - 41.67\pi + 125 \times \frac{\pi}{3} = E_A$$

$$E_{max} = E_A + 125\pi$$

$$E_{min} = E_A - 41.67\pi$$

$$J_f = \frac{E_{max} - E_{min}}{\delta\omega_m^2} = \frac{E_A + 125\pi - (E_A - 41.67\pi)}{0.1 \times \left(\frac{\pi \times 60}{30}\right)^2} = 132.7\text{kg} \cdot \text{m}^2$$

## 12.5 Aperiodic Speed Fluctuation and Regulation in a Machine

When the resistance changes irregularly, and the driving force can not adjust itself, this may cause a fluctuation of speed of the machine. It is called aperiodic fluctuation of speed.

The method of regulation of speed is that a governor is mounted in the machine. For example, in a diesel-electric unit, the diesel engine is the prime mover which drives the electric generator, and the generator supplies electric current. If the power consumed is reduced, the rotational speed of the diesel engine must be reduced relevantly, and this can be done by controlling the oil supply system with a governor. Fig. 12-12 shows a diagram of a governor.

The governor shaft 1 is rotated by the engine $W_1$ and a sleeve 2 carried on the shaft is positioned

**解**　作一条代表 $M_d$、平行于 $\varphi$ 轴的直线，在一个周期内与 $M$ 轴、$M_r$ 及周期末端线的交点为 $A$、$B$、$C$、$D$、$E$、$F$，见图 12-11b。

在一个运转周期内，等效驱动力矩 $M_d$ 所做的功与等效阻抗力矩 $M_r$ 所做的功相等。

$$M_d \times 2\pi = 600 \times \frac{\pi}{4} + 600 \times \frac{\pi}{6}$$

$$M_d = 125\text{N} \cdot \text{m}$$

设周期开始点的动能为 $E_A$，则其余各点的动能分别为

$$E_B = E_A + \Delta E_1 = E_A + 125\pi$$

$$E_C = E_B - \Delta E_2 = E_A + 125\pi - (600 - 125) \times \frac{\pi}{4} = E_A + 6.25\pi$$

$$E_D = E_C + \Delta E_3 = E_A + 6.25\pi + 125 \times \frac{\pi}{4} = E_A + 37.5\pi$$

$$E_E = E_D - \Delta E_4 = E_A + 37.5\pi - (600 - 125) \times \frac{\pi}{6} = E_A - 41.67\pi$$

$$E_F = E_E + \Delta E_5 = E_A - 41.67\pi + 125 \times \frac{\pi}{3} = E_A$$

$$E_{max} = E_A + 125\pi$$

$$E_{min} = E_A - 41.67\pi$$

将 $E_{max}$、$E_{min}$ 代入式（12-21）得

$$J_f = \frac{E_{max} - E_{min}}{\delta\omega_m^2} = \frac{E_A + 125\pi - (E_A - 41.67\pi)}{0.1 \times \left(\frac{\pi \times 60}{30}\right)^2}\text{kg} \cdot \text{m}^2 = 132.7\text{kg} \cdot \text{m}^2$$

## 12.5　非周期性速度波动及其调节

有些机器的稳定运转过程中，驱动力所做的功突然大于阻抗力所做的功，或者阻抗力所做的功突然大于驱动力所做的功，两者在一个运转周期内做的功不再相等，破坏了稳定运转的平衡条件。使得机器主轴的速度突然加速或减速。这样的速度波动没有周期性，因此，不能用安装飞轮的方法进行速度波动的调节。

例如：在内燃机驱动的发电机组中，若用电负荷突然减小，则导致发电机组中的阻抗力也随之减小，而内燃机提供的驱动力矩没有改变，发电机转子的转速升高。用电负荷的继续减小，将导致发电机转子的转速继续升高，有可能发生飞车事故。反之，若用电负荷突然增大，则导致发电机组中的阻抗力也随之增大，而内燃机提供的驱动力矩没有改变，发电机转子的转速降低。用电负荷的继续增大，将导致发电机转子的转速继续降低，直致发生停车事故。因此，必须研究这种非周期性速度波动的调节方法。

机械运转的平衡条件受到破坏，会导致机械系统的运转速度发生非周期性的变化。安装飞轮已不能调节这种速度波动。在机械系统中安装调速装置可以重新建立一种平衡关系，称这种装置为调速器。调速器的种类很多，常用的有机械式调速器和电子式调速器。以下简单介绍机械式调速器的工作原理。

图 12-12 所示为内燃机驱动的发电机组中的机械式离心调速器工作示意图，$W_1$ 为内燃机，$W_2$ 为发电机，6 为弹簧，3、5、7 为杆件，套筒 2 对称空套在主轴 1 上。

当与内燃机 $W_1$ 相连接的主轴 1 的转度增加时，安装在杆件 5 末端的重球 4 所产生的离心惯性力 $F$ 使杆件 3 张开，并带动套筒 2 往上移动。再通过连杆机构 $AOBCD$ 中的杆件 $CD$

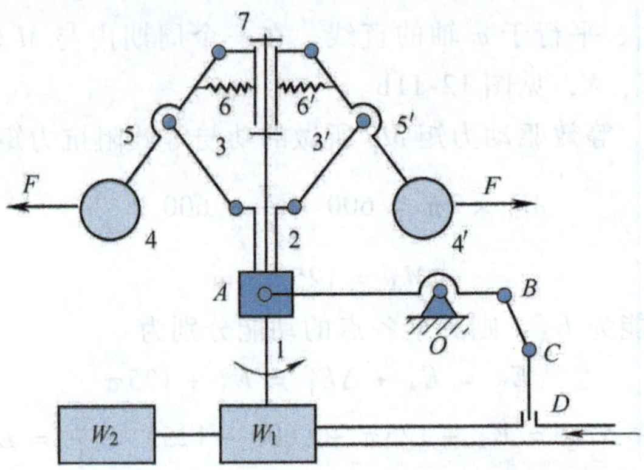


Fig. 12-12 Centrifugal governor（离心调速器）

1—shaft（主轴） 2—sleeve（套筒） 3、5、7—rod（杆件）

4—ball（重球） 6—spring（弹簧）

$W_1$—diesel engine（内燃机） $W_2$—electric generator（发电机）

by the ball-crank-lever. When the rotating speed of the shaft increases, the centrifugal forces acting on the two balls increase too, the sleeve will move up. The oil supply will be reduced, and the electric power generated will be reduced due to the reduction of the rotating speed. The fluctuation of speed caused by internal load is regulated by the governor.

动作，减小油路的流通面积，从而减小了内燃机的驱动力。套筒经过多次的振荡后，停留在固定位置，从而建立起新的平衡关系。反之，由于外载荷的突然增加而造成机械主轴转速下降时，调速器中的重球所受的离心惯性力也随之减小。重球往里靠近，套筒下移，油路开口增加。进油量的增加导致内燃机的驱动力矩增加，当与外载荷平衡时，套筒经过几次振荡后停留在固定位置，被打破的平衡关系重新建立起来。

不同的机械，使用的调速器种类也不相同。在风力发电机中，要随风力的强弱调整叶片的角度，达到调整风力发电机主轴转速的目的。水力发电机中，调速器安装在水轮机中，通过调整水轮机叶轮的角度，改变进水的流量，达到调整发电机主轴转速的目的。

关于调速器的详细原理与设计可参阅有关调速器的专业书籍。

# Chapter 13

## Balance of Machinery

# 第13章

## 机械的平衡设计

## 13.1 Introduction

**1. Balance Purposes**

A rotating shaft or rotor will experience centrifugal forces if its center of mass does not lie exactly on the rotating centerline. The centrifugal force exerted on the frame by moving machine member will be time varying and impart vibratory motion to the frame. This vibration and accompanying noise can produce human discomfort, alter the desired machine performance or may cause failure of the rotor or the support.

The purpose of balance is to reduce unbalance to an acceptable level and possibly to eliminate it entirely.

**2. Classification of Balance**

There are three types of balance.

(1) Balance of rigid rotors When a rotor is rotating about its own centerline of rotation at an angular velocity, the deformation of the rotor is small and can be negligible, and it is said to be the rigid rotor, otherwise, it is a flexible rotor. In this chapter we only deal with the rigid rotors.

Fig. 13-1 shows a rotating rigid rotor at a constant angular velocity. There are two inertia forces on the two parallel planes and the following conditions must be satisfied. The sum of the inertia forces must be zero and the sum of the moments must also be zero.

$$\sum F_i = 0, \sum F_i b = 0$$

If the distance $b$ is small, usually $b/d \leqslant 0.2$, the inertia moment causing a bending of the shaft can be negligible. This unbalance is fairly easy to protect, for example, if the rotor was mounted horizontally on knife-edge bearings, the rotor would always seek the static position due to gravity. So it is said to be in static balance. Another name for static balance is single-plane balance. It means that the masses which are generating the inertia forces are nearly in the same plane.

If the distance $b$ is large, usually $b/d > 0.2$, the inertia moment must also be required, such an unbalance is called dynamic balance. Another name for dynamic balance is sometimes two-plane balance. It requires that two criteria must be met. The sum of the forces must be zero (static balance) and the sum of the moment must be zero.

(2) Balance of flexible rotors When a rotor is rotating about its own centerline of rotation at an angular velocity, the deformation of the rotor can not be negligible, it is said to be a flexible rotor.

(3) Balance of linkages The rotating links of a linkage, such as crank and rockers, can be individually balanced by the rotating balance methods. The coupler is in complex motion and has no fixed pivot, thus its mass center is always in motion, and the inertia force of the link has variable magnitude and sense. We can not attach a mass to the link for balancing it. The global mass center of the linkage normally will change position as the linkage moves. So balance of a linkage is more difficult than balance of rotors. If we can somehow force this global mass center to be stationary in the frame, we will have a state of balance for the overall linkage.

## 13.1　机械平衡概述

**1. 研究机械平衡的目的**

机械运转时，运动构件会产生大小和方向做周期性变化的惯性力，并引起机械及其基础产生强迫振动。如果其振幅较大，或其频率接近系统的固有频率，会导致机械的工作性能和可靠性下降、零件材料内部疲劳损伤加剧，从而使机械设备遭到破坏，甚至危及人员的安全。

研究机械平衡的目的就是根据惯性力的变化规律，进行平衡设计和平衡实验，消除或减少构件所产生的惯性力，减轻机械振动，降低噪声，提高机械系统的工作性能和使用寿命。

**2. 机械平衡的种类**

组成机械的构件按照运动方式可分为三种：做定轴转动的构件、往复移动的构件和做平面复合运动的构件。由于构件的结构及运动形式的不同，产生的惯性力和平衡方法也不同。

在平衡技术中，把做定轴转动的构件称为转子。转子分为刚性转子和挠性转子两种。

（1）刚性转子的平衡　工作转速低于一阶临界转速，转子本身弹性变形可以忽略不计时，称其为刚性转子。

图 13-1 所示的转子直径为 $d$ 、宽度为 $b$，左右两面各有一个相位相反的不平衡质量 $m$。其平衡条件为：

$$\sum F_i = 0, \quad \sum F_i b = 0$$

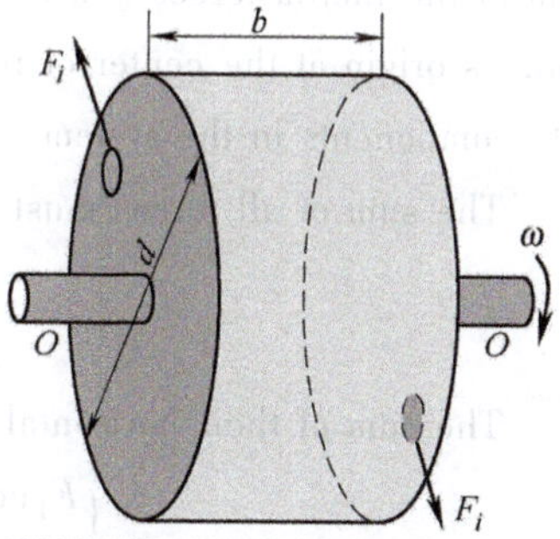


Fig. 13-1　Centrifugal forces of the rotor（转子的惯性力）

若转子的宽度 $b$ 很小，当 $b/d \leqslant 0.2$ 时，可不考虑转子宽度的影响，两面惯性力产生的力偶矩可忽略不计。其质量可认为分布在同一平面内。其不平衡现象在静止状态即可表现出来，称为静不平衡。这类转子可以只进行静平衡设计。静平衡设计仅需在一个平面内对转子的惯性力进行平衡，又称单面平衡。

当转子的宽度 $b$ 很大，且 $b/d > 0.2$ 时，转子的宽度不能忽略，惯性力偶矩的影响也不能忽略不计。这类长圆柱形转子的不平衡现象在静止时不易显示出来，只有在运转过程中才会出现明显的不平衡特征，称为动不平衡。这类转子需要进行动平衡设计。动平衡设计需对惯性力和惯性力偶矩进行平衡，需要在转子的左右两个平面内进行平衡设计，又称双面平衡。

（2）挠性转子的平衡　有些长圆柱形的转子在高速运转过程中，转子本身会发生明显的弯曲变形，产生动挠度，从而使其惯性力显著增大，称这类发生弹性变形的转子为挠性转子。挠性转子的平衡原理是基于弹性梁的横向振动理论，其平衡设计可参阅相关书籍。

（3）机构的平衡　当机构中含有往复移动的构件和平面运动的构件时，因构件质心位置随机构运动发生变化，故质心处的加速度与惯性力也随构件的运动而变化，所以不能针对某一构件进行平衡，只能对整个机构进行研究，通过合理设计，设法使各运动构件惯性力的合力和合力偶作用在机架上，故此类平衡问题又称为机械在机座上的平衡。

## 13.2 Balance Design of Rigid Rotors

The process of designing in order to reduce unbalance to an acceptable level and possibly eliminate it entirely is called balance design.

**1. Static Balance of Rigid Rotors**

The unbalanced forces of a rigid rotor are due to the acceleration of masses in the rotor. The requirement for static balance is that the sum of all forces in the rotating rotor must be zero.

Fig. 13-2 shows a rigid rotor rotating with a constant angular velocity of $\omega$. A number of masses, such as three, are depicted by point masses at different radii in the same transverse plane. If $m_1$, $m_2$, $m_3$ are the masses rotating at radii $r_1$, $r_2$, $r_3$ respectively in the same plane, then each mass produces a centrifugal force acting radically outwards from the axis of rotation.

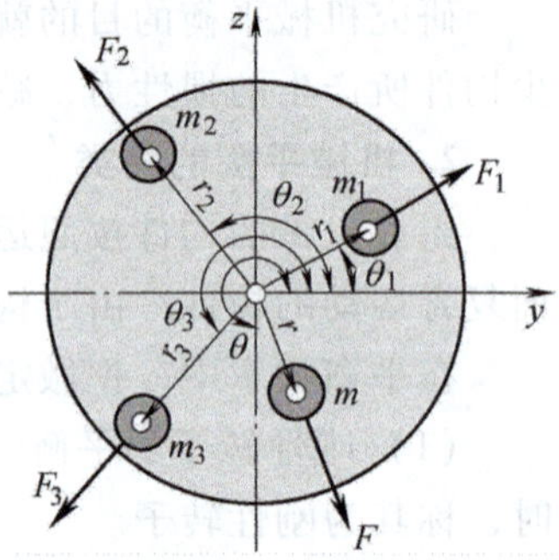

Fig. 13-2 Static balance of rigid rotor(刚性转子的静平衡)

Let $m$ be the mass which must be added at some radius $r$ in order to produce equilibrium. Static balance will be produced if the sum of the inertia forces is zero. We will set up a coordinate system with its origin at the center of rotation and resolve the inertia forces into components in the system.

The sum of all forces must be zero for static balance. The inertia forces can be in equilibrium.

$$\sum_{i=1}^{n} F_i + F = 0 \tag{13-1}$$

The sum of their horizontal components and vertical components must be zero, thus:

$$\begin{cases} F_1\cos\theta_1 + F_2\cos\theta_2 + F_3\cos\theta_3 + F\cos\theta = 0 \\ F_1\sin\theta_1 + F_2\sin\theta_2 + F_3\sin\theta_3 + F\sin\theta = 0 \end{cases} \tag{13-2}$$

The centrifugal force $F_i = m_i r_i \omega^2$, take it into equation(13-2), and then:

$$\begin{cases} m_1 r_1\cos\theta_1 + m_2 r_2\cos\theta_2 + m_3 r_3\cos\theta_3 + mr\cos\theta = 0 \\ m_1 r_1\sin\theta_1 + m_2 r_2\sin\theta_2 + m_3 r_3\sin\theta_3 + mr\sin\theta = 0 \end{cases} \tag{13-3}$$

Where $mr$ is called product of mass and radius.

The equation (13-3) can be written as follows in simplified form.

$$\sum_{i=1}^{n} m_i r_i \cos\theta_i + mr\cos\theta = 0 \tag{13-4}$$

$$\sum_{i=1}^{n} m_i r_i \sin\theta_i + mr\sin\theta = 0 \tag{13-5}$$

Where, $\theta_i$ represents angular orientation with respect to the $y$ axis.

The equation can be solved either mathematically or graphically. To solve it mathematically, square and add the equations (13-4) and (13-5).

$$mr = \sqrt{\left(\sum_{i=1}^{n} m_i r_i \cos\theta_i\right)^2 + \left(\sum_{i=1}^{n} m_i r_i \sin\theta_i\right)^2} \tag{13-6}$$

## 13.2　刚性转子的平衡设计

在转子的设计阶段，必须对其进行平衡计算，以检查其惯性力和惯性力矩是否平衡。若不平衡，则需在结构上采取措施，消除不平衡惯性力的影响，这一过程称为转子的平衡设计。

**1. 刚性转子的静平衡设计**

宽径比 $b/d \leqslant 0.2$ 的圆盘状转子需要进行静平衡设计。设计原理是转子上各不平衡质量所产生的离心惯性力与所加配重（或所减配重）产生的离心惯性力的合力为零，利用力平衡方程解出应加或应减配重的大小与方位。

转子离心惯性力的计算公式为：$F = mr\omega^2$。转子离心惯性力的静平衡条件为

$$\sum_{i=1}^{n} F_i + F = 0 \tag{13-1}$$

图 13-2 所示的转子上，已知三个不平衡质量的大小分别为 $m_1$、$m_2$、$m_3$；相对直角坐标系的方位为 $r_1$、$\theta_1$，$r_2$、$\theta_2$，$r_3$、$\theta_3$，设应加平衡质量为 $m$，坐标方位为 $r$、$\theta$。所加平衡质量 $m$ 产生的惯性力与三个不平衡质量 $m_1$、$m_2$、$m_3$ 所产生的惯性力的合力为零时，其质量中心位于回转轴线上，实现了转子的静平衡。

根据力系平衡方程可有

$$\begin{cases} F_1\cos\theta_1 + F_2\cos\theta_2 + F_3\cos\theta_3 + F\cos\theta = 0 \\ F_1\sin\theta_1 + F_2\sin\theta_2 + F_3\sin\theta_3 + F\sin\theta = 0 \end{cases} \tag{13-2}$$

设第 $i$ 个构件上的惯性力为 $F_i$，其值为 $F_i = m_i r_i \omega^2$，代入式（13-2）中得

$$\begin{cases} m_1 r_1\cos\theta_1 + m_2 r_2\cos\theta_2 + m_3 r_3\cos\theta_3 + mr\cos\theta = 0 \\ m_1 r_1\sin\theta_1 + m_2 r_2\sin\theta_2 + m_3 r_3\sin\theta_3 + mr\sin\theta = 0 \end{cases} \tag{13-3}$$

写成通式为

$$\sum_{i=1}^{n} m_i r_i\cos\theta_i + mr\cos\theta = 0 \tag{13-4}$$

$$\sum_{i=1}^{n} m_i r_i\sin\theta_i + mr\sin\theta = 0 \tag{13-5}$$

当转子以角速度 $\omega$ 转动时，决定离心惯性力的大小与方位的只有质径积 $mr$。在平衡技术中，常称 $mr$ 为不平衡量。

联立求解式（13-4）与式（13-5），可求转子上应加的质径积 $mr$ 为

$$mr = \sqrt{\left(\sum_{i=1}^{n} m_i r_i\cos\theta_i\right)^2 + \left(\sum_{i=1}^{n} m_i r_i\sin\theta_i\right)^2} \tag{13-6}$$

选定所加平衡质量的位置半径 $r$ 后，即可确定所加的平衡质量 $m$。

We can either select a value for $m$ and value for the necessary radius $r$ at which it should be placed, or choose a desired radius and solve for the mass that must be placed there. The angle $\theta$ is as follows:

$$\theta = \arctan \frac{-\sum_{i=1}^{n}(m_i r_i \sin\theta_i)}{-\sum_{i=1}^{n}(m_i r_i \cos\theta_i)} \tag{13-7}$$

The signs of the numerator and denominator of the function identify the quadrant of the angle.

| numerator | + | + | - | - |
|---|---|---|---|---|
| denominator | + | - | - | + |
| quadrant | I | II | III | IV |

**Example 13-1** A circular disk mounted on a shaft carries three attached masses of 4kg($m_1$), 3kg($m_2$) and 2.5kg($m_3$) at radial distance of 75mm, 85mm and 50mm and at the angular positions of 45°, 135° and 240° respectively. The angular positions are measured counterclockwise from the reference line along the $y$ axis. Determine the amount of the counter-mass at a radial distance of 75mm required for the static balance.

**Solution**

Fig. 13-3 shows the various masses, angular positions and their radial distances. According to the equation (13-6) and the given data, we have

$$mr = \sqrt{\left(\sum_{i=1}^{n} m_i r_i \cos\theta_i\right)^2 + \left(\sum_{i=1}^{n} m_i r_i \sin\theta_i\right)^2}$$

$m_1 = 4\text{kg}, r_1 = 75\text{mm}, \theta_1 = 45°; m_2 = 3\text{kg}, r_2 = 85\text{mm}, \theta_2 = 135° m_3 = 2.5\text{kg}; r_3 = 50\text{mm}, \theta_3 = 240°, r = 75\text{mm}$

Put the above data into the formula, then

$$mr = \sqrt{(m_1 r_1 \cos\theta_1 + m_2 r_2 \cos\theta_2 + m_3 r_3 \cos\theta_3)^2 + (m_1 r_1 \sin\theta_1 + m_2 r_2 \sin\theta_2 + m_3 r_3 \sin\theta_3)^2}$$

$$mr = 285.8\text{kgmm}$$

$$m = 3.81\text{kg}$$

The angle $\theta$ is as follows.

$$\theta = \arctan \frac{-\sum_{i=1}^{n}(m_i r_i \sin\theta_i)}{-\sum_{i=1}^{n}(m_i r_i \cos\theta_i)}$$

$$= \arctan \frac{-(m_1 r_1 \sin\theta_1 + m_2 r_2 \sin\theta_2 + m_3 r_3 \sin\theta_3)}{-(m_1 r_1 \cos\theta_1 + m_2 r_2 \cos\theta_2 + m_3 r_3 \cos\theta_3)} = \arctan \frac{-284.2}{-(-30.68)} = \arctan(-9.26)$$

$$= 276°12'$$

$\theta$ lies in the fourth quadrant.

**2. Dynamic Balance of Rigid Rotors**

The most general case of distribution of masses on a rigid rotor is that in which the masses lie in various transverse planes as shown in Fig. 13-4. The rotor revolves with a uniform angular velocity

整理式（13-4）与式（13-3），可求平衡质量 $m$ 所在方位角 $\theta$ 为

$$\theta = \arctan \frac{-\sum_{i=1}^{n}(m_i r_i \sin\theta_i)}{-\sum_{i=1}^{n}(m_i r_i \cos\theta_i)} \tag{13-7}$$

根据式（13-7）中分子与分母的正负号可区别方位角所在的象限。

| 分子 | + | + | - | - |
|---|---|---|---|---|
| 分母 | + | - | - | + |
| 象限 | Ⅰ | Ⅱ | Ⅲ | Ⅳ |

**例 13-1** 图 13-3 所示的盘状转子上附有三个不平衡质量，其质量大小与方位为：$m_1=4\text{kg}$，$m_2=3\text{kg}$，$m_3=2.5\text{kg}$，$r_1=75\text{mm}$，$r_1=75\text{mm}$，$r_3=50\text{mm}$，$\theta_1=45°$，$\theta_2=135°$，$\theta_3=240°$；进行静平衡设计时，设配重半径 $r=75\text{mm}$，求所加配重的大小与方位。

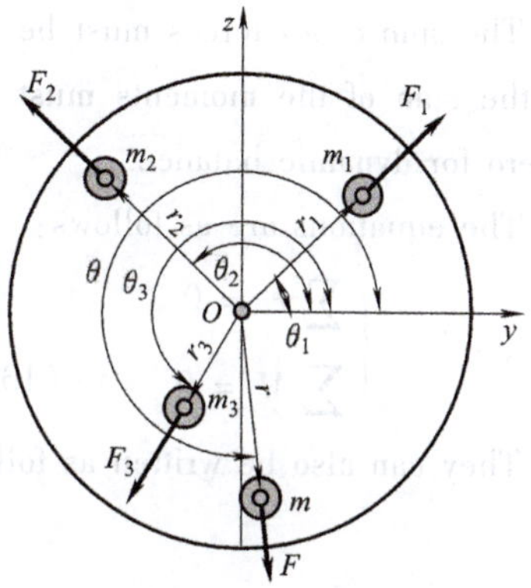


Fig. 13-3 Static balance of thd ditsk rotor

（盘状转子的静平衡设计）

**解** 把已知不平衡质量的大小、向径与方位角代入式（13-6）中，并求解所加配重 $m$ 如下。

$m_1=4\text{kg}$，$r_1=75\text{mm}$，$\theta_1=45°$；$m_2=3\text{kg}$，$r_2=85\text{mm}$，$\theta_2=135°$；$m_3=2.5\text{kg}$；$r_3=50\text{mm}$，$\theta_3=240°$ $r=75\text{mm}$

$$mr = \sqrt{\left(\sum_{i=1}^{n} m_i r_i \cos\theta_i\right)^2 + \left(\sum_{i=1}^{n} m_i r_i \sin\theta_i\right)^2}$$

$$mr = \sqrt{(m_1 r_1 \cos\theta_1 + m_2 r_2 \cos\theta_2 + m_3 r_3 \cos\theta_3)^2 + (m_1 r_1 \sin\theta_1 + m_2 r_2 \sin\theta_2 + m_3 r_3 \sin\theta_3)^2}$$

$$mr = 285.8\text{kg}\cdot\text{mm}$$

$$m = 3.81\text{kg}$$

再把上述已知数据代入式（13-7），可求解配重所在方位角 $\theta$。

$$\theta = \arctan \frac{-\sum_{i=1}^{n}(m_i r_i \sin\theta_i)}{-\sum_{i=1}^{n}(m_i r_i \cos\theta_i)}$$

$$= \arctan \frac{-(m_1 r_1 \sin\theta_1 + m_2 r_2 \sin\theta_2 + m_3 r_3 \sin\theta_3)}{-(m_1 r_1 \cos\theta_1 + m_2 r_2 \cos\theta_2 + m_3 r_3 \cos\theta_3)} = \arctan \frac{-284.2}{-(-30.68)} = \arctan(-9.26)$$

$$= 276°12'$$

进行静平衡设计时，也可以在所求平衡质量位置的反方向减重，具体情况要根据转子的结构确定。

**2. 刚性转子的动平衡设计**

对于宽径比 $b/d>0.2$ 的圆柱状转子，不能忽略转子的宽度，转子上不平衡质量不能视为集中在一个平面内，而是分布在多个平面内，所产生的离心惯性力不在同一回转平面，因

$\omega$, and $m_1$, $m_2$, $m_3$ are the masses attached to the rotor in planes 1, 2, 3 respectively and at radii $r_1$, $r_2$, $r_3$. Choose two parallel transverse planes Ⅰ and Ⅱ, and the distance between them is $l$, so that the distances of the plane 1, 2 and 3 from plane Ⅰ are $l_1$, $l_2$, $l_3$ respectively.

We can balance the rotor through the addition of two counter-masses, $m_{\mathrm{I}}$, $m_{\mathrm{II}}$, placed in transverse plane I and plane II at radii $r_{\mathrm{I}}$, $r_{\mathrm{II}}$.

The sum of all forces must be zero and the sum of the moments must also be zero for dynamic balance.

The equations are as follows:

$$\begin{cases} \sum F = 0 \\ \sum M = 0 \end{cases} \qquad (13\text{-}8)$$

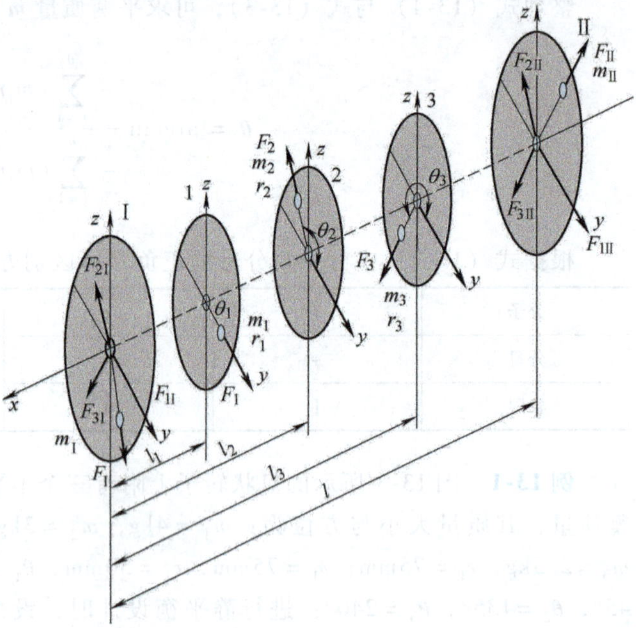


Fig. 13-4 Dynamic balance of rigid rotor (转子的动平衡)

They can also be written as follows:

$$\begin{aligned} &\sum F = 0 \\ &\sum M_{\mathrm{I}} = 0 \\ &\sum M_{\mathrm{II}} = 0 \end{aligned} \qquad (13\text{-}9)$$

Where $\sum M_{\mathrm{I}}$ is the sum of moments of inertia forces about plane I; $\sum M_{\mathrm{II}}$ is the sum of moments of inertia forces about plane Ⅱ.

Then equation (13-9) may be written as:

$$\sum_{i=1}^{3} m_i r_i + m_{\mathrm{I}} r_{\mathrm{I}} + m_{\mathrm{II}} r_{\mathrm{II}} = 0 \qquad (13\text{-}10)$$

$$\sum_{i=1}^{3} m_i r_i l_i + m_{\mathrm{II}} r_{\mathrm{II}} l = 0 \qquad (13\text{-}11)$$

$$\sum_{i=1}^{3} m_i r_i (l - l_i) + m_{\mathrm{I}} r_{\mathrm{I}} l = 0 \qquad (13\text{-}12)$$

The sum of their horizontal components and vertical components must be zero, thus equation (13-11) may be written as:

$$\sum_{i=1}^{3} m_i r_i l_i \cos\theta_i + m_{\mathrm{II}} r_{\mathrm{II}} l \cos\theta_{\mathrm{II}} = 0 \qquad (13\text{-}13)$$

$$\sum_{i=1}^{3} m_i r_i l_i \sin\theta_i + m_{\mathrm{II}} r_{\mathrm{II}} l \sin\theta_{\mathrm{II}} = 0 \qquad (13\text{-}14)$$

Squaring and adding the equation (13-13) and (13-14), we have:

而形成轴面惯性力偶，且该力偶的作用方位随转子的回转而变化。对这类转子进行动平衡设计时，要求转子各偏心质量产生的惯性力和惯性力偶矩同时得以平衡。

刚性转子的动平衡条件为

$$\begin{cases} \sum F = 0 \\ \sum M = 0 \end{cases} \tag{13-8}$$

在图 13-4 所示的转子中，设已知偏心质量 $m_1$、$m_2$、$m_3$ 分别位于 1、2、3 平面内，方位分别为 $r_1$、$\theta_1$，$r_2$、$\theta_2$，$r_3$、$\theta_3$。当转子以等角速度 $\omega$ 旋转时，所产生的惯性力 $F_1$、$F_2$、$F_3$ 形成一个空间力系。

$$F_1 = m_1 r_1 \omega^2,\ F_2 = m_2 r_2 \omega^2,\ F_3 = m_3 r_3 \omega^2$$

为求解方便，刚性转子的动平衡方程可写为

$$\begin{aligned} \sum F &= 0 \\ \sum M_{\mathrm{I}} &= 0 \\ \sum M_{\mathrm{II}} &= 0 \end{aligned} \tag{13-9}$$

式中，$\sum M_{\mathrm{I}}$ 为各面惯性力对平衡面Ⅰ的力矩之和；$\sum M_{\mathrm{II}}$ 为各面惯性力对平衡面Ⅱ的力矩之和。

将式（13-9）写成质径积的形式为

$$\sum_{i=1}^{3} m_i r_i + m_{\mathrm{I}} r_{\mathrm{I}} + m_{\mathrm{II}} r_{\mathrm{II}} = 0 \tag{13-10}$$

$$\sum_{i=1}^{3} m_i r_i l_i + m_{\mathrm{II}} r_{\mathrm{II}} l = 0 \tag{13-11}$$

$$\sum_{i=1}^{3} m_i r_i (l - l_i) + m_{\mathrm{I}} r_{\mathrm{I}} l = 0 \tag{13-12}$$

由于式（13-10）含有四个未知数，可利用式（13-11）和式（13-12）求解。

再把式（13-11）写成 $y$ 轴和 $z$ 轴的投影方程，则有

$$\sum_{i=1}^{3} m_i r_i l_i \cos\theta_i + m_{\mathrm{II}} r_{\mathrm{II}} l \cos\theta_{\mathrm{II}} = 0 \tag{13-13}$$

$$\sum_{i=1}^{3} m_i r_i l_i \sin\theta_i + m_{\mathrm{II}} r_{\mathrm{II}} l \sin\theta_{\mathrm{II}} = 0 \tag{13-14}$$

从式（13-13）和式（13-14）中，可求解面Ⅱ的质径积和所在相位角

$$m_{\mathrm{II}} r_{\mathrm{II}} = \frac{\left[\left(\sum_{i=1}^{n} l_i m_i r_i \cos\theta_i\right)^2 + \left(\sum_{i=1}^{n} l_i m_i r_i \sin\theta_i\right)^2\right]^{1/2}}{l} \tag{13-15}$$

$$\theta_{\mathrm{II}} = \arctan \frac{-\sum_{i=1}^{n} l_i m_i r_i \sin\theta_i}{-\sum_{i=1}^{n} l_i m_i r_i \cos\theta_i} \tag{13-16}$$

$$m_{\mathrm{II}} r_{\mathrm{II}} = \frac{\left[\left(\sum_{i=1}^{n} l_i m_i r_i \cos\theta_i\right)^2 + \left(\sum_{i=1}^{n} l_i m_i r_i \sin\theta_i\right)^2\right]^{1/2}}{l} \tag{13-15}$$

Dividing equation (13-14) by equation (13-13), we have:

$$\theta_{\mathrm{II}} = \arctan \frac{-\sum_{i=1}^{n} l_i m_i r_i \sin\theta_i}{-\sum_{i=1}^{n} l_i m_i r_i \cos\theta_i} \tag{13-16}$$

Equation (13-12) can be written as:

$$\sum_{i=1}^{n} l m_i r_i \cos\theta_i - \sum_{i=1}^{n} l_i m_i r_i \cos\theta_i + l m_{\mathrm{I}} r_{\mathrm{I}} \cos\theta_{\mathrm{I}} = 0 \tag{13-17}$$

$$\sum_{i=1}^{n} l m_i r_i \sin\theta_i - \sum_{i=1}^{n} l_i m_i r_i \sin\theta_i + l m_{\mathrm{I}} r_{\mathrm{I}} \sin\theta_{\mathrm{I}} = 0 \tag{13-18}$$

Squaring and adding equation (13-17) and equation (13-18), we have:

$$m_{\mathrm{I}} r_{\mathrm{I}} = \frac{\left[\left(\sum_{i=1}^{n} l m_i r_i \cos\theta_i - \sum_{i=1}^{n} l_i m r_i \cos\theta_i\right)^2 + \left(\sum_{i=1}^{n} l m_i r_i \sin\theta_i - \sum_{i=1}^{n} l_i m_i r_i \sin\theta_i\right)^2\right]^{1/2}}{l} \tag{13-19}$$

Dividing equation (13-18) by equation (13-17), we have

$$\theta_{\mathrm{I}} = \arctan \frac{-\sum_{i=1}^{n} l m_i r_i \sin\theta_i + \sum_{i=1}^{n} l_i m_i r_i \sin\theta_i}{-\sum_{i=1}^{n} l m_i r_i \cos\theta_i + \sum_{i=1}^{n} l_i m_i r_i \cos\theta_i} \tag{13-20}$$

**Example 13-2** A rotating rotor carries three unbalanced masses of 4kg, 3kg and 2.5kg at radial distance of 75mm, 85mm and 50mm and at the angular positions of 45 °, 135 °and 240 °respectively. The second and the third masses are in the planes at 200 mm and 375mm from the plane of the first mass. The angular positions are measured counter-clockwise from the reference line along $y$-axis and viewing the shaft from the first mass end.

The shaft length is 800mm between bearings and the distance between the plane of the first mass and the bearing at that end is 225mm.

Determine the amount of the counter-masses in planes at 75mm from the bearings for the complete balance of the rotor. The first counter-mass is to be in a plane between the first mass and the left bearing. The second counter-mass is to be in a plane between the third mass and the right bearing. Fig. 13-5 shows the planes of the counter-masses as well as the planes of the unbalanced masses.

**Solution**

Plane $I$ is to be taken as the reference plane and the distance of the unbalanced planes are as follows:

$m_1 = 4\text{kg}$, $r_1 = 75\text{mm}$, $\theta_1 = 45°$; $m_2 = 3\text{kg}$, $r_2 = 85\text{mm}$, $\theta_2 = 135°$; $m_3 = 2.5\text{kg}$, $r_3 = 50\text{mm}$, $\theta_3 = 240°$

$l_1 = 225 - 75 = 150\text{mm}$

同理，将式（13-12）写成投影方程为

$$\sum_{i=1}^{n} lm_i r_i \cos\theta_i - \sum_{i=1}^{n} l_i m_i r_i \cos\theta_i + lm_{\mathrm{I}} r_{\mathrm{I}} \cos\theta_{\mathrm{I}} = 0 \tag{13-17}$$

$$\sum_{i=1}^{n} lm_i r_i \sin\theta_i - \sum_{i=1}^{n} l_i m_i r_i \sin\theta_i + lm_{\mathrm{I}} r_{\mathrm{I}} \sin\theta_{\mathrm{I}} = 0 \tag{13-18}$$

联立求解，得

$$m_{\mathrm{I}} r_{\mathrm{I}} = \frac{\left[\left(\sum_{i=1}^{n} lm_i r_i \cos\theta_i - \sum_{i=1}^{n} l_i m r_i \cos\theta_i\right)^2 + \left(\sum_{i=1}^{n} lm_i r_i \sin\theta_i - \sum_{i=1}^{n} l_i m_i r_i \sin\theta_i\right)^2\right]^{1/2}}{l} \tag{13-19}$$

$$\theta_{\mathrm{I}} = \arctan \frac{-\sum_{i=1}^{n} lm_i r_i \sin\theta_i + \sum_{i=1}^{n} l_i m_i r_i \sin\theta_i}{-\sum_{i=1}^{n} lm_i r_i \cos\theta_i + \sum_{i=1}^{n} l_i m_i r_i \cos\theta_i} \tag{13-20}$$

任何动不平衡的刚性转子，无论具有多少个偏心质量，以及分布于多少个回转平面内，都可以在任选的两个平衡基面分别加上或减去一个适当的配重，使转子得到完全的平衡。

**例 13-2** 对一长圆柱形转子进行动平衡设计。三个不平衡平面的不平衡质量及方位如图 13-5 所示。该转子两端轴承之间的距离为 800mm，两选定的平衡面各距轴承 75mm，左端不平衡面 1 距左轴承距离为 225mm，不平衡面 2、3 距不平面 1 的距离分别为 200mm 和 375mm。平衡面 I 加重半径 $r_{\mathrm{I}}$ =75mm，平衡面 II 加重半径 $r_{\mathrm{II}}$ =40mm，求应加在平衡面 I 和 II 处的配重大小与方位。

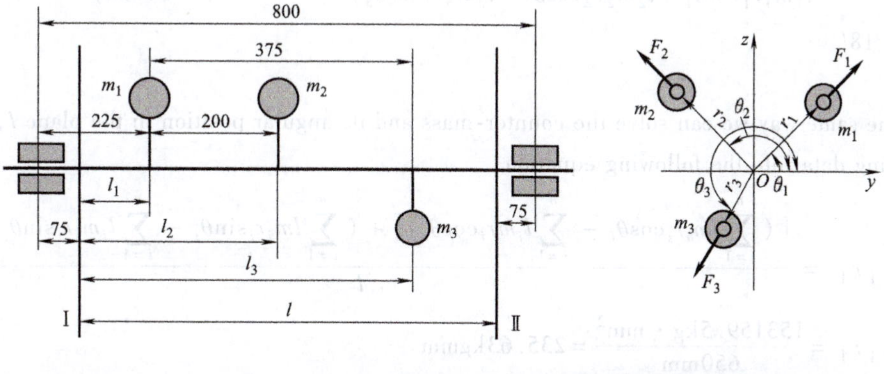


Fig. 13-5 Masses contribution of rigidcylindrical rotor（刚性圆柱转子的质量分布）

**解** 以平衡面 I 为参考平面，求解各不平衡面的轴向尺寸。

$$l_1 = 225\text{mm} - 75\text{mm} = 150\text{mm}$$

$$l_2 = 200\text{mm} + 150\text{mm} = 350\text{mm}$$

$$l_3 = 375\text{mm} + 150\text{mm} = 525\text{mm}$$

$$l = 800\text{mm} - 2 \times 75\text{mm} = 650\text{mm}$$

$m_1$ = 4kg，$r_1$ = 75mm，$\theta_1$ = 45°；$m_2$ = 3kg，$r_2$ = 85mm，$\theta_2$ = 135°；$m_3$ = 2.5kg，$r_3$ =

$l_2 = 200 + 150 = 350\text{mm}$

$l_3 = 375 + 150 = 525\text{mm}$

$l = 800 - 2 \times 75 = 650\text{mm}$

Put the above data into the formula 13-15, then

$$m_{\text{II}} r_{\text{II}} = \frac{[(\sum_{i=1}^{n} l_i m_i r_i \cos\theta_i)^2 + (\sum_{i=1}^{n} l_i m_i r_i \sin\theta_i)^2]^{1/2}}{l}$$

$$mr = \sqrt{(m_1 r_1 \cos\theta_1 + m_2 r_2 \cos\theta_2 + m_3 r_3 \cos\theta_3)^2 + (l_1 m_1 r_1 \sin\theta_1 + l_2 m_2 r_2 \sin\theta_2 + l_3 m_3 r_3 \sin\theta_3)^2}$$
$$= 285.8\text{kg} \cdot \text{mm}$$

$$m_{\text{II}} r_{\text{II}} = \frac{\sqrt{(l_1 m_1 r_1 \cos\theta_1 + l_2 m_2 r_2 \cos\theta_2 + l_3 m_3 r_3 \cos\theta_3)^2 + (l_1 m_1 r_1 \sin\theta_1 + l_2 m_2 r_2 \sin\theta_2 + l_3 m_3 r_3 \sin\theta_3)^2}}{l}$$

$$= \frac{74567.8994\text{kg} \cdot \text{mm}^2}{650\text{mm}} = 114.7198\text{kg} \cdot \text{mm}$$

We can select a value for $m$ or a value for the necessary radius.

$r_{\text{II}} = 40\text{mm}$, then

$$m_{\text{II}} = \frac{114.7198\text{kg} \cdot \text{mm}}{40\text{mm}} = 2.86799\text{mm}$$

The angle $\theta_{\text{II}}$ is as follows.

$$\theta_{\text{II}} = \arctan \frac{-\sum_{i=1}^{n} l_i m_i r_i \sin\theta_i}{-\sum_{i=1}^{n} l_i m_i r_i \cos\theta_i}$$

$$= \arctan \frac{-(l_1 m_1 r_1 \sin\theta_1 + l_2 m_2 r_2 \sin\theta_2 + l_3 m_3 r_3 \sin\theta_3)}{-(l_1 m_1 r_1 \cos\theta_1 + l_2 m_2 r_2 \cos\theta_2 + l_3 m_3 r_3 \cos\theta_3)} = \arctan \frac{-38096}{-(-64102)} = \arctan(-0.594)$$

$$= 329°18'$$

In the same way we can solve the counter-mass and its angular position in the plane $I$, only put the knowing data into the following equation.

$$m_{\text{I}} r_{\text{I}} = \frac{[(\sum_{i=1}^{n} l m_i r_i \cos\theta_i - \sum_{i=1}^{n} l_i m r_i \cos\theta_i)^2 + (\sum_{i=1}^{n} l m_i r_i \sin\theta_i - \sum_{i=1}^{n} l_i m_i r_i \sin\theta_i)^2]^{1/2}}{l}$$

$$m_{\text{I}} r_{\text{I}} = \frac{153159.5\text{kg} \cdot \text{mm}^2}{650\text{mm}} = 235.63\text{kgmm}$$

we suppose that $r_{\text{I}} = 75\text{mm}$, then

$m_{\text{I}} = 3.14\text{kg}$

The angular position in plane $I$ can be solved as follows:

$$\theta_{\text{I}} = \arctan \frac{-\sum_{i=1}^{n} l m_i r_i \sin\theta_i + \sum_{i=1}^{n} l_i m_i r_i \sin\theta_i}{-\sum_{i=1}^{n} l m_i r_i \cos\theta_i + \sum_{i=1}^{n} l_i m_i r_i \cos\theta_i}$$

50mm，$\theta_3 = 240°$

把上述数据代入式（13-15）并求解得

$$m_{\mathrm{II}} r_{\mathrm{II}} = \frac{\left[\left(\sum_{i=1}^{n} l_i m_i r_i \cos\theta_i\right)^2 + \left(\sum_{i=1}^{n} l_i m_i r_i \sin\theta_i\right)^2\right]^{1/2}}{l}$$

$$mr = \sqrt{(m_1 r_1 \cos\theta_1 + m_2 r_2 \cos\theta_2 + m_3 r_3 \cos\theta_3)^2 + (l_1 m_1 r_1 \sin\theta_1 + l_2 m_2 r_2 \sin\theta_2 + l_3 m_3 r_3 \sin\theta_3)^2}$$

$$= 285.8\text{kg} \cdot \text{mm}$$

$$m_{\mathrm{II}} r_{\mathrm{II}} = \frac{\sqrt{(l_1 m_1 r_1 \cos\theta_1 + l_2 m_2 r_2 \cos\theta_2 + l_3 m_3 r_3 \cos\theta_3)^2 + (l_1 m_1 r_1 \sin\theta_1 + l_2 m_2 r_2 \sin\theta_2 + l_3 m_3 r_3 \sin\theta_3)^2}}{l}$$

$$= \frac{74567.8994\text{kg} \cdot \text{mm}^2}{650\text{mm}} = 114.7198\text{kg} \cdot \text{mm}^2$$

$$r_{\mathrm{II}} = 40\text{mm}$$

$$m_{\mathrm{II}} = \frac{114.7198\text{kg} \cdot \text{mm}}{40\text{mm}} = 2.86799\text{kg}$$

平衡面Ⅱ配重的方位角如下：

$$\theta_{\mathrm{II}} = \arctan \frac{-\sum_{i=1}^{n} l_i m_i r_i \sin\theta_i}{-\sum_{i=1}^{n} l_i m_i r_i \cos\theta_i}$$

$$= \arctan \frac{-(l_1 m_1 r_1 \sin\theta_1 + l_2 m_2 r_2 \sin\theta_2 + l_3 m_3 r_3 \sin\theta_3)}{-(l_1 m_1 r_1 \cos\theta_1 + l_2 m_2 r_2 \cos\theta_2 + l_3 m_3 r_3 \cos\theta_3)} = \arctan \frac{-38096}{-(-64102)} = \arctan(-0.594)$$

$$= 329°18'$$

同理，可求解平衡面Ⅰ的配重大小与方位。

$$m_{\mathrm{I}} r_{\mathrm{I}} = \frac{\left[\left(\sum_{i=1}^{n} l m_i r_i \cos\theta_i - \sum_{i=1}^{n} l_i m r_i \cos\theta_i\right)^2 + \left(\sum_{i=1}^{n} l m_i r_i \sin\theta_i - \sum_{i=1}^{n} l_i m_i r_i \sin\theta_i\right)^2\right]^{1/2}}{l}$$

$$m_{\mathrm{I}} r_{\mathrm{I}} = \frac{153159.5\text{kg} \cdot \text{mm}^2}{650\text{mm}} = 235.63\text{kg} \cdot \text{mm}$$

$$m_{\mathrm{I}} = 3.14\text{kg}$$

$$\theta_{\mathrm{I}} = \arctan \frac{-\sum_{i=1}^{n} l m_i r_i \sin\theta_i + \sum_{i=1}^{n} l_i m_i r_i \sin\theta_i}{-\sum_{i=1}^{n} l m_i r_i \cos\theta_i + \sum_{i=1}^{n} l_i m_i r_i \cos\theta_i}$$

$$= \arctan \frac{-(l(m_1 r_1 \sin\theta_1 + m_2 r_2 \sin\theta_2 + m_3 r_3 \sin\theta_3) + (l_1 m_1 r_1 \sin\theta_1 + l_2 m_2 r_2 \sin\theta_2 + l_3 m_3 r_3 \sin\theta_3)}{-(l(m_1 r_1 \cos\theta_1 + m_2 r_2 \cos\theta_2 + m_3 r_3 \cos\theta_3) + (l_1 m_1 r_1 \cos\theta_1 + l_2 m_2 r_2 \cos\theta_2 + l_3 m_3 r_3 \cos\theta_3)}$$

$$= \arctan \frac{-225.62}{-67.96}$$

$$= 253°12'$$

$$=\arctan\frac{-(l(m_1r_1\sin\theta_1+m_2r_2\sin\theta_2+m_3r_3\sin\theta_3))+(l_1m_1r_1\sin\theta_1+l_2m_2r_2\sin\theta_2+l_3m_3r_3\sin\theta_3)}{-(l(m_1r_1\cos\theta_1+m_2r_2\cos\theta_2+m_3r_3\cos\theta_3))+(l_1m_1r_1\cos\theta_1+l_2m_2r_2\cos\theta_2+l_3m_3r_3\cos\theta_3)}$$

$$=\arctan\frac{-225.62}{-67.96}$$

$$=253°12'$$

## 13.3 Balance Test of Rigid Rotors

Though care is taken in the design of rotors of a machine to eliminate any unbalance, some residual unbalance still will be left in the finished rotor. This may happen due to slight variation in the density of the material or inaccuracies in the casting, forged or machining. Since the centrifugal force and couple vary as the square of the speed, even the small errors may lead to serious troubles at high speed of rotation. Thus there is need for a means to measure and correct the imbalance in rotating system.

The balancing machines may be used to measure the static unbalance or dynamic unbalance or both. A balance machine is able to indicate whether a rotor is in balance or not and if it is not, then it measures the unbalance by indicating its magnitude and location.

**1. Static Balance Test**

If the distance $b$ of a rotor is small, usually $b/d \leqslant 0.2$, the inertia moment causing a bending of the shaft can be neglected. Static balance machines are used for rotors of small axial dimensions such as fans, gears, belt wheels and impellers, etc.

Fig. 13-6 shows a rigid rotor 1 with the shaft laid on the horizontal parallel ways 2. By gravity $G$, the rotor will roll until the center of the rotor gravity lies on the lowest position.

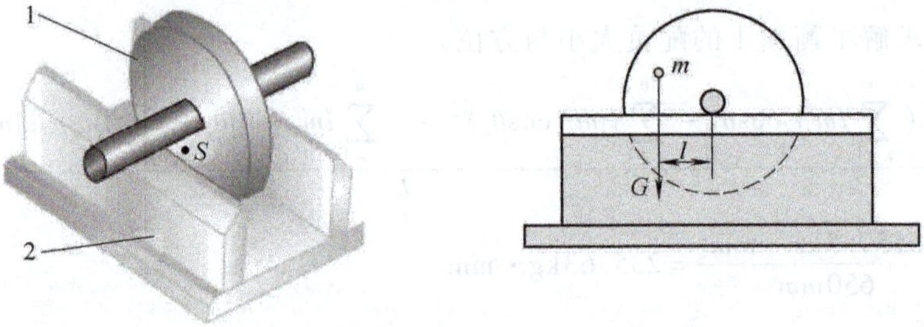


Fig. 13-6 Static balance test (静平衡实验)

1—rigid rotor (待平衡转子) 2—horizontal parallel ways (刀口状平衡架)

A temporary mass is placed upon the rotor in opposite direction. The trial-and-error method may be used for static balance. If the rotor is in static balance, the rotor will not roll regardless of the angular position of the rotor. Then either a permanent mass or masses are attached to the rotor, or the proper amount of mass may be removed at a position diametrically opposite in order to produce the same effect.

**2. Dynamic Balance Test**

For dynamic balance of a rotor, two balance of counter-masses are required to be used in any

## 13.3 刚性转子的平衡试验

经过平衡设计的转子实现了理论上的平衡。但是，由于制造、装配误差及材质不均匀等原因，实际生产出来的转子在运转时还会出现不平衡现象，这种不平衡在设计阶段是无法消除的，需要通过试验来确定不平衡量的大小和方位，然后利用增加或去除质量的方法予以平衡。

**1. 刚性转子的静平衡试验**

对于宽径比 $b/d \leqslant 0.2$ 的刚性转子，可进行静平衡实验。静平衡实验设备比较简单，一般采用带有两根平行导轨的静平衡架。为减小轴颈与导轨之间的摩擦，导轨端口形状常做成刀口状和圆弧状。图 13-6 为静平衡试验示意图，1 为待平衡转子，2 为刀口状平衡架。

将一个具有偏心质量的圆盘状转子放在静平衡支架上，偏心质量对其转动中心会产生一个重力矩 $Gl$，并驱动转子转动，直到重心位于正下方才会停止。进行静平衡实验时，首先调整好支架的水平状态，然后将转子轴颈放置在支架的一端，轻轻使转子向另一端滚动，待其静止时，在正上方作一标记，然后使转子反方向滚动。若转子仍在上次位置附近静止，则说明在该位置时的质心位于转子轴线的下方。在其上方加一配重或在其下方减一配重，再反复试验，直到该转子在任意位置都能静止，说明转子的重心与其回转轴线趋于重合。

由于轴颈和支架之间的摩擦会影响平衡的精度，所以重要的圆盘状转子还要在动平衡机上进行静平衡。

**2. 刚性转子的动平衡试验**

对于宽径比 $b/d > 0.2$ 的刚性转子，需进行动平衡试验。刚性转子的动平衡试验要在动平衡机上进行。转子不平衡而产生的离心惯性力和惯性力偶矩，将使转子的支承产生强迫振动，转子支承处振动的强弱反映了转子的不平衡情况。各类动平衡机的工作原理都是通过测量转子支承处的振动强度和相位来测定转子不平衡量的大小与方位的。通过测量两个支承处的振动就可以知道两平面的平衡结果。图 13-7 所示为工业动平衡机。

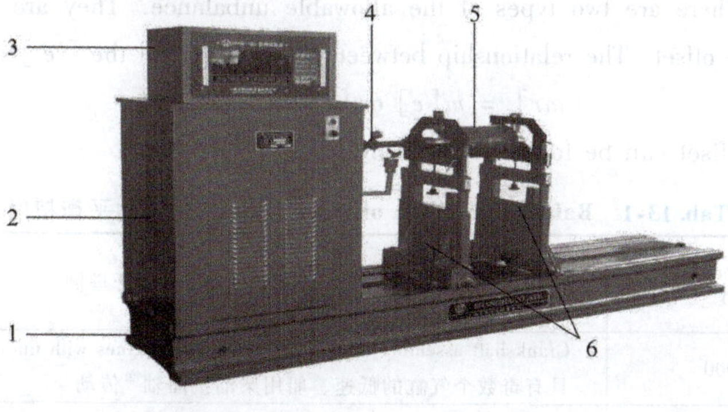


Fig. 13-7 Dynamic balance machine in industry（工业动平衡机）
1—base（底座） 2—power box（动力箱） 3—computer system（计算机系统）
4—spindle（主轴） 5—rotor（转子） 6—carriage（支承架）

安装在底座 1 上的动力箱 2 为主轴 4 提供动力，驱动转子 5 转动。转子 5 安装在前后支承架 6 上。转子的振动传递到支承架上，由传感器将支承架上的振动信号经过选频、整形、放大等信号处理后，经计算机系统 3 显示出转子不平衡质量的大小与方位。在转子的两个平衡面进行加重或减重后，再重新进行平衡，直到满足平衡精度为止。

two correct planes. Dynamic balance is achieved by adding or removing masses in these two planes. This requires a dynamic balance machine. Fig. 13-5 shows a common type of dynamic balance machine, which is used in industry.

The power box 2 and two carriages 6 are mounted on the base 1. The rotor 5 is supported on the two carriages, and it is connected to the spindle 4 of the power box. The magnitudes and the angular positions of the unbalance masses of the rotor can be indicated by the computer system 3.

Many balance machines are based on the principle of vibrations. When the rotor is rotating on the carriages, the centrifugal forces will cause the carriages to do transverse vibration. Two transducers are each mounted in a carriage and they measure the vibration of the carriages. The electromagnetic transducers deliver voltage proportional to the amplitude of the carriages. This voltage is amplified electronically and delivered to circuitry which can compute its peak magnitude and the phase angle of that peak with respect to some reference signal. The reference signal is supplied by a benchmark generator.

The signals are sent to the computer for processing and computation of the needed balance masses and locations. The mass-radius product needed in the correction planes on each side of the rotor can then be calculated. After the correction radius is determined, the balance masse and angular position may be calculated for each correction plane.

In heavy machinery like turbines and generators, it is not possible to balance the rotors by mounting them in the balance machine. In such case, the balance has to be done under normal condition on its own bearings. This is called field balance.

**3. Balance Precision**

After a rigid rotor has been balanced by using a balance machine, the centerline of mass of the rotor will be coincident with the centerline of rotation of the rotor theoretically, but in practice, they can not be coincident completely. An offset between the centerline of mass and the centerline of rotation of the rotor always exists. To assure the balance precision, the actual unbalance must be less or equal to the allowable unbalance. There are two types of the allowable unbalance. They are allowable mass-radius product and allowable offset. The relationship between the $[mr]$ and the $[e]$ is as follows:

$$[mr] = m[e] \text{ or } [e] = [mr]/m$$

The allowable offset can be found in the Tab. 13-1.

**Tab. 13-1 Balance precision of rigid rotors（转子的平衡精度）**

| Precisions 精度等级 | $A = \frac{[e]\omega}{1000}$ ① mm/s | Rotors 典型转子举例 |
|---|---|---|
| A4000 | 4000 | Crankshaft assembly of slow marine diesel engines with uneven number of cylinders<br>具有奇数个气缸的低速、船用柴油机曲轴②传动 |
| A1600 | 1600 | Crankshaft assembly of large two-cycle engines<br>刚性安装的大型两冲程发动机曲轴传动装置 |
| A630 | 630 | Crankshaft assembly of large four-cycle engines<br>大型四冲程发动机曲轴传动 |
| A250 | 250 | Crankshaft assembly of four-cycle engines at high speed<br>高速四缸柴油机曲轴传动装置 |
| A100 | 100 | Crankshaft assembly of six-cycle engines at high speed, cars, trucks, locomotives<br>六缸或六缸以上的高速柴油机曲轴传动装置，汽车和机车用发动机整机 |

（续）

| Precisions 精度等级 | $A=\frac{[e]\omega}{1000}$ ① mm/s | Rotors 典型转子举例 |
|---|---|---|
| A40 | 40 | Car wheels, driving shafts, crankshaft assembly of six-cycle engines at high speed<br>汽车轮、轮缘、轮组、转动轴，弹性安装的六缸或六缸以上的高速四冲程发动机曲轴传动装置，汽车和机车用发动机的曲轴传动装置 |
| A16 | 16 | Individual engine components<br>特殊要求的传动轴（螺旋桨轴、万向传动轴），破碎机械及农用机械的零部件，汽车和机车用发动机的特殊部件，有特殊要求的六缸或六缸以上发动机的曲轴传动装置 |
| A6.3 | 6.3 | Rotors of working machines, fans, flywheels, machine tools<br>作业机械的回转零件，船用主汽轮机的齿轮，风扇，航空燃气轮机转子部件，泵的叶轮，离心机鼓轮，机床及一般机械的回转零部件，普通电机转子，特殊要求的发动机回转零部件 |
| A2.5 | 2.5 | Cars and steam turbines, machine tool spindle, small electric armatures<br>燃汽轮机和汽轮机的转子部件，刚性汽轮发电机的转子，透平压缩机转子，机床主轴和驱动部件，特殊要求的大型和中型电机转子，小型电机转子，透平驱动泵 |
| A1.0 | 1.0 | Tape recorder and phonograph drivers, grinding machine drivers, micro-motors<br>磁带记录仪及录音机驱动部件，磨床驱动部件，特殊要求的微型电机转子 |
| A0.4 | 0.4 | Spindles of precision machine tools, grinding wheels, motor rotors, gyroscopes<br>精密磨床的主轴，砂轮，电机转子，陀螺仪 |

注：① $\omega$ /（rad/s）；[$e$] /（μm）

Crankshaft devices is included Crankshaft, flywheel, clutch, belt wheel, and etc.

② 曲轴传动装置包括曲轴、飞轮、离合器、带轮、减振器、连杆回转部分等的组件

**3. 刚性转子的平衡精度**

经过平衡试验的转子还会存在一些残存的不平衡量。只要转子的残存不平衡量小于许用不平衡量就可以满足工作要求。

转子的许用不平衡量有两种表示方法，即质径积表示法和偏心距表示法。转子的许用不平衡质径积用 [$mr$] 表示，转子质心距离回转轴线的许用偏心距以 [$e$] 表示。两者的关系为

$$[e]=[mr]/m$$

偏心距是一个与转子质量无关的绝对量，而质径积是与转子质量有关的相对量。通常，对于具体给定的转子，用许用不平衡质径积较好，因为它直观，便于平衡操作，缺点是不能反映转子和平衡机的平衡精度。而为了便于比较，在衡量转子平衡的优劣或衡量平衡的精度时，用许用偏心距较好。

关于转子的许用不平衡量，目前我国尚未制定统一的标准。表 13-1 所列为国际标准化组织（ISO）制定的刚性转子平衡精度标准。

表 3-1 中给出各种典型转子的平衡精度与对应的许用不平衡量，可供参考使用。使用该表时，可首先选定平衡精度 $A$，然后依据表中公式计算出许用偏心距 [$e$]，再计算许用质径积，判别平衡结果是否满足平衡精度要求。

许用偏心距为

$$[e]=\frac{1000}{\omega}\times A$$

When using the allowable offset [$e$], the follows shoud be noted:

1) For static balance. The allowable offset [ $e$ ] may be used directly in the balance plane.

2) For dynamic balance. The allowable offset [ $e$ ] means that it is the allowable offset in the plane which passes through the center of the mass, and this value must be resolved into two correction planes. Fig. 13-8 shows a rigid rotor in which point $S$ is the center of mass. The plane Ⅰ and plane Ⅱ are correction planes respectively.

The allowable unbalance of the two correction planes are as follows:

$$[mr]_{\mathrm{I}} = m[e]\frac{b}{a+b}, [mr]_{\mathrm{II}} = m[e]\frac{a}{a+b}$$

## 13.4 Balance of Planar Mechanisms

We have known that all the inertia forces acting on the moving links in a mechanism can be replaced by a single total inertia force and a single total moment of inertia which act on the frame of the mechanism. If the resultant of all the inertia forces and the resultant of all the inertia couples acting on the frame are zero, there are no shaking forces and shaking couples. The mechanism runs smoothly. We will discover that in most mechanisms, by adding appropriate balance masses, the shaking forces and the shaking couples can be reduced.

The reciprocating unbalance is caused by the inertia forces associated with translating mass. Its effects are very evident in machines such as piston engines and compressors. Balance of reciprocate machines is more difficult than balance of rotors.

Fig. 13-9a shows a slider-crank mechanism. The countermass $m_{22}$ is mounted at the extension line of the connecting rod, so that the mass center of the rod and the slider is located at the pivot $B$. The countermass $m_{11}$ is mounted at the extension line of the crank, so that the mass center of the crank is located at the pivot $A$. The total mass center of the mechanism is stationary at the fixed pivot $A$, and the acceleration of pivot $A$ is zero. Thus the mechanism has been balanced completely.

But sometimes it is usually not practically to eliminate them completely; usually we only balance a part of shaking force by adding appropriate balance mass. The partial balance of the mechanism can be obtained as shown in Fig. 13-9b.

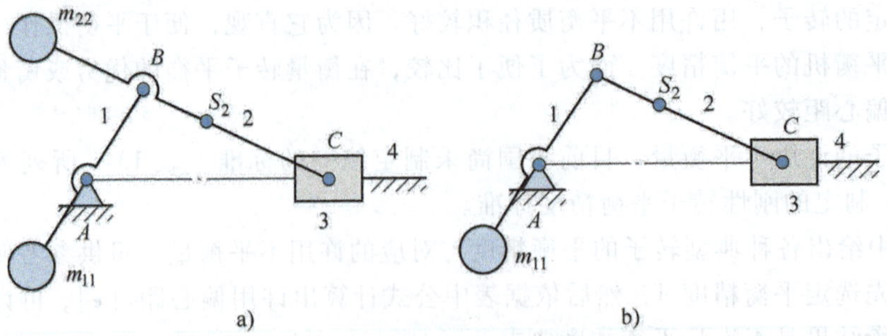


Fig. 13-9 Balance of linkages (机构的平衡)

许用质径积为

$$[mr] = m[e]。$$

进行静平衡试验时，由于转子的不平衡质量与配重均在同一个平面内，故转子质心与回转中心之间的最大距离应控制在许用偏心距之内，其许用质径积为 $m[e]$。

进行动平衡试验时，先求出质心所在平面的许用偏心距 $[e]$，再求许用质径积 $m\ [e]$，然后求出两个平衡基面内的许用质径积。以图 13-8 所示转子为例，设转子质量为 $m$，质心位于点 $S$ 处，平衡基面Ⅰ、Ⅱ上的许用质径积分别为

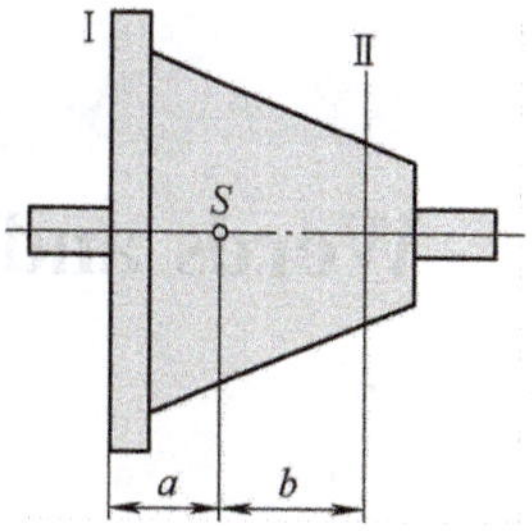


Fig. 13-8 Distribution of the allowed mass-radius product（许用质径积的分配）

$$[mr]_{\mathrm{I}} = m[e]\frac{b}{a+b},[mr]_{\mathrm{II}} = m[e]\frac{a}{a+b}$$

进行平衡试验时，两个平面的剩余不平衡量分别与上述值相比较即可知道平衡效果。

## 13.4 平面机构的平衡简介

### 1. 平面机构的平衡条件

机构运动时，各运动构件所产生的惯性力可以合成为一个通过机构质心 $S$ 处的总惯性力和一个总惯性力偶矩，这个总惯性力和总惯性力偶矩全部由基座承受。因此，为了消除机构在基座上引起的动压力，就必须设法平衡这个总惯性力和总惯性力偶矩。故机构平衡的条件是作用在机构质心的总惯性力和总惯性力偶矩分别为零，即

$$\sum m_i a_S = 0, \sum M_i = 0$$

式中，$\sum m_i$ 为机构中各构件的质量和；$a_S$ 为机构总质心处的加速度；$\sum M_i$ 为机构中各构件的惯性力矩之和。

实际平衡中，总惯性力偶矩对基座的影响应当与外加驱动力矩和阻抗力矩一并研究（因这三者都将作用到基座上），但是由于驱动力矩和阻抗力矩与机械的工作性质有关，单独平衡惯性力偶矩往往没有意义，故本章只讨论总惯性力的平衡问题。

机构的总质量 $\sum m_i$ 不可能为零，若使机构惯性力得以平衡，机构惯性力平衡的条件只有满足机构的总质心处的加速度 $a_S=0$。满足 $a_S=0$ 的条件是机构总质心静止不动或做匀速直线运动。由于机构在运动过程中总质心的运动轨迹为封闭曲线，总质心不可能做匀速直线运动，故机构惯性力的平衡条件只有总质心静止不动。

### 2. 平面机构平衡方法

对平面机构进行平衡时，可利用在运动构件上加减配重的方法，使机构总质心位于机架上并静止不动。图 13-9a 所示的曲柄滑块机构中，在连杆 $BC$ 延长线上安装配重 $m_{22}$，使连杆与滑块的质心位于曲柄和连杆的铰链点 $B$ 处，在曲柄的延长线上安装配重 $m_{11}$，使曲柄 1 的质心位于点 $A$，这样机构总质心位于固定点，对应总质心处的加速度为零。

通过在构件上加减配重，可调整机构总质心的位置，使总质心在机架上静止不动。因为过大的配重会增加机构质量，也会增大机构尺寸，故有时也采取平衡部分惯性力的方法。图 13-9b 所示曲柄滑块机构中，仅在曲柄相反方向上加装一个配重，可平衡部分惯性力，但减小了机构的尺寸与质量。

# Appendix 附录一
# Words and Phrases in Both Chinese and English
# 中英名词及短语对照

## Chapter 1 Introduction 绪论

双语教材：bilingual textbook
机械原理：theory of machines and mechanisms
机械学：mechanology
机构学：mechanism
机械设计及理论：machine design and theory of machine
机械：machinery
机器：machine
机构：mechanism
内燃机：internal-combustion engine
活塞：piston
曲轴：crankshaft
构件：link，member
零件：part，element
螺母：nut
螺栓：bolt
垫片：washer
轴承：bearing
轴瓦：bearing bush
连杆：coupler，connecting rod
机械设计的步骤：mechanical design procedure
微型机械（微米级）：micromachine
小型机械：minimachine
机器组成：constitution of machine
机械运动学：kinematics of machinery
机械动力学：dynamics of machinery
过程：process
机械设计：machine design
国际机械原理联合会：international federation for theory of machines and mechanisms
单缸四冲程内燃机：one-cylinder four-stroke cycle combustion engine
吸气行程：intake stroke
压缩行程：compression stroke
压力行程：power stroke
排气行程：exhaust stroke

## Chapter 2 Structural Analysis of Planar Mechanisms 平面机构的结构分析

机构组成：constitution of mechanism
结构分析：structure analysis
原动机：prime machine，prime power
原动件：driving link，input link
从动件：driven link，output link
机架：frame，fixed link
连架杆：link connected with frame，side link
相对运动：relative motion
接触：contact
运动副：kinematical pair
运动副的分类：classification of kinematical pairs
低副：lower pair
高副：higher pair
转动副：turning pair，revolute pair
移动副：sliding pair，prismatic pair
滚滑高副：sliding-turning pair
二副杆：binary link
三副杆：ternary link

平面机构：plane mechanism
空间机构：spatial mechanism
运动链：kinematical chain
平面运动链：plane chain
空间运动链：spatial chain
闭链：closed chain
开链：unclosed chain，open chain
自由度：degree of freedom，mobility
约束：constraint
虚约束：overconstraint，redundant constraint
局部自由度：partial degree of freedom，passive degree of freedom，redundant degree of freedom
复合铰链：multiple pin joints，compound hinges
机构运动简图：schematic diagram of mechanism，kinematic sketch
颚式破碎机：jaw crusher machine
长度比例尺：length scale
刚性构件：rigid link
柔性构件：flexible link
活动构件：moving link
高副低代：replacement of higher pair by lower pair
机构的结构分析：structure analysis of mechanism
杆组：link group
Ⅱ级杆组：class Ⅱ link group
Ⅲ级杆组：class Ⅲ link group
复杂机构：complex mechanism
等效机构：equivalent mechanism
平面机构自由度计算公式：gruebler's criterion for degree of freedom of plane mechanism
空间机构自由度计算公式：kutzback criterion for degree of freedom of spatial mechanism

## Chapter 3 Kinematic Analysis of Planar Mechanisms
## 平面机构的运动分析

图解法：graphical method
相对运动图解法：graphical method of relative motion
瞬心法：velocity analysis by instantaneous centers
解析法：analytical method
运动分析：kinematic analysis
线位移：displacement，linear displacement
线速度：velocity，linear velocity
线加速度：acceleration，linear acceleration
转角：angle
角位移：angular displacement
角速度：angular velocity
角加速度：angular acceleration
位移方程：displacement equation
速度方程：relative velocity equation
加速度方程：relative acceleration equation
绝对速度：absolute velocity
相对速度：relative velocity
牵连速度：converted velocity，velocity of the base point
法向速度：normal velocity
切向速度：tangential velocity
哥式加速度：coriolis component of acceleration，coriolis acceleration
位移分析：displacement analysis
速度分析：velocity analysis
速度影像：velocity image
加速度影像：acceleration image
加速度分析：acceleration analysis
重合点：coincident points
瞬心：instant center，instantaneous center
速度瞬心：velocity center
绝对瞬心：absolute center
相对瞬心：relative center
瞬心多边形：instantaneous center polygon
用运动副直接连接的两构件瞬心：prime center
三心定理：aronhold-kennedy theorem

极点：pole
速度比例尺：velocity scale
加速度比例尺：acceleration scale
速度多边形：velocity polygon
加速度多边形：acceleration polygon
矢量：vector
标量：scalar
大小与方向：magnitude and direction
直角坐标系：rectangular coordinate
封闭矢量环：closed vector loop
矩阵：matrix
线形代数方程：linear algebraic equation
线性方程组：linear equations
非线性方程组：non-linear equations
数学模型：mathematical model
计算机辅助设计：computer aided design（CAD）
计算机程序：computer program

## Chapter 4 Force Analysis of Planar Mechanisms
## 平面机构的力分析

力分析：force analysis
静力分析：static analysis
动力分析：dynamic analysis
静力平衡：static equilibrium
作用力：applied force
力矩：moment
驱动力：driving force
阻抗力：resistance force
重力：gravity
惯性力：inertia force
约束反力：reactive force，reaction
分离体：free-body diagram
二力构件：two-force member
三力构件：three-force member
力作用线：action line of force
平衡方程：equation of equilibrium
摩擦：friction
摩擦因数：coefficient of friction
当量摩擦因数：equivalent coefficient of friction
干摩擦：dry friction
平面摩擦：friction on plane
斜面摩擦：friction on inclined plane
矩形螺纹摩擦：friction on square-threaded screw
三角螺纹摩擦：friction on v-threaded screw
滑动摩擦：sliding friction
接触表面：contacting surface
磨损：wear
摩擦力：friction force
摩擦角：friction angle
当量摩擦角：equivalent friction angle
总反力：total resultant force
轴颈摩擦：pin friction，friction in journal bearing
摩擦圆：friction circle
达朗伯原理：d'alember's principle
惯性力矩：inertia torque，moment of inertia
质心：center of mass，mass center
转动惯量：mass moment of inertia
离心力：centrifugal force
水平力：horizontal force
切向力：tangential force
法向力：normal force
自锁：self-locking
机械效率：mechanical efficiency
自锁机构：self-locking mechanism
输入功率：input power
输出功率：output power

## Chapter 5 Synthesis of Planar Linkages
## 平面连杆机构及其设计

运动综合：kinematic synthesis
精确综合：precision synthesis

近似综合：approximate synthesis
型综合：type synthesis
数综合：number synthesis
尺度综合：dimensional synthesis
精度：precision
结构误差：structural error
按行程速度变化系数综合四杆机构：synthesis of four-bar linkage for coefficient of travel speed variation
按连杆的对应位置综合四杆机构：synthesis of four-bar linkage for specified coupler positions
按连架杆的对应位置综合四杆机构：synthesis of four-bar linkage for coordination of angular positions
按连杆曲线综合四杆机构：synthesis of four-bar linkage for prescribed coupler path
刚体导引：guiding a body through a number of prescribed positions
函数发生：function generation, coordination of the positions of the output and input link
轨迹生成：path generation, guiding a point along a prescribed path
连杆机构：linkage
平面连杆机构：plane linkage
空间连杆机构：spatial linkage
四杆机构：four-bar linkage
曲柄摇杆机构：crank-rocker linkage
双曲柄机构：double-crank linkage
双摇杆机构：double-rocker linkage
曲柄滑块机构：slider-crank linkage
曲柄摇块机构：rocking-block linkage
摆动导杆机构：rocking guide-bar linkage
转动导杆机构：rotating guide-bar linkage
移动导杆机构：sliding guide-bar linkage
正弦机构：scotch-yoke linkage, sine linkage
正切机构：rapson's slide linkage, tangent linkage
双滑块机构：double slider linkage
双转块机构：double rotating block linkage
平行四边形机构：parallel-crank four-bar linkage
曲柄存在的条件：grashoff' law
压力角：pressure angle
传动角：transmission angle
极位夹角：acute angle corresponding crank positions when the rocker is at two extreme positions
极限位置：extreme position, limiting position
行程速度变化系数：coefficient of travel speed variation, advance to return time ratio
急回特性：quick-return characteristics
急回运动：quick-return motion
急回机构：quick-return mechanism
死点：dead point
机构演化：evolution of mechanism
机构变异：mutation of mechanism
机架变换：inversion of mechanism
移动：translation
定轴转动：rotation about a fixed axis
一般平面运动：general plane motion

## Chapter 6 Design of Cam Mechanisms
## 凸轮机构及其设计

凸轮：cam
高速凸轮：high-speed cam
凸轮机构：cam mechanism
从动件：follower
滚子：roller
滚子半径：radius of roller
滚子中心：center of roller
平底长度：length of flat-face
直动从动件：translating follower
摆动从动件：oscillating follower
滚子从动件：roller follower
尖顶从动件：knife-edge follower

平底从动件：flat-faced follower，flat follower
曲底从动件：spherical-faced follower
对心直动从动件盘形凸轮机构：disk cam with radial translating follower
对心直动滚子从动件盘形凸轮机构：disk cam with radial translating roller follower
对心直动尖顶从动件盘形凸轮机构：disk cam with radial translating knife-edge follower
偏置直动从动件盘形凸轮机构：disk cam with offset translating follower
偏置直动滚子从动件盘形凸轮机构：disk cam with offset translating roller follower
偏置直动尖顶从动件盘形凸轮机构：disk cam with offset translating knife-edge follower
摆动尖顶从动件盘形凸轮机构：disk cam with oscillating knife-edge follower
摆动平底从动件盘形凸轮机构：disk cam with oscillating flat follower
摆动滚子从动件盘形凸轮机构：disk cam with oscillating roller follower
直动滚子从动件圆柱凸轮机构：cylindrical cam with translating roller follower
摆动滚子从动件圆柱凸轮机构：cylindrical cam with oscillating roller follower
运动规律：follower motion
多项式运动规律：polynomial motion
一阶多项式运动规律：first-degree polynomial motion
二阶多项式运动规律：second-degree polynomial motion
3-4-5 次多项式运动规律：3-4-5 polynomial motion
等速运动规律：uniform motion，straight-line motion
等加速、等减速运动规律：parabolic motion，constant acceleration motion
三角函数运动规律：trigonometric function motion
摆线运动规律：cycloidal motion
简谐运动规律：simple harmonic motion
正弦运动规律：sine acceleration motion，sine motion
余弦运动规律：cosine acceleration motion，cosine motion
组合运动规律：combination of follower motion
位移线图：displacement diagram
位移方程：displacement equation
速度线图：velocity diagram
加速度线图：acceleration diagram
盘形凸轮：disk cam，plate cam
移动凸轮：translating cam，wedge cam
等径凸轮：conjugate yock radical cam
等宽凸轮：yock radical cam，constant-breadth cam
共轭凸轮：conjugate cam
圆柱凸轮：cylindrical cam
圆锥凸轮：conical cam
反转法原理：principle of inversion
包络线：envelope
实际廓线基圆：base circle
偏距：offset
偏距圆：offset circle
理论廓线基圆：prime circle
理论廓线：pitch curve
实际廓线：cam profile
曲率：curvature
凸轮廓线的曲率半径：radius of curvature of cam profile
凸轮廓线的曲率中心：center of curvature of cam profile
运动失真：motion distortion，undercutting
尖点：sharp point，cusp
外凸曲线：convex curve
内凹曲线：concave curve
滚子中心点：trace point
开始点：initial point
升程：rise travel
回程：return travel
行程：stroke
从动件位移：follower displacement

凸轮转角：cam angle
停顿：dwell
升程运动角：rise angle，angle of ascent
回程运动角：return angle，angle of descent
休止角：repose angle，angle of dwell
刀具中心轨迹：locus of cutter center
刚性冲击：rigid impulse
柔性冲击：flexible impulse

## Chapter 7 Design of Gear Mechanisms

## 齿轮机构及其设计

齿轮机构：gear mechanism
平行轴齿轮：parallel gears
相交轴齿轮：intersecting gears
交错轴齿轮：skew gears
圆柱齿轮：cylindrical gear
锥齿轮：bevel gear
蜗轮：warm gear
蜗杆：warm
直齿圆柱齿轮：spur gear
斜齿圆柱齿轮：helical gear
齿廓曲线：tooth profile
齿廓啮合基本定律：fundamental law of toothed gearing，law of gearing
发生线：generating line
渐开线：involute
渐开线性质：involute property
渐开线函数：involute function
渐开线方程：involute equation
中心距：distance of centers
标准中心距：standard distance of centers
实际安装中心距：mounted distance of centers
节点：pitch point
节圆：pitch circle
分度圆：reference circle，standard pitch circle
基圆：base circle
齿顶圆：addendum circle
齿根圆：dedendum circle
齿数：number of teeth
模数：module
齿距：circular pitch
基节：base pitch
法节：normal pitch
齿厚：tooth thickness
齿槽宽：width of space，space width
任意圆周上的齿厚：tooth thickness along an arbitrary circle
齿顶高：addendum
齿根高：dedendum
全齿高：whole depth
顶隙：clearance
侧隙：backlash
齿顶高系数：coefficient of addendum
顶隙系数：coefficient of clearance
基圆直径：diameter of base circle
分度圆直径：diameter of reference circle
节圆直径：diameter of pitch circle
齿顶圆直径：diameter of addendum circle
齿根圆直径：diameter of dedendum circle
公法线长度：length of common normal
小齿轮：pinion
外齿轮：external gear
齿条：rack
内齿轮：internal gear，ring gear
啮合：mesh，engaging
啮合点：engaging point，meshing point
啮合线：line of action
啮合角：meshing angle，angle of obliquity
正确啮合的条件：condition of correctly engaging
重合度：contact ratio
滑动系数：slip ratio
齿轮加工：forming of gear teeth，gear manufacture
成形法：form milling
范成法：generating
盘形刀具：disk milling cutter

指形刀具：finger cutter
齿条刀具：rack-shaped cutter
干涉：interference
根切：undercutting
不发生根切的最少齿数：minimum number of teeth to avoid undercutting
标准齿轮：standard gear
变位齿轮：nonstandard gear，modified gear
变位系数：coefficient of offset
变位：offset
正变位：positive offset
负变位：negative offset
正变位齿轮：positive offset modified gear
负变位齿轮：negative offset modified gear
不发生根切的最小变位系数：smallest coefficient of offset to avoid undercutting
分度圆分离系数：separated coefficient of reference circle
齿顶高降低系数：shortening coefficient of addendum
无侧隙啮合方程：equation of engagement with zero backlash
分度圆分离方程：equation of separation of reference circle
基圆柱：base cylinder
渐开面：involute surface
发生面：generating plane
渐开螺旋面：involute helicoid
螺旋：thread
螺旋角：helix angle
左旋：left-hand
右旋：right-hand
法面：normal plane
端面：transverse plane，plane of rotation
法向参数：normal parameter
端面参数：transverse parameter
法向压力角：normal angle
端面压力角：transverse angle
法向模数：normal module
端面模数：transverse module
法向基节：normal base pitch
端面基节：transverse base pitch
法向齿厚：normal tooth thickness
端面齿厚：transverse tooth thickness
法向齿顶高系数：coefficient of normal addendum
端面齿顶高系数：coefficient of transverse addendum
法向顶隙系数：coefficient of normal radical clearance
端面顶隙系数：coefficient of transverse radical clearance
法向重合度：normal contact ratio
端面重合度：transverse contact ratio
当量齿轮：equivalent spur gear
当量齿数：number of teeth of the equivalent spur gear
法向力：normal force
径向力：radial force
轴向力：axial force
人字齿轮：herringbone gear
螺旋齿轮：crossed helical gear
阿基米德蜗杆：achimedes worm
蜗杆直径系数；quotient of warm diameter
主截面：main section
螺纹升角：lead angle
直齿锥齿轮：straight-tooth bevel gear
曲齿锥齿轮：spiral bevel gear
球面渐开线：spherical involute
轴角：shaft angle
齿顶圆锥：face cone
节圆锥：pitch cone
齿根圆锥：root cone
节锥角：pitch cone angle
锥距：cone distance
锥顶：common apex
顶锥角：face angle
根锥角：root angle
齿顶角：addendum angle
齿根角：dedendum angle

背锥：back cone
平面锥齿轮：crown gear
准双曲面锥齿轮：hypoid gear

## Chapter 8 Gear Trains
## 轮系及其设计

轮系：gear train
定轴轮系：ordinary gear train
单式轮系：simple gear train
复式轮系：compound gear train
回归轮系：reverted gear train
周转轮系：epicyclic gear train
差动轮系：differential gear train
行星轮系：planetary gear train
混合轮系：compound epicyclic gear train, combined gear train
串联轮系：tandem combined gear train
封闭轮系：closed combined gear train
传动比：angular velocity ratio, speed ratio
太阳轮：sun gear
行星轮：planet gear
系杆：planet carrier, arm
转化机构：inversion gear train, converted gear train
轮系效率：efficiency of gear train
少齿差行星传动：planetary gear train with small tooth difference
摆线针轮：cycloidal-pin wheel
摆线齿轮：cycloidal gear
谐波传动：harmonic drive
波数：number of wave
波发生器：wave generator
刚轮：rigid circular spline
柔轮：flex spline, flexible wheel
惰轮：idle gear
平动齿轮传动：parallel move gearing
减速器：reducer

## Chapter 9 Introduction of Screws, Hook's Couplings and Intermittent Mechanisms
## 螺旋机构、万向联轴器、间歇运动机构简介

螺旋传动：screw drive
螺旋机构：screw mechanism
往复移动：rectilinear motion
差动螺旋机构：differential screw mechanism
滚珠螺旋丝杠机构：roller screw mechanism
万向联轴器：universal joints, hook's coupling
单万向联轴器：single universal joint
十字叉：cross piece
叉头：yoke
双万向联轴器：double universal joints
间歇运动机构：intermittent motion mechanism
自动机械：automatic machinery
间歇转动：intermittent rotation
棘轮机构：ratchet mechanism
棘轮：ratchet wheel, ratchet
棘爪：pawl
无声棘轮：silent ratchet wheel
槽轮机构：geneva wheel mechanism
锁紧盘：locking plate
内槽轮机构：inverse geneva wheel mechanism
运动系数：action coefficient
蜂窝煤机：beehive coal tool machine
不完全齿轮机构：intermittent gear mechanism
分度凸轮机构：indexing cam mechanism
圆柱形分度凸轮机构：cylindrical cam indexing mechanism
蜗杆形分度凸轮机构：worm-cam indexing mechanism

## Chapter 10 Spatial Mechanisms and Robotic Mechanisms
## 空间连杆机构及机器人机构概述

空间机构：spatial mechanism
空间串联机构：spatial mechanism in series, spatial tandem mechanism
空间并联机构：spatial mechanism in parallel, spatial parallel mechanism
空间运动副：spatial kinematic pair
球面副：spherical pair
圆柱副：cylintrical pair
螺旋副：helical pair
球销副：slotted spherical pair
Ⅰ类副：class Ⅰ kinematic pair
Ⅱ类副：class Ⅱ kinematic pair
Ⅲ类副：class Ⅲ kinematic pair
Ⅳ类副：class Ⅳ kinematic pair
Ⅴ类副：class Ⅴ kinematic pair
分类：classification
机器人：robot
机械手：manipulator
串联机器人：tandem robot
并联机器人：parallel robot
闭链机器人：closed chain robot
开链机器人：unclosed chain robot
平面机器人：planer robot
空间机器人：spatial robot
正解：normal solution
逆解：inverse solution

## Chapter 11 Design of Mechanism systems
## 机构系统设计

方案：project, scheme
运动系统：kinematic scheme, kinematic system
机电产品：electromechanical product
现代设计：modern design
优化设计：optimal design
机械运动方案：project of mechanical system
机构系统：system of mechanisms, mechanism system
传动机构：transmission mechanism
工作执行机构：operating mechanism
控制系统：control system
电动机：electromotor
运动循环图：motion circulation diagram
运动协调设计：design of motion coordination, design of motion harmonization
概念设计：conceptual design
机构的组合：combination of mechanisms
创新：Innovation, creation
创新设计：innovating design, creative design
基本机构：elementary mechanism, Basic mechanism
复杂机构：complex mechanism
串联：serial connection, Tandem connection
并联：parallel connection, connection in parallel
叠加组合：superposed connection
封闭连接：closed connecting
牛头刨床：shaper
压床：press machine
压力机：punching machine

## Chapter 12 Fluctuation and Regulation in Speed of Machines
## 机械系统的运转及速度波动的调节

机械系统：machinery system
机械的运转：operating of machinery
速度波动：fluctuation in speed
速度波动的调节：regulation in speed

调速器：governor
飞轮：flywheel
飞轮设计：flywheel design
公式：formula
运转不均匀系数：coefficient of fluctuation in speed
最大角速度：maximum angular speed
最小角速度：minimum angular speed
平均角速度：average angular speed
能量：energy
动能：kinetic energy
最大动能：maximum kinetic energy
最小动能：minimum kinetic energy
驱动力做功：work done by driving forces
阻抗力做功：work done by resistance forces
盈功：increment work
亏功：decrement work
动能方程；equation of kinetic energy
瞬时功率：instantaneous power
稳定运转周期：period of steady motion
平均速度：average speed
制动器：brake
等效构件：equivalent link
等效力：equivalent force
等效力矩：equivalent moment
等效驱动力：equivalent driving force
等效驱动力矩：equivalent driving moment
等效阻抗力：equivalent resistance force
等效阻抗力矩：equivalent resistance moment
等效质量：equivalent mass
等效转动惯量：equivalent moment of inertia
飞轮矩：moment of gyration，flywheel characteristic
微分方程：differential equation
积分方程：integral equation

## Chapter 13　Balance of Machinery

## 机械的平衡

平衡：balance，equilibrium
机械平衡：balance of machinery
静平衡：static balance
动平衡：dynamic balance
不平衡：unbalance
转子：rotor
刚性转子：rigid rotor
挠性转子：flexible rotor
动平衡机：dynamic balance machine
平行平面：parallel plane
校正平面：correcting plane
平衡实验：balance test
平衡精度：balance precision
许用不平衡量：allowable unbalance，allowable amount of unbalance
许用偏心距：allowable offset
转子的质量：mass of rotor
转子的重量：gravity of rotor
平衡质量：balance mass
质径积：mass-radius product
残存的不平衡量：residual unbalance
现场平衡：field balance
振动力：shaking force
振动力矩：shaking couple
机构平衡：balance of mechanism
部分平衡：partial balance
完全平衡：complete balance

# Appendix 附录二
# Common symbols and formulas pronunciation
# 符号与公式读法

1. 数字读法（Writing and saying numbers）

20 以内的数字读法很容易，这里仅列举 20 以上的数字读法。

（1）20 以上整数读法（Writing and saying integers over twenty）

20 以上的整数读相应的基数词。例如：

35，thirty-five；67，sixty-seven；71，seventy-one，etc.

（2）100 以上整数读法（Writing and saying integers over hundred）

百位数与十位数之间加 and，基数词 + hundred + and + 基数词。例如：

101，one hundred and one；329，three hundred and twenty nine；856，eight hundred and fifty six

（3）1，000 以上整数读法（Writing and saying integers over thousand）

基数词 + thousand + 基数词。例如：

1，100，one thousand one hundred；5，682，five thousand six hundred and eighty two

33，423，thirty three thousand four hundred and twenty three

（4）10，000 以上的整数（Writing and saying integers over ten thousand）

10，000（一万），ten thousand；100，000（十万），one hundred thousand；

1，000，000（一百万），one million；10，000，000（一千万），ten million；

100，000，000（一亿），one hundred million

（5）亿以上的大数读法（Writing and saying integers over one hundred million）

英美有不同的读法（英国/美国）。例如：

十亿，one thousand million/one billion；百亿，ten thousand million/ten billion；

千亿，one hundred thousand million/one hundred billion；万亿，one thousand billion/one trillion

（6）小数读法（Writing and saying decimals）

小数点读 point，小数点前后数字读相应的基数词；美式英语中的零 0 读 zero，英式英语中的零 0 读 nought。例如：

0.67，zero point six seven（美式）/nought point six seven（英式）；79.3，seventy-nine point three；

423.08，four hundred and twenty three point zero eight；

3.14159，three point one four one five nine（小数点后位数过多时，可直接读对应的基数词）

（7）分数读法（Writing and saying fractions）

分数有两种读法，可根据自己的阅读习惯选择。

1）分子读基数词，当分子大于 1 时，分母读序数词复数。

2）分子读基数词 + over + 分母读基数词。

例如：

1/2，one half（习惯读法）；1/3，one third（习惯读法）；1/4，one quarter/one fourth；

2/3，two thirds/two over three；3/4，three fourths/three quarters/three over four；

1/10，one tenth/a tenth；1/100，one hundredth/a hundredth

1/1000，one thousandth/a thousandth；19/56，nineteen fifty-sixths/nineteen over fifty-six；$\frac{2}{3}$，two thirds/two over three；$4\frac{2}{3}$，four and two thirds/four and two over three

复杂分数读法如下：

31/144，thirty-one over one hundred and forty -four/thirty-one over one four four；

7/1234，seven over one thousand two hundred and thirty-four/seven over one two three four

（8）百分数读法（Writing and saying percentages）

百分数读法为：基数词 + percent；千分数读法为：基数词 + permille。例如：

5%，five percent；15%，fifteen percent；86%，eighty six percent；

0.2%，zero point two percent；0.3%，decimal three percent；4‰，four per mille

（9）序数词读法（Writing and saying ordinal numbers）

1st，first；2nd，second；3rd，third；4th，fourth；5th，fifth；…9th，ninth；10th，tenth；11th，eleventh；12th，twelfth；…19th，nineteenth；21，twenty first；22，twenty-second；23，twenty-third；24，twenty-fourth；

38th，thirty-eighth

2. 标点符号读法（Writing and saying punctuation）

仅在阅读课文时才读标点符号。

句号，dot；逗号，comma；冒号，colon；分号，semicolon；问号，question mark；

圆括号，brackets；方括号，square brackets；大括号，braces

3. 数学运算符号读法（Writing and saying mathematical expressions）

加号（+），plus；20 +5，twenty plus five；$x+y$，$x$ plus $y$ 等。

减号（-），minus；23 -7，twenty three minus seven；$x-y$，$x$ minus $y$ 等。

正号（+），positive；+5，positive five；$+x$，positive $x$ 等。

负号（-），negative；-5，negative five；$-y$，negative $y$ 等。

乘号（×），times or multiplied by；3 ×5，three times five/three multiplied by five 等。

除号（÷），divided by；9 ÷4，nine divided by four；over，a/b，a over b

$\frac{a-b}{c+d}$，$a$ minus $b$ divided by $c$ plus $d$

等号（=），is；equals；is equal to

大于号（>），is greater than；小于号（<），is less than

约等于号（≈），is approximately equal to or approximately equals

小于等于号（≤），equals to or less than；大于等于号（≥），equals to or greater than

4. 矢量叉积与点积读法（Writing and saying vector cross product and dot product）

叉积，cross product：A × B，the cross product of vector A and B，vector A cross B

点积，dot product：A. B，the dot product of vector A and B，vector A dot B

5. 逻辑运算读法（Writing and saying logic operations）

x ∈ A，x belongs to A or x is an element（or a member）of A；

A ⊂ B，A is contained in B or A is a subset of B；A ⊃ B，A contains B or B is a subset of A

6. 指数表达式（Writing and saying exponent）

一个数或代数式的平方和立方有固定读法，其他指数采用“to the power of”。例如：

$3^2$，three squared；$x^2$，$x$ squared；$5^3$，five cubed；$x^3$，$x$ cubed；$6^{10}$，six to the power of ten；$y^n$，$y$ to the power of $n$；$y^{n-1}$，$y$ to the power of $n$ minus one；$x_{B1}^2$，$x$ squared sub $B$ one

7. 根号读法（Writing and saying $n$th root of）

开平方与开立方有固定读法，其他采用：the $n$th root of。例如：

$\sqrt{x}$，square root of $x$；$\sqrt[3]{x}$，cube root of $x$；$\sqrt[5]{x}$，fifth root of $x$；$\sqrt[n]{x}$，$n$th root of $x$；

$x^{\frac{1}{n}}$或$\sqrt[n]{x}$：$x$ to the power of one over $n$ or the $n$th root of $x$；

8. 三角函数读法（Writing and saying trigonometric function）

$\sin\alpha$，sine$\alpha$；$\cos\varphi$，cosine$\varphi$；$\tan\alpha$，tangent$\alpha$；$\cot\alpha$，cotangent$\alpha$；

$\arctan\dfrac{a}{b}$，arctangent $a$ over $b$

9. 上下角标读法（Writing and saying superscripts and subscripts）

上下角标有两种读法：一是读出上角标或下角标；二是类似汉语的读法，不读出角标。

读法 1：上角标读 superscript or sup，下角标读 subscript or sub。例如：

$F^n$，$F$ superscript $n$/$F$ sup $n$；$F_3$，$F$ subscript three/$F$ sub three；

$v_{B1}$，$v$ sub $B$ one；$F_{32}$，$F$ sub three two；$v_{B1}^n$，$v$ sup $n$ sub $B$ one；$a_{B1}^n$，$a$ sup $n$ sub $B$ one

读法 2：类似汉语的读法，省略角标，口语时常用。例如：

$F^n$，$F$ $n$；$F_3$，$F$ three；$v_{B1}$，$v$ $B$ one；$F_{32}$，$F$ three two；

$v_{B1}^n$，$v$ $n$ $B$ one/$v$ $B$ one $n$；$a_{B1}^n$，$a$ $n$ $B$ one/$a$ $B$ one $n$

10. 撇号读法（Writing and saying apostrophe）

无角标时，先读字母或数字 + prime；有角标时，先读撇号，再读角标；带有平方时，后读平方。例如：

$b'$，$b$ prime；$b''$，$b$ double prime；$b'''$，$b$ triple prime；$b_1''$：$b$ double prime sub one；

$b''^2$，$b$ double prime squared

11. 微分读法（Writing and saying differentials）

$dx$，differential of $x$ 或直读 dee $x$；$\Delta x$，increment of $x$ 或直读 delta $x$；

$f(x)$，the function $f$ of $x$；$\dfrac{dy}{dx}$，the first derivative of $y$ with respect to $x$；

$\dfrac{d^2y}{dx^2}$，the second derivative of $y$ with respect to $x$；$\dfrac{d^ny}{dx^n}$，the $n$th derivative of $y$ with respect to $x$；

$y'$，the first derivative of $y$ with respect to $x$；$y''$，the second derivative of $y$ with respect to $x$；

$f'''(x)$，the third derivative of $f$ with respect to $x$；$f^{(4)}(x)$，the fourth derivative of $f$ with respect to $x$；

$\partial f/\partial x_1$，the partial derivative of $f$ with respect to $x$ sub one；

$\partial^2 f/\partial x_1^2$，the second partial derivative of $f$ with respect to $x$ sub one；

$\dot{y}$，$y$ dot，the first derivative of $y$ with respect to $t$；

$\ddot{y}$，$y$ two dots，the second derivative of $y$ with respect to $t$；

12. 积分的读法（Writing and saying integrals）

$\int f(x)\,dx$：integral of the function of $x$；$\int_a^b f(x)\,dx$：integral of the function of $x$ from $a$ to $b$

13. 矩阵的读法（Writing and saying matrix）

$[A]^T$：transpose matrix of $A$

$[A)]^{-1}$：inverse matrix of $A$

$[A] \cdot [x] = [B]$，matrix $A$ times $x$ equals $B$

较复杂矩阵运算，不需详细读出每个矩阵元素，一般采用简单读法。例如：

$$\begin{bmatrix} -l_2\sin\varphi_2 & l_3\sin\varphi_3 \\ l_2\cos\varphi_2 & -l_3\cos\varphi_3 \end{bmatrix}\begin{bmatrix} \omega_2 \\ \omega_3 \end{bmatrix} = \begin{bmatrix} l_1\omega_1\sin\varphi_1 \\ l_1\omega_1\cos\varphi_1 \end{bmatrix}$$

The left matrix times the right unknown matrix equals the constant matrix.

14. 不等式读法（Writing and saying inequalities）

$x > y$，$x$ is greater than $y$；$x \geqslant y$，$x$ is greater than or equal to $y$；

$x < y$，$x$ is less than $y$；$x \leqslant y$，$x$ is less than or equal to $y$；

$0 < x < 1$，zero is less than $x$ is less than one；$x$ is between 0 and 1；

$0 \leqslant x \leqslant 1$，zero is less than or equal to $x$，and $x$ is less than or equal to one；

$a \leqslant x \leqslant b$，$a$ is less than or equal to $x$，and $x$ is less than or equal to $b$

15. 求和的读法（Writing and saying summations）

$\sum_1^n F_i$，the summation $F$ sub $i$ from one to $n$ or sigma $F_i$，$i$ from one to $n$.

$\sum_1^3 m_i r_i$，the sigma of $m$ sub $i$ times $r$ sub $i$ from one to three or sigma $m$ $i$ times $r$ $i$，$i$ from one to three.

16. 公式读法（Writing and saying formulas）

（1）简单公式（Simple algebraic expression）

简单公式可按照字母、数字、符号直读即可，括号可以省略不读。例如：

$x+y=6$，$x$ plus $y$ is six. $x-y=3$，$x$ minus $y$ is three. /$x$ minus $y$ equals three.

$x+y=124$，$x$ plus $y$ is one hundred and twenty four。/$x$ plus $y$ equals one hundred and twenty four.

$x-y=1.29$，$x$ minus $y$ is one point two nine. /$x$ minus $y$ equals one point two nine.

$a+b-c-d=0$，$a$ plus $b$ minus $c$ minus $d$ is zero. /$a$ plus $b$ minus $c$ minus $d$ is equal to zero.

$(x+y)^2=m$，$x$ plus $y$ all squared is $m$. /$x$ plus $y$ squared is equal to $m$.

$x^2+y^2=n$，$x$ squared plus $y$ squared is $n$. /$x$ squared plus $y$ squared equals $n$.

$xy=5$，$x$ times $y$ equals five. /$x$ is multiplied by $y$ is five.

$\frac{a+b}{c+d}=5$，$a$ plus $b$ divided by the sum of $c$ and $d$ equals five. /$a$ plus $b$ divided by $c$ plus $d$ is equal to five.

$(x/y)^3=1.24$，$x$ over $y$ all cubic equals one point two four.

$2:3$，the ratio of two to three

$z_1:z_2:z_3=a:b:c$，the ratio of $z$ sub one，$z$ sub two and $z$ sub three equals the ratio of $a$ $b$ $c$. the ratio of $z$ one，$z$ two and $z$ three equals the ratio of $a$ $b$ and $c$.（省略角标读法）

$\lim_{x\to a} f(x)=b$，the limit of function $f$ $x$ is $b$ as $x$ approaches $a$

$25\times(267-225)=1050$，twenty-five times left bracket two hundred and sixty-seven minus two hundred and twenty－five right bracket equals one thousand and fifty.（直读法）

twenty-five times two hundred and sixty-seven minus two hundred and twenty-five equals one thousand and fifty.（简读法）

（2）机械原理常用公式读法列举（Reading examples in the book）

1）公式不过于复杂时，可直接按简单公式的读法直接读出。当公式较长时，上下角标读法可省略。例如：

① $F = 3n - 2p_l - p_h$, three times $n$ minus two times $p$ sub $l$ minus $P$ sub $h$ equals $F$.

three times $n$ minus two times $p$ $l$ minus $P$ $h$ equals $F$.（可省略下角标）

three $n$ minus two $p$ $l$ minus $p$ $h$ equals $F$.（可省略乘号和下角标，教学中常用此类读法）

② $N = \frac{k(k-1)}{2} = \frac{4 \times (4-1)}{2} = 6$, $N$ equals $k$ times $k$ minus one over two equals six.

③ $v_C = v_B + v_{CB}$, $v$ sub $C$ equals $v$ sub $B$ plus $v$ sub $CB$, or $v$ $C$ equals $v$ $B$ plus $v$ $CB$.

④ $v_{B2} = v_{B1} + v_{B2B1}$, $v$ sub $B$ two equals $v$ sub $B$ one plus $v$ sub $B$ two $B$ one, or $v$ $B$ two equals $v$ $B$ one plus $v$ $B$ two $B$ one.

⑤ $a_{B2} = a_{B1} + a^k{}_{B2B1} + a^r{}_{B2B1}$, $a$ $B$ sub two equals $a$ $B$ sub one plus $a$ sup $k$ sub $B$ two $B$ one plus $a$ sup $r$ sub $B$ two $B$ one.

$a$ $B$ two equals $a$ $B$ one plus $a$ $k$ $B$ two $B$ one plus $a$ $r$ $B$ two $B$ one.

⑥ $\omega_2 = \frac{v_{CB}}{L_{BC}} = \frac{\mu_v bc}{L_{BC}}$, Omega sub two equals $v$ sub $CB$ over $L$ sub $BC$ and equals the velocity scale $\mu$ sub $v$ times the length of $bc$ over the length of $BC$.

$\omega$ two equals $v$ $CB$ over $L$ $BC$ and equals $\mu$ $v$ times $bc$ over $LBC$.（省略角标）

⑦ $a_C^t = c''c'\mu_a$, $a$ sup $t$ sub $c$ equals $c$ double prime $c$ prime times $\mu$ sub $a$/$a$ $c$ $t$ equals $c$ double prime $c$ prime times $\mu$ $a$/The tangential acceleration at point $C$ equals the $c$ double prime $c$ prime times $\mu$ sub $a$.

⑧ $F_{i2} = -m_2 a_{s2}$, the force $F$ sub $i$ two equals negative $m$ sub two times $a$ sub $s$ two. /$F$ $i$ two equals negative $m$ two times $a$ $s$ two.

⑨ $M_{s2} = -J_{S2}\alpha_2$, the moment $M$ sub $s$ two equals negative $J$ sub $s$ two times $\alpha$ sub two. / $M$ $s$ two equals negative $J$ $s$ two times $\alpha$ two.

⑩ $[A] . [F_{ij}] = [B]$, the matrix $A$ multiplied by matrix $F$ $ij$ equals matrix $B$.

⑪ $F_d = G\tan(\alpha + \varphi)$, $F$ sub $d$ equals $G$ divided by tangent bracket $\alpha$ plus $\varphi$ bracket closed. /$Fd$ equals $G$ divide by tangent $\alpha$ plus $\varphi$.

⑫ $\mathrm{d}F = f\mathrm{d}N$, the differential of $F$ equals $f$ divided by the differential of $N$.

⑬ $\theta = 180° \frac{K-1}{K+1}$, $\theta$ equals one hundred and eighty degrees times the value of $K$ minus one over $K$ plus one. /$\theta$ equals one hundred and eighty degrees times $K$ minus one over $K$ plus one.

⑭ $\cos\gamma = \frac{b^2 + c^2 - a^2 - d^2 + 2ad\cos\varphi}{2bc}$, cosine $\gamma$ equals $b$ squared plus $c$ squared mines $a$ squared minus $d$ squared plus two $a$ $d$ cosine $\varphi$ over two $b$ $c$.

⑮ $s = c_0 + c_1\varphi + c_2\varphi^2 + c_3\varphi^3 + \cdots + c_n\varphi^n$, $s$ equals $c$ sub zero plus $c$ sub *one* times $\varphi$ plus $c$ sub two times $\varphi$ squared plus $c$ sub three times $\varphi$ cubed plus until to $c$ sub $n$ times $\varphi$ to the power of $n$. /$s$ is equal to $c$ zero plus $c$ one times $\varphi$ plus $c$ two times $\varphi$ squared plus $c$ three times $\varphi$ cubed plus until $c$ $n$ times $\varphi$ to the power of $n$.

⑯ $F = 6n - (P_1 + 2P_2 + 3P_3 + 4P_4 + 5P_5) = 6n - \sum iP_i$, $F$ equals six times $n$ minus $P$ sub one plus two $P$ sub two plus three $P$ sub three plus four $P$ sub four plus five $P$ sub five , and equals six times $n$ minus the sum of $i$ times $P$ sub $i$. /$F$ is equal to six $n$ minus $p$ one plus two $p$ two plus three $p$ three plus four $p$ four plus five $p$ five, and $F$ equals six $n$ minus the sum of $i$ $P$ sub $i$.

⑰ $J\frac{d\omega}{dt}+\frac{\omega^2}{2}\frac{dJ}{d\varphi}=M=M_d-M_r$, $J$ times the first derivative of $\omega$ with respect to t plus $\omega$ squared over two times the first derivative of $J$ with respect to $\varphi$ equals $M$, and is equal to $M$ d minus $M$ $r$.

⑱ $\rho=\frac{(1+y'^2)^{3/2}}{y''}$, $\rho$ equals bracket one plus $y$ prime squared bracket to the power of three over two over $y$ double prime. /$\rho$ is equal to one plus $y$ prime squared to the power of three second divided by $y$ double prime.

⑲ $\sum_{i=1}^{n}F_i+F=0$, the sum of $F$ sub $i$ from $i$ equals one to $n$ plus $F$ is zero.

⑳ $(x_{Bi}-x_A)^2+(y_{Bi}-y_A)^2=a^2$, $x$ sub $Bi$ minus $x$ sub $A$ all squared plus $y$ sub $Bi$ minus $y$ sub $A$ all squared equals $a$ squared. /$x$ $Bi$ minus $xA$ all squared plus $y$ $Bi$ minus $y$ $A$ all squared equals $a$ squared.（无角标）

㉑ $s=\frac{2h}{\Phi^2}\varphi^2$, $s$ is equal to two $h$ divided by capital $\Phi$ squared times $\varphi$ squared.

㉒ $s=h-\frac{2h}{\Phi^2}(\Phi-\varphi)^2$, $s$ is equal to $h$ minus two $h$ divided by capital $\Phi$ squared times capital $\Phi$ minus $\varphi$ all squared.

㉓ $s=h\left(\frac{10}{\Phi^3}\varphi^3-\frac{15}{\Phi^4}\varphi^4+\frac{6}{\Phi^5}\varphi^5\right)$, $s$ is equal to $h$ times bracket ten divided by capital $\Phi$ to the power of three times $\varphi$ to the power of three minus fifteen divided by capital $\Phi$ to the power of four times $\varphi$ to the power of four plus six divided by capital $\Phi$ to the power of five times $\varphi$ to the power of five bracket.

㉔ $\tan\alpha=\frac{\frac{ds}{d\varphi}}{\sqrt{r_0^2-e^2}+s}$, tangent $\alpha$ is equal to the derivative of $s$ with respect to $\varphi$ divided by square root of $r$ zero squared minus $e$ squared plus $s$.

㉕ $\frac{1}{2}J\omega^2-\frac{1}{2}J_0\omega_0^2=\int_{\varphi_0}^{\varphi}Md\varphi$, one half $J$ omega squared minus one half $J$ zero omega zero squared is equal to the integral of $M$ with respect to phi from phi zero to phi.

㉖ $\omega=\sqrt{\frac{J_0}{J}\omega_0^2+\frac{2}{J}\int_{\varphi_0}^{\varphi}Md\varphi}$, omega is equal to square root of $J$ zero over $J$ times omega zero squared plus two over $J$ times the integral of $M$ with respect to phi from phi zero to phi.

㉗ $\sum_{i=1}^{n}m_ir_i\cos\theta_i+mr\cos\theta=0$, the sum of $m$ $i$ $r$ $i$ times cosine theta $i$ for $i$ from one to $n$ plus $m$ $r$ cosine theta is equal to zero.

㉘ $mr=\sqrt{\left(\sum_{i=1}^{n}m_ir_i\cos\theta_i\right)^2+\left(\sum_{i=1}^{n}m_ir_i\sin\theta_i\right)^2}$, $m$ $r$ is equal to the root of sum of $m$ $i$ times $r$ $i$ times cosine theta $i$ squared for $i$ from one to $n$ plus sum of $m$ $i$ $r$ $i$ times sine theta $i$ squared for $i$ from one to $n$.

㉙ $\theta=\arctan\frac{-\sum_{i=1}^{n}(m_ir_i\sin\theta_i)}{-\sum_{i=1}^{n}(m_ir_i\cos\theta_i)}$, theto is equal to arctangent negative sum of $m$ $i$ $r$ $i$ sine theta $i$

for $i$ from one to $n$ divided by negative sum of $m$ $i$ $r$ $i$ cosine theta $i$ for $i$ from one to $n$.

2）公式中的同类运算符号过多时，可采用简化读法。例如：

$M_f=\int_{r_1}^{r_2}\mathrm{d}M_f=\int_{r_1}^{r_2}2\pi fp\rho^2\mathrm{d}\rho$ , the $M$ sub $f$ equals the integral of dee $M$ sub $f$ from $r$ sub one to $r$ sub two and equals the integral of product of *two*, $\pi$, $f$, $p$, $\rho$ squared, d$\rho$. /$M$ $f$ is equal to the integral of the function of $M$ $f$ from $r$ one to $r$ two, and it is equal to the integral of two $\pi$ $f$, $p$, $\rho$ squared differential of with respect $\rho$ from $r$ one to $r$ two.

3）表达式过于复杂时，可不必全部读出。例如：

$U=f\ [\ (x_P-d)\ \cos\gamma+y\sin\gamma]\ (x_P^2+y_P^2+e^2-a^2)\ \ -ex_P\ [(x_P-d)^2+y_P^2+f^2-c^2]$

The value $U$ equals the following expression…

4）有时可不按照公式中字母或数字直读，可按照公式的含义去读。例如：

① $v_C=v_B+v_{CB}$, the absolute velocity at point $C$ equals the absolute velocity at point $B$ plus the velocity of point $C$ relative to $B$.

② $v_{B2}=v_{B1}+v_{B2B1}$, the velocity at point $B$ sub two equals the velocity at point $B$ sub one plus the relative velocity at the coincident point $B$ sub two and $B$ sub one. /The velocity at point $B$ two equals the velocity at point $B$ one plus the relative velocity at the coincident point $B$ two and $B$ one.（省略角标）

③ $a_{B2}=a_{B1}+a^k{}_{B2B1}+a^r{}_{B2B1}$, the acceleration at point $B$ sub two equals the acceleration at point $B$ sub one plus the Coriolis acceleration of the coincident point $B$ sub two $B$ sub one plus the relative acceleration of point $B$ sub two $B$ sub one.

④ $i_{1k}=\dfrac{\omega_1}{\omega_k}=(-1)^m\dfrac{z_2\cdots z_k}{z_1\cdots z_{k-1}}$, the ratio $i$ sub one $k$ equals $\omega$ sub one over $\omega$ sub $k$ and also equals negative one to the power of $m$ times the product of driven gear tooth numbers over the product of driving gear tooth numbers. /The ratio $i$ one $k$ equals $\omega$ one over $\omega$ $k$ and also equals negative one to the power of $m$ times the product of driven gear tooth numbers over the product of driving gear tooth numbers.

⑤ $i_{15}=i_{12}i_{23}i_{3'4}i_{4'5}$, the ratio $i$ sub one five equals the product of gear ratios in the train.

⑥ $J_f=\dfrac{E_{\max}-E_{\min}}{\delta\omega^2{}_m}$, the moment of inertia of the flywheel equals the difference between the maximum energy and minimum energy over the product of $\delta$ and $\omega$ squared sub $m$.

注：

1. 按照公式中的数字、字母、符号、角标直读，显得有些冗长、枯燥，所以口语中经常省略括号、上下角标。较长的公式一般不读，用简化方式读即可。

2. 有些符号、表达式、公式给出了几种不同的读法，供读者参考。

3. 由于数学公式种类和数量过多，本附录仅列出部分公式。但其读法具有普遍性，也适合阅读其他科技英语中的数学公式。

# References 参考文献


[1] Robert L. Norton. Design of Machinery [M]. New York: McGraw-Hill Inc, 2002.

[2] S S Rattan. Theory of Machines [M]. New York: Tata McGraw-Hill Education Prlvate Limited, 2009.

[3] J E Shigley. Theory of Machine and Mechanisms [M]. New York: McGraw-Hill Inc, 1980.

[4] Reuleux F. Kinematics of machinery [M]. London: Dover Publications Inc, 1963.

[5] Erdman A G. Sandor G N. Mechanism Design [M]. New York: Prentice-Hall International Inc, 1984.

[6] Uicker, Shigley. Theory of Machine and Mechanisms [M]. London: Oxford University Press, 2003.

[7] Hamilton H. Mabi. Mechanisms and Dynamics of Machinery [M]. New York: John Wiley and Sons Inc, 1987.

[8] R C Hibbeler. Engineering Mechanics [M]. New York: Prentice-hall International Inc, 1995.

[9] Wilson, Sadler, Michels. Kinematics and Dynamics of Machinery [M]. New York: Harper & Row, 1983.

[10] Barnacle, Walker. Mechanics of Machines [M]. London: Oxford Pergamon, 1965.

[11] George H. Martin . Kinematics and Dynamics of Machines [M]. New York: McGraw-Hill Inc, 1982.

[12] Jerimy. Hirschhorm. Kinematics and Dynamics of Plane Mechanisms [M]. New York: McGraw-Hill Inc, 1962.

[13] John R. Zimmerman. Elementary Kinematics of Mechanisms [M]. New York: John Wiley &Sons, 1975.

[14] Ye Zhonghe, Lan Zhaohui. Mechanisms and Machine Theory [M]. Beijing: Higher Education Press, 2001.

[15] 张春林. 机械原理 [M]. 北京：高等教育出版社，2006.

[16] 张春林. 高等机构学 [M]. 北京：北京理工大学出版社，2006.

[17] 张春林. 机械创新设计 [M]. 北京：机械工业出版社，2007.

[18] 申永胜. 机械原理 [M]. 北京：清华大学出版社，1999.

[19] 张　策. 机械原理与机械设计 [M]. 北京：机械工业出版社，2004.

[20] 孙　桓，陈作谋，葛文杰. 机械原理 [M]. 北京：高等教育出版社，2006.

[21] 郑文纬，吴克坚. 机械原理 [M]. 北京：高等教育出版社，1997.

[22] 张春林，余跃庆. 机械原理教学参考书 [M]. 北京：高等教育出版社，2009.

[23] 邹慧君，张春林，李杞仪. 机械原理 [M]. 北京：高等教育出版社，2006.

[24] 黄茂林，秦伟. 机械原理 [M]. 北京：机械工业出版社，2002.